High Voltage Engineering

Fundamentals

High Voltage Engineering

Fundamentals

Second edition

E. Kuffel
Dean Emeritus,
University of Manitoba,
Winnipeg, Canada

W.S. Zaengl
Professor Emeritus,
Electrical Engineering Dept.,
Swiss Federal Institute of Technology,
Zurich, Switzerland

J. Kuffel
Manager of High Voltage and Current Laboratories,
Ontario Hydro Technologies,
Toronto, Canada

AMSTERDAM • BOSTON • HEIDELBERG • LONDON • NEW YORK • OXFORD
PARIS • SAN DIEGO • SAN FRANCISCO • SINGAPORE • SYDNEY • TOKYO
Newnes is an imprint of Elsevier

Newnes is an imprint of Elsevier
Linacre House, Jordan Hill, Oxford OX2 8DP, UK
30 Corporate Drive, Suite 400, Burlington, MA 01803, USA

First edition published by Pergamon Press 1984
Reprinted 1986
Second edition published by Butterworth Heinemann 2000
Reprinted 2001, 2003, 2004, 2005, 2006, 2007, 2008

British Library Cataloguing in Publication Data
A catalogue record for this book is available from the British Library

Library of Congress Cataloging-in-Publication Data
A catalog record for this book is available from the Library of Congress

ISBN: 978-0-7506-3634-6

Ttransferred to Digital Printing in 2009

Contents

Preface to second edition xi

Preface to first edition xv

Chapter 1 Introduction 1

1.1 Generation and transmission of electric energy 1

1.2 Voltage stresses 3

1.3 Testing voltages 5
1.3.1 Testing with power frequency voltages 5
1.3.2 Testing with lightning impulse voltages 5
1.3.3 Testing with switching impulses 6
1.3.4 D.C. voltages 6
1.3.5 Testing with very low frequency voltage 7

References 7

Chapter 2 Generation of high voltages 8

2.1 Direct voltages 9
2.1.1 A.C. to D.C. conversion 10
2.1.2 Electrostatic generators 24

2.2 Alternating voltages 29
2.2.1 Testing transformers 32
2.2.2 Series resonant circuits 40

2.3 Impulse voltages 48
2.3.1 Impulse voltage generator circuits 52
2.3.2 Operation, design and construction of impulse generators 66

2.4 Control systems 74

References 75

Chapter 3 Measurement of high voltages 77

3.1 Peak voltage measurements by spark gaps 78
3.1.1 Sphere gaps 79
3.1.2 Reference measuring systems 91

3.1.3 Uniform field gaps 92
3.1.4 Rod gaps 93

3.2 Electrostatic voltmeters 94

3.3 Ammeter in series with high ohmic resistors and high ohmic resistor voltage dividers 96

3.4 Generating voltmeters and field sensors 107

3.5 The measurement of peak voltages 109
3.5.1 The Chubb–Fortescue method 110
3.5.2 Voltage dividers and passive rectifier circuits 113
3.5.3 Active peak-reading circuits 117
3.5.4 High-voltage capacitors for measuring circuits 118

3.6 Voltage dividing systems and impulse voltage measurements 129
3.6.1 Generalized voltage generation and measuring circuit 129
3.6.2 Demands upon transfer characteristics of the measuring system 132
3.6.3 Fundamentals for the computation of the measuring system 139
3.6.4 Voltage dividers 147
3.6.5 Interaction between voltage divider and its lead 163
3.6.6 The divider's low-voltage arm 171

3.7 Fast digital transient recorders for impulse measurements 175
3.7.1 Principles and historical development of transient digital recorders 176
3.7.2 Errors inherent in digital recorders 179
3.7.3 Specification of ideal A/D recorder and parameters required for h.v. impulse testing 183
3.7.4 Future trends 195

References 196

Chapter 4 Electrostatic fields and field stress control 201

4.1 Electrical field distribution and breakdown strength of insulating materials 201

4.2 Fields in homogeneous, isotropic materials 205
4.2.1 The uniform field electrode arrangement 206
4.2.2 Coaxial cylindrical and spherical fields 209
4.2.3 Sphere-to-sphere or sphere-to-plane 214
4.2.4 Two cylindrical conductors in parallel 218
4.2.5 Field distortions by conducting particles 221

4.3 Fields in multidielectric, isotropic materials 225
4.3.1 Simple configurations 227
4.3.2 Dielectric refraction 232
4.3.3 Stress control by floating screens 235

4.4 Numerical methods 241
4.4.1 Finite difference method (FDM) 242

4.4.2 Finite element method (FEM) 246
4.4.3 Charge simulation method (CSM) 254
4.4.4 Boundary element method 270

References 278

Chapter 5 Electrical breakdown in gases 281

5.1 Classical gas laws 281
5.1.1 Velocity distribution of a swarm of molecules 284
5.1.2 The free path λ of molecules and electrons 287
5.1.3 Distribution of free paths 290
5.1.4 Collision-energy transfer 291

5.2 Ionization and decay processes 294
5.2.1 Townsend first ionization coefficient 295
5.2.2 Photoionization 301
5.2.3 Ionization by interaction of metastables with atoms 301
5.2.4 Thermal ionization 302
5.2.5 Deionization by recombination 302
5.2.6 Deionization by attachment–negative ion formation 304
5.2.7 Mobility of gaseous ions and deionization by diffusion 308
5.2.8 Relation between diffusion and mobility 314

5.3 Cathode processes – secondary effects 316
5.3.1 Photoelectric emission 317
5.3.2 Electron emission by positive ion and excited atom impact 317
5.3.3 Thermionic emission 318
5.3.4 Field emission 319
5.3.5 Townsend second ionization coefficient γ 321
5.3.6 Secondary electron emission by photon impact 323

5.4 Transition from non-self-sustained discharges to breakdown 324
5.4.1 The Townsend mechanism 324

5.5 The streamer or 'Kanal' mechanism of spark 326

5.6 The sparking voltage–Paschen's law 333

5.7 Penning effect 339

5.8 The breakdown field strength (E_b) 340

5.9 Breakdown in non-uniform fields 342

5.10 Effect of electron attachment on the breakdown criteria 345

5.11 Partial breakdown, corona discharges 348
5.11.1 Positive or anode coronas 349
5.11.2 Negative or cathode corona 352

5.12 Polarity effect – influence of space charge 354

5.13 Surge breakdown voltage–time lag 359

5.13.1 Breakdown under impulse voltages 360
5.13.2 Volt–time characteristics 361
5.13.3 Experimental studies of time lags 362

References 365

Chapter 6 Breakdown in solid and liquid dielectrics 367

6.1 Breakdown in solids 367
6.1.1 Intrinsic breakdown 368
6.1.2 Streamer breakdown 373
6.1.3 Electromechanical breakdown 373
6.1.4 Edge breakdown and treeing 374
6.1.5 Thermal breakdown 375
6.1.6 Erosion breakdown 381
6.1.7 Tracking 385

6.2 Breakdown in liquids 385
6.2.1 Electronic breakdown 386
6.2.2 Suspended solid particle mechanism 387
6.2.3 Cavity breakdown 390
6.2.4 Electroconvection and electrohydrodynamic model of dielectric breakdown 391

6.4 Static electrification in power transformers 393

References 394

Chapter 7 Non-destructive insulation test techniques 395

7.1 Dynamic properties of dielectrics 395
7.1.1 Dynamic properties in the time domain 398
7.1.2 Dynamic properties in the frequency domain 404
7.1.3 Modelling of dielectric properties 407
7.1.4 Applications to insulation ageing 409

7.2 Dielectric loss and capacitance measurements 411
7.2.1 The Schering bridge 412
7.2.2 Current comparator bridges 417
7.2.3 Loss measurement on complete equipment 420
7.2.4 Null detectors 421

7.3 Partial-discharge measurements 421
7.3.1 The basic PD test circuit 423
7.3.2 PD currents 427
7.3.3 PD measuring systems within the PD test circuit 429
7.3.4 Measuring systems for apparent charge 433
7.3.5 Sources and reduction of disturbances 448
7.3.6 Other PD quantities 450
7.3.7 Calibration of PD detectors in a complete test circuit 452

7.3.8 Digital PD instruments and measurements 453

References 456

Chapter 8 Overvoltages, testing procedures and insulation coordination 460

8.1 The lightning mechanism 460
8.1.1 Energy in lightning 464
8.1.2 Nature of danger 465

8.2 Simulated lightning surges for testing 466

8.3 Switching surge test voltage characteristics 468

8.4 Laboratory high-voltage testing procedures and statistical treatment of results 472
8.4.1 Dielectric stress–voltage stress 472
8.4.2 Insulation characteristics 473
8.4.3 Randomness of the appearance of discharge 473
8.4.4 Types of insulation 473
8.4.5 Types of stress used in high-voltage testing 473
8.4.6 Errors and confidence in results 479
8.4.7 Laboratory test procedures 479
8.4.8 Standard test procedures 484
8.4.9 Testing with power frequency voltage 484
8.4.10 Distribution of measured breakdown probabilities (confidence in measured $P(V)$) 485
8.4.11 Confidence intervals in breakdown probability (in measured values) 487

8.5 Weighting of the measured breakdown probabilities 489
8.5.1 Fitting of the best fit normal distribution 489

8.6 Insulation coordination 492
8.6.1 Insulation level 492
8.6.2 Statistical approach to insulation coordination 495
8.6.3 Correlation between insulation and protection levels 498

8.7 Modern power systems protection devices 500
8.7.1 MOA – metal oxide arresters 500

References 507

Chapter 9 Design and testing of external insulation 509

9.1 Operation in a contaminated environment 509

9.2 Flashover mechanism of polluted insulators under a.c. and d.c. 510
9.2.1 Model for flashover of polluted insulators 511

9.3 Measurements and tests 512
9.3.1 Measurement of insulator dimensions 513

9.3.2 Measurement of pollution severity 514
9.3.3 Contamination testing 517
9.3.4 Contamination procedure for clean fog testing 518
9.3.5 Clean fog test procedure 519
9.3.6 Fog characteristics 520

9.4 Mitigation of contamination flashover 520
9.4.1 Use of insulators with optimized shapes 520
9.4.2 Periodic cleaning 520
9.4.3 Grease coating 521
9.4.4 RTV coating 521
9.4.5 Resistive glaze insulators 521
9.4.6 Use of non-ceramic insulators 522

9.5 Design of insulators 522
9.5.1 Ceramic insulators 523
9.5.2 Polymeric insulators (NCI) 526

9.6 Testing and specifications 530
9.6.1 In-service inspection and failure modes 531

References 531

Index 533

Preface to Second Edition

The first edition as well as its forerunner of Kuffel and Abdullah published in 1970 and their translations into Japanese and Chinese languages have enjoyed wide international acceptance as basic textbooks in teaching senior undergraduate and postgraduate courses in High-Voltage Engineering. Both texts have also been extensively used by practising engineers engaged in the design and operation of high-voltage equipment. Over the years the authors have received numerous comments from the text's users with helpful suggestions for improvements. These have been incorporated in the present edition. Major revisions and expansion of several chapters have been made to update the continued progress and developments in high-voltage engineering over the past two decades.

As in the previous edition, the principal objective of the current text is to cover the fundamentals of high-voltage laboratory techniques, to provide an understanding of high-voltage phenomena, and to present the basics of high-voltage insulation design together with the analytical and modern numerical tools available to high-voltage equipment designers.

Chapter 1 presents an introduction to high-voltage engineering including the concepts of power transmission, voltage stress, and testing with various types of voltage. Chapter 2 provides a description of the apparatus used in the generation of a.c., d.c., and impulse voltages. These first two introductory chapters have been reincorporated into the current revision with minor changes.

Chapter 3 deals with the topic of high-voltage measurements. It has undergone major revisions in content to reflect the replacement of analogue instrumentation with digitally based instruments. Fundamental operating principles of digital recorders used in high-voltage measurements are described, and the characteristics of digital instrumentation appropriate for use in impulse testing are explained.

Chapter 4 covers the application of numerical methods in electrical stress calculations. It incorporates much of the contents of the previous text, but the section on analogue methods has been replaced by a description of the more current boundary element method.

Chapter 5 of the previous edition dealt with the breakdown of gaseous, liquid, and solid insulation. In the new edition these topics are described in two chapters. The new Chapter 5 covers the electrical breakdown of gases. The breakdown of liquid and solid dielectrics is presented in Chapter 6 of the current edition.

Chapter 7 of the new text represents an expansion of Chapter 6 of the previous book. The additional areas covered comprise a short but fundamental introduction to dielectric properties of materials, diagnostic test methods, and non-destructive tests applicable also to on-site monitoring of power equipment. The expanded scope is a reflection of the growing interest in and development of on-site diagnostic testing techniques within the electrical power industry. This area represents what is perhaps the most quickly evolving aspect of high-voltage testing. The current drive towards deregulation of the power industry, combined with the fact that much of the apparatus making up the world's electrical generation and delivery systems is ageing, has resulted in a pressing need for the development of in-service or at least on-site test methods which can be applied to define the state of various types of system assets. Assessment of the remaining life of major assets and development of maintenance practices optimized both from the technical and economic viewpoints have become critical factors in the operation of today's electric power systems. Chapter 7 gives an introduction and overview of the fundamental aspects of on-site test methods with some practical examples illustrating current practices.

Chapter 8 is an expansion of Chapter 7 from the previous edition. However, in addition to the topics of lightning phenomena, switching overvoltages and insulation coordination, it covers statistically based laboratory impulse test methods and gives an overview of metal oxide surge arresters. The statistical impulse test methods described are basic tools used in the application of insulation coordination concepts. As such, an understanding of these methods leads to clearer understanding of the basis of insulation coordination. Similarly, an understanding of the operation and application of metal oxide arresters is an integral part of today's insulation coordination techniques.

Chapter 9 describes the design, performance, application and testing of outdoor insulators. Both ceramic and composite insulators are included. Outdoor insulators represent one of the most critical components of transmission and distribution systems. While there is significant experience in the use of ceramic insulators, composite insulators represent a relatively new and quickly evolving technology that offers a number of performance advantages over the conventional ceramic alternative. Their use and importance will continue to increase and therefore merits particular attention.

The authors are aware of the fact that many topics also relevant to the fundamentals of high-voltage engineering have again not been treated. But every textbook about this field will be a compromise between the limited space available for the book and the depth of treatment for the selected topics. The inclusion of more topics would reduce its depth of treatment, which should be good enough for fundamental understanding and should stimulate further reading.

The authors would like to express their thanks to Professors Yuchang Qiu of X'ian Jaotong University, Stan. Grzybowski of Mississippi State University, Stephen Sebo of Ohio State University for their helpful suggestions in the selection of new material, Ontario Power Technologies for providing help in the preparation of the text and a number of illustrations and Mrs Shelly Gerardin for her skilful efforts in scanning and editing the text of the first edition. Our special thanks go to Professor Yuchang Qiu for his laborious proof reading of the manuscript.

Finally we would like to express our personal gratitude to Mr Peter Kuffel and Dr Waldemar Ziomek for their invaluable help in the process of continued review and preparation of the final manuscript and illustrations.

Preface to First Edition

The need for an up-to-date textbook in High Voltage Engineering fundamentals has been apparent for some time. The earlier text of Kuffel and Abdullah published in 1970, although it had a wide circulation, was of somewhat limited scope and has now become partly outdated.

In this book an attempt is made to cover the basics of high voltage laboratory techniques and high voltage phenomena together with the principles governing design of high voltage insulation.

Following the historical introduction the chapters 2 and 3 present a comprehensive and rigorous treatment of laboratory, high voltage generation and measurement techniques and make extensive references to the various international standards.

Chapter 4 reviews methods used in controlling electric stresses and introduces the reader to modern numerical methods and their applications in the calculation of electric stresses in simple practical insulations.

Chapter 5 includes an extensive treatment of the subject of gas discharges and the basic mechanisms of electrical breakdown of gaseous, liquid and solid insulations.

Chapter 6 deals with modern techniques for discharge detection and measurement. The final chapter gives an overview treatment of systems overvoltages and insulation coordination.

It is hoped the text will fill the needs of senior undergraduate and graduate students enrolled in high voltage engineering courses as well as junior researchers engaged in the field of gas discharges. The in-depth treatment of high voltage techniques should make the book particularly useful to designers and operators of high voltage equipment and utility engineers.

The authors gratefully acknowledge Dr. M. M. Abdullah's permission to reproduce some material from the book *High Voltage Engineering*, Pergamon Press, 1970.

E. Kuffel, W.S. Zaengal
March 1984

Chapter 1
Introduction

1.1 Generation and transmission of electric energy

The potential benefits of electrical energy supplied to a number of consumers from a common generating system were recognized shortly after the development of the 'dynamo', commonly known as the generator.

The first public power station was put into service in 1882 in London (Holborn). Soon a number of other public supplies for electricity followed in other developed countries. The early systems produced direct ccurrent at low-voltage, but their service was limited to highly localized areas and were used mainly for electric lighting. The limitations of d.c. transmission at low-voltage became readily apparent. By 1890 the art in the development of an a.c. generator and transformer had been perfected to the point when a.c. supply was becoming common, displacing the earlier d.c. system. The first major a.c. power station was commissioned in 1890 at Deptford, supplying power to central London over a distance of 28 miles at 10 000 V. From the earliest 'electricity' days it was realized that to make full use of economic generation the transmission network must be tailored to production with increased interconnection for pooling of generation in an integrated system. In addition, the potential development of hydroelectric power and the need to carry that power over long distances to the centres of consumption were recognized.

Power transfer for large systems, whether in the context of interconnection of large systems or bulk transfers, led engineers invariably to think in terms of high system voltages. Figure 1.1 lists some of the major a.c. transmission systems in chronological order of their installations, with tentative projections to the end of this century.

The electric power (P) transmitted on an overhead a.c. line increases approximately with the surge impedance loading or the square of the system's operating voltage. Thus for a transmission line of surge impedance Z_L ($\cong 250\,\Omega$) at an operating voltage V, the power transfer capability is approximately $P = V^2/Z_L$, which for an overhead a.c. system leads to the following results:

V (kV)	400	700	1000	1200	1500
P (MW)	640	2000	4000	5800	9000

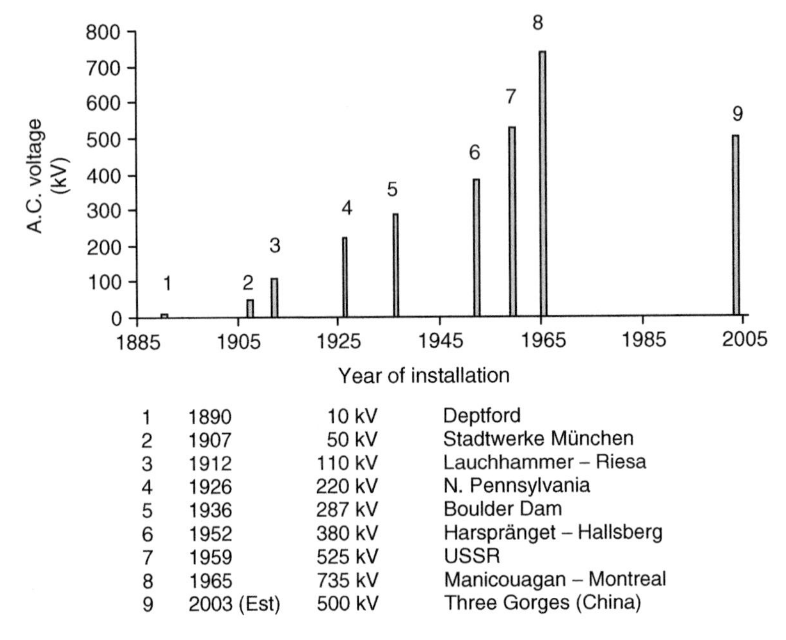

Figure 1.1 *Major a.c. systems in chronological order of their installations*

The rapidly increasing transmission voltage level in recent decades is a result of the growing demand for electrical energy, coupled with the development of large hydroelectric power stations at sites far remote from centres of industrial activity and the need to transmit the energy over long distances to the centres. However, environmental concerns have imposed limitations on system expansion resulting in the need to better utilize existing transmission systems. This has led to the development of Flexible A.C. Transmission Systems (FACTS) which are based on newly developing high-power electronic devices such as GTOs and IGBTs. Examples of FACTS systems include Thyristor Controlled Series Capacitors and STATCOMS. The FACTS devices improve the utilization of a transmission system by increasing power transfer capability.

Although the majority of the world's electric transmission is carried on a.c. systems, high-voltage direct current (HVDC) transmission by overhead lines, submarine cables, and back-to-back installations provides an attractive alternative for bulk power transfer. HVDC permits a higher power density on a given right-of-way as compared to a.c. transmission and thus helps the electric utilities in meeting the environmental requirements imposed on the transmission of electric power. HVDC also provides an attractive technical and economic solution for interconnecting asynchronous a.c. systems and for bulk power transfer requiring long cables.

Table 1.1 summarizes a number of major HVDC schemes in order of their in-service dates. Figure 1.2 provides a graphic illustration of how HVDC transmission voltages have developed. As seen in Figure 1.2 the prevailing d.c. voltage for overhead line installations is 500 kV. This 'settling' of d.c. voltage has come about based on technical performance, power transfer requirements, environmental and economic considerations. Current trends indicate that d.c. voltage levels will not increase dramatically in the near future.

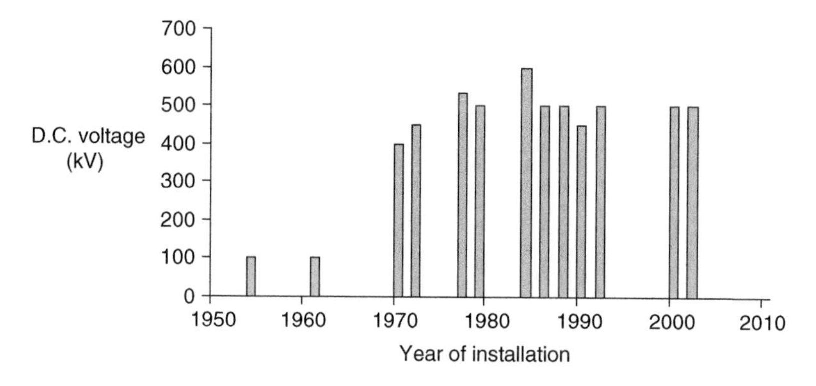

Figure 1.2 *Major d.c. systems in chronological order of their installations*

1.2 Voltage stresses

Normal operating voltage does not severely stress the power system's insulation and only in special circumstances, for example under pollution conditions, may operating voltages cause problems to external insulation. Nevertheless, the operating voltage determines the dimensions of the insulation which forms part of the generation, transmission and distribution equipment. The voltage stresses on power systems arise from various overvoltages. These may be of external or internal origin. External overvoltages are associated with lightning discharges and are not dependent on the voltage of the system. As a result, the importance of stresses produced by lightning decreases as the operating voltage increases. Internal overvoltages are generated by changes in the operating conditions of the system such as switching operations, a fault on the system or fluctuations in the load or generations.

Their magnitude depends on the rated voltage, the instance at which a change in operating conditions occurs, the complexity of the system and so on. Since the change in the system's conditions is usually associated with switching operations, these overvoltages are generally referred to as switching overvoltages.

Table 1.1 *Major HVDC schemes*

Scheme	*Year*	*Power (MW)*	*D.C. voltage (kv)*	*Line or cable length (km)*	*Location*
Gottland 1	1954	20	±100	96	Sweden
English Channel	1961	160	±100	64	England–France
Pacific Intertie	1970	1440	±400	1362	USA
Nelson River 1	1972	1620	±450	892	Canada
Eel River	1972	320	2 × 80	Back to back	Canada
Cabora Bassa	1978	1920	±533	1414	Mozambique–South Africa
Nelson River 2	1978	900	±250	930	Canada
	1985	1800	±500		
Chateauguay	1984	1000	2 × 140	Back to back	Canada
Itaipu 1	1984	200	±300	785	Brazil
	1985	1575			
	1986	2383	±600		
Intermountain	1986	1920	±500	784	USA
Cross Channel	1986	2000	2 × ±270	72	England–France
Itaipu 2	1987	3150	±600	805	Brazil
Gezhouba–Shanghai	1989	600	500	1000	China
	1990	1200	±500		
Fenno-Skan	1989	500	400	200	Finland–Sweden
Rihand-Delhi	1991	1500	±500	910	India
Hydro Quebec–New England	1990	2000	±450	1500	Canada–USA
Baltic Cable	1994	600	450	250	Sweden–Germany
Tian Guang	2000 (est)	1800	±500	960	China
Three Gorges	2002 (est)	3000	±500	–	China

Source: HVDC Projects Listing, D.C. & Flexible A.C. Transmission Subcommittee of the IEEE Transmission and Distribution Committee, Working Group on HVDC, and Bibliography and Records, January 1998 Issue.

In designing the system's insulation the two areas of specific importance are:

(i) determination of the voltage stresses which the insulation must withstand, and
(ii) determination of the response of the insulation when subjected to these voltage stresses.

The balance between the electric stresses on the insulation and the dielectric strength of this insulation falls within the framework of insulation coordination and will be discussed in Chapter 8.

1.3 Testing voltages

Power systems equipment must withstand not only the rated voltage (V_m), which corresponds to the highest voltage of a particular system, but also overvoltages. Accordingly, it is necessary to test h.v. equipment during its development stage and prior to commissioning. The magnitude and type of test voltage varies with the rated voltage of a particular apparatus. The standard methods of measurement of high-voltage and the basic techniques for application to all types of apparatus for alternating voltages, direct voltages, switching impulse voltages and lightning impulse voltages are laid down in the relevant national and international standards.

1.3.1 Testing with power frequency voltages

To assess the ability of the apparatus's insulation withstand under the system's power frequency voltage the apparatus is subjected to the 1-minute test under 50 Hz or 60 Hz depending upon the country. The test voltage is set at a level higher than the expected working voltage in order to be able to simulate the stresses likely to be encountered over the years of service. For indoor installations the equipment tests are carried out under dry conditions only. For outdoor equipment tests may be required under conditions of standard rain as prescribed in the appropriate standards.

1.3.2 Testing with lightning impulse voltages

Lightning strokes terminating on transmission lines will induce steep rising voltages in the line and set up travelling waves along the line and may damage the system's insulation. The magnitude of these overvoltages may reach several thousand kilovolts, depending upon the insulation. Exhaustive measurements and long experience have shown that lightning overvoltages are characterized by short front duration, ranging from a fraction of a microsecond to several tens of microseconds and then slowly decreasing to zero. The standard impulse voltage has been accepted as an aperiodic impulse that reaches its peak value in 1.2 μsec and then decreases slowly (in about 50 μsec) to half its peak value. Full details of the waveshape of the standard impulse voltage together with the permitted tolerances are presented in Chapter 2, and the prescribed test procedures are discussed in Chapter 8.

In addition to testing equipment, impulse voltages are extensively used in research laboratories in the fundamental studies of electrical discharge mechanisms, notably when the time to breakdown is of interest.

1.3.3 Testing with switching impulses

Transient overvoltages accompanying sudden changes in the state of power systems, e.g. switching operations or faults, are known as switching impulse voltages. It has become generally recognized that switching impulse voltages are usually the dominant factor affecting the design of insulation in h.v. power systems for rated voltages of about 300 kV and above. Accordingly, the various international standards recommend that equipment designed for voltages above 300 kV be tested for switching impulses. Although the waveshape of switching overvoltages occurring in the system may vary widely, experience has shown that for flashover distances in atmospheric air of practical interest the lowest withstand values are obtained with surges with front times between 100 and 300 μsec. Hence, the recommended switching surge voltage has been designated to have a front time of about 250 μsec and half-value time of 2500 μsec. For GIS (gas-insulated switchgear) on-site testing, oscillating switching impulse voltages are recommended for obtaining higher efficiency of the impulse voltage generator Full details relating to generation, measurements and test procedures in testing with switching surge voltages will be found in Chapters 2, 3 and 8.

1.3.4 D.C. voltages

In the past d.c. voltages have been chiefly used for purely scientific research work. Industrial applications were mainly limited to testing cables with relatively large capacitance, which take a very large current when tested with a.c. voltages, and in testing insulations in which internal discharges may lead to degradation of the insulation under testing conditions. In recent years, with the rapidly growing interest in HVDC transmission, an increasing number of industrial laboratories are being equipped with sources for producing d.c. high voltages. Because of the diversity in the application of d.c. high voltages, ranging from basic physics experiments to industrial applications, the requirements on the output voltage will vary accordingly. Detailed description of the various main types of HVDC generators is given in Chapter 2.

1.3.5 Testing with very low-frequency voltage

In the earlier years when electric power distribution systems used mainly paper-insulated lead covered cables (PILC) on-site testing specifications called for tests under d.c. voltages. Typically the tests were carried out at $4-4.5V_0$. The tests helped to isolate defective cables without further damaging good cable insulation. With the widespread use of extruded insulation cables of higher dielectric strength, the test voltage levels were increased to $5-8V_0$. In the 1970s premature failures of extruded dielectric cables factory tested under d.c. voltage at specified levels were noted(1). Hence on-site testing of cables under very low frequency (VLF) of ~0.1 Hz has been adopted. The subject has been recently reviewed(1,2).

References

1. Working Group 21.09. After-laying tests on high voltage extruded insulation cable systems, *Electra*, No. 173 (1997), pp. 31–41.
2. G.S. Eager *et al.* High voltage VLF testing of power cables, *IEEE Trans Power Delivery*, **12**, No. 2 (1997), pp. 565–570.

Chapter 2

Generation of high voltages

A fundamental knowledge about generators and circuits which are in use for the generation of high voltages belongs to the background of work on h.v. technology.

Generally commercially available h.v. generators are applied in routine testing laboratories; they are used for testing equipment such as transformers, bushings, cables, capacitors, switchgear, etc. The tests should confirm the efficiency and reliability of the products and therefore the h.v. testing equipment is required to study the insulation behaviour under all conditions which the apparatus is likely to encounter. The amplitudes and types of the test voltages, which are always higher than the normal or rated voltages of the apparatus under test, are in general prescribed by national or international standards or recommendations, and therefore there is not much freedom in the selection of the h.v. testing equipment. Quite often, however, routine testing laboratories are also used for the development of new products. Then even higher voltages might be necessary to determine the factor of safety over the prospective working conditions and to ensure that the working margin is neither too high nor too low. Most of the h.v. generator circuits can be changed to increase the output voltage levels, if the original circuit was properly designed. Therefore, even the selection of routine testing equipment should always consider a future extension of the testing capabilities.

The work carried out in research laboratories varies considerably from one establishment to another, and the type of equipment needed varies accordingly. As there are always some interactions between the h.v. generating circuits used and the test results, the layout of these circuits has to be done very carefully. The classes of tests may differ from the routine tests, and therefore specially designed circuits are often necessary for such laboratories. The knowledge about some fundamental circuits treated in this chapter will also support the development of new test circuits.

Finally, high voltages are used in many branches of natural sciences or other technical applications. The generating circuits are often the same or similar to those treated in the following sections. It is not the aim, however, of this introductory text to treat the broad variations of possible circuits, due to space limitation. Not taken into account are also the differing problems of electrical power generation and transmission with high voltages of a.c. or d.c., or the pure testing technique of h.v. equipment, the procedures of which may be found in relevant standards of the individual equipment. Power generation and transmission problems are treated in many modern books, some of which are listed within the bibliography of an earlier report.[1]*

This chapter discusses the generation of the following main classes of voltages: direct voltages, alternating voltages, and transient voltages.

2.1 Direct voltages

In h.v. technology direct voltages are mainly used for pure scientific research work and for testing equipment related to HVDC transmission systems. There is still a main application in tests on HVAC power cables of long length, as the large capacitance of those cables would take too large a current if tested with a.c. voltages (see, however, 2.2.2: Series resonant circuits). Although such d.c. tests on a.c. cables are more economical and convenient, the validity of this test suffers from the experimentally obtained stress distribution within the insulating material, which may considerably be different from the normal working conditions where the cable is transmitting power at low-frequency alternating voltages. For the testing of polyethylene h.v. cables, in use now for some time, d.c. tests are no longer used, as such tests may not confirm the quality of the insulation.[50]

High d.c. voltages are even more extensively used in applied physics (accelerators, electron microscopy, etc.), electromedical equipment (X-rays), industrial applications (precipitation and filtering of exhaust gases in thermal power stations and the cement industry; electrostatic painting and powder coating, etc.), or communications electronics (TV, broadcasting stations). Therefore, the requirements on voltage shape, voltage level, and current rating, short- or long-term stability for every HVDC generating system may differ strongly from each other. With the knowledge of the fundamental generating principles it will be possible, however, to select proper circuits for a special application.

In the International Standard IEC 60-1[2] or IEEE Standard. 4-1995[3] the value of a direct test voltage is defined by its arithmetic mean value, which will be designated as $\overline{V}$. Therefore, this value may be derived from

$$\overline{V} = \frac{1}{T}\int_0^T V(t)\,\mathrm{d}t. \tag{2.1}$$

where T equals a certain period of time if the voltage $V(t)$ is not constant, but periodically oscillating with a frequency of $f = 1/T$. Test voltages as applied to test objects then deviate periodically from the mean value. This means that

* Superscript numbers are to References at the end of the chapter.

a ripple is present. The amplitude of the ripple, δV, is defined as half the difference between the maximum and minimum values, or

$$\delta V = 0.5(V_{max} - V_{min}). \tag{2.2}$$

The ripple factor is the ratio of the ripple amplitude to the arithmetic mean value, or $\delta V/\overline{V}$. For test voltages this ripple factor should not exceed 3 per cent unless otherwise specified by the appropriate apparatus standard or be necessary for fundamental investigations.

The d.c. voltages are generally obtained by means of rectifying circuits applied to a.c. voltages or by electrostatic generation. A treatment of the generation principles according to this subdivision is appropriate.

2.1.1 A.C. to D.C. conversion

The rectification of alternating currents is the most efficient means of obtaining HVDC supplies. Although all circuits in use have been known for a long time, the cheap production and availability of manifold solid state rectifiers has facilitated the production and application of these circuits fundamentally. Since some decades, there is no longer a need to employ valves, hot cathode gas-filled valves, mercury pool or corona rectifiers, or even mechanical rectifiers within the circuits, for which the auxiliary systems for cathode heating, etc., have always aggravated their application. The state of the art of such earlier circuits may be found in the work of Craggs and Meek,[4] which was written in 1954. All rectifier diodes used now adopt the Si type, and although the peak reverse voltage is limited to less than about 2500 V, rectifying diode units up to tens and hundreds of kVs can be made by series connections if appropriate means are applied to provide equal voltage distribution during the non-conducting period. One may treat and simulate, therefore, a rectifier within the circuits – independently of the voltage levels – simply by the common symbol for a diode.

The theory of rectifier circuits for low voltages and high power output is discussed in many standard handbooks. Having the generation of high d.c. voltages in mind, we will thus restrict the treatment mainly to single-phase a.c. systems providing a high ratio of d.c. output to a.c. input voltage. As, however, the power or d.c. output is always limited by this ratio, and because very simple rectifier circuits are in use, we will treat only selected examples of the many available circuits.

Simple rectifier circuits

For a clear understanding of all a.c. to d.c. conversion circuits the *single-phase half-wave rectifier* with voltage smoothing is of basic interest (Fig. 2.1(a)). If we neglect the leakage reactance of the transformer and the small internal

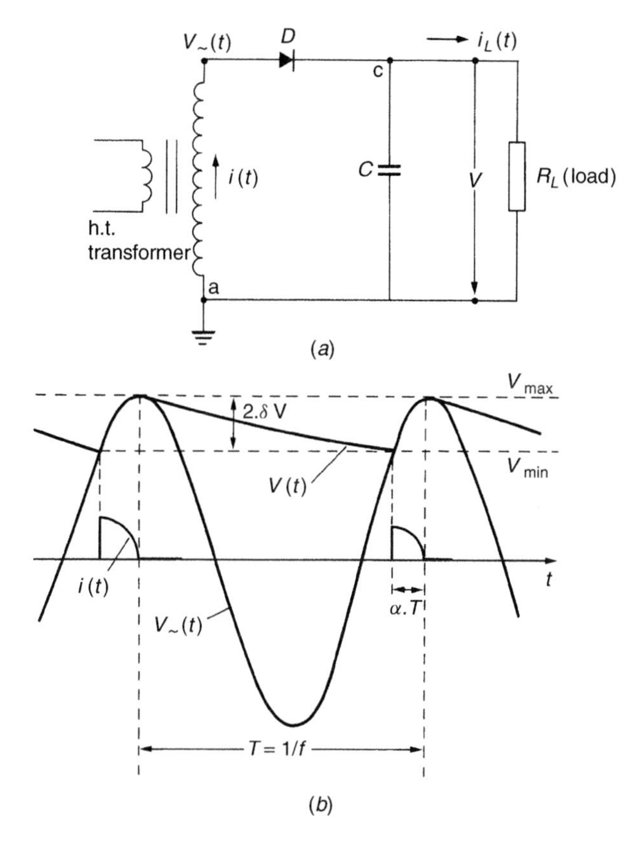

Figure 2.1 *Single-phase half-wave rectifier with reservoir capacitance C. (a) Circuit. (b) Voltages and currents with load R_L*

impedance of the diodes during conduction – and this will be done throughout unless otherwise stated – the reservoir or smoothing capacitor C is charged to the maximum voltage $+V_{max}$ of the a.c. voltage $V_{\sim}(t)$ of the h.t. transformer, when D conducts. This is the case as long as $V < V_{\sim}(t)$ for the polarity of D assumed. If $I = 0$, i.e. the output load being zero ($R_L = \infty$), the d.c. voltage across C remains constant ($+V_{max}$), whereas $V_{\sim}(t)$ oscillates between $\pm V_{max}$. The diode D must be dimensioned, therefore, to withstand a peak reverse voltage of $2V_{max}$.

The output voltage V does not remain any more constant if the circuit is loaded. During one period, $T = 1/f$ of the a.c. voltage a charge Q is transferred to the load R_L, which is represented as

$$Q = \int_T i_L(t)\,\mathrm{d}t = \frac{1}{R_L}\int_T V(t)\,\mathrm{d}t = IT = \frac{I}{f}. \tag{2.3}$$

I is therefore the mean value of the d.c. output $i_L(t)$, and $V(t)$ the d.c. voltage which includes a ripple as shown in Fig. 2.1(b). If we introduce the ripple factor δV from eqn (2.2), we may easily see that $V(t)$ now varies between

$$V_{max} \geq V(t) \geq V_{min}; \quad V_{min} = V_{max} - 2(\delta V). \tag{2.4}$$

The charge Q is also supplied from the transformer within the short conduction time $t_c = \alpha T$ of the diode D during each cycle. Therefore, Q equals also to

$$Q = \int_{\alpha T} i(t)\,dt = \int_T i_L(t)\,dt. \tag{2.5}$$

As $\alpha T \ll T$, the transformer and diode current $i(t)$ is pulsed as shown idealized in Fig. 2.1(b) and is of much bigger amplitudes than the direct current $i_L \cong I$. The ripple δV could be calculated exactly for this circuit based upon the exponential decay of $V(t)$ during the discharge period $T(1-\alpha)$. As, however, for practical circuits the neglected voltage drops within transformer and rectifiers must be taken into account, and such calculations are found elsewhere,[3] we may assume that $\alpha = 0$. Then δV is easily found from the charge Q transferred to the load, and therefore

$$Q = 2\delta VC = IT; \quad \delta V = \frac{IT}{2C} = \frac{I}{2fC}. \tag{2.6}$$

This relation shows the interaction between the ripple, the load current and circuit parameter design values f and C. As, according to eqn (2.4), the mean output voltage will also be influenced by δV, even with a constant a.c. voltage $V_{\sim}(t)$ and a lossless rectifier D, no load-independent output voltage can be reached. The product fC is therefore an important design factor.

For h.v. test circuits, a sudden voltage breakdown at the load ($R_L \to 0$) must always be taken into account. Whenever possible, the rectifiers should be able to carry either the excessive currents, which can be limited by fast, electronically controlled switching devices at the transformer input, or they can be protected by an additional resistance inserted in the h.t. circuit. The last method, however, increases the internal voltage drop.

Half-wave rectifier circuits have been built up to voltages in the megavolt range, in general by extending an existing h.v. testing transformer to a d.c. current supply. The largest unit has been presented by Prinz,[5] who used a 1.2-MV cascaded transformer and 60-mA selenium-type solid state rectifiers with an overall reverse voltage of 3.4 MV for the circuit. The voltage distribution of this rectifier, which is about 12 m in length, is controlled by sectionalized parallel capacitor units, which are small in capacitance value in comparison with the smoothing capacitor C (see Fig. 2.14). The size of such circuits, however, would be unnecessarily large for pure d.c. supplies.

The other disadvantage of the single-phase half-wave rectifier concerns the possible saturation of the h.v. transformer, if the amplitude of the direct current is comparable with the nominal alternating current of the transformer. The biphase half-wave (or single-phase full-wave) rectifier as shown in Fig. 2.2 overcomes this disadvantage, but it does not change the fundamental efficiency, considering that two h.v. windings of the transformer are now available. With reference to the frequency f during one cycle, now each of the diodes D_1 and D_2 is conducting for one half-cycle with a time delay of $T/2$. The ripple factor according to eqn (2.6) is therefore halved. It should be mentioned that the real ripple will also be increased if both voltages $V_{1\sim}$ and $V_{2\sim}$ are not exactly equal. If $V_{2\,max}$ would be smaller than $(V_{1\,max} - 2\delta V)$ or V_{min}, this h.v. winding would not charge the capacitance C. The same effect holds true for multiphase rectifiers, which are not treated here.

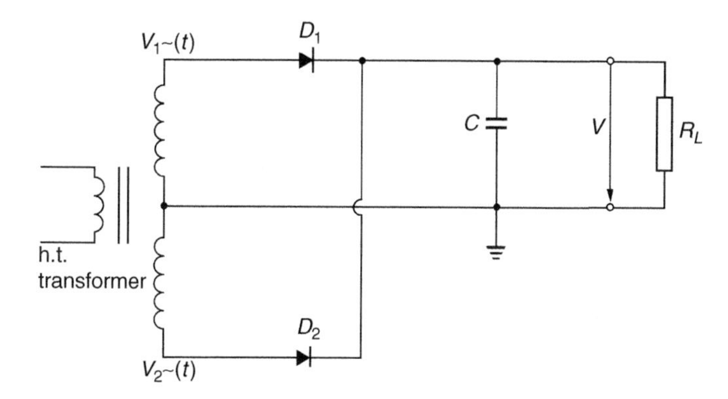

Figure 2.2 *Biphase half-wave rectifier circuit with smoothing capacitor C*

Thus single-phase full-wave circuits can only be used for h.v. applications if the h.t. winding of the transformer can be earthed at its midpoint and if the d.c. output is single-ended grounded. More commonly used are single-phase voltage doublers, a circuit of which is contained in the voltage multiplier or d.c. cascade of Fig. 2.6, see stage 1. Although in such a circuit grounding of the h.v. winding is also not possible, if asymmetrical d.c. voltages are produced, the potential of this winding is fixed. Therefore, there is no danger due to transients followed by voltage breakdowns.

Cascade circuits

The demands from physicists for very high d.c. voltages forced the improvement of rectifying circuits quite early. It is obvious that every multiplier circuit in which transformers, rectifiers and capacitor units have only to withstand a fraction of the total output voltage will have great advantages. Today there are many standard cascade circuits available for the conversion of modest a.c. to high d.c. voltages. However, only few basic circuits will be treated.

In 1920 Greinacher, a young physicist, published a circuit[6] which was improved in 1932 by Cockcroft and Walton to produce high-energy positive ions.[7] The interesting and even exciting development stages of those circuits have been discussed by Craggs and Meek.[4] To demonstrate the principle only, an n-stage single-phase cascade circuit of the 'Cockcroft–Walton type', shown in Fig. 2.3, will be presented.

HV output open-circuited: $I = 0$. The portion $0 - n' - V(t)$ is a half-wave rectifier circuit in which C'_n charges up to a voltage of $+V_{\max}$ if $V(t)$ has reached the lowest potential, $-V_{\max}$. If C_n is still uncharged, the rectifier D_n conducts as soon as $V(t)$ increases. As the potential of point n' swings up to $+V_{2\max}$ during the period $T = 1/f$, point n attains further on a steady potential of $+2V_{\max}$ if $V(t)$ has reached the highest potential of $+V_{\max}$. The part $n' - n - 0$ is therefore a half-wave rectifier, in which the voltage across D'_n can be assumed to be the a.c. voltage source. The current through D_n that

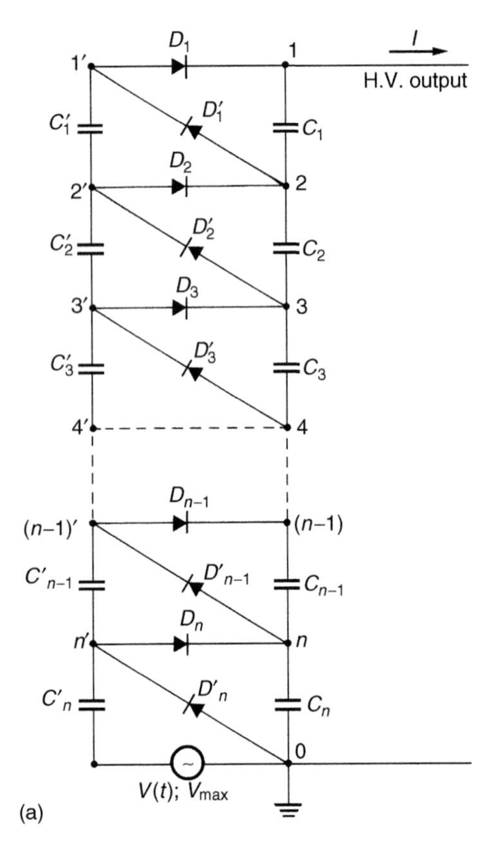

Figure 2.3 *(a) Cascade circuit according to Cockroft–Walton or Greinacher. (b) Waveform of potentials at the nodes, no load*

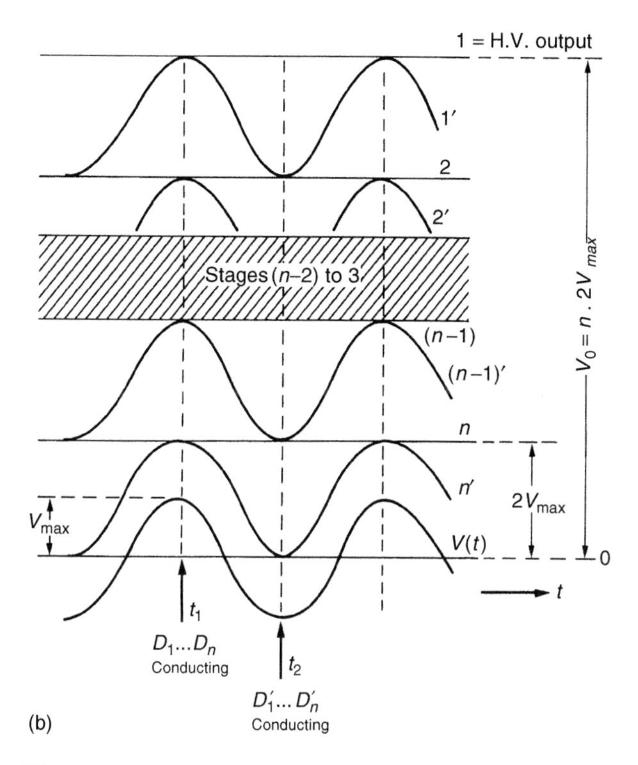

Figure 2.3 *(continued)*

charged the capacitor C_n was not provided by D'_n, but from $V(t)$ and C'_n. We assumed, therefore, that C'_n was not discharged, which is not correct. As we will take this into consideration for the loaded circuit, we can also assume that the voltage across C_n is not reduced if the potential n' oscillates between zero and $+2V_{\max}$. If the potential of n', however, is zero, the capacitor C'_{n-1} is also charged to the potential of n, i.e. to a voltage of $+2V_{\max}$. The next voltage oscillation of $V(t)$ from $-V_{\max}$ to $+V_{\max}$ will force the diode D_{n-1} to conduct, so that also C_{n-1} will be charged to a voltage of $+2V_{\max}$.

In Fig. 2.3(b) the steady state potentials at all nodes of the circuit are sketched for the circuit for zero load conditions. From this it can be seen, that:

- the potentials at the nodes $1', 2' \ldots n'$ are oscillating due to the voltage oscillation of $V(t)$;
- the potentials at the nodes $1, 2 \ldots n$ remain constant with reference to ground potential;
- the voltages across all capacitors are of d.c. type, the magnitude of which is $2V_{\max}$ across each capacitor stage, except the capacitor C'_n which is stressed with $V_{\max}$ only;

- every rectifier D_1, $D'_1 \ldots D_n$, D'_n is stressed with $2V_{max}$ or twice a.c. peak voltage; and
- the h.v. output will reach a maximum voltage of $2nV_{max}$.

Therefore, the use of several stages arranged in this manner enables very high voltages to be obtained. The equal stress of the elements used is very convenient and promotes a modular design of such generators. The number of stages, however, is strongly limited by the current due to any load. This can only be demonstrated by calculations, even if ideal rectifiers, capacitors and an ideal a.c. voltage source are assumed.

Finally it should be mentioned that the lowest stage n of the cascade circuit (Fig. 2.3(a)) is the Cockcroft–Walton voltage doubler. The a.c. voltage source $V(t)$ is usually provided by an h.t. transformer, if every stage is built for high voltages, typically up to about 300 kV. This source is always symmetrically loaded, as current is withdrawn during each half-cycle (t_1 and t_2 in Fig. 2.3(b)). The voltage waveform does not have to be sinusoidal: every symmetrical waveform with equal positive and negative peak values will give good performance. As often high-frequency input voltages are used, this hint is worth remembering.

H.V. output loaded: $I > 0$. If the generator supplies any load current I, the output voltage will never reach the value $2nV_{max}$ as shown in Fig. 2.3(b). There will also be a ripple on the voltage, and therefore we have to deal with two quantities: the voltage drop ΔV_0 and the peak-to-peak ripple $2\delta V$. The sketch in Fig. 2.4 shows the shape of the output voltage and the definitions of

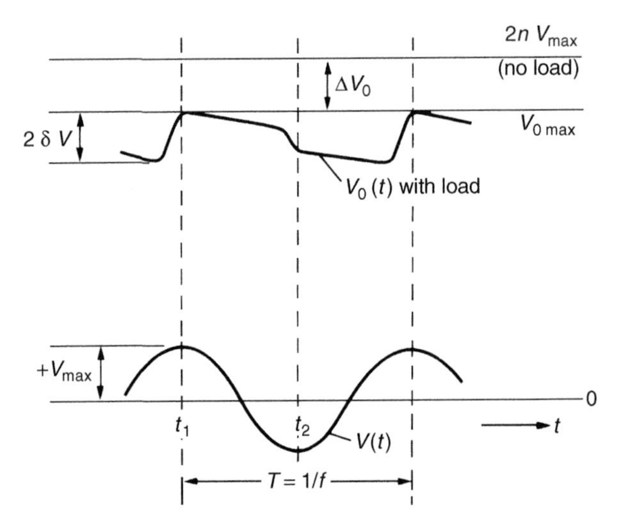

Figure 2.4 *Loaded cascade circuit, definitions of voltage drop ΔV_0 and ripple δV*

ΔV_0 and $2\delta V$. The time instants t_1 and t_2 are in agreement with Fig. 2.3(b). Therefore, the peak value of V_o is reached at t_1, if $V(t)$ was at $+V_{\max}$ and the rectifiers $D_1 \ldots D_n$ just stopped to transfer charge to the 'smoothing column' $C_1 \ldots C_n$. After that the current I continuously discharges the column, interrupted by a sudden voltage drop shortly before t_2: this sudden voltage drop is due to the conduction period of the diodes $D'_1 \ldots D'_n$, during which the 'oscillating column' $C'_1 \ldots C'_n$ is charged.

Now let a charge q be transferred to the load per cycle, which is obviously $q = I/f = IT$. This charge comes from the smoothing column, the series connection of $C_1 \ldots C_n$. If no charge would be transferred during T from this stack via $D'_1 \ldots D'_n$ to the oscillating column, the peak-to-peak ripple would merely be

$$2\delta V = IT \sum_{i=1}^{n} (1/C_i).$$

As, however, just before the time instant t_2 every diode $D'_1 \ldots D'_n$ transfers the same charge q, and each of these charges discharges all capacitors on the smoothing column between the relevant node and ground potential, the total ripple will be

$$\delta V = \frac{1}{2f}\left(\frac{1}{C_1} + \frac{2}{C_2} + \frac{3}{C_3} + \cdots \frac{n}{C_n}\right). \tag{2.7}$$

Thus in a cascade multiplier the lowest capacitors are responsible for most ripple and it would be desirable to increase the capacitance in the lower stages. This is, however, very inconvenient for h.v. cascades, as a voltage breakdown at the load would completely overstress the smaller capacitors within the column. Therefore, equal capacitance values are usually provided, and with $C = C_1 = C_2 \ldots C_n$, eqn (2.7) is

$$\delta V = \frac{I}{fC} \times \frac{n(n+1)}{4}. \tag{2.7a}$$

To calculate the total voltage drop ΔV_0, we will first consider the stage n. Although the capacitor C'_n at time t_1 will be charged up to the full voltage $V_{\max}$, if ideal rectifiers and no voltage drop within the a.c.-source are assumed, the capacitor C_n will only be charged to a voltage

$$(V_{C_n})_{\max} = 2V_{\max} - \frac{nq}{C'_n} = 2V_{\max} - \Delta V_n$$

as C_n has lost a total charge of (nq) during a full cycle before and C'_n has to replace this lost charge. At time instant t_2, C_n transfers the charge q to C'_{n-1}

equal amounts q to $C'_{n-2}, \ldots C'_2, C'_1$ and q to the load during T. Therefore, C'_{n-1} can only be charged up to a maximum voltage of

$$(V_{C'_{n-1}})_{\max} = \left(2V_{\max} - \frac{nq}{C'_n}\right) - \frac{nq}{C_n}$$

$$= (V_{C_n})_{\max} - \frac{nq}{C_n}.$$

As the capacitor C'_{n-1} will be charged up to this voltage minus $(n-1)q/c'_{n-1}$, etc., one can easily form the general rules for the total voltage drop at the smoothing stack $C_1 \ldots C_n$

If all the capacitors within the cascade circuit are equal or

$$C_1 = C'_1 = C_2 = C'_2 = \ldots C_n = C'_n = C,$$

then the voltage drops across the individual stages are

$$\Delta V_n = (q/c)n;$$

$$\Delta V_{n-1} = (q/c)[2n + (n-1)];$$

$$\vdots$$

$$\Delta V_1 = (q/c)[2n + 2(n-1) + 2(n-2) + \ldots + 2 \times 2 + 1]. \tag{2.8}$$

By summation, and with $q = I/f$, we find

$$\Delta V_0 = \frac{1}{fC}\left(\frac{2n^3}{3} + \frac{n^2}{2} - \frac{n}{6}\right). \tag{2.9}$$

Thus the lowest capacitors are most responsible for the total ΔV_0 as is the case of the ripple, eqn (2.7). However, only a doubling of C'_n is convenient, since this capacitor has to withstand only half the voltage of the other capacitors; namely $V_{\max}$. Therefore, ΔV_n decreases by an amount of $0.5\,nq/c$, which reduces ΔV of every stage by the same amount, thus n times. Hence,

$$\Delta V_0 = \frac{1}{fC}\left(\frac{2n^3}{3} - \frac{n}{6}\right). \tag{2.10}$$

For this case and $n \geq 4$ we may neglect the linear term and therefore approximate the maximum output voltage by

$$V_{0\max} \cong 2nV_{\max} - \frac{I}{fC} \times \frac{2n^3}{3}. \tag{2.11}$$

For a given number of stages, this maximum voltage or also the mean value $V_0 = (V_{0\max} - \delta V)$ will decrease linearly with the load current I at constant frequency, which is obvious. For a given load, however, V_0 may rise initially with the number of stages n, but reaches an optimum value and even decreases if n is too large. Thus – with respect to constant values of I V_{max}, f and C – the highest value can be reached with the 'optimum' number of stages, obtained by differentiating eqn (2.11) with respect to n. Then

$$n_{opt} = \sqrt{\frac{V_{max} f C}{I}} \tag{2.12}$$

For a generator with $V_{max} = 100\,\text{kV}$, $f = 500\,\text{Hz}$, $C = 7\,\mu\text{F}$ and $I = 500\,\text{mA}$, $n_{opt} = 10$. It is, however, not desirable to use the optimum number of stages, as then $V_{0\max}$ is reduced to 2/3 of its maximum value ($2nV_{max}$). Also the voltage variations for varying loads will increase too much.

The application of this circuit to high power output, which means high products of IV_0 is also limited by eqns (2.9) and (2.11), in which again the large influence of the product fC can be seen. An increase of supply frequency is in general more economical than an increase of the capacitance values; small values of C also provide a d.c. supply with limited stored energy, which might be an essential design factor, i.e. for breakdown investigations on insulating materials. A further advantage is related to regulation systems, which are always necessary if a stable and constant output voltage V_0 is required. Regulation can be achieved by a measurement of V_0 with suitable voltage dividers (see Chapter 3, section 3.6.4) within a closed-loop regulation system, which controls the a.c. supply voltage $V(t)$. For fast response, high supply frequencies and small stored energy are prerequisites.

For tall constructions in the MV range, the circuit of Fig. 2.3(a) does not comprise all circuit elements which are influencing the real working conditions. There are not only the impedances of the diodes and the supply transformer which have to be taken into consideration; stray capacitances between the two capacitor columns and capacitor elements to ground form a much more complex network. There are also improved circuits available by adding one or two additional 'oscillating' columns which charge the same smoothing stack. This additional column can be fed by phase-shifted a.c. voltages, by which the ripple and voltage drop can further be reduced. For more details see reference 8.

Cascade generators of Cockcroft–Walton type are used and manufactured today worldwide. More information about possible constructions can be found in the literature[9,10] or in company brochures. The d.c. voltages produced with this circuit may range from some 10 kV up to more than 2 MV, with current ratings from some 10 μA up to some 100 mA. Supply frequencies of 50/60 Hz are heavily limiting the efficiency, and therefore higher frequencies up to about 1000 Hz (produced by single-phase alternators) or some 10 kHz (produced by electronic circuits) are dominating.

Also for this kind of generators, voltage reversal can be performed by a reversal of all diodes. For some special tests on components as used for HVDC transmission, a fast reversal of the d.c. voltages is necessary. This can be done with special mechanical arrangements of the diodes, as published by W. Hauschild *et al.*[50,51] Figure 2.5 shows such a unit for a d.c. voltage up to

Figure 2.5 *A Cockroft–Walton d.c. generator for voltages up to 900 kV/10 mA with fast polarity reversal at ETH Zurich (courtesy HIGH VOLT, Dresden, Germany)*

900 kV. Here, also the general structure of the Cockroft–Walton circuit can be identified.

Voltage multiplier with cascaded transformers

The multiple charge transfer within the cascade circuit of the Cockcroft–Walton type demonstrated the limitations in d.c. power output. This disadvantage can be reduced if single- or full-wave rectifier systems, each having its own a.c. power source, are connected in series at the d.c. output only. Then the a.c. potentials remain more or less at d.c. potentials. Although there are many modifications possible, the principle that will be demonstrated here is based upon a very common circuit, which is shown in Fig. 2.6. Every transformer per stage consists of an l.v. primary (1), h.v. secondary (2), and l.v. tertiary winding (3), the last of which excites the primary winding of the next upper stage. As none of the h.v. secondary windings is on ground potential, a d.c. voltage insulation within each transformer (T_1, T_2, etc.) is necessary, which can be subdivided within the transformers. Every h.v. winding feeds two half-wave rectifiers, which have been explained before. Although there

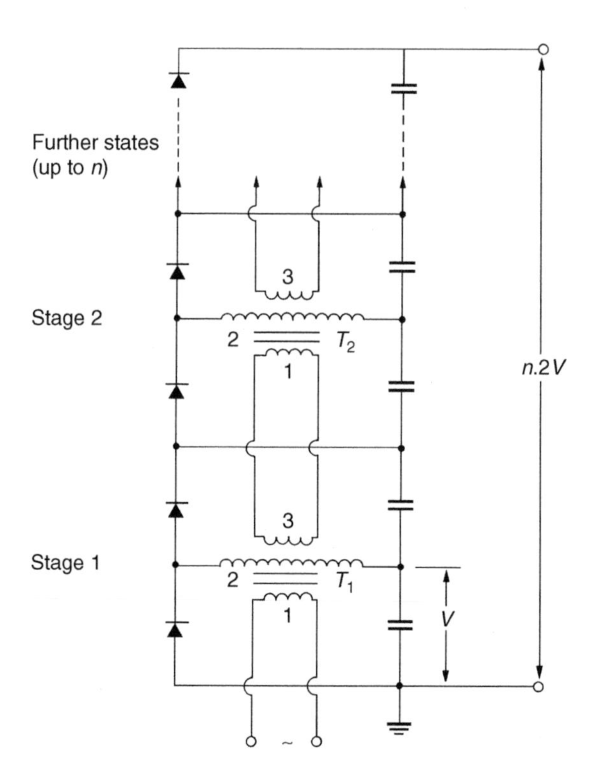

Figure 2.6 *D.C. cascade circuit with cascaded transformers*

are limitations as far as the number of stages is concerned, as the lower transformers have to supply the energy for the upper ones, this circuit, excited with power frequency, provides an economical d.c. power supply for h.v. testing purposes with moderate ripple factors and high power capabilities.

The 'Engetron' circuit (Deltatron)

A very sophisticated cascade transformer HVDC generator circuit was described by Enge in a US Patent.[11] Although such generators might be limited in the power output up to about 1 MV and some milliamperes, the very small ripple factors, high stability, fast regulation and small stored energies are essential capabilities of this circuit.

The circuit is shown in Fig. 2.7. It consists primarily of a series connection of transformers, which do not have any iron core. These transformers are coupled by series capacitors C_s which compensate most of the stray inductance.

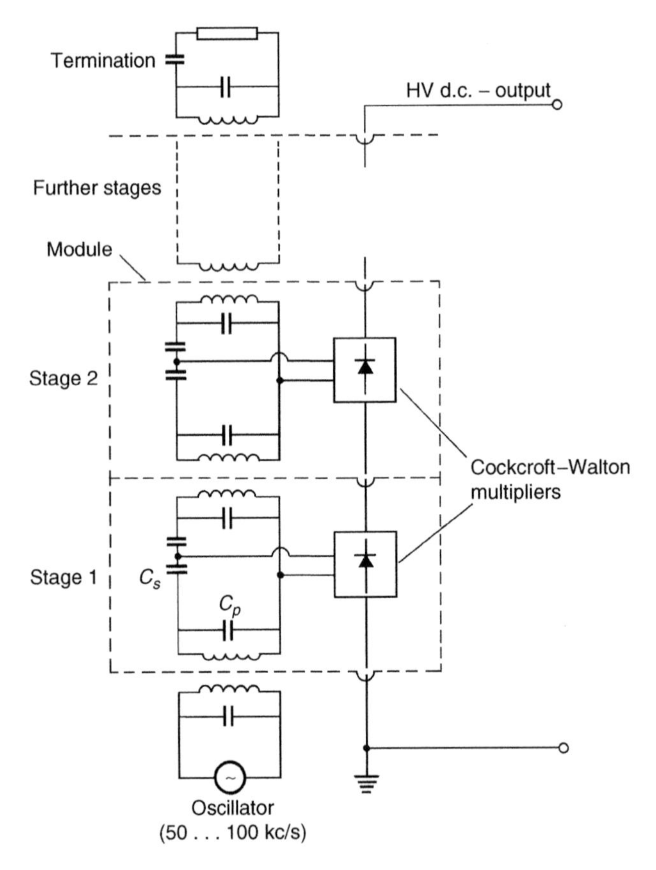

Figure 2.7 *The 'Engetron' or Deltatron principle*

of the transformers. In addition to this, to every primary and secondary winding a capacitor C_p is connected in parallel, which provides an overcompensation of the magnetizing currents. The whole chain of cascaded transformers is loaded by a terminating resistor; thus the network acts similarly to a terminated transmission line along which the a.c. voltage remains nearly constant and has a phase shift between input (high-frequency power supply) and output (termination). The transformers, therefore, are not used to increase the a.c. voltage.

It is now possible to connect to every stage indicated as usual Cockcroft–Walton cascade circuit, with only a small input voltage (some kV), producing, however, output voltages of some 10 kV per stage. The storage columns of these Cockcroft–Walton cascades are then directly series connected, providing the high d.c. output voltage for the whole cascade transformer HVDC generator unit. Typically up to about 25 stages can be used, every stage being modular constructed. As these modules are quite small, they can be stacked in a cylindrical unit which is then insulated by SF_6. Not shown in Fig. 2.7 is the voltage regulation system, which is controlled by a parallel mixed R-C voltage divider and a high-frequency oscillator, whose frequency ranges from 50 to 100 kHz. As for these high frequencies the capacitors within the Cockcroft–Walton circuits can be very small, and the energy stored is accordingly low; regulation due to load variations or power voltage supply variations is very fast (response time typically about 1 msec). The small ripple factor is not only provided by the storage capacitor, but also by the phase-shifted input voltages of the cascade circuits. Amongst the disadvantages is the procedure to change polarity, as all modules have to be reversed.

Summary and concluding remarks to 2.1.1

It has been shown that all a.c. to d.c. voltage conversion systems could be classed between the circuits of Figs 2.1 and 2.3, if single-phase a.c. voltages are converted into d.c. voltages. A high d.c. to a.c. voltage ratio can only be gained with a high product of a.c. frequency and energy stored in the smoothing capacitors, as they have to store electrical energy within each cycle, during which the a.c. power is oscillating. If, therefore, the d.c. output should be very stable and continuous, a high product (fC) is necessary. A reduction of stored energy is possible if the a.c. power is not only provided at ground potential, this means if a.c. power is injected into the circuits at different potential levels. The savings, therefore, can be made either on the a.c. or d.c. side. The large variety of possible circuits and technical expenditure is always strongly related to the 'quality' of the d.c. power needed, this means to the stability and the ripple of the output voltage.

2.1.2 Electrostatic generators

Electrostatic generators convert mechanical energy directly into electrical energy. In contrast to electromagnetic energy conversion, however, electrical charges are moved in this generator against the force of electrical fields, thus gaining higher potential energies and consuming mechanical energy. All historical electrostatic machines, such as the Kelvin water dropper or the Wimshurst machine, are therefore forerunners of modern generators of this type. A review of earlier machines may be found in reference 12.

Besides successful developments of 'dust generators' presented by Pauthenier *et al.*[(13)] the real breakthrough in the generation of high and ultra-high d.c. voltages is linked with Van de Graaff, who in 1931 succeeded with the development of electrostatic belt-driven generators.[(14)] These generators are in common use today in nuclear physics research laboratories. Figure 2.8 demonstrates the principle of operation, which is described in more detail in reference 4. Charge is sprayed onto an insulating moving belt by means of corona discharge points (or direct contact) which are at some 10 kV from earth potential. The belt, the width of which may vary widely (some cm up to metres), is driven at about 15–30 m/sec by means of a motor and the charge is conveyed to the upper end where it is removed from the belt by discharging points connected to the inside of an insulated metal electrode through which

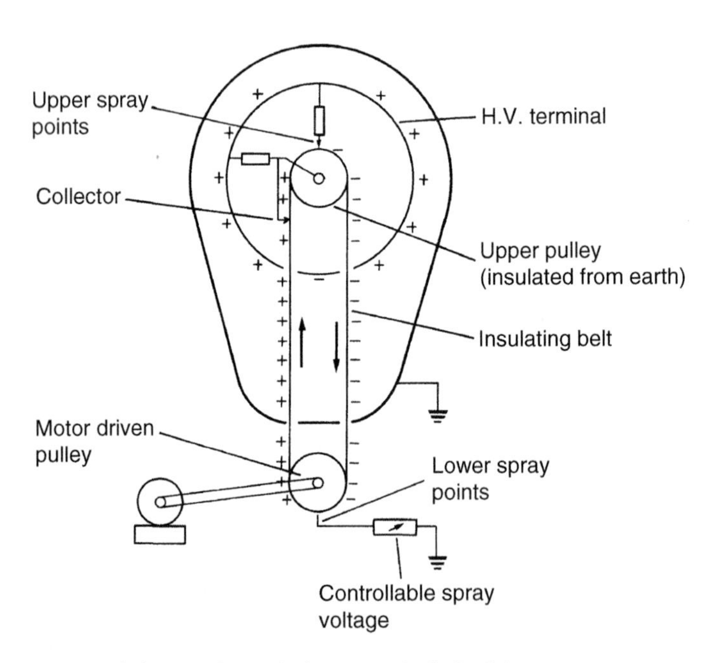

Figure 2.8 *Outline of electrostatic belt-driven generator*

the belt passes. The entire equipment is usually enclosed in an earthed metal tank filled with insulating compressed gases of good performance such as air, mixtures of N_2–CO_2, Freon 12 (CCl_2, F_2) or SF_6. For simple applications the metal tank can be omitted, so that the insulation is provided by atmospheric air only.

The potential of the h.v. terminal at any instant is $V = Q/C$ above earth, where Q is the charge stored and C is the capacitance of the h.v. electrode to ground. The potential of the terminal rises at a rate given by $\mathrm{d}V/\mathrm{d}t = I/C$, where

$$I = \hat{S}bv \tag{2.13}$$

is the net charging current to the terminal. In this equation, $\hat{S}$ is the charge density at the belt in coulombs/m^2, b its width in m, and v the belt speed in m/sec. In practice, $\mathrm{d}V/\mathrm{d}t$ may reach a value of 1 MV/sec and it appears that the final potential of the h.v. electrode would be infinite in the absence of any mechanism of charge loss. Equilibrium is in practice established at a potential such that the charging current equals the discharge current which includes load currents – also due to voltage dividers, leakage currents and corona losses, if present – and by voltage regulating systems which are based on voltage measurement and the controllable spray unit.

While the h.v. terminal electrode can easily be shaped in such a way that local discharges are eliminated from its surface, the field distribution between this electrode and earth along the fast moving belt is of greatest importance. The belt, therefore, is placed within properly shaped field grading rings, the grading of which is provided by resistors and sometimes additional corona discharge elements.

The lower spray unit, shown in Fig. 2.8, may consist of a number of needles connected to the controllable d.c. source so that the discharge between the points and the belt is maintained. The collector needle system is placed near the point where the belt enters the h.v. terminal.

A self-inducing arrangement is commonly used for spraying on the down-going belt charges of polarity opposite to that of the h.v. terminal. The rate of charging of the terminal, for a given speed of the belt, is therefore doubled. To obtain a self-charging system, the upper pulley is connected to the collector needle and is therefore maintained at a potential higher than that of the h.v. terminal. The device includes another system of points (shown as upper spray points in Fig. 2.8) which is connected to the inside of the h.v. terminal and is directed towards the pulley at the position shown. As the pulley is at a higher positive potential, the negative charges of the corona at the upper spray points are collected by the belt. This neutralizes any remaining positive charges on the belt and leaves any excess negative charges which travel down with it and are neutralized at the lower spray points.

For a rough estimation of the current I which can be provided by such generators, we may assume a homogeneous electrical field E normal to the belt running between the lower spray points and the grounded lower pulley. As $E = D/\varepsilon_0 = \hat{S}/\varepsilon_0$, D being the flux density, ε_0 the permittivity and $\hat{S}$ the charge density according to eqn (3.13) deposited at the belt, with $\varepsilon_0 = 8.85 \times 10^{-12}$ As/Vm, the charge density cannot be larger than about 2.7×10^{-5} As/m^2 if $E = 30$ kV/cm. For a typical case the belt speed might be $v = 20$ m/sec and its width $b = 1$ m. The charging current according to eqn (2.13) is then $I \cong 540\,\mu$A. Although with sandwiched belts the output current might be increased as well as with self-inducing arrangements mentioned above, the actual short-circuit currents are limited to not more than a few mA with the biggest generators.

The main advantages of belt-driven electrostatic generators are the high d.c. voltages which can easily be reached, the lack of any fundamental ripple, and the precision and flexibility, though any stability of the voltage can only be achieved by suitable stabilizing devices. Then voltage fluctuations and voltage stability may be in the order down to 10^{-5}.

The shortcomings of these generators are the limited current output, as mentioned above, the limitations in belt velocity and its tendency for vibrations, which aggravates an accurate grading of the electrical fields, and the maintenance necessary due to the mechanically stressed parts.

The largest generator of this type was set into operation at Oak Ridge National Laboratory.[15] A view of this tandem-type heavy ion accelerator is shown in Fig. 2.9. This generator operates with 25 MV, and was tested up to internal flashovers with about 31 MV.

For h.v. testing purposes only a limited amount of generators are in use due to the limited current output. A very interesting construction, however, comprising the Van de Graaff generator as well as a coaxial test arrangement for testing of gases, is used at MIT[16] by Cooke. This generator, with an output of about 4 MV, may be controlled to provide even very low frequency a.c. voltages.

The disadvantages of the belt-driven generators led Felici to develop electrostatic machines with insulating cylindrical rotors which can sustain perfectly stable movement even at high speeds. The schematic diagram of such a machine[17] is shown in Fig. 2.10. To ensure a constant narrow air gap, the stator is also made in the form of a cylinder. If the stator is a perfect insulator, ions are deposited on its surface which tend to weaken the field. In order to avoid such ion screening, a slight conductivity has to be provided for the stator and resistivities in the range 10^{11}–$10^{13}\,\Omega$/cm have been found satisfactory.

The overall efficiency of the machine is higher than 90 per cent and the life expectancies are only limited by mechanical wearing of the bearings, provided the charge density on the rotor surface is kept within limits which depend upon the insulating material employed. Epoxy cylinders have a practically unlimited

Figure 2.9 *25-MV electrostatic tandem accelerator (Oak Ridge National Laboratory)*

life if the density remains sufficiently low. Unlike the rectifier circuit, the cylindrical generator delivers a smooth and continuous current without any ripple.

Sames of France have built two-pole generators of the Felici type. They give an output of 600 kV at 4 mA and are suitable for use with particle accelerator, electrostatic paint spray equipment, electrostatic precipitator, X-ray purposes and testing h.v. cables. A cross-sectional view of the generator

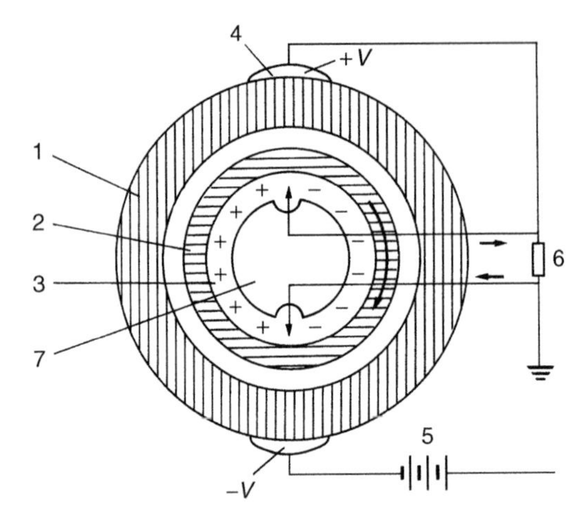

Figure 2.10 *Diagrammatic cross-section of the Felici generator. (1) Cylindrical stator. (2) Insulating rotor. (3) Ionizer. (4) Contact metallic segments. (5) Auxiliary generator. (6) Load. (7) Stationary insulating core* ($-V = 30$ *kV*; $+V = 200$ *kV*)

is shown in Fig. 2.11. The distinctive features include a cylindrical pressure vessel enclosing the generator, the rotor of which is driven at 3000 rpm by means of an induction motor. Ions from an exciting source are sprayed onto the rotor at the charging poles and are transported to the output poles with a consequent rise of potential. The transfer of charge takes place by means of thin blades placed a short distance from the rotor, and in the absence of any rubbing contact the efficiency of the machine is about 90 per cent. The characteristics of the 600-kV generator are such that the fluctuation in the voltage is less than 10^{-4} per cent and the voltage drop at full load current of 4 mA is only 500 V. For a 5 per cent variation in the main voltage, the generator voltage remains within 10^{-5} per cent.

The main applications of these 'rotating barrel' generators are in physics as well as in different areas of industrial applications, but rarely in h.v. testing. The maximum voltages are limited to less than 750 kV.

Finally, another type of electrostatic generator is the vacuum-insulated 'varying capacitance machine', first discussed in detail by Trump(18) and recently again investigated by Philp.(19) This machine provides a high voltage in the range up to about 1 MV and/or high power in the range of megawatts. The high efficiency, however, could only be reached by high field gradients within the generator, which up to now can only be obtained theoretically by assuming the possible high E values in vacuum. It is, however, doubtful whether the stresses necessary can be reached within the large electrode areas

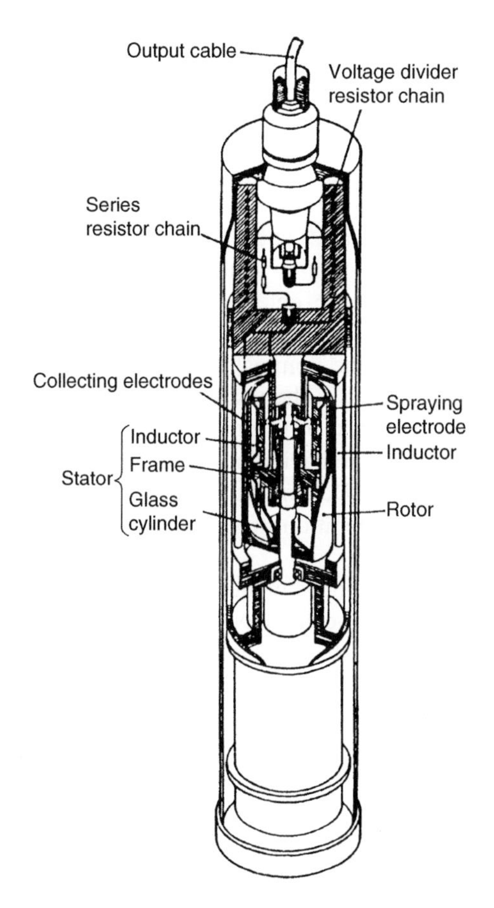

Figure 2.11 *Sames electrostatic generator*

present in such generators, and therefore only a reference to this type of generator might be useful.

2.2 Alternating voltages

As electric power transmission with high a.c. voltages predominates in our transmission and distribution systems, the most common form of testing h.v. apparatus is related to high a.c. voltages. It is obvious then that most research work in electrical insulation systems has to be carried out with this type of voltage.

In every laboratory HVAC supplies are therefore in common use. As far as the voltage levels are concerned, these may range from about 10 kV r.m.s. only up to more than 1.5 MV r.m.s. today, as the development of transmission voltages up to about 1200 kV has proceeded for many years. For routine testing, the voltage levels for power-frequency testing are always related to the highest r.m.s. phase-to-phase voltage V_m of power transmission systems. This 'rated power-frequency short duration withstand voltage' V_t is different for different apparatus used within the transmission systems and also dependent upon the type of insulation coordination applied (see Chapter 8). For $V_m < 300\,\text{kV}$, the ratio V_t/V_m is up to about 1.9 and may decrease with higher values of V_m. If, nevertheless, higher nominal voltages for the a.c. testing supplies are foreseen, the necessity for the determination of safety factors are most responsible for this fact.

In general, all a.c. voltage tests are made at the nominal power frequency of the test objects. Typical exceptions are related to the testing of iron-cored windings, i.e. potential or instrument transformers, or to fundamental studies on insulating materials or systems. For iron-cored windings, the frequency has to be raised to avoid saturation of the core. Depending upon the type of testing equipment used, the methods for the generation of variable-frequency voltages might be expensive.

A fundamental design factor for all a.c. testing supplies is an adequate control system for a continuous regulation of the high output voltages. In general, this will be performed by a control of the primary or l.v. input of the voltage step-up systems. It is not the aim of this chapter to deal with the details of these systems. Some hints related to the different methods will be given in section 2.4.

Although power transmission systems are mostly of three-phase type, the testing voltages are usually single-phase voltages to ground. The waveshapes must be nearly pure sinusoidal with both half-cycles closely alike, and according to the recommendations[2,3] the results of a high-voltage test are thought to be unaffected by small deviations from a sinusoid if the ratio of peak-to-r.m.s. values equals $\sqrt{2}$ within ± 5 per cent, a requirement which can be assumed to be met if the r.m.s. value of the harmonics does not exceed 5 per cent of the r.m.s. value of the fundamental. The r.m.s. value is for a cycle of T

$$V_{(\text{r.m.s.})} = \sqrt{\frac{1}{T}\int_0^T V^2(t)\,\mathrm{d}t}.$$

The nominal value of the test voltage, however, is defined by its peak value divided by $\sqrt{2}$, i.e. $V_{\text{peak}}/\sqrt{2}$. The reason for this definition can be found in the physics of breakdown phenomena in most of the insulating materials, with the breakdown mainly following the peak voltages or the highest values of field strength.

Testing of h.v. apparatus or h.v. insulation always involves an application of high voltages to capacitive loads with low or very low power dissipation only. In general, power dissipation can be completely neglected if the nominal power output of the supply is determined. If C_t is the capacitance of the equipment or sample under test, and V_n the nominal r.m.s. voltage of the h.v. testing supply, the nominal KVA rating P_n may be calculated from the design formula

$$P_n = kV_n^2 \omega C_t \qquad (2.14)$$

in which the factor $k \geq 1$ accounts for additional capacitances within the whole test circuit and some safety factor. Examples for additional capacitances are h.v. electrodes and connections between test object and voltage source, which might have large diameters and dimensions to avoid heavy discharges or even partial discharges, or measurement devices as, e.g., capacitor voltage dividers or sphere gaps frequently incorporated within the test circuit. This safety factor k might range from only about 2 for very high voltages of $\geq$1 MV, and may increase to higher values for lower nominal voltages, as over-dimensioning is economically possible. The capacitance of test equipment C_t may change considerably, depending upon the type of equipment. Typical values are:

Simple post or suspension insulators	some 10 pF
Bushings, simple and graded	~100–1000 pF
Potential transformers	~200–500 pF
Power transformers	
<1000 kVA	~1000 pF
>1000 kVA	~1000–10 000 pF
H.V. power cables:	
Oil-paper impregnated	~250–300 pF/m
Gaseous insulated	~60 pF/m
Metal clad substation, SF_6 insulated	~1000–>10 000 pF

One may calculate the nominal currents $I_n = P_n/V_n$ from eqn (2.14) for different test voltages, different C_t values as shown above, and proper safety factors k. From such estimations it may be seen that these currents may range from some 10 mA for testing voltages of 100 kV only, up to amperes in the megavolt range. Although these currents are not high and the nominal power is moderate, many efforts are necessary to keep the test equipment as small as possible, as the space is limited and expensive within any h.v. laboratory. Frequently the equipment will be used also for field testing. Then the portability and transportation calls for lightweight equipment. Some facilities are possible by the fact that most of the test voltages are only of short duration. The nominal ratings are, therefore, often related to short time periods of 15 min. Due to the relatively large time constants for the thermal temperature rise, no sophisticated cooling systems are in general necessary within the voltage testing supplies.

A final introductory remark is related to the necessity that all supplies can withstand sudden voltage breakdowns of the output voltage. The stress to the windings and coils accompanied by the breakdown events is usually not related to the short-circuit currents and thus the magnetic forces within the windings, as those currents are not large either; more frequently it is the stray potential distribution between the windings which will cause insulation failures. One may also provide proper damping resistors between h.v. testing supply and the test equipment to reduce the rate of the sudden voltage drop and to avoid any overvoltages within the test circuit caused by interruptions of the breakdown phenomena. Nominal values of such damping resistors between 10 and 100 kΩ will usually not influence the test conditions. These resistors, however, are expensive for very high voltages and it should be checked whether the a.c. voltage supply can withstand the stresses without the damping resistors.

Most of the above remarks are common to the two main methods for the generation of high a.c. testing voltages: transformers and resonant circuits.

2.2.1 Testing transformers

The power frequency single-phase transformer is the most common form of HVAC testing apparatus. Designed for operation at the same frequency as the normal working frequency of the test objects (i.e., 60 or 50 Hz), they may also be used for higher frequencies with rated voltage, or for lower frequencies, if the voltages are reduced in accordance to the frequency, to avoid saturation of the core.

From the considerations of thermal rating, the kVA output and the fundamental design of the iron core and windings there is not a very big difference between a testing and a single-phase power transformer. The differences are related mainly to a smaller flux density within the core to avoid unnecessary high magnetizing currents which would produce higher harmonics in the voltage regulator supplying the transformer, and to a very compact and well-insulated h.v. winding for the rated voltage. Therefore, a single-phase testing unit may be compared with the construction of a potential transformer used for the measurement of voltage and power in power transmission systems.

For a better understanding of advanced circuits, the fundamental design of such 'single unit testing transformers' will be illustrated. Figure 2.12(a) shows the well-known circuit diagram. The primary winding '2' is usually rated for low voltages of ≤ 1 kV, but might often be split up in two or more windings which can be switched in series or parallel (not shown here) to increase the regulation capabilities. The iron core '1' is fixed at earth potential as well as one terminal of each of the two windings. Simplified cross-sections of two possible constructions for the unit itself are given in Figs 2.12(b) and (c). In both cases the layout arrangement of core and windings is basically the same. Figure 2.12(b), however, shows a grounded metal tank unit, for which an h.v. bushing '6' is necessary to bring the high voltage out of the tank '5'. Instead of a bushing, a coaxial cable could also be used if this improves

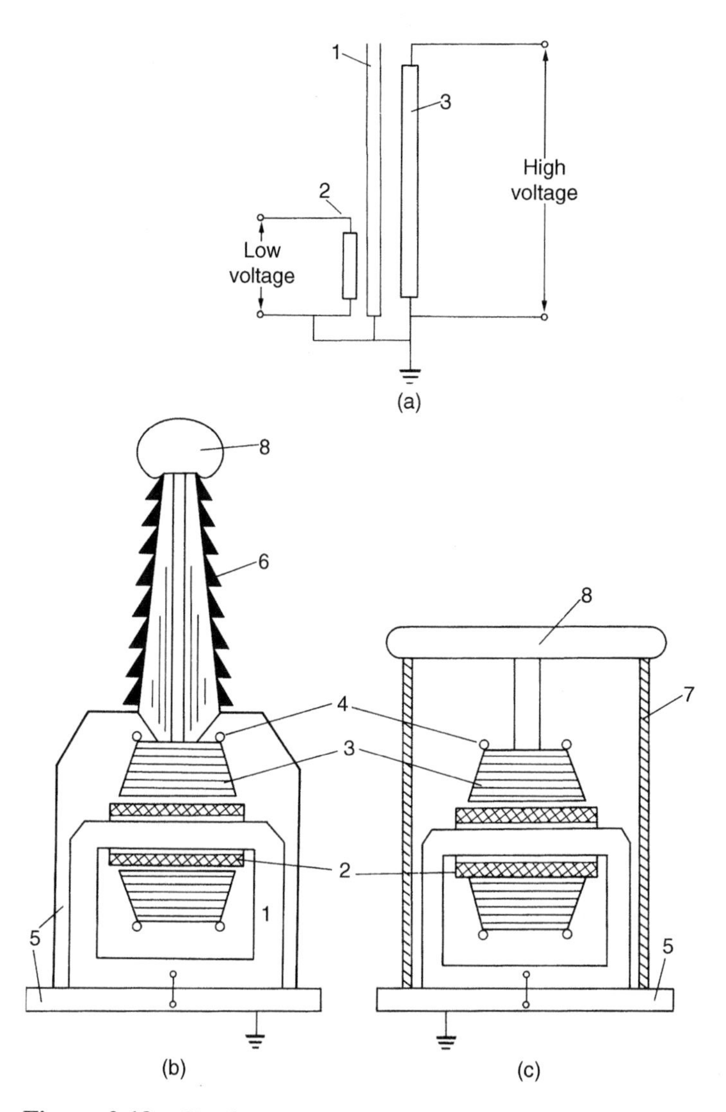

Figure 2.12 *Single unit testing transformers. (a) Diagram. (b & c) different construction units. (1) Iron core. (2) Primary l.v. or exciting winding. (3) Secondary h.v. winding. (4) Field grading shield. (5) Grounded metal tank and base. (6) H.V. bushing. (7) Insulating shell or tank. (8) H.V. electrode*

the connection between testing transformer and test object. In Fig. 2.12(c) the active part of the transformer is housed within an isolating cylinder '7' avoiding the use of the bushing. This construction reduces the height, although the heat transfer from inside to outside is aggravated. In both cases the vessels would be filled with high-quality transformer oil, as most of the windings are oil-paper insulated.

The sectional view of the windings shows the primary winding close to the iron core and surrounded by the h.v. winding '3'. This coaxial arrangement reduces the magnetic stray flux and increases, therefore, the coupling of both windings. The shape of the cross-sectional view of winding no. 3 is a hint to the usual layout of this coil: the beginning (grounded end) of the h.v. winding is located at the side close to the core, and the end close to a sliced metal shield, which prevents too high field intensities at h.v. potential. Between both ends the single turns are arranged in layers, which are carefully insulated from each other by solid materials (kraft paper sheets for instance). Adjacent layers, therefore, form coaxial capacitors of high values, and if those capacitances are equal – produced by the reduced width of the single layers with increasing diameters – the potential distribution for transient voltages can be kept constant. By this procedure, the trapezoidal shape of the cross-section is originated.

It may well be understood that the design of the h.v. winding becomes difficult if voltages of more than some 100 kV must be produced within one coil. Better constructions are available by specialized techniques, mainly by 'cascading' transformers.

The first step in this technique is to place two h.v. windings on one iron core, to join both windings in series and to connect this junction with the core.[20] For illustration, the circuit diagram is shown in Fig. 2.13 in combination with a simplified cross-section of the active part. The arrangement could still be treated as a single unit transformer, as only one core exists. The mid-point of the h.v. winding is connected to the core and to a metal tank, if such a tank is used as a vessel. The cross-section shows that the primary winding '2' is, however, placed now around the first part '3a' of the whole h.t. winding, whose inner layer, which is at half-potential of the full output voltage, is connected to the core. There are two additional windings, '4a' and '4b', rated for low voltages, which act as compensating windings. These are placed close to the core and reduce the high leakage reactance between '3b' and the primary '2'. Often an exciting winding '5', again a winding rated for low voltages as the primary winding, is also available. This exciting winding is introduced here as it will be needed for the cascading of transformers. Note that this winding is at the full output potential of the transformer.

Although no vessel is shown in which such a unit would be immersed, it can easily be understood that for metal tank construction (see Fig. 2.12(b)) two h.v. bushings are now necessary. The tank itself must be insulated from earth for half-output voltage. This typical view for testing transformers can be seen in Fig. 2.14. If, however, insulating tanks are employed, this internal layout may not necessarily be recognized from outside.

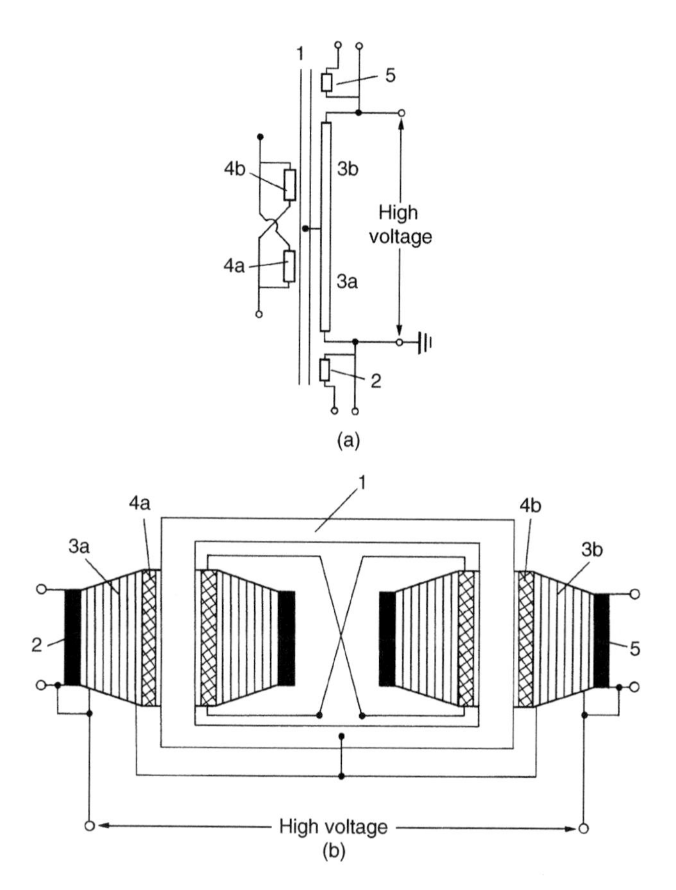

Figure 2.13 *Single unit testing transformer with mid-point potential at core: Diagram (a) and cross-section (b). (1) Iron core. (2) Primary winding. (3a & b) High-voltage windings. (4a & b) compensating windings. (5) Exciting winding*

Cascaded transformers

For voltages higher than about 300 to 500 kV, the cascading of transformers is a big advantage, as the weight of a whole testing set can be subdivided into single units and therefore transport and erection becomes easier. A review of earlier constructions is given in reference 4.

A prerequisite to apply this technique is an exciting winding within each transformer unit as already shown in Fig. 2.13. The cascading principle will be illustrated with the basic scheme shown in Fig. 2.15. The l.v. supply is connected to the primary winding '1' of transformer I, designed for an h.v. output of V as are the other two transformers. The exciting winding

Figure 2.14 *Testing transformer for 1200 kV r.m.s. comprising three single unit transformers according to Fig. 2.13, with metallic tanks and bushings (High Voltage Laboratory, Technical University of Munich, Germany). (Note. Suspended at ceiling and connected with transformer is a selenium-type rectifier with a reverse voltage of 3.4 MV, see ref. 5.)*

'3' supplies the primary of the second transformer unit II; both windings are dimensioned for the same low voltage, and the potential is fixed to the high potential V. The h.v. or secondary windings '2' of both units are series connected, so that a voltage of 2 V is produced hereby. The addition of the stage III needs no further explanation. The tanks or vessels containing the

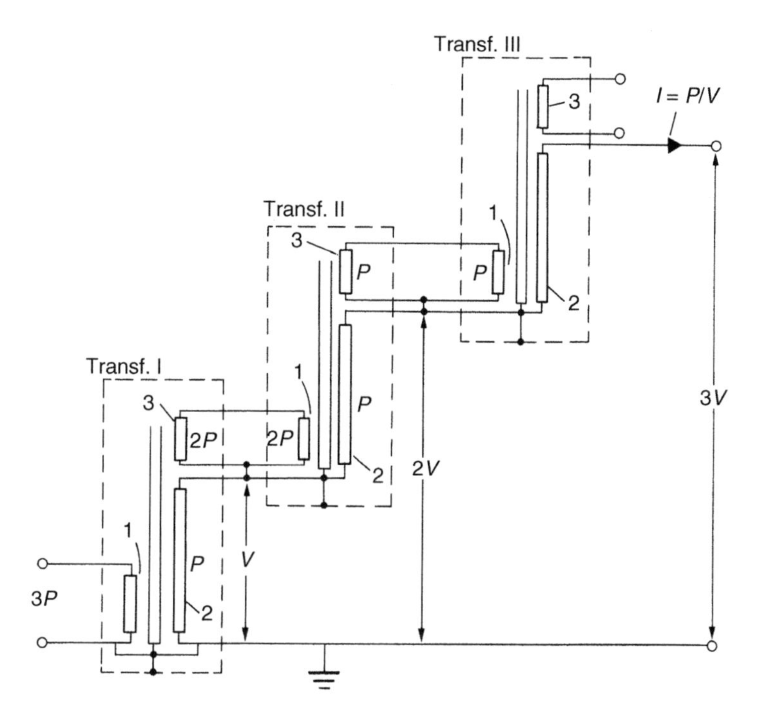

Figure 2.15 *Basic circuit of cascaded transformers. (1) Primary windings. (2) Secondary h.t. windings. (3) Tertiary exciting windings*

active parts (core and windings) are indicated by dashed lines only. For a metal tank construction and the non-subdivided h.v. winding assumed in this basic scheme, the core and tank of each unit would be tapped to the l.v. terminal of each secondary winding as indicated. Then the tank of transformer I can be earthed; the tanks of transformers II and III are at high potentials, namely V and 2 V above earth, and must be suitably insulated. Through h.t. bushings the leads from the exciting coils '3' as well as the tappings of the h.v. windings are brought up to the next transformer. If the h.v. windings of each transformer are of mid-point potential type (see Fig. 2.13), the tanks are at potentials of 0.5 V, 1.5 V and 2.5 V respectively, as shown in Fig. 2.14. Again, an insulating shell according to Fig. 2.12 could avoid the h.t. bushings, rendering possible the stacking of the transformer units as shown in Fig. 2.16.

The disadvantage of transformer cascading is the heavy loading of primary windings for the lower stages. In Fig. 2.15 this is indicated by the letter P, the product of current and voltage for each of the coils. For this three-stage cascade the output kVA rating would be $3P$, and therefore each of the h.t. windings '2' would carry a current of $I = P/V$. Also, only the primary winding of transformer III is loaded with P, but this power is drawn from the exciting winding of transformer II. Therefore, the primary of this second stage is loaded with $2P$. Finally, the full power $3P$ must be provided by the primary of transformer I. Thus an adequate dimensioning of the primary and exciting coils is necessary. As for testing of insulation, the load is primarily a capacitive one, a compensation of this capacitive load by l.v. reactors, which are in parallel

Figure 2.16 *(a) Cascaded testing transformers with insulating shell construction (courtesy IREQ, Canada)*

Figure 2.16 *(b) Cascaded testing transformers with metal tanks and coolers. Total voltage 3000 kV, 4 A (courtesy HIGH-VOLT Dresden, Germany)*

to the primary windings, is possible. As these reactors must be switched in accordance to the variable load, however, one usually tries to avoid this additional expense. It might also be necessary to add tuned filters to improve the waveshape of the output voltage, that is to reduce higher harmonics.[(21)]

Without any compensation, the overloading of the lower stage transformers introduces a relatively high internal impedance of the whole cascade circuit. In a simplified equivalent circuit of each transformer unit, which consists of a three-windings-type, we may define leakage or stray reactances X for each winding, the primary X_p, the h.t. winding X_h and the exciting winding X_e. Neglecting losses within the windings and magnetizing currents, the somewhat simplified calculation of the resultant reactance X_{res} of a cascade unit with n transformers having the individual reactances X_{pv}, X_{hv} and X_{ev} shows

$$X_{\text{res}} = \sum_{v=1}^{n} [X_{hv} + (n - \nu)^2 X_{ev} + (n + 1 - \nu)^2 X_{pv}]. \tag{2.15}$$

(All reactances related to same voltage.)

Assuming three equal transformer units, the equation leads to a resultant reactance of

$$X_{\text{res}} = 3X_h + 5X_e + 14X_p$$

instead of only $3(X_h + X_e + X_p)$ which might be expected.

Cascaded transformers are the dominating HVAC testing units in all large testing laboratories. In Fig. 2.16(a) the 2.4-MV cascade of the Quebec Hydro Research and Testing Laboratory can be seen. Here, each of the six 600 kV single units are of insulating case type and the two lower stages consist of four units to avoid overloading for full rated current. The world's largest a.c. testing station at WEI Istra near Moscow, Russia, is equipped with a cascaded testing transformer rated for 3 MV, 12 MVA, which is shown in Fig. 2.16(b). This very large unit was designed and built by the former firm TuR, Dresden, Germany, and commissioned by Siemens.

A final remark relates to the effect that for all transformers the output voltage will increase with load, as this is formed by capacitors. The equivalent circuit of a transformer loaded by capacitors forms a series resonant circuit, which is shown in Fig. 2.17 and will be used to introduce the resonant circuits for testing purposes. With nominal load, the exciting frequency is well below resonance frequency, so that the voltage increase is only about proportional to the load. If the testing transformer, however, is switched to a primary voltage higher than about half the rated voltage, the output voltage will oscillate with resonance frequency, and the amplitude may easily become higher than the rated voltage. The impedance of the voltage regulators used must also be taken into account for quantitative calculations.

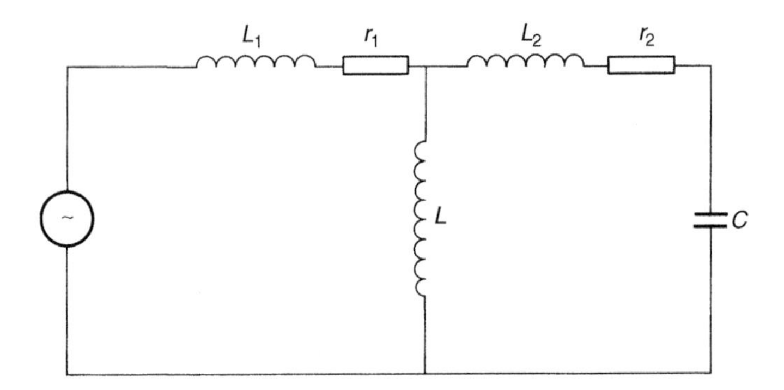

Figure 2.17 *Equivalent circuit of straight test set consisting of a transformer and test capacitance*

2.2.2 Series resonant circuits

The tuned series resonant h.v. testing circuit arose as a means of overcoming the accidental and unwanted resonance to which the more conventional test sets are more prone. If we consider a conventional 'straight' test set such as the first unit in Fig. 2.15 used in, say, testing a capacitor C, then its equivalent circuit will be that shown in Fig. 2.17. In this circuit $(r_1 + j\omega L_1)$ represents the impedances of the voltage regulator and the transformer primary. ωL represents the transformer shunt impedance which is usually large compared with L_1 and L_2 and can normally be neglected. $(r_2 + j\omega L_2)$ represents the impedance of the transformer secondary. $1/\omega C$ represents the impedance of the load.

If by chance $\omega(L_1 + L_2) = l/\omega C$, accidental resonance occurs. At supply frequency the effect can be extremely dangerous, as the instantaneous voltage application can be of the order of 20 times the intended high voltage. This has given rise to some vicious explosions during cable testing. The greatest possibility of this occurring is when testing at the maximum limit of current and relatively low voltages, i.e. high capacitive load. Unfortunately the inductance of most of the supply regulators varies somewhat over its range, so that resonance does not necessarily occur when the voltages are at their low switch-on value, but rather suddenly at the higher voltage range.

Resonance of a harmonic can similarly occur, as harmonic currents are present due to the transformer iron core. In recent years, also the power supply voltages contain a large amount of harmonics due to the still increasing application of power electronics. These resonances are not quite so disastrous, but third harmonics have been observed of greater amplitude than the fundamental, and even the thirteenth harmonic can give a 5 per cent ripple on the voltage waveform. This form of harmonic resonance causes greater voltage distortion than other effects and occurs insidiously at particular capacitance loads, usually unnoticed by conventional instrumentation.

With the series resonant set, however, the resonance is controlled at fundamental frequency and no unwanted resonance can therefore occur.

Historically, in the period 1935–45, power engineers were increasingly aware of the potentialities of tuned circuits. It was not, however, until the late 1940s that engineers at Ferranti, England, and Standard Telefon of Kabelfabrik, Norway, combined to make this a practical proposition culminating in a 600-kV, 2400-kVA a.c. testing equipment completed in 1950, although an earlier version of a resonance transformer for supplying X-ray equipment has been described by Charlton.(22)

The development of this technique will be demonstrated based upon some circuits shown in Fig. 2.18. In each circuit, the capacitance C_t represents the almost pure capacitive load of the test objects and a constant power supply frequency is assumed. In Fig. 2.18(a) a continuously variable inductance (reactor) is connected to the l.v. winding of a step-up transformer, whose secondary winding is rated for the full test voltage. By this means, the impedance of the reactor is converted to the h.v. side. If the inductance of the reactor is tuned to match the impedance of the capacitive load, the idle power of the load is completely compensated. The step-up transformer, however, has to carry the full load current, which is a disadvantage of this circuit. The same disadvantage applies to the circuit of Fig. 2.18(b), although no special means are necessary to cascade two or more units. The inductors are

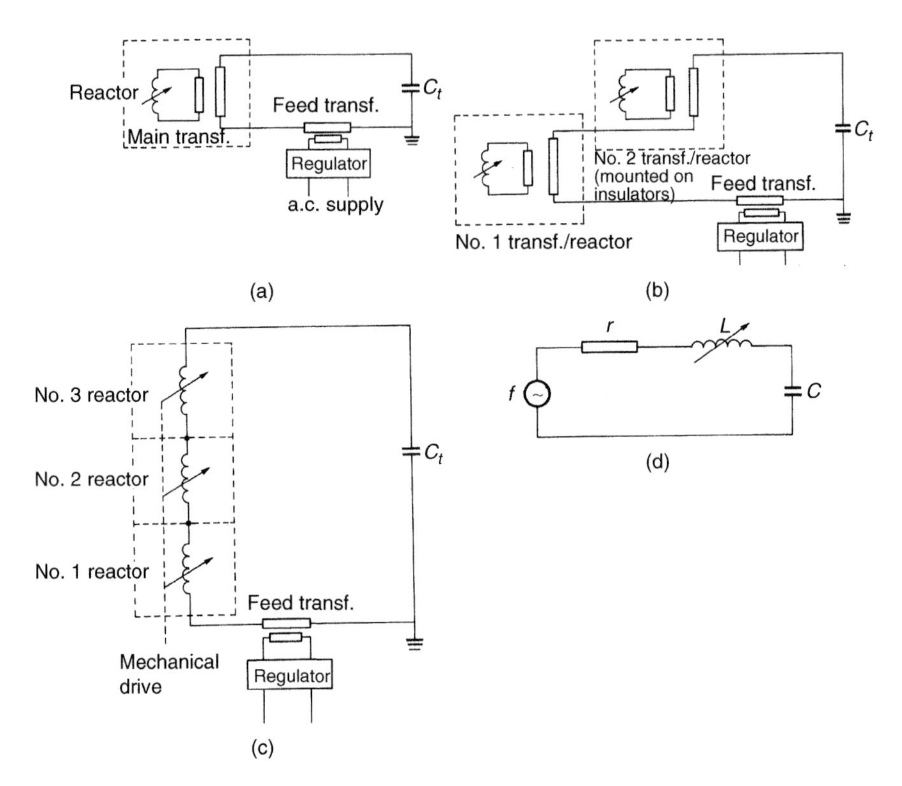

Figure 2.18 *Series resonant circuit for transformer/reactor. (a) Single transformer/reactor. (b) Two or more units in series. (c) Simplified diagram of s.r. circuit for h.t. reactor units in series. (d) Equivalent diagram of s.r. circuits*

designed for high-quality factors $Q = \omega L/R$ within the limits of the inductance variation. The feed transformer therefore injects the losses of the circuits only.

These types of s.r. circuits have been produced mainly up to about the late 1960s, since it was not possible to design continuously variable reactors for high voltages. Then, as described in reference 23, a new technique was developed with split iron cores, the gap of which is variable. This technique now provides h.t. continuously variable reactors up to about 300 kV per unit. Thus the testing step-up transformers can be omitted, as indicated in Fig. 2.18(c). The inductance of these h.t. reactors may be changed by up to 10 to 20 times, offering the opportunity to tune the circuits with capacitances C_t, which vary within the same order.

The equivalent circuit diagram for all these circuits is simply a low damped s.r. circuit sketched in Fig. 2.18(d). Because the equations of such circuits are well known, detailed designs will not be discussed here. It should be emphasized that the high output voltage may best be controlled by a continuously variable a.c. supply voltage, i.e. by a voltage regulator transformer, if the circuit was tuned before. The feed transformers are rated for the nominal currents of the inductors; the voltage rating can be as low as V/Q, if V is the full output voltage and Q the worst quality factor of the whole circuit. The additional advantages of the s.r. circuits may be summarized as follows.

Additional advantages of the series resonant circuit

1. The voltage waveshape is improved not only by the elimination of unwanted resonances, but also by attenuation of harmonics already in the power supply. A realistic figure for the amplification of the fundamental voltage amplitude on resonance is between 20 and 50 times for power frequencies of 50/60 Hz. Higher harmonic voltages are divided in the series circuit with a decreasing proportion across the capacitive load. It is easily seen that harmonics in the supply become insignificant.
2. The power required from the supply is lower than the kVA in the main test circuit. It represents only about 5 per cent of the main kVA with a unity power factor.
3. If a failure of the test specimen occurs, no heavy power arc will develop, as only the load capacitance will be discharged. This is of great value to the cable industry where a power arc can sometimes lead to the dangerous explosion of the cable termination. It has also proved invaluable for development work as the weak part of the test object is not completely destroyed. Additionally, as the arc is self-extinguishing due to this voltage collapse, it is possible to delay the tripping of the supply circuit.
4. The series or parallel operation of h.t. reactor or l.t. reactor/h.t. transformer units is simple and very efficient. Any number of units may be put in series without the high impedance problems associated with a cascaded testing transformer group (see eqn (2.15)). Equal voltage distributions for series connections are easily provided by a proper control of the individual reactor impedances. For heavy current testing it is possible to parallel the reactor or reactor/transformer units, even if the impedances are different, merely by controlling each associated reactance.
5. Various degrees of sophistication are possible concerning auto-tuning devices keeping the set in tune, if supply frequency or load capacitance varies during a long-term test, or concerning auto-voltage control.

Figure 2.19 shows cascaded h.t. reactors for an s.r. circuit according to Fig. 2.18(c). From this figure a further advantage may be seen not mentioned before, namely the reduction in the size and weight of such units in comparison to testing transformers. For testing transformers typically, a specific weight of about 10 to 20 kg/kVA (not including the necessary regulating and control equipment) can be assumed. According to reference 24 and Fig. 2.18(c), this

Figure 2.19 *2.2 MV series resonant circuit (Hitachi Research Laboratory, supplied by Hipotronix, Brewster, USA)*

weight for oil-insulated continuously variable h.v. reactors can be reduced to about 3 to 6 kg/kVA for a power frequency of 60 Hz.

For field testing of cables, large rotating machines or metal-clad gas-insulated substations (GIS), a still further reduction of weight and size of the testing equipment is very desirable. This goal was reached by the development[25] of a different kind of series resonant circuit, for which chokes with constant inductances are used. As the load capacitance cannot be kept within narrow limits, the supply frequency must thus be continuously variable to achieve resonance. This disadvantage, however, may be eliminated by the novel features, which may be briefly explained by the schematic diagram sketched in Fig. 2.20. An exciter supply, connected to the l.v. mains, excites the s.r. circuit with a variable frequency; this supply is therefore designed as a controlled frequency converter, which are standard devices nowadays. This converter supplies again only the losses of the testing circuit, which are usually less than about 2 per cent of the reactive power of C_t, if frequencies equal or higher than 50 Hz are used. The chokes can easily be designed for such high-quality factors, which increase with frequency up to some 100 Hz. In Fig. 2.20 only one nominal inductance L_n of the h.v. reactor is indicated; this inductance, however, might be provided by any number of chokes in series and/or in parallel. C_t represents the test object and other shunt capacitances, e.g. capacitor voltage dividers or some frequency-adjusting capacitor units, if a specified testing frequency f must be achieved. Due to the resonance condition, this frequency is always

$$f = \frac{1}{2\pi\sqrt{L_n C_t}}. \tag{2.16}$$

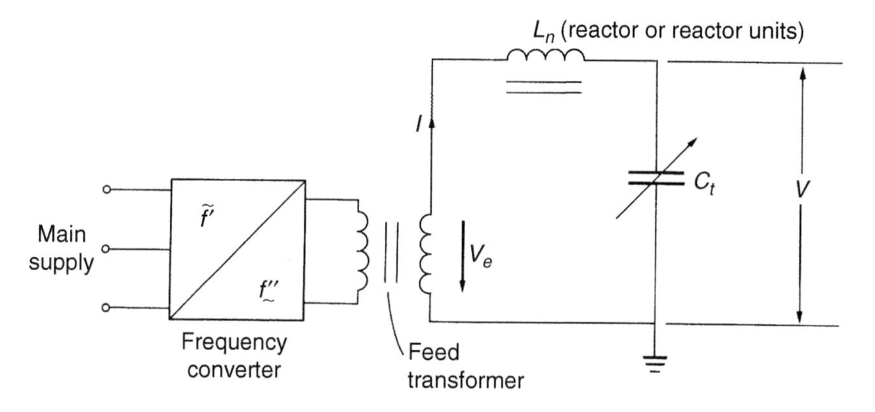

Figure 2.20 *Schematic diagram of s.r. test circuit with variable test frequency*

The nominal inductance L_n will predominantly be designed according to a nominal capacitance $C_n = C_t$ which is the highest capacitance that can be tested with the full rated voltage $V = V_n$ of the circuit, and a nominal frequency f_n, which is the lowest frequency within this rated voltage. With the above equation we thus obtain

$$L_n = \frac{1}{(2\pi)^2 f_n^2 C_n}. \qquad (2.17)$$

A further criterion for the design of the choke is the maximum or nominal current $I = I_n$, which either overheats the coil or saturates the iron core, if any. As the losses are very small, we may neglect $R \ll \omega L_n$ within the whole frequency range; I_n may thus directly be derived from the voltage drop across L_n, which is nearly the full rated voltage V_n, or from the fact that for all frequencies or every cycle the magnetic energy of the choke is equivalent to the electric energy stored within the test specimen. Thus

$$I_n = \frac{V_n}{2\pi f_n L_n} = V_n \sqrt{\frac{C_n}{L_n}}. \qquad (2.18)$$

These three equations are used to demonstrate the normalized operating characteristics of the circuit. For test objects with capacitance values C_t different from C_n, the resulting testing frequency f will also be different from f_n. The variation of the frequency then becomes, according to eqn (2.16),

$$\frac{f}{f_n} = \sqrt{\frac{C_n}{C_t}} = \frac{1}{\sqrt{C_t/C_n}}. \qquad (2.19)$$

For $C_t \leq C_n$, the reactor L_n can be used up to the full rated voltage V_n. Although the frequency increases according to eqn (2.19), the load current will always be lower than I_n. Ohm's law or eqn (2.18) can be used to derive the relationship of the normalized current for $C_t \leq C_n$,

$$\frac{I}{I_n} = \frac{f_n}{f} = \sqrt{\frac{C_t}{C_n}}. \qquad (2.20)$$

For $C_t > C_n$, this circuit may still and conveniently be applied, if the testing voltage $V = V_t$ is decreased to keep the current at its nominal value I_n. As the current I is always proportional to the testing voltage, we may extend eqn (2.20) to

$$\frac{I}{I_n} = \frac{V_t}{V_n} = \sqrt{\frac{C_t}{C_n}} \qquad (2.20a)$$

and apply this equation to show the necessary reduction of the testing voltage for $C_t > C_n$, if we limit I to I_n:

$$\frac{V_t}{V_n} = \frac{1}{\sqrt{C_t/C_n}}. \qquad (2.21)$$

The normalized operating conditions given by eqns (2.19), (2.20) and (2.21) are illustrated in Fig. 2.21. Whereas for quite small test specimens the test frequency f may conveniently be limited by the addition of a permanent h.t. capacitor as, e.g., a capacitor voltage divider; the relatively modest variation of this frequency for large capacitors under test will improve the flexibility of applications, i.e. for the testing of power cables with a.c. voltages. The actual limitations in testing of very large test specimens with lower voltages than V_n are given by the reduction of Q for too low frequencies, and the frequency for which the exciter transformer saturates.

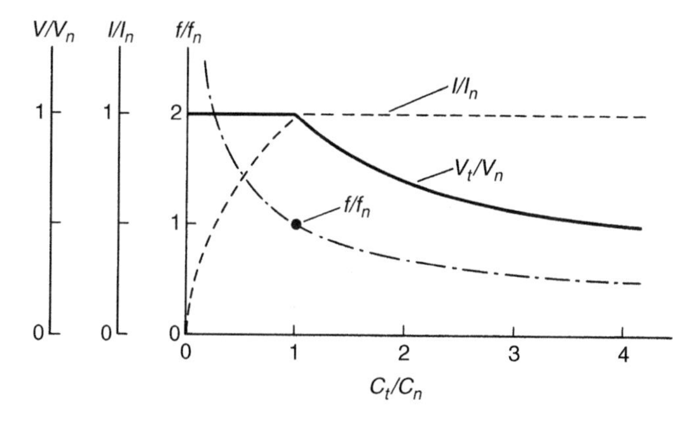

Figure 2.21 *Operating characteristics of circuit according to Fig. 2.20*

The prototype reactor described in reference 25 was designed for $V_n = 200\,\text{kV}$, $I_n = 6\,\text{A}$, $f_n = 100\,\text{Hz}$. The total weight is about 300 kg only, yielding a specific weight of 0.25 kg/kVA for the nominal frequency. The reactor has a cylindrical bar iron core, which is at half-potential of the subdivided h.t. winding; this winding is coaxially placed across the core. The construction provides excellent and high Q values between 50 and 150 within a frequency range of 50 to more than 1000 Hz. Thus a very small a.c. testing supply is available which can easily be handled and conveyed. A further advantage of this circuit is obviously related to the cheap generation of frequencies higher than power frequencies, which may be used for ageing tests. Since about 1980, powerful series resonance circuits with variable frequencies were being used more often,[26,52] as it was recognized that at least for on-site tests the influence of frequencies, which are not very far from the nominal working frequency, is of minor importance. Apart from tests on GIS, the main application is now related to on-site tests of polymeric cables, for which d.c. voltage tests are no longer applied due to well-known reasons.[50,51] Figure 2.22 shows the test set-up comprising 12 modular reactor units of

Figure 2.22 *Modular reactors of a series resonant circuit with variable frequency during an on-site test of very long polymeric cables (courtesy FKH, Zurich, Switzerland)*

200 kV/6 A each as used during on-site a.c. voltage tests of 110 kV XLPE cables with a length of about 7 km.

Whereas s.r. circuits are still used less in h.v. laboratories than testing transformers, especially designed resonant circuits have often been applied in conjunction with X-ray sets even for voltages in the MV range.[(4)]

2.3 Impulse voltages

As explained in detail in Chapter 8, disturbances of electric power transmission and distribution systems are frequently caused by two kinds of transient voltages whose amplitudes may greatly exceed the peak values of the normal a.c. operating voltage.

The first kind are lightning overvoltages, originated by lightning strokes hitting the phase wires of overhead lines or the busbars of outdoor substations. The amplitudes are very high, usually in the order of 1000 kV or more, as every stroke may inject lightning currents up to about 100 kA and even more into the transmission line;[27] each stroke is then followed by travelling waves, whose amplitude is often limited by the maximum insulation strength of the overhead line. The rate of voltage rise of such a travelling wave is at its origin directly proportional to the steepness of the lightning current, which may exceed 100 kA/μsec, and the voltage levels may simply be calculated by the current multiplied by the effective surge impedance of the line. Too high voltage levels are immediately chopped by the breakdown of the insulation and therefore travelling waves with steep wave fronts and even steeper wave tails may stress the insulation of power transformers or other h.v. equipment severely. Lightning protection systems, surge arresters and the different kinds of losses will damp and distort the travelling waves, and therefore lightning overvoltages with very different waveshapes are present within the transmission system.

The second kind is caused by switching phenomena. Their amplitudes are always related to the operating voltage and the shape is influenced by the impedances of the system as well as by the switching conditions. The rate of voltage rise is usually slower, but it is well known that the waveshape can also be very dangerous to different insulation systems, especially to atmospheric air insulation in transmission systems with voltage levels higher than 245 kV.[28]

Both types of overvoltages are also effective in the l.v. distribution systems, where they are either produced by the usual, sometimes current-limiting, switches or where they have been transmitted from the h.v. distribution systems. Here they may often cause a breakdown of electronic equipment, as they can reach amplitudes of several kilovolts, and it should be mentioned that the testing of certain l.v. apparatus with transient voltages or currents is a need today.[29] Such tests also involve 'electromagnetic compatibility (EMC) tests', which will not be discussed here.

Although the actual shape of both kinds of overvoltages varies strongly, it became necessary to simulate these transient voltages by relatively simple means for testing purposes. The various national and international standards define the impulse voltages as a unidirectional voltage which rises more or less rapidly to a peak value and then decays relatively slowly to zero. In the relevant IEC Standard 60,[2] widely accepted today through national committees,[3] a distinction is made between lightning and switching impulses, i.e. according to the origin of the transients. Impulse voltages with front durations varying from less than one up to a few tens of microseconds are, in general, considered as lightning impulses. Figure 2.23(a) shows the shape for such a 'full' lightning impulse voltage as well as sketches for the same voltage chopped at the tail (Fig. 2.23(b)) or on the front (Fig. 2.23(c)), i.e. interrupted by a

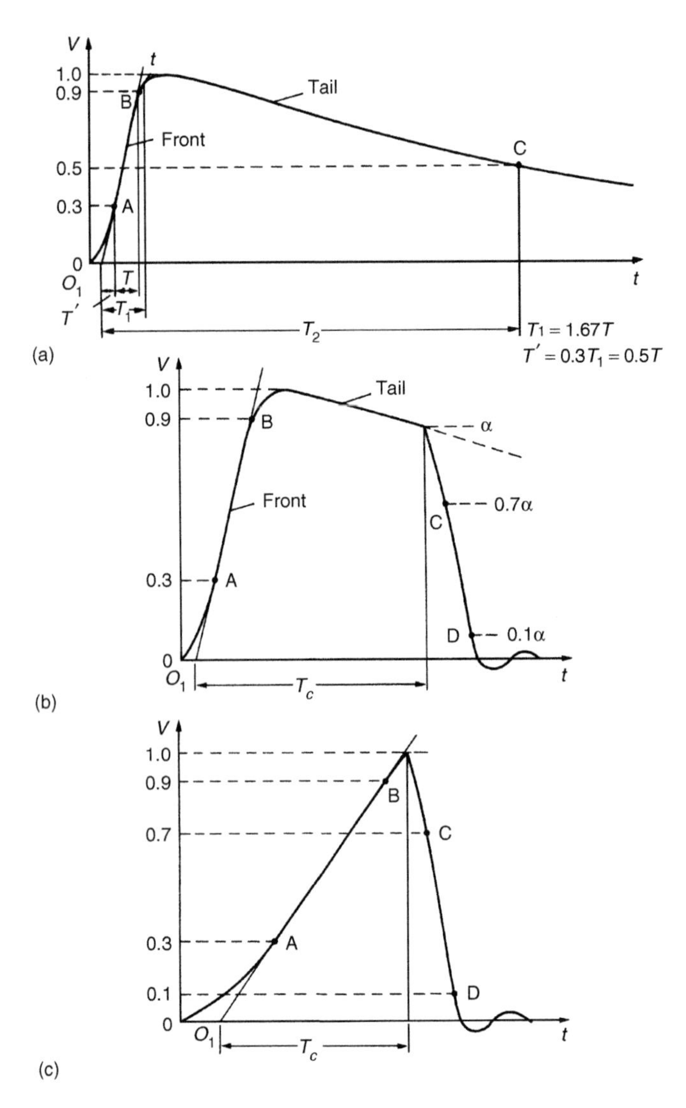

Figure 2.23 *General shape and definitions of lightning impulse (LI) voltages. (a) Full LI. (b) LI chopped on the tail. (c) LI chopped on the front. T_1: front time. T_2: time to half-value. T_c: time to chopping. O_1: virtual origin*

disruptive discharge. Although the definitions are clearly indicated, it should be emphasized that the 'virtual origin' O_1 is defined where the line AB cuts the time axis. The 'front time' T_1, again a virtual parameter, is defined as 1.67 times the interval T between the instants when the impulse is 30 per cent and 90 per cent of the peak value for full or chopped lightning impulses.

For front-chopped impulses the 'time to chopping' T_c is about equal to T_1. The reason for defining the point A at 30 per cent voltage level can be found in most records of measured impulse voltages. It is quite difficult to obtain a smooth slope within the first voltage rise, as the measuring systems as well as stray capacitances and inductances may cause oscillations. For most applications, the (virtual) front time T_1 is 1.2 µs, and the (virtual) time to half-value T_2 is 50 µs. In general the specifications[(2)] permit a tolerance of up to ±30 per cent for T_1 and ±20 per cent for T_2. Such impulse voltages are referred to as a T_1/T_2 impulse, and therefore the 1.2/50 impulse is the accepted standard lightning impulse voltage today. Lightning impulses are therefore of very short duration, mainly if they are chopped on front. Due to inherent measurement errors (see Chapter 3, section 3.6) and uncertainties in the evaluation the 'time parameters' T_1, T_2 and T_c or especially the time difference between the points C and D (Figs 2.23(b) and (c)) can hardly be quantified with high accuracy.

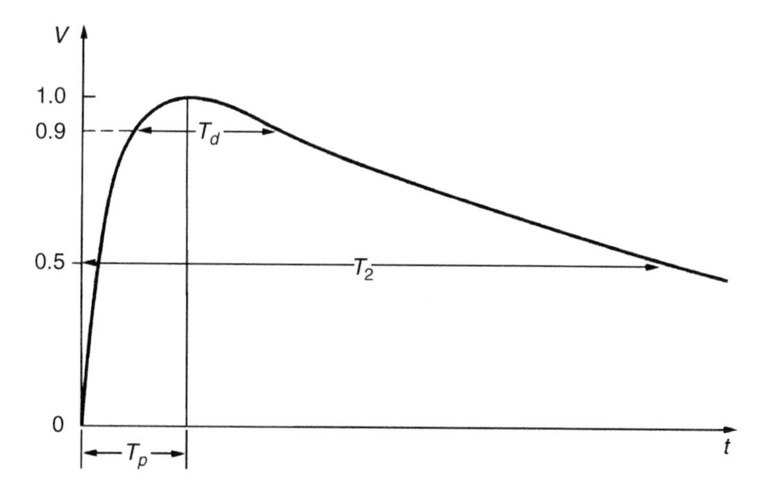

Figure 2.24 *General shape of switching impulse voltages. T_p: time to peak. T_2: time to half-value. T_d: time above 90 per cent*

Figure 2.24 illustrates the slope of a switching impulse. Whereas the time to half-value T_2 is defined similarly as before, the time to peak T_p is the time interval between the *actual* origin and the instant when the voltage has reached its maximum value. This definition could be criticized, as it is difficult to establish the actual crest value with high accuracy. An additional parameter is therefore the time T_d, the time at 90 per cent of crest value. The different definitions in comparison to lightning impulses can be understood if the time scale is emphasized: the standard switching impulse has time parameters (including tolerances) of

$$T_p = 250\,\mu s \pm 20\%$$

$$T_2 = 2500\,\mu s \pm 60\%$$

and is therefore described as a 250/2500 impulse. For fundamental investigations concerning the insulation strength of long air gaps or other apparatus, the time to peak has to be varied between about 100 and 1000 μs, as the breakdown strength of the insulation systems may be sensitive upon the voltage waveshape.[28]

2.3.1 *Impulse voltage generator circuits*

The introduction to the full impulse voltages as defined in the previous section leads to simple circuits for the generation of the necessary waveshapes. The rapid increase and slow decay can obviously be generated by discharging circuits with two energy storages, as the waveshape may well be composed by the superposition of two exponential functions. Again the load of the generators will be primarily capacitive, as insulation systems are tested. This load will therefore contribute to the stored energy. A second source of energy could be provided by an inductance or additional capacitor. For lightning impulses mainly, a fast discharge of pure inductor is usually impossible, as h.v. chokes with high energy content can never be built without appreciable stray capacitances. Thus a suitable fast discharge circuit will always consist essentially of two capacitors.

Single-stage generator circuits

Two basic circuits for single-stage impulse generators are shown in Fig. 2.25. The capacitor C_1 is slowly charged from a d.c. source until the spark gap G breaks down. This spark gap acts as a voltage-limiting and voltage-sensitive switch, whose ignition time (time to voltage breakdown) is very short in comparison to T_1. As such single-stage generators may be used for charging voltages from some kV up to about 1 MV, the sphere gaps (see Chapter 3, section 3.1) will offer proper operating conditions. An economic limit of the charging voltage V_0 is, however, a value of about 200 to 250 kV, as too large diameters of the spheres would otherwise be required to avoid excessive inhomogeneous field distributions between the spheres. The resistors R_1, R_2 and the capacitance C_2 form the waveshaping network. R_1 will primarily damp the circuit and control the front time T_1. R_2 will discharge the capacitors and therefore essentially control the wavetail. The capacitance C_2 represents the full load, i.e. the object under test as well as all other capacitive elements which are in parallel to the test object (measuring devices; additional load capacitor to avoid large variations of T_1/T_2, if the test objects are changed). No inductances are assumed so far, and are neglected in the first fundamental analysis, which is also necessary to understand multistage generators. In general this approximation is permissible, as the inductance of all circuit elements has to be kept as low as possible.

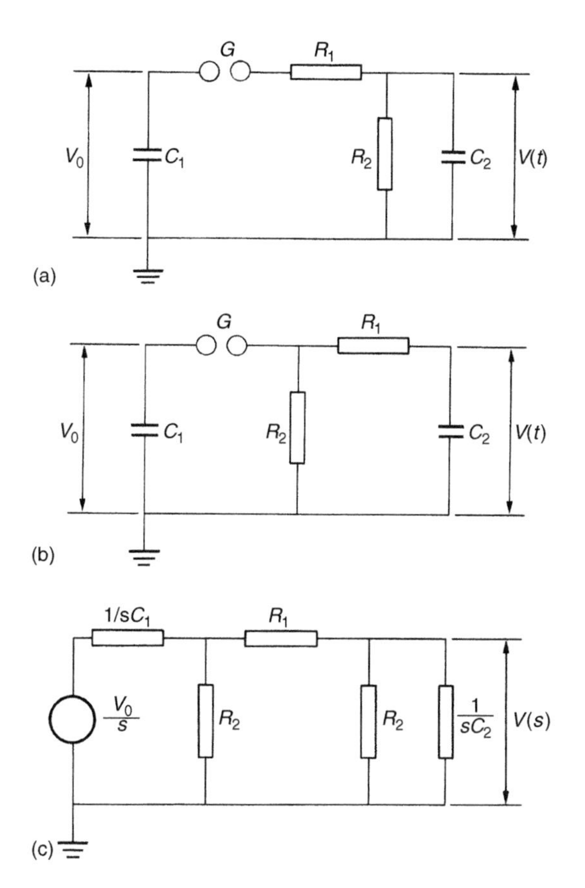

Figure 2.25 *Single-stage impulse generator circuits (a) and (b). C_1: discharge capacitance. C_2: load capacitance. R_1: front or damping resistance. R_2: discharge resistance. (c) Transform circuit*

Before starting the analysis, we should mention the most significant parameter of impulse generators. This is the maximum stored energy

$$W = \tfrac{1}{2}C_1(V_{0_{\max}})^2 \tag{2.22}$$

within the 'discharge' capacitance C_1. As C_1 is always much larger than C_2, this figure determines mainly the cost of a generator.

For the analysis we may use the Laplace transform circuit sketched in Fig. 2.25(c), which simulates the boundary condition, that for $t \leq 0$ C_1 is charged to V_0 and for $t > 0$ this capacitor is directly connected to the wave-shaping network. For the circuit Fig. 2.25(a) the output voltage is thus given by the expression

$$V(s) = \frac{V_0}{s} \frac{Z_2}{Z_1 + Z_2},$$

where

$$Z_1 = \frac{1}{C_1 s} + R_1;$$

$$Z_2 = \frac{R_2/C_2 s}{R_2 + 1/C_2 s}.$$

By substitution we find

$$V(s) = \frac{V_0}{k} \frac{1}{s^2 + as + b} \tag{2.23}$$

where

$$\begin{aligned} a &= \left(\frac{1}{R_1 C_1} + \frac{1}{R_1 C_2} + \frac{1}{R_2 C_2}\right); \\ b &= \left(\frac{1}{R_1 R_2 C_1 C_2}\right); \\ k &= R_1 C_2. \end{aligned} \tag{2.24}$$

For circuit Fig. 2.25(b) one finds the same general expression (eqn (2.23)), with the following constants; however,

$$\begin{aligned} a &= \left(\frac{1}{R_1 C_1} + \frac{1}{R_1 C_2} + \frac{1}{R_2 C_1}\right); \\ &\left.\begin{aligned} b &= \left(\frac{1}{R_1 R_2 C_1 C_2}\right); \\ k &= R_1 C_2. \end{aligned}\right\} \text{ as above} \end{aligned} \tag{2.25}$$

For both circuits, therefore, we obtain from the transform tables the same expression in the time domain:

$$V(t) = \frac{V_0}{k} \frac{1}{(\alpha_2 - \alpha_1)} [\exp(-\alpha_1 t) - \exp(-\alpha_2 t)] \tag{2.26}$$

where α_1 and α_2 are the roots of the equation $s^2 + as + b = 0$, or

$$\alpha_1, \alpha_2 = \frac{a}{2} \mp \sqrt{\left(\frac{a}{2}\right)^2 - b}. \tag{2.27}$$

The output voltage $V(t)$ is therefore the superposition of two exponential functions of different signs. According to eqn (2.27), the negative root leads to a larger time constant $1/\alpha_1$ than the positive one, which is $1/\alpha_2$. A graph of the expression (eqn (2.26)) is shown in Fig. 2.26, and a comparison with Figs 2.23 and 2.24 demonstrates the possibility to generate both types of impulse voltages with these circuits.

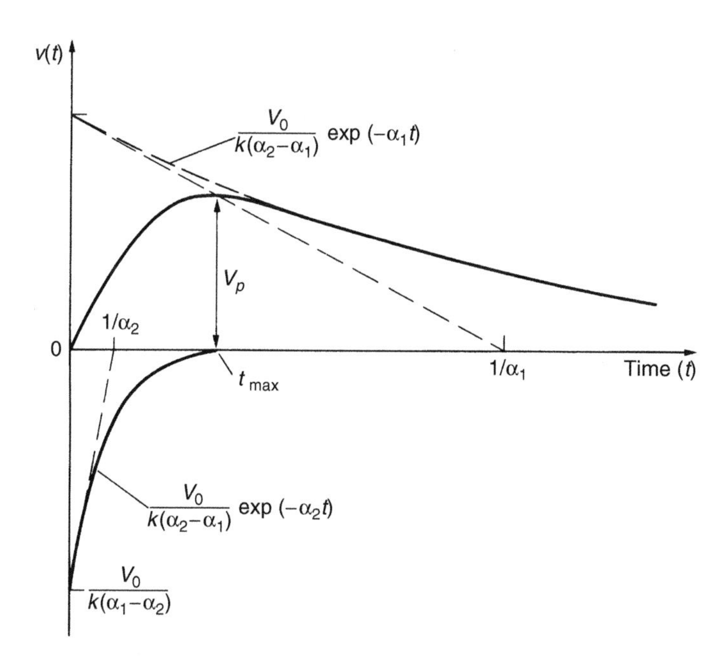

Figure 2.26 *The impulse voltage wave and its components according to circuits in Fig. 2.25*

Although one might assume that both circuits are equivalent, a larger difference may occur if the voltage efficiency, η, is calculated. This efficiency is defined as

$$\eta = \frac{V_p}{V_0}; \tag{2.28}$$

V_p being the peak value of the output voltage as indicated in Fig. 2.26. Obviously this value is always smaller than 1 or 100 per cent. It can be calculated by finding t_{max} from $dV(t)/dt = 0$; this time for the voltage $V(t)$ to rise to its peak value is given by

$$t_{max} = \frac{\ln(\alpha_2/\alpha_1)}{(\alpha_2 - \alpha_1)}. \tag{2.29}$$

Substituting this equation into eqn (2.26), one may find

$$\eta = \frac{(\alpha_2/\alpha_1)^{-[(\alpha_2/\alpha_1-\alpha_1)]} - (\alpha_2/\alpha_1)^{-[(\alpha_2/\alpha_2-\alpha_1)]}}{k(\alpha_2 - \alpha_1)}. \tag{2.30}$$

For a given impulse shape T_1/T_2 or T_p/T_2 of the impulse voltages the values of α_1 and α_2 must be equal. The differences in efficiency η can only be due, therefore, to differences in the value of $k = R_1C_2$ for both circuits. We may first calculate this term for the circuit Fig. 2.25(b), which has always a higher efficiency for a given ratio of C_2/C_1, as during the discharge the resistors R_1 and R_2 do not form a voltage-dividing system. The product R_1C_2 is found by eqn (2.27) by forming

$$\alpha_1 \cdot \alpha_2 = b$$
$$\alpha_1 + \alpha_2 = a \tag{2.31}$$

and by the substitution of a and b from eqn (2.25). Then we obtain

$$k = R_1C_2 = \frac{1}{2}\left(\frac{\alpha_2+\alpha_1}{\alpha_2\cdot\alpha_1}\right)\left[1 - \sqrt{1 - 4\frac{\alpha_2\cdot\alpha_1}{(\alpha_2+\alpha_1)^2}\left(1+\frac{C_2}{C_1}\right)}\right]. \tag{2.32}$$

For $C_2 \le C_1$, which is fulfilled in all practical circuits, and with $\alpha_2 \gg \alpha_1$ for all normalized waveshapes, one may simplify this equation to

$$k \cong \frac{1 + C_2/C_1}{(\alpha_2 + \alpha_1)}. \tag{2.33}$$

The substitution of this expression in eqn (2.30) finally results in

$$\eta = \frac{C_1}{(C_1 + C_2)} = \frac{1}{1 + (C_2/C_1)} \tag{2.34}$$

if again the inequality $\alpha_2 \gg \alpha_1$ is taken into account. The voltage efficiency for this circuit will therefore rise continuously, if (C_2/C_1) decreases to zero. Equation (2.34) indicates the reason why the discharge capacitance C_1 should be much larger than the load C_2.

Less favourable is the circuit Fig. 2.25(a). The calculation of η may be based upon the substitution of α_1 and α_2 in eqn (2.30) from eqn (2.27), and a treatment of the ratio $R_1/R_2 = f(C_2/C_1)$, which increases heavily with decreasing values of C_2/C_1. With minor approximations and the inequality $\alpha_2 \gg \alpha_1$ one may find the result

$$\eta \cong \frac{C_1}{(C_1 + C_2)} \frac{R_2}{(R_1 + R_2)} = \frac{1}{(1 + C_2/C_1)} \frac{1}{(1 + R_1/R_2)}. \qquad (2.35)$$

The comparison with eqn (2.34) shows the decrease in η due to an additional factor. As the ratio R_1/R_2 is dependent upon the waveshape, the simple dependency from (C_2/C_1) only is lost. For a 1.2/50 μs impulse and similar impulse voltages the rapid increase of R_1/R_2 leads to a decrease of η for $C_2/C_1 \lessapprox 0.1$; therefore, the efficiency moves through an optimum value and decreases for high C_2/C_1 values as well as for small ones. One could even show that for very small C_2/C_1 ratios this circuit will fail to work.

In practice, both circuits are in use, often, however, in mixed and modified form. If resistive h.v. dividers are placed in parallel to the test object, their resistor value may contribute to discharge the circuits. The front resistor R_1 is often subdivided, mainly in multistage generators treated later on. Nevertheless, the dependency of the voltage efficiency factors η is displayed in Fig. 2.27 for the standard lightning impulse voltage 1.2/50 μsec as well as for some other waveshapes. More information is available in the literature.[30]

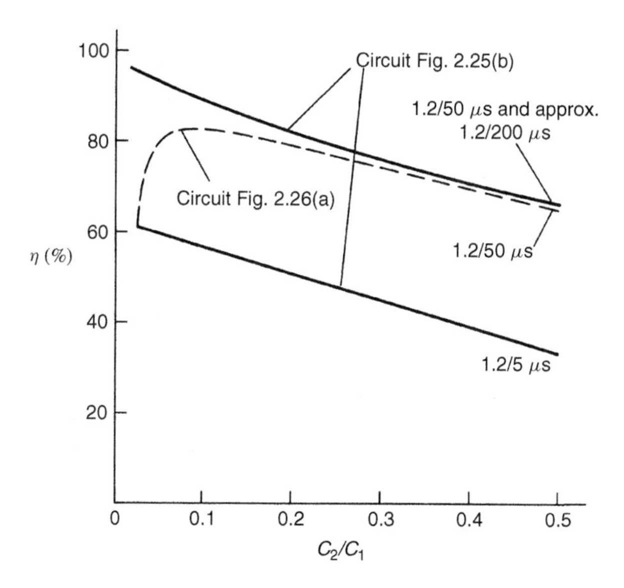

Figure 2.27 *Voltage efficiency factors η in dependency of the capacitance ratio C_2/C_1 for lightning impulses T_1/T_2*

Dimensioning of circuit elements. The common task is to find the resistor values for R_1 and R_2, as C_2 and C_1 are known in general. For larger generators, the discharge capacitors are always given and dimensioned for a good efficiency (see eqns (2.34) and (2.35)) within a certain range of C_2. This total load capacitance can easily be measured if it is not known in advance. The unknown resistance values can then be calculated using eqn (2.31) and the circuit-dependent values for a and b due to eqns (2.24) and (2.25). The result will be for:

Circuit Fig. 2.25(a):

$$R_1 = \frac{1}{2C_1}\left[\left(\frac{1}{\alpha_1}+\frac{1}{\alpha_2}\right) - \sqrt{\left(\frac{1}{\alpha_1}+\frac{1}{\alpha_2}\right)^2 - \frac{4(C_1+C_2)}{\alpha_1\alpha_2 \cdot C_2}}\right]. \tag{2.36}$$

$$R_2 = \frac{1}{2(C_1+C_2)}\left[\left(\frac{1}{\alpha_1}+\frac{1}{\alpha_2}\right) + \sqrt{\left(\frac{1}{\alpha_1}+\frac{1}{\alpha_2}\right)^2 - \frac{4(C_1+C_2)}{\alpha_1\alpha_2 C_2}}\right]. \tag{2.37}$$

Circuit Fig. 2.25(b):

$$R_1 = \frac{1}{2C_2}\left[\left(\frac{1}{\alpha_1}+\frac{1}{\alpha_2}\right) - \sqrt{\left(\frac{1}{\alpha_1}+\frac{1}{\alpha_2}\right)^2 - \frac{4(C_1+C_2)}{\alpha_1\alpha_2 C_1}}\right]. \tag{2.38}$$

$$R_2 = \frac{1}{2(C_1+C_2)}\left[\left(\frac{1}{\alpha_1}+\frac{1}{\alpha_2}\right) + \sqrt{\left(\frac{1}{\alpha_1}+\frac{1}{\alpha_2}\right)^2 - \frac{4(C_1+C_2)}{\alpha_1\alpha_2 C_1}}\right]. \tag{2.39}$$

All these equations contain the time constants $1/\alpha_1$ and $1/\alpha_2$, which depend upon the waveshape. There is, however, no simple relationship between these time constants and the times T_1, T_2 and T_p as defined in the national or international recommendations, i.e. in Figs 2.23 and 2.24. This relationship can be found by applying the definitions to the analytical expression for $V(t)$, this means to eqn (2.26). The relationship is irrational and must be computed numerically. The following table shows the result for some selected waveshapes:

T_1/T_2 (μs)	T_p/T_2 (μs)	$1/\alpha_1$ (μs)	$1/\alpha_2$ (μs)
1.2/5	–	3.48	0.80
1.2/50	–	68.2	0.405
1.2/200	–	284	0.381
250/2500	–	2877	104
–	250/2500	3155	62.5

The standardized nominal values of T_1 and T_2 are difficult to achieve in practice, as even for fixed values of C_1 the load C_2 will vary and the exact values for R_1 and R_2 according to eqns (2.38) and (2.39) are in general not available. These resistors have to be dimensioned for the rated high voltage of the generator and are accordingly expensive. The permissible tolerances for T_1 and T_2 are therefore necessary and used to graduate the resistor values. A recording of the real output voltage $V(t)$ will in addition be necessary if the admissible impulse shape has to be testified.

Another reason for such a measurement is related to the value of the test voltage as defined in the recommendations.[2,3] This magnitude corresponds to the crest value, if the shape of the lightning impulse is smooth. However, oscillations or an overshoot may occur at the crest of the impulse. If the frequency of such oscillations is not less than 0.5 MHz or the duration of overshoot not over 1 μsec, a 'mean curve' (see Note below) should be drawn through the curve. The maximum amplitude of this 'mean curve' defines the value of the test voltage. Such a correction is only tolerated, provided their single peak amplitude is not larger than 5 per cent of the crest value. Oscillations on the front of the impulse (below 50 per cent of the crest value) are tolerated, provided their single peak amplitude does not exceed 25 per cent of the crest value. It should be emphasized that these tolerances constitute the permitted differences between specified values and those actually recorded by measurements. Due to measuring errors the true values and the recorded ones may be somewhat different.

Note. With the increasing application of transient or digital recorders in recording of impulse voltages it became very obvious that the definition of a 'mean curve' for the evaluation of lightning impulse parameters of waveforms with oscillations and/or overshoot, as provided by the standards,[2,3] is insufficient. Any software, written to evaluate the parameters, needs clear instructions which are not yet available. As this matter is still under consideration (by CIGRE Working Group 33.03) and a revision of the current standards may provide solutions, no further comments to this problem are given.

The origin of such oscillations or the overshoot can be found in measuring errors (heavily oscillating 'step response', see Chapter 3, section 3.6) as well as by the inductances within every branch of the circuit or the stray capacitances, which will increase with the physical dimensions of the circuit. As far as inductances are concerned, a general rule for the necessary critical damping of single-stage or – with less accuracy – of multistage generators can easily be demonstrated by Fig. 2.28. If individual inductances L_1, L_2 are considered within the discharge circuit as indicated in Fig. 2.28(a), a second order differential equation determines the output voltage across the load capacitance C_2. However, such an equivalent circuit cannot be exact, as additional circuits related to stray capacitances are not taken into account. Thus we may only combine the total inductance within the C_1–C_2 circuit to single inductance L, as shown in Fig. 2.28(b), and neglect the positions of the tail resistors, which have no big influence. This reduces the circuit to a simple damped series resonant circuit, and the critical resistance $R = R_1$ for the circuit to be non-oscillatory is given by the well-known equation

$$R_1 \cong R = 2\sqrt{\frac{L}{C}} \tag{2.40}$$

where

$$\frac{1}{C} = \frac{1}{C_1} + \frac{1}{C_2}$$

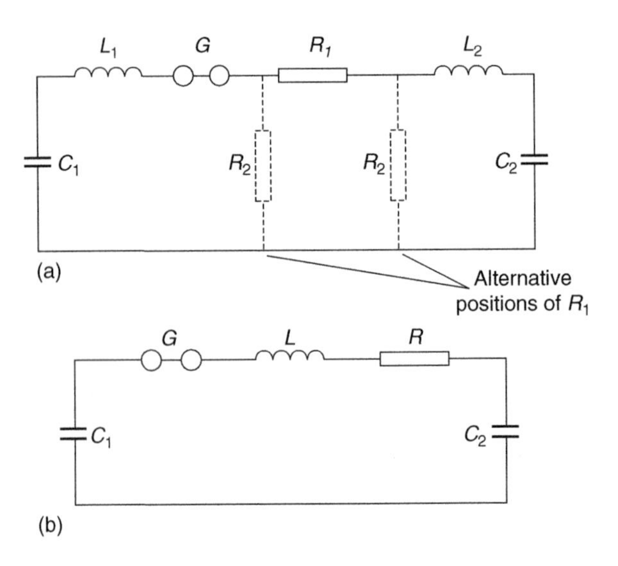

Figure 2.28 *Simplified circuit of impulse generator and load. Circuit showing alternative positions of the wave tail control resistance. (b) Circuit for calculation of wave front oscillations*

This equation is in general suitable for predicting the limiting values for the front resistor R_1. The extremely tedious analytical analysis of circuits containing individual inductances is shown elsewhere.[31–34] Computer programs for transients may also be used to find the origin of oscillations, although it is difficult to identify good equivalent circuits.

Multistage impulse generator circuits

The difficulties encountered with spark gaps for the switching of very high voltages, the increase of the physical size of the circuit elements, the efforts necessary in obtaining high d.c. voltages to charge C_1 and, last but not least, the difficulties of suppressing corona discharges from the structure and leads during the charging period make the one-stage circuit inconvenient for higher voltages.

In order to overcome these difficulties, in 1923 Marx[35] suggested an arrangement where a number of condensers are charged in parallel through high ohmic resistances and then discharged in series through spark gaps. There are many different, although always similar, multistage circuits in use. To demonstrate the principle of operation, a typical circuit is presented in Fig. 2.29 which shows the connections of a six-stage generator. The d.c. voltage charges the equal stage capacitors C'_1 in parallel through the high value charging resistors R' as well as through the discharge (and also charging)

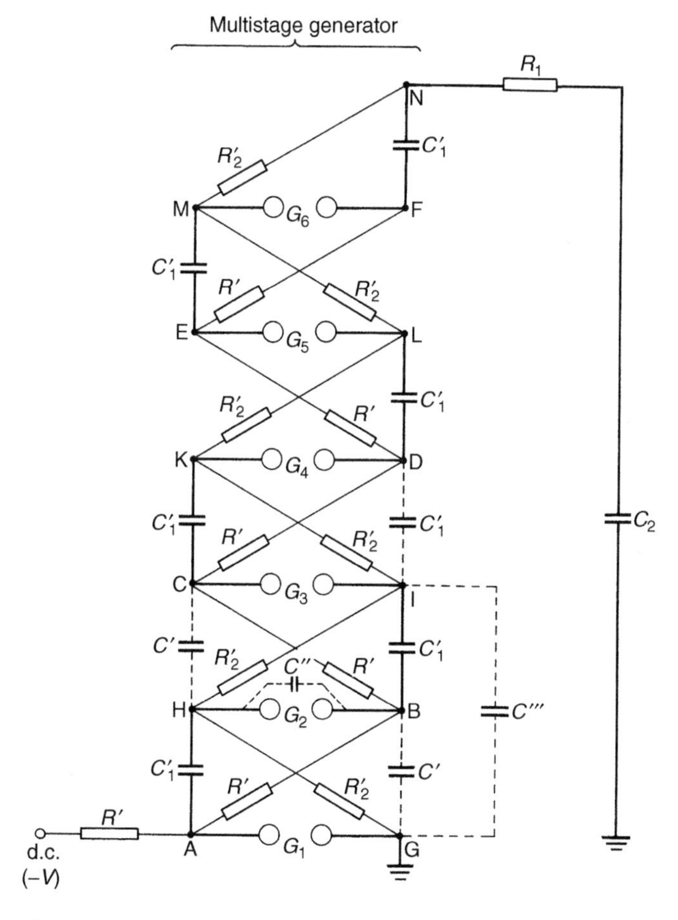

Figure 2.29 *Basic circuit of a six-stage impulse generator (Marx generator)*

resistances R_2', which are much smaller than the resistors R' and are comparable with R_2 in Fig. 2.25. At the end of the relatively long charging period (typically several seconds up to 1 minute), the points $A, B, \ldots, F$ will be at the potential of the d.c. source, e.g. $-V$ with respect to earth, and the points $G, H, \ldots, N$ will remain at the earth potential, as the voltage drop during charging across the resistors R_2' is negligible. The discharge or firing of the generator is initiated by the breakdown of the lowest gap G_1 which is followed by a nearly simultaneous breakdown of all the remaining gaps. According to the traditional theory, which does not take into account the stray capacitances indicated by the dotted lines, this rapid breakdown would be caused by high overvoltages across the second and further gaps: when the first gap fires, the potential at point A changes rapidly from $-V$ to zero, and thus the point H increases its potential to $+V$. As the point B still would remain at the charging potential, $-V$, thus a voltage of $2V$ would appear across G_2. This high overvoltage would therefore cause this gap to break down and the potential at point I would rise to $+2V$, creating a potential difference of $3V$ across gap G_3, if again the potential at point C would remain at the charging potential. This traditional interpretation, however, is wrong, since the potentials B and C can – neglecting stray capacitances – also follow the adjacent potentials of the points A and B, as the resistors R' are between. We may only see up to now that this circuit will give an output voltage with a polarity opposite to that of the charging voltage.

In practice, it has been noted that the gap G_2 must be set to a gap distance only slightly greater than that at which G_1 breaks down; otherwise it does not operate. According to Edwards, Husbands and Perry[31] for an adequate explanation one may assume the stray capacitances C', C'' and C''' within the circuit. The capacitances C' are formed by the electrical field between adjacent stages; C''' has a similar meaning across two stages. C'' is the capacitance of the spark gaps. If we assume now the resistors as open circuits, we may easily see that the potential at point B is more or less fixed by the relative magnitudes of the stray capacitances. Neglecting C' between the points H and C and taking into account that the discharge capacitors C_1' are large in comparison to the stray capacitances, point B can be assumed as mid-point of a capacitor voltage divider formed by C'' and C'/C'''. Thus the voltage rise of point A from $-V$ to zero will cause the potential B to rise from $-V$ to a voltage of

$$V_B = -V + V\left(\frac{C''}{C' + C'' + C'''}\right) = -V\left(\frac{C' + C'''}{C' + C'' + C'''}\right).$$

Hence the potential difference across G_2 becomes

$$V_{G2} = +V - (-V_B) = V\left(1 + \frac{C' + C'''}{C' + C'' + C'''}\right).$$

If C'' equals zero, the voltage across G_2 will reach its maximum value $2V$. This gap capacitance, however, cannot be avoided. If the stage capacitances C' and C''' are both zero, V_{G2} will equal V, and a sparking of G_2 would not be possible. It is apparent, therefore, that these stray capacitances enhance favourable conditions for the operation of the generator. In reality, the conditions set by the above equations are approximate only and are, of course, transient, as the stray capacitances start to discharge via the resistors. As the values of C' to C''' are normally in the order of some 10 pF only, the time constants for this discharge may be as low as 10^{-7} to 10^{-8} sec.

Thus the voltage across G_2 appears for a short time and leads to breakdown within several tens of nanoseconds. Transient overvoltages appear across the further gaps, enhanced also by the fact that the output terminal N remains at zero potential mainly, and therefore additional voltages are built up across the resistor R'_2. So the breakdown continues and finally the terminal N attains a voltage of $+6V$, or nV, if n stages are present.

The processes associated with the firing of such generators are even more sophisticated. They have been thoroughly analysed and investigated experimentally.[31,36,37]

In practice for a consistent operation it is necessary to set the distance for the first gap G_1 only slightly below the second and further gaps for earliest breakdown. It is also necessary to have the axes of the gaps in one vertical plane so that the ultraviolet illumination from the spark in the first gap irradiates the other gaps. This ensures a supply of electrons released from the gap to initiate breakdown during the short period when the gaps are subjected to the overvoltage. If the first gap is not electronically triggered, the consistency of its firing and stability of breakdown and therefore output voltage is improved by providing ultraviolet illumination for the first gap. These remarks indicate only a small part of the problems involved with the construction of spark gaps and the layout of the generator. Before some of these additional problems are treated, we shall treat more realistic Marx circuits as used for the explanations so far.

In Fig. 2.29, the wavefront control resistor R_1 is placed between the generator and the load only. Such a single 'external' front resistor, however, has to withstand for a short time the full rated voltage and therefore is inconveniently long or may occupy much space. This disadvantage can be avoided if either a part of this resistance is distributed or if it is completely distributed within the generator. Such an arrangement is illustrated in Fig. 2.30, in which in addition the series connection of the capacitors C'_1 and gaps (as proposed originally by Goodlet[38]) is changed to an equivalent arrangement for which the polarity of the output voltage is the same as the charging voltage. The charging resistors R' are always large compared with the distributed resistors R'_1 and R'_2, and R'_2 is made as small as is necessary to give the required time to halve-value T_2. Adding the external front resistor R''_1 helps to damp oscillations otherwise

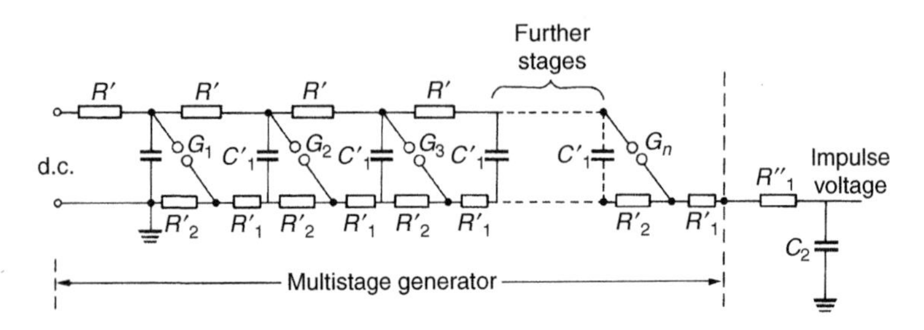

Figure 2.30 *Multistage impulse generator with distributed discharge and front resistors. R_2': discharge resistors. R_1': internal front resistors. R_1'': external front resistor*

excited by the inductance and capacitance of the external leads between the generator and the load, if these leads are long. It may be readily seen that this circuit can be reduced to the single-stage impulse generator circuit shown in Fig. 2.25(b). If the generator has fired, the total discharge capacitance C_1 may be calculated as

$$\frac{1}{C_1} = \sum^{n} \frac{1}{C_1'};$$

the effective front resistance R_1 as

$$R_1 = R_1'' + \sum^{n} R_1';$$

and the effective discharge resistance R_2 – neglecting the charging resistances R' – as

$$R_2 = nR_2' = \sum^{n} R_2';$$

where n is the number of stages.

The consistent firing of such circuits could be explained as for the generator of Fig. 2.29. For both generator circuits, the firing is aggravated if the resistances R_2' have relatively low values. According to eqns (2.22) and (2.39) such low values appear with generators of high energy content and/or short times to half-value, T_2. Then the time constant for discharging the stray capacitances to ground C''' (Fig. 2.29) will be too low and accordingly the overvoltages for triggering the upper stages too short. By additional means providing high resistance values within the firing period, this disadvantage can be avoided.[39]

Special circuits for generating switching impulse voltages

The common impulse generator circuits discussed so far are well capable of producing standard switching impulses with adequate voltage efficiency η, if the circuit is well designed and the ratio C_2/C_1 is kept adequately small.

Other methods, however, have taken advantage of utilizing testing transformers to step up the amplitudes from impulse voltages also. One such circuit is shown in Fig. 2.31.[2] An initially charged capacitor C_1 is discharged into the waveshaping circuit R_1, C_2 as well as into the l.v. winding of the transformer. The elements R_1 and C_2 or other suitable components, in the dotted rectangle, may be used to control the waveshape. The wave tail is not only controlled by the resistive voltage divider included, but also from the main inductance of the transformer equivalent circuit. The time to crest T_p is even without R_1, C_2 limited by the series inductance of the transformer, L_s, which forms a series resonant circuit in combination with C_1 and the load capacitance C_2'. Neglecting any losses within the circuit, the voltage across the test object would therefore start with a $(1 - \cos \omega t)$ function, and as $T_p \approx T/2 = \pi/f_r$, f_r being the resonance frequency, the time to crest is approximately

$$T_{\text{cr}} \cong \pi\sqrt{L_s C}$$

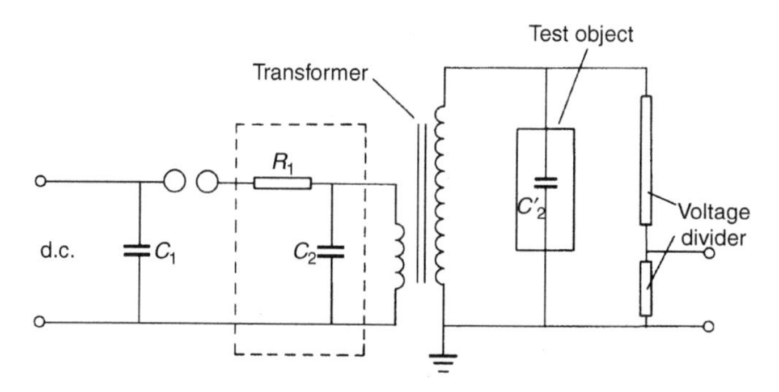

Figure 2.31 *Circuit for generation of switching impulses using a transformer*

where, neglecting transformer ratio,

$$C = \frac{C_1 C_2'}{C_1 + C_2'}.$$

In general, low values of T_p are difficult to achieve, as L_s is quite large and also the capacitance of the h.v. winding of the transformer contributes to the load C_2'. Further problems arise with transient oscillations within the transformer windings, mainly if cascaded testing transformers are used.[40]

The physical phenomena of disruptive discharges in long air gaps as well as in other insulating systems are often related to the front of switching impulses only. Therefore, with switching surges it is not always necessary to produce double-exponential waveshapes as recommended in Fig. 2.24. In fact, many investigations are sometimes made with unusual shapes of high voltages, and one can establish many circuits for mainly oscillating voltages, whose variety cannot be treated here. Only the common impulse voltage circuit with a strongly increased inductance will demonstrate this variety, as such circuits came into use for field testing of GIS.[41,42] The principle of such circuits is demonstrated in Figs 2.32(a) and (b). If the front resistor R_1 in Fig. 2.25(a) is replaced by a series inductance, the circuit of Fig. 2.32(a) results, which was first described for the generation of high switching impulse voltages up to 500 kV by Bellaschi and Rademacher.[43] A typical waveshape of the output voltage is also included; it may easily be calculated, since the circuit is a damped series resonant circuit only. The advantage of such a circuit is the nearly doubling of the output voltage, if $C_2 \ll C_1$, in comparison to the charging voltage V_0. Also a proper damping does not decrease the amplitude of the oscillation very much, and therefore the first increase of the voltage may be used as the front of a switching impulse. This fundamental circuit can be applied to multistage impulse generators, in which the front resistors are replaced by h.t. chokes.[42]

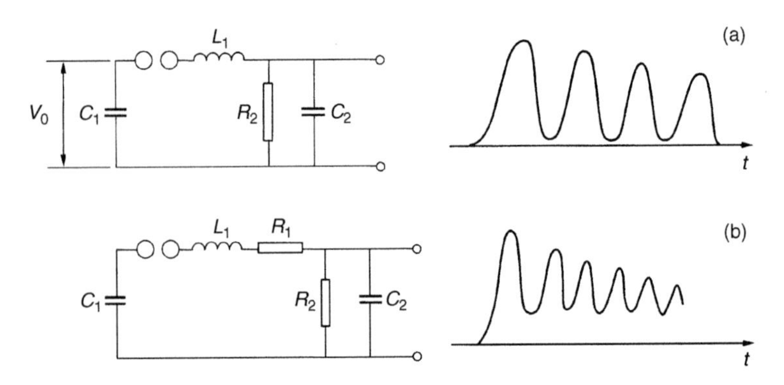

Figure 2.32 *Circuits for the generation of oscillating switching impulses*

A modular design of such generators offers the opportunity for easy transportation and erection on field site.

The second circuit of Fig. 2.32 uses an additional resistor R_1 in series to L_1, and is therefore simply a pure impulse voltage circuit with a high inductance within the discharge circuit. Therefore, unidirectional damped oscillations are produced. Also this circuit can be used for high voltages, as originally shown by Kojema and Tanaka.[44]

2.3.2 Operation, design and construction of impulse generators

Although the main aim of this book is not concerned with detailed information about the design and construction of h.v. equipment, some additional remarks are necessary for a better understanding of the problems involved with the operation and use of impulse generators. The advice given within this chapter is mainly related to section 2.3.1 and concerns multistage generators.

Every generator needs a d.c. power supply to charge the discharge capacitance C_1. This supply may simply consist of an h.t. transformer and rectifiers providing unidirectional currents, as the voltage smoothing is made by C_1. The d.c. supply should primarily act as a current source, so that the charging time can be controlled. The charging times should not be shorter than 5 to 10 sec, as every voltage application to an object under test may lead to prestressing effects within the insulation, influencing the withstand or breakdown strength. Much longer charging times or time intervals between successive voltage applications may be necessary, depending upon the material tested. For the rough-controlled charging, the d.c. supply is usually only voltage controlled by a voltage regulator at the primary of the h.t. transformer. Manufacturers nowadays provide thyristor-controlled charging supplies with current-limiting output. By this method, a programmed charging of generators is possible to reach equal charging times for all levels of impulse voltages.

The layout of the construction of multistage impulse generators is largely governed by the type of capacitors involved. Oil-paper-insulated capacitors of low inductance and high capability for fast discharging are in common use; mineral oil is often replaced by special fluids, providing higher permittivity to increase the capacitance per volume.

For earlier constructions predominantly capacitor units have been used, having the dielectric assembled in an insulating cylinder of porcelain or varnished paper with plane metal end-plates. This construction provided the obvious advantage that the stages of capacitors could be built up in the vertical columns, each stage being separated from the adjacent one by supports of the same or similar form as the capacitor units without dielectric. Such a construction is illustrated by the generator shown in Fig. 2.33. The disadvantage relates to the difficult replacement of failing capacitor units, and therefore this originally preferred construction is not much used today. New designs prefer complete modular constructions with capacitor units within insulating cylinders or vessels, or within metal tanks and bushings. This design originated from improvements in the capacitor dielectrics, which could reduce the size of the capacitors significantly even for voltages up to 100 kV per unit or more. To improve the consistency in firing the spark gaps, especially in cases of large generators with many stages, the gaps are mounted inside a cylindrical housing in which the pressure can be controlled, and hence the gaps do not require adjusting. Such a construction is shown in Fig. 2.34 for a 20-stage, 4-MV, 200-kJ indoor generator. Besides such indoor constructions, many of the generators for very high impulse voltages are used under outdoor conditions for the direct testing of outdoor material and to avoid the use of too large laboratories. To eliminate the detrimental influence of weather conditions

Figure 2.33 *2.4-MV impulse generator UMIST*

on electrical insulation as well as the mechanical influences (corrosion, etc.), most of these generators are housed in huge insulating cylinders providing for the full insulation of the output voltage and providing the opportunity to have the generator itself under air conditioning.

Besides the charging resistances R', all of the waveshaping resistors should be placed in such a way that they can easily be exchanged and replaced, as they must be changed often to ensure the waveshapes necessary.

Figure 2.34 *Indoor impulse generator, 20-stage, 4-MV, 200-kJ, with encapsulated sphere gaps (courtesy of CEPEL, Rio de Janero, manufactured by Haefely)*

The resistors may be composed of wire, liquids or composite resistive materials. Although the high heat capacity involved with liquid and composite resistors would give preference to these resistor types, the instability of the resistance values is of big disadvantage. Therefore highly non-inductively wound wire resistors are best for the front and discharge resistors R_1 and

R_2 of the circuits. Wavefront resistors are quite satisfactory if their L/R value is less than about 0.1 μsec.

The spark gaps are usually mounted on horizontal arms and the setting of the gaps is adjusted by a remotely controlled motor in conjunction with an indicator. This remark is, of course, related to gaps working in open air only. Sometimes encapsulated and pressurized gaps are used, for which the breakdown voltage is controlled by the gas pressure. The use of proper gas mixtures gives good switching performance.(45)

Each generator should have a device to earth the capacitors when it is not in operation. Due to relaxation phenomena, d.c.-operated capacitors can build up high voltages after a short-time short-circuit.

All leads and electrodes within the generator should be dimensioned properly to avoid too heavy corona discharges during the charging period. During the short time of discharge and therefore impulse generation, partial discharges cannot be prevented. A complete immersion of the generators in improved insulation materials, as mineral-oil or high-pressure gases, could reduce the size effectively; such solutions, however, are only used for special purposes.(46,47)

Finally, some explanations refer to the tripping and synchronization of the operation of impulse generators. The simple method of tripping the generators by non-triggered sphere gaps suffers from the disadvantage that the exact instant of firing is not predictable. Furthermore, the presence of unavoidable dust can cause irregular operation of multistage generators due to the following main reason: dust particles are likely to be attracted to the spheres stressed with d.c. voltages during charging, and the breakdown voltage can strongly be reduced by these particles.(48) If dust is randomly deposited on the lowest gap (G_1 in Fig. 2.29 or 2.30), the dispersion of the d.c. breakdown voltage of this gap increases and thus the output voltage will not be stable. To overcome this irregular firing of generators, which even with triggered gaps may occur, a protection against dust may be provided. As separate enclosures of the gaps with any insulating material prevents any ultraviolet illumination from reaching the other gaps, only a common enclosure of all gaps is satisfactory. Early investigations demonstrating those effects have been described by Edwards *et al.*(31) The common enclosure of all gaps provides, today, a reliable method to ensure stable tripping (see Fig. 2.34).

A stable self-tripping by a fast mechanical closure of the first gap is a simple means to avoid self-firing due to dust particles; however, this method cannot be used to synchronize the impulse voltage with other events within a very short time interval of a microsecond or less. The early need for synchronization arose from the necessity to initiate the time sweep of oscilloscopes used for the voltage recording. Later on, analogue impulse oscilloscopes have been manufactured with built-in time delays, so that a pre-trigger may not be necessary. Today, triggered firing of impulse generators would not be even required as transient recorders are available (see Chapter 3) which replace the analogue impulse oscilloscopes and which provide a continuous pre-trigger system. However, as such recorders may still not yet be in common use and since controlled or triggered switching of high voltages has so many applications in different fields, a brief review of controlled switching devices will be presented.

There are many factors and properties which have to be considered if controlled switching of voltages has to be achieved. The most essential factors include:

- the magnitude of the switched voltage (some kV up to MV),
- the magnitude of the control voltage or signal,
- the time delay between control signal and final stage of the switching,
- the jitter of the time delay,
- the conductivity of the switch in open and closed position,
- the inductance of the switch,
- the magnitude of the current switched,
- the repetition or recurrence frequency, and finally,
- the number of switching operations admissible.

For voltages higher than about 10 kV no solid state electronic element is able to operate. Up to some 10 kV, different types of thyratrons may be used, especially those with heated cathode and hydrogen content. Special types of thyratrons may switch voltages up to 100 kV and currents up to 20 kA with time delays of about 10 μsec and very low jitter down to a few nanoseconds. These elements are expensive, however, and the application is aggravated by the energy supply for the heated cathode, if the cathode is at high potential. Thus, for voltages higher than 100 kV only spark gaps are nearly unlimited in application, if they are properly controlled.

The physical mechanism responsible for the very fast transition of the resistance value of a spark gap begins with the streamer breakdown of the insulating gas and finishes with the arc, which has unlimited current-carrying capabilities. The time-dependent resistance of a gap may be calculated from the well-known 'spark laws' due to Toepler, Rompe-Weizel or Braginskii.[49] Investigations show that the oldest law associated with Toepler may conveniently be used for the computation of this time-dependent resistance, $R(t)$. If $i(t)$ is the current flowing in the gap, this resistance is given as

$$R(t) = \frac{k_T d}{\int_0^t i(t)\,dt}$$

where d is the gap distance and k_T the 'Toepler spark constant'. The integration may be started ($t = 0$) by a finite, not too small, current, due to the early beginning of the spark formation. The values of k_T are not real constants; they are slightly dependent upon the gas involved and the field strength within the gap before breakdown. For air, many measurements have yielded a value of $k_T \cong 0.5 \times 10^{-4} \pm 20$ per cent Vs/cm. The above relationship may be applied to a discharge circuit consisting of a discharge capacitor C in series with a resistance R and a homogeneous spark gap to calculate the current $i(t)$ and the time-dependent voltage drop across the gap. If then a time to breakdown T_b is defined as the time from the 90 to the 10 per cent instant values of the decreasing voltage, for $C \gtrapprox 2\,\text{nF}$, $R \gtrapprox 100\,\Omega$ one may derive the dependency

$$T_b \cong 13 \times 10^6 \frac{k_T}{E},$$

where T_b is in ns, k_T in Vs/cm and E in kV/cm. Thus for short switching times high field strength E before breakdown is necessary. Such high values can be achieved by pressurizing the gap, as the breakdown strength will increase about proportionally with the gas pressure (see Chapter 5, section 5.5). Also in air at atmospheric pressure switching times of about 20 ns will be reached for voltages up to some 100 kV.

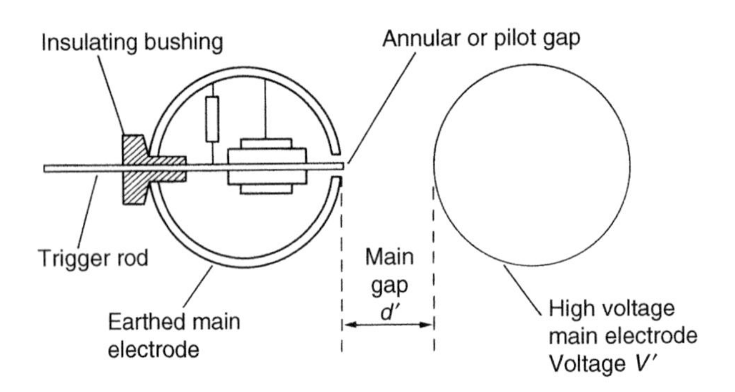

Figure 2.35 *The Trigatron spark gap*

The development of triggered and therefore controlled spark gaps cannot be discussed in detail. Only the principle will be considered using the arrangement displayed in Fig. 2.35, which provides good operating characteristics. This arrangement, known as 'Trigatron', consists essentially of a three-electrode gap. The main electrodes – indicated as h.v. and earthed electrodes – may consist of spheres, hemispheres or other nearly homogeneous electrode configurations. A small hole is drilled into the earthed electrode into which a metal rod projects. The annular gap between the rod and the surrounding sphere is typically about 1 mm. The metal rod or trigger electrode forms the third electrode, being essentially at the same potential as the drilled electrode, as it is connected to it through a high resistance, so that the control or tripping pulse can be applied between these two electrodes. For this special arrangement, a glass tube is fitted across the rod and is surrounded by a metal foil connected to the potential of the main electrode. The function of this tube is to promote corona discharges around the rod as this causes photoionization in the pilot gap, if a tripping impulse is applied to the rod. Due to this photoionization primary electrons are available in the annular gap which start the breakdown without appreciable time delay. The glass tube (or a tube of different solid insulation material, such as epoxy resin) may also fill the annular gap, so that the rod as well as the tube with its face is flush with the outside surface of the sphere. Thus a surface discharge is caused by the tripping pulse.

If a voltage V stresses the main gap, which is lower than the peak voltage at which self-firing occurs, this main gap will break down at a voltage even appreciably lower than the self-firing voltage V_s, if a tripping pulse is applied. The Trigatron requires a pulse of some kilovolts, typically $\leq 10\,\text{kV}$, and the tripping pulse should have a steep front with steepness $\gtrapprox 0.5\,\text{kV/nsec}$ to keep the jitter of the breakdown as small as possible. The first essential operating characteristic refers to the voltage operating limits, at which a steady operation or switching is possible. Such a characteristic is sketched in Fig. 2.36, where the operating voltage V, the voltage across the main gap, is shown in dependency of the main gap distance. The upper operation limit is identical with the self-firing voltage as defined earlier; the lower operating limit is that at which still a steady operation or breakdown is obtained with a predetermined jitter, for instance $\leq 100\,\text{ns}$, or time delay. Such a characteristic is clearly dependent

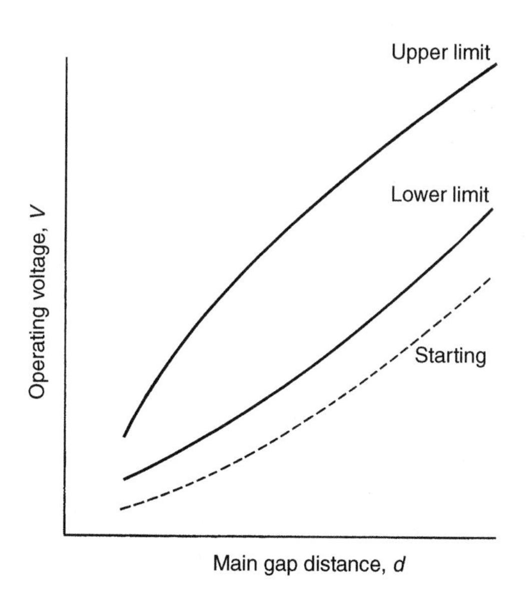

Figure 2.36 *Sketch of an operating characteristic of a Trigatron*

upon the detailed construction of the Trigatron. These characteristics are also polarity sensitive, mainly if the field distribution within the gap is not very homogeneous; the polarity sensitivity refers also to the polarity of the tripping pulse, which should always be of opposite polarity of the main voltage which had to be switched.

The physical mechanism which causes the main gap to break down is fundamentally understood, although it might be quite complex in detail. Indeed, it is recognized today that two types of mechanism are active. For small spacings d and a given tripping voltage V, the breakdown may directly be initiated by the distortion and enhancement of the electrical field between trigger electrode and the opposite main electrode, leading to a direct breakdown between these two electrodes. The arc then commutates from the larger electrode to the drilled electrode for the main current. The second type of breakdown takes place for larger gap distances. The trigger pulse causes a breakdown of the annular or pilot gap, and the large amount of charge carriers of all types available after sparking will initiate the breakdown of the main gap.

2.4 Control systems

In sections 2.1 to 2.3 only the basic principles for the generation of high d.c, a.c. and impulse voltages have been described. In general, the specific circuit and its components are only a part of a whole test system which is at least supplemented by sophisticated measuring (see Chapter 3) and control devices. In the simplest case of 'manual operation', measurement and control are performed by an operator and all results are recorded and further evaluated by hand. With the availability of quite cheap and extremely efficient PCs it is now easy to fulfil the demands for partly or fully automatic test sequences as well as to perform automatic recording and evaluation of test results. Even older available voltage generators can, in general, be updated with the necessary hard- and software. New types of h.v. generating systems are always prepared and equipped with the necessary interfaces for full automatic operation.

The main problem in designing such computer-aided control, measuring and evaluation systems is related to the hard electromagnetic environment present during high voltage tests, for which breakdown phenomena emitting very fast and strong electromagnetic transients are common. Industrial PCs of high 'electromagnetic compatibility' must thus be used as control computers, if a control room is not additionally shielded. As electronic measuring devices are also quite sensitive to the electromagnetic environment, larger and well-shielded control racks containing all items related to automatic testing, e.g. computer, monitor, keyboard, printer, transient recorders, etc., can be used. No limitations are set by applying special software for running the tests as prescribed by special test procedures, by recording individual results and evaluating test sequences by, e.g., statistical procedures, or for presenting test records. Of big advantage is the application of interface systems which are also commonly used in other industrial control systems.

Additional information is usually provided in the pamphlets or websites of the manufacturers of high-voltage generators.

References

1. Report of the Task Force on Power System Textbooks of PEEC/IEEE; Electric Power Systems Textbooks. *Trans. IEEE.* PAS **100** (1981), pp. 4255–4262.
2. IEC Publication 60: *High-voltage test techniques.* Part 1: General definitions and test requirements, Second Edition, 1989-11. Part 2: Measuring Systems, Second Edition, 1994-11.
3. IEEE Std 4-1995. *IEEE Standard Techniques for High-voltage Testing.*
4. J.D. Craggs and J.M. Meek. *High Voltage Laboratory Technique.* Butterworth, London, 1954.
5. H. Prinz. *Feuer, Blitz und Funke.* F. Bruckmann-Verlag, Munich, 1965.
6. H. Greinacher. *Erzeugung einer Gleichspannung vom vielfachen Betrag einer Wechselspannung ohne Transformator.* Bull. *SEV* **11** (1920), p. 66.
7. J.D. Cockcroft and E.T.S. Walton. Experiments with high velocity ions. *Proc. Roy. Soc. London,* Series A, **136** (1932), pp. 619–630.
8. K. Hammer and K. Kluge. Besonderheiten bei der Entwicklung von Gleich- spannungspruefanlagen mit grossen Abgabestroemen. *Elektrie* **35** (1981), pp. 127–131.
9. H.P.J. Brekoo and A. Verhoeff. *Phil. Tech. Rev.* **23** (1962), p. 338.
10. M. Wagstaff. *Direct Current* **7** (1962), p. 304.
11. H.A. Enge. Cascade transformer high voltage generator. US Patent No. 3,596,167 (July 1971).
12. N.J. Felici. *Elektrostatische Hochspannungs-Generatoren.* Verlag G. Braun, Karlsruhe, 1957.
13. M. Pauthenier and M. Moreau-Hanot. *J. de Phys. et le Radium* **8** (1937), p. 193.
14. Van Atta *et al.* The design, operation and performance of the Round Hill electrostatic generator. *Phys. Rev.* **49** (1936), p. 761.
15. Holifield Heavy Ion Research Facility. Pamphlet of Oak Ridge National Laboratory, Oak Ridge/Tenn., USA, 1981.
16. MIT. Study of gas dielectrics for cable insulation. EPRI Report No. EL 220 (October 1977).
17. N.J. Felici. *Direct Current* **1** (1953), p. 122.
18. J.G. Trump. *Elect. Eng.* **66** (1947), p. 525.
19. S.F. Philps. The vacuum-insulated, varying capacitance machine. *Trans. IEEE.* **EI 12** (1977), p. 130.
20. E.T. Norris and F.W. Taylor. *J. IEE.* **69** (1931), p. 673.
21. W. Müller. Untersuchung der Spannungskurvenform von Prüftranformatoren an einem Modell. *Siemens-Zeitschrift* **35** (1961), pp. 50–57.
22. E.E. Charlton, W.F. Westendorp, L.E. Dempster and G. Hotaling. *J. Appl. Phys.* **10** (1939), p. 374.
23. R. Reid. High voltage resonant testing. IEEE PES Winter Meeting 1974, Conf. Paper C74 038-6.
24. R. Reid. New method for power frequency testing of metal clad gas insulated substations and larger rotary machines in the field. World Electrotechn. Congress, Moscow 1977, Section 1, Report 29.
25. F. Bernasconi, W.S. Zaengl and K. Vonwiller. A new HV-series resonant circuit for dielectric tests. 3rd Int. Symp. on HV Engg., Milan, Report 43.02,1979.
26. W.S. Zaengl *et al.* Experience of a.c. voltage tests with variable frequency using a lightweight on-site s.r. device. CIGRE-Session 1982, Report 23.07.

27. R.H. Golde. *Lightning*, Vols I and II. Academic Press, London/New York/San Francisco, 1977.
28. Les Renardieres Group. Positive discharges in long air gaps at Les Renardieres. *Electra* No. **53**, July 1977.
29. F.A. Fisher and F.D. Martzloff. Transient control levels, a proposal for insulation coordination in low-voltage systems. *Trans. IEEE* **PAS 95** (1976), pp. 120–129.
30. O. Etzel and G. Helmchen. Berechnung der Elemente des Stossspannungs-Kreises für die Stossspannungen 1,2/50, 1,2/5 und 1,2/200. *ETZ-A* **85** (1964), pp. 578–582.
31. F.S. Edwards, A.S. Husbands and F.R. Perry. *Proc. IEE* **981** (1951), p. 155.
32. A. Vondenbusch. Ein allgemeines Berechnungsverfahren fuer Stossschaltungen mit voneinander unabhaengigen Energiespeichern. Ph.D. Thesis, TH Aachen, 1968.
33. J.R. Eaton and J.P. Gebelein. *GE Rev.* **43** (1940), p. 322.
34. J.L. Thomason. *Trans. AIEE* **53** (1934), p. 169.
35. Marx, E. Deutsches Reichspatent (German Patent) No. 455933.
36. A. Rodewald. Ausgleichsvorgaenge in der Marxschen Vervielfachungsschaltung nach der Zuendung der ersten Schaltfunkenstrecke. *Bull. SEV* **60** (1969), pp. 37–44.
37. F. Heilbronner. Das Durchzünden mehrstufiger Stossgeneratoren. *ETZ-A* **92** (1971), pp. 372–376.
38. B.L. Goodlet. *J. IEE* **67** (1929), p. 1177, and British Patent No. 344 862.
39. A. Rodewald and K. Feser. The generation of lightning and switching impulse voltages in the UHV region with an improved Marx circuit. *Trans. IEEE* **PAS 93** (1974), pp. 414–420.
40. D. Kind and H. Wehinger. Transients in testing transformers due to the generation of switching voltages. *Trans. IEEE* **PAS 97** (1978), pp. 563–568.
41. W. Boeck, D. Kind and K.H. Schneider. Hochspannungspruefungen von SF_6-Anlagen vor Ort. *ETZ-A* **94** (1973), pp. 406–412.
42. K. Feser. High voltage testing of metal-enclosed, gas-insulated substations on-site with oscillating switching impulse voltages. *Gaseous Dielectrics II*, Pergamon Press (1980), pp. 332–340 (Proc. of the 2nd Int. Symp. on Gas. Diel., Knoxville/Tenn., USA).
43. P.L. Bellaschi and L.B. Rademacher. *Trans. AIEE* **65** (1946), p. 1047.
44. K. Kojema and S. Tanaka. *JIEE Japan* **83** (1963), p. 42.
45. J.M. Christensen *et al.* A versatile automatic 1.2 MV impulse generator. 2nd Int. Symp. on High Voltage Engg., Zurich, 1975, paper 2.1-03.
46. F. Brändlin, K. Feser and H. Sutter. Eine fahrbare Stossanlage für die Prüfung von gekapselten SF_6-isolierten Schaltanlagen. *Bull. SEV* **64** (1973), pp. 113–119.
47. F. Jamet and G. Thomer. *Flash Radiography*, Elsevier, Amsterdam, 1976.
48. T.E. Allibone and J.C. Saunderson. The influence of particulate matter on the breakdown of large sphere-gaps. Third Int. Symp. on Gaseous Diel., Knoxville/Tenn., USA, 1982. (*Gaseous Dielectrics III*, Pergamon Press, 1982, pp. 293–299.)
49. K. Moeller. Ein Beitrag zur experimentellen Ueberprüfung der Funkengesetze von Toepler, Rompe-Weizel und Braginskii. *ETZ-A* **92** (1971), pp. 37–42.
50. H. Frank, W. Hauschild, I. Kantelberg and H. Schwab. *HV DC Testing Generator for Short-Time Reversal on Load.* 4th Int. Symposium on High Voltage Engineering (ISH), Athens 1983, paper 5105.
51. W. Schufft and Y. Gotanda. *A new DC Voltage Test System with Fast Polarity Reversal.* 10th Int. Symposium on High Voltage Engineering (ISH), Montreal 1997, Vol. 4, pp. 37–40.
52. W. Hauschild *et al. Alternating voltage on-site testing of XLPE cables: The parameter selection of frequency-tuned resonant test systems.* 10th Int. Symposium on High Voltage Engineering (ISH), Montreal 1977, Vol. 4, pp. 75–78.

Chapter 3

Measurement of high voltages

Measurement of high voltages – d.c., a.c. or impulse voltages – involves unusual problems that may not be familiar to specialists in the common electrical measurement techniques. These problems increase with the magnitude of the voltage, but are still easy to solve for voltages of some 10 kV only, and become difficult if hundreds of kilovolts or even megavolts have to be measured. The difficulties are mainly related to the large structures necessary to control the electrical fields, to avoid flashover and sometimes to control the heat dissipation within the circuits.

This chapter is devoted to the measurement of voltages applied for the testing of h.v. equipment or in research. Voltage-measuring methods used within the electric power transmission systems, e.g. instrument transformers, conventional or non-conventional ones, are not discussed. Such methods are summarized in specialized books as, for instance, reference 2, distributed publications,[3,4]* or a summary given in reference 1. An introduction into some measuring methods related to non-destructive insulation testing is presented separately (Chapter 7), and a brief reference related to the measurement of electrical fields is included in Chapter 4, section 4.4.

The classification of the measuring methods by sections according to the type of voltages to be measured would be difficult and confusing. A basic principle of quantifying a voltage may cover all kinds of voltage shapes and thus it controls the classification. The essential part of a measuring system relates also to the elements or apparatus representing the individual circuit elements. These could be treated separately, but a preferred treatment is within the chapter, in which special problems first arise. Due to space limitation no constructional details are given, but the comments referring to such problems should carefully be noted. The classification used here could introduce difficulties in selecting proper methods for the measurement of given voltages. Therefore, at this point a table is included (Table 3.1) which correlates the methods treated within the corresponding sections to the type of voltages to be measured.

* Superscript numbers are to references at the end of the chapter.

Table 3.1 *(Note '+' means 'in combination with')*

Type of voltage / *Quantity*	*d.c. voltages*	*a.c. voltages*	*Impulse voltages*
Mean value	3.3 3.4 3.6.4	Not Applicable	Not Applicable
r.m.s. value	3.2 3.6.4 + 3.2	3.2 3.6.4 + 3.2	Not Applicable
Crest values	3.1 3.4 (special des.) 3.3 (divider) + ripple by CRO or 3.7 (see also 3.6.4)	3.1 3.4 (special des.) 3.5.1 3.5.2 + 3.5.3 (see also 3.6.4)	3.1 3.6.4 (special) + 3.5.3 or 3.7
Voltage shape	3.4 (special des.) or 3.6.4 + CRO or 3.7	3.4 (special des.) or 3.6.4 + CRO or 3.7	3.6.4 (special) + CRO or 3.7

3.1 Peak voltage measurements by spark gaps

Simple spark gaps insulated by atmospheric air can be used to measure the amplitude of a voltage above about 10 kV. The complex mechanism of this physical effect, often employed in protecting equipment from overvoltages (protection gaps), is treated in Chapter 5. Although spark gaps for measurement purposes might be applied following given rules and recommendations only, a misuse can be avoided through an adequate study of the physical phenomena. As the fast transition from an either completely insulating or still highly insulating state of a gap to the high conducting arc state is used to determine a voltage level, the disruptive discharge does not offer a direct reading of the voltage across the gap. A complete short-circuit is the result of a spark, and therefore the voltage source must be capable to allow such a short-circuit, although the currents may and sometimes must be limited by resistors in series with the gap. Strictly speaking, spark gaps according to sections 3.1.1 and 3.1.3 can be considered as approved calibration devices with a limited accuracy, i.e known measuring uncertainty, but with a high reliability. Because of their high reliability and simplicity, spark gaps will probably never completely disappear from h.v. laboratories. More accurate and easier-to-use devices incorporating electronic circuits are generally applied for routine measurements. But these circuits are often sensitive to the electromagnetic effects and may sometimes fail to work. A regular calibration of such devices against approved spark gaps thus eliminates the possibility of large measuring errors and awkward consequences.

The geometry of a spark gap is a decisive factor for its application. For some decades the international and also national standards recommend the sphere gap (section 3.1.1) and now also the rod/rod gap for approved voltage measurements, as their reliability are best confirmed. The uniform field gaps (section 3.1.3) are merely included here to demonstrate their disadvantages and to save the beginner troublesome experiments.

3.1.1 Sphere gaps

Two adjacent metal spheres of equal diameters whose separation distance is limited, as discussed later, form a sphere gap for the measurement of the peak value of either d.c., a.c. or both kinds of impulse voltages. The ability to respond to peak values of voltages, if the duration of the peak region is not too short in time ($\gtrsim 1-3\,\mu$sec), is governed by a short statistical time lag, i.e. the waiting time for an electron to appear to initiate an electron avalanche and breakdown streamer, and an equally short formative time lag required for the voltage breakdown or fast current increase within the breakdown channel (see Fig. 5.42). The limitation in gap distance provides a fairly homogeneous field distribution so that no predischarge or corona appears before breakdown; the formative time lags are, therefore, also short. The permanent presence of primary or initiatory electrons within the regions of maximum field gradients to start critical avalanches within a short time lag is of great importance. The electrical field distribution within the high field regions must sufficiently be controlled by the geometry of the electrode and the air density as well as its composition must be known. Air is composed of various types of molecules which will influence the breakdown voltage. All these influences can be accounted for by the well-known breakdown criteria of gases (see Chapter 5) besides the primary electron impact, whose presence is a prerequisite.

All instructions as given in the still relevant IEC Recommendation[5] or National Standards[6] in detail can be related to these effects. The two standardized arrangements for the construction of the sphere gaps are shown in Figs 3.1(a) and 3.1(b). It should be noted also that in the horizontal arrangement one sphere must be earthed.

These figures contain most of the instructions necessary to define the geometry, except for values A and B which require some explanation. These two parameters define clearances such as to maintain the field distribution between the points on the two spheres that are closest to each other (sparking

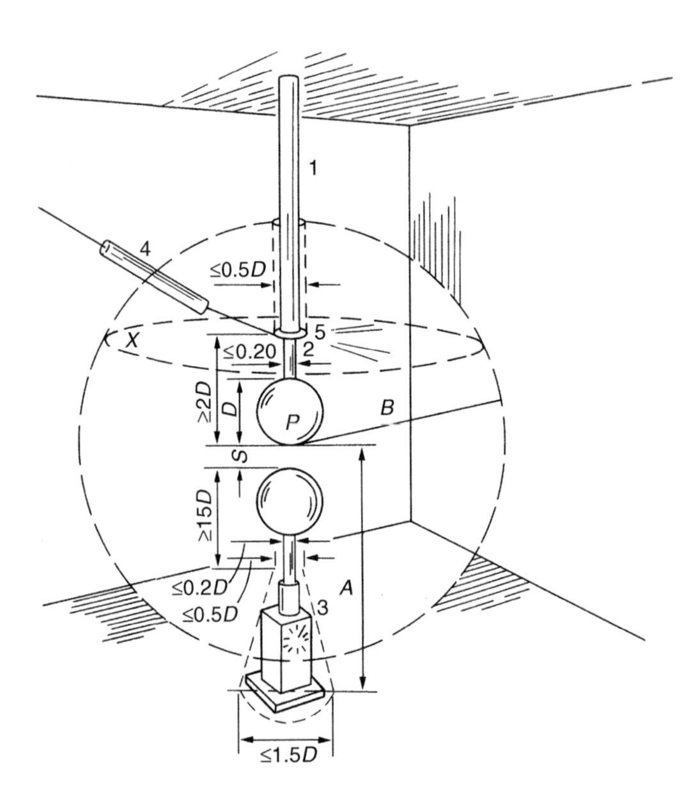

Figure 3.1(a) *Vertical sphere gap. 1. Insulating support. 2. Sphere shank. 3. Operating gear, showing maximum dimensions. 4. High-voltage connection with series resistor. 5. Stress distributor, showing maximum dimensions. P. Sparking point of h.v. sphere. A. Height of P above ground plane. B. Radius of space free from external structures. X. Item 4 not to pass through this plane within a distance B from P. Note: The figure is drawn to scale for a 100-cm sphere gap at radius spacing. (Reproduced from ref. 5)*

points) within narrow limits. The height of the sparking point P above the horizontal ground plane, which can be a conducting network in or on the floor of the laboratory, or a conducting surface on the support in which the sphere gap is placed, must be within given limits related to the sphere diameter D. To be accepted as a standard measuring device, a minimum clearance B around the sphere must also be available, within which no extraneous objects (such as walls, ceilings, transformer tanks, impulse generators) or supporting framework for the spheres are allowed. Table 3.2 gives the required clearances. Related to the accuracy of the field distribution are also requirements for the construction of the spheres and their shanks. The most important rules are reproduced partly:

Tolerances on size, shape and surface of spheres and their shanks

The spheres shall be carefully made so that their surfaces are smooth and their curvature is as uniform as possible. The diameter shall nowhere differ by more than 2 per cent from the nominal value. They should be reasonably free from surface irregularities in the region of the sparking points. This region is defined by a circle such as would be drawn on the spheres by a pair of dividers set to an opening of $0.3D$ and centred on the sparking point. The freedom from surface irregularities shall be checked by adequate measuring devices (for more details see reference 5 or 6).

The surfaces of the spheres in the neighbourhood of the sparking points shall be free from any trace of varnish, grease or other protective coating. They shall be clean and dry, but need not to be polished. If the spheres become

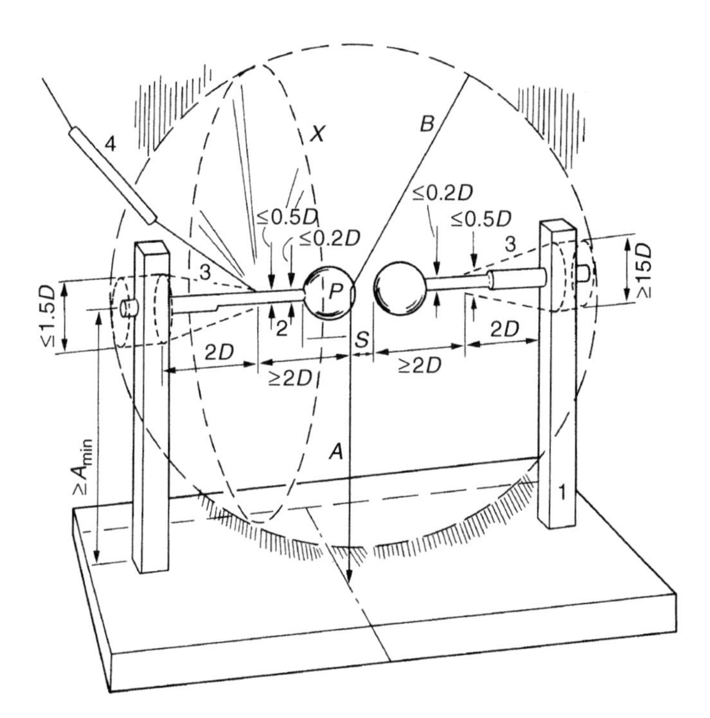

Figure 3.1(b) *Horizontal sphere gap. 1. Insulating support. 2. Sphere shank. 3. Operating gear, showing maximum dimensions. 4. High-voltage connection with series resistor. P. Sparking point of h.v. sphere. A. Height of P above ground plane. B. Radius of space free from external structures. X. Item 4 not to pass through this plane within a distance B from P. Note: The figure is drawn to scale for a 25-cm sphere gap at a radius spacing. (Reproduced from ref. 5).*

Table 3.2 *Clearance around the spheres*

Sphere diameter D (mm)	*Minimum value of A*	*Maximum value of A*	*Minimum Value of B*
62.5	7D	9D	14S
125	6	8	12
250	5	7	10
500	4	6	8
750	4	6	8
1000	3.5	5	7
1500	3	4	6
2000	3	4	6

excessively roughened or pitted in use, they shall be refinished or replaced. For relative air humidity exceeding 90 per cent, moisture may condense on the surface and the measurement will then cease to be accurate.

The sphere shanks shall be reasonably in line and the shanks of the h.v. sphere shall be free from sharp edges or corners, but the diameter of the shank shall not exceed $0.2D$ over a length D. If a stress distributor is used at the end of the shank, its greatest dimension shall be in accordance with Fig. 3.1.

Disruptive discharge voltages

If these and all otherwise recommended conditions are fulfilled, a sphere gap of diameter D and spacing S will spark at a peak voltage whose value will be close to the nominal values shown in Tables 3.3 and 3.4. These 'calibration data', related to the atmospheric reference conditions (temperature 20°C; air pressure 101.3 kPa or 760 mmHg) and the kind and polarity of voltage applied, are a result of joint international measurements within the period 1920 to about 1955; a summary of this research work is found in the bibliography of reference 6.

Note. For every sphere diameter the sparking voltage is a non-linear function of the gap distance, which is mainly due to the increasing field inhomogeneity and only less to the physics of breakdown. All table values could well be simulated by polynominals of order 6 or even less. Note also, that many table values are the result of only linear interpolation between points which have been the result of actual measurements.

For d.c. voltages the measurement is generally subject to larger errors, caused by dust or fibres in the air.[23,24] In this case the results are considered to have an estimated uncertainty of ±5 per cent provided that the spacing is less than $0.4D$ and excessive dust is not present.

Table 3.3

(PART 1) Sphere gap with one sphere grounded

Peak values of disruptive discharge voltages (50% for impulse tests) are valid for:

alternating voltages,
negative lightning impulse voltages,
negative switching impulse voltages,
direct voltages of either polarity.

Atmospheric reference conditions: 20°C and 101.3 kPa

Sphere gap spacing (mm)	*Voltage, kV peak*		
	Sphere diameter (cm)		
	6.25	*12.5*	*25*
5	17.2	16.8	
10	31.9	31.7	
15	45.5	45.5	
20	58.5	59.0	
25	69.5	72.5	72.5
30	79.5	85.0	86
35	(87.5)	97.0	99
40	(95.0)	108	112
45	(101)	119	125
50	(107)	129	137
55	(112)	138	149
60	(116)	146	161
65		154	173
70		(161)	184
80		(174)	206
90		(185)	226
100		(195)	244
110		(203)	261
120		(212)	275
125		(214)	282
150			(314)
175			(342)
200			(366)
225			(385)
250			(400)

(*continued overleaf*)

Table 3.3 *(continued)*

(PART 2) Sphere gap with one sphere grounded

	Voltage, kV peak				
	Sphere diameter (cm)				
Sphere gap spacing (mm)	*50*	*75*	*100*	*150*	*200*
50	138	138	138	138	
75	202	203	203	203	203
100	263	265	266	266	266
125	320	327	330	330	330
150	373	387	390	390	390
175	420	443	443	450	450
200	460	492	510	510	510
250	530	585	615	630	630
300	(585)	665	710	745	750
350	(630)	735	800	850	855
400	(670)	(800)	875	955	975
450	(700)	(850)	945	1050	1080
500	(730)	(895)	1010	1130	1180
600		(970)	(1110)	1280	1340
700		(1025)	(1200)	1390	1480
750		(1040)	(1230)	1440	1540
800			(1260)	(1490)	1600
900			(1320)	(1580)	1720
1000			(1360)	(1660)	1840
1100				(1730)	(1940)
1200				(1800)	(2020)
1300				(1870)	(2100)
1400				(1920)	(2180)
1500				(1960)	(2250)
1600					(2320)
					(2320)
1700					(2370)
1800					(2410)
1900					(2460)
2000					(2490)

Note. The figures in parentheses, which are for spacing of more than $0.5D$, will be within ±5 per cent if the maximum clearances in Table 3.2 are met. On errors for direct voltages, see text.

Table 3.4

(PART 1) Sphere gap with one sphere grounded

Peak values of disruptive discharge voltages (50% values) are valid for:

positive lightning impulses,
positive switching impulses,
direct voltages of either polarity.

Atmospheric reference conditions: 20°C and 101.3 kPa

Sphere gap spacing (mm)	*Voltage, kV peak*		
	Sphere diameter (cm)		
	6.25	*12.5*	*25*
5	17.2	16.8	–
10	31.9	31.7	31.7
15	45.9	45.5	45.5
20	59	59	59
25	71.0	72.5	72.7
30	82.0	85.5	86
35	(91.5)	98.0	99
40	(101)	110	112
45	(108)	122	125
50	(115)	134	138
55	(122)	145	151
60	(127)	155	163
65		(164)	175
70		(173)	187
80		(189)	211
90		(203)	233
100		(215)	254
110		(229)	273
120		(234)	291
125		(239)	299
150			(337)
175			(368)
200			(395)
225			(416)
250			(433)

(*continued overleaf*)

Table 3.4 *(continued)*

(PART 2) Sphere gap with one sphere grounded

Sphere gap spacing (mm)	*Voltage, kV peak*				
	Sphere diameter (cm)				
	50	*75*	*100*	*150*	*200*
50	138	138	138	138	138
75	203	202	203	203	203
100	263	265	266	266	266
125	323	327	330	330	330
150	380	387	390	390	390
175	432	447	450	450	450
200	480	505	510	510	510
250	555	605	620	630	630
300	(620)	695	725	745	750
350	(670)	770	815	858	860
400	(715)	(835)	900	965	980
450	(745)	(890)	980	1060	1090
500	(775)	(940)	1040	1150	1190
600		(1020)	(1150)	1310	1380
700		(1070)	(1240)	(1430)	1550
750		(1090)	(1280)	(1480)	1620
800			(1310)	(1530)	1690
900			(1370)	(1630)	1820
1000			(1410)	(1720)	1930
1100				(1790)	(2030)
1200				(1860)	(2120)
1300				(1930)	(2200)
1400				(1980)	(2280)
1500				(2020)	(2350)
1600					(2410)
1700					(2470)
1800					(2510)
1900					(2550)
2000					(2590)

Note. The figures in parentheses, which are for spacing of more than $0.5D$, will be within ±5 per cent if the maximum clearances in Table 3.2 are met.

For a.c. and impulse voltages, the tables are considered to be 'accurate' (to have an estimated uncertainty) within ±3 per cent for gap lengths up to $0.5D$. The tables are not valid for impulses below 10 kV and gaps less than $0.05D$ due to the difficulties to adjust the gap with sufficient accuracy. Values for spacing larger than $0.5D$ are regared with less accuracy and, for that reason, are shown in parentheses.

Remarks on the use of the sphere gap

The sphere gap represents a capacitance, which may form a series resonant circuit with its leads. Heavy predischarges across a test object will excite superimposed oscillations that may cause erratic breakdown. To avoid excessive pitting of the spheres, protective series resistances may be placed between test object and sphere gap, whose value may range from 0.1 to 1 MΩ for d.c. and a.c. power frequency voltages. For higher frequencies, the voltage drop would increase and it is necessary to reduce the resistance. For impulse voltages such protective resistors should not be used or should not exceed a value of 500 Ω (inductance less than 30 μH).

The disruptive discharge values of Tables 3.3 and 3.4 apply to measurements made without irradiation other than random ionization already present, except in

- the measurement of voltages below 50 kV peak, irrespective of the sphere diameters,
- the measurement of voltages with spheres of 125 mm diameter and less, whatever the voltage.

Therefore, for measurements under these conditions, additional irradiation is recommended and is essential if accurate and consistent results are to be obtained, especially in the case of impulse voltages and small spacing (see also below). For irradiation a quartz tube mercury vapour lamp having a minimum rating of 35 W and a current of at least 1 A is best applicable. Irradiation by capsules containing radioactive materials having activities not less than 0.2 mCi (7,4 10^6 Bq) and preferably of about 0.6 mCi (22,2 10^6 Bq), inserted in the h.v. sphere near the sparking points, needs precautions in handling the radioactive materials.

The application of spark gaps is time consuming. The procedure usually consists of establishing a relation between a high voltage, as measured by the sphere gap, and the indication of a voltmeter, an oscilloscope, or other device connected in the control circuit of the equipment. Unless the contrary can be shown, this relation ceases to be valid if the circuit is altered in any respect other than a slight change of the spacing of the spheres. The voltage measured by the sphere gap is derived from the spacing. The procedure in establishing the relationship varies with the type of voltage to be measured, as follows: for the measurement of direct and alternating voltages, the voltage shall be applied with an amplitude low enough not to cause disruptive discharge during the switching transient and it is then raised sufficiently slowly for the l.v. indicator to be read accurately at the instant of disruptive discharge of the gap. Alternatively, a constant voltage may be applied across the gap and the spacing between the spheres slowly reduced until disruptive discharge occurs.

If there is dust or fibrous material in the air, numerous low and erratic disruptive discharges may occur, especially when direct voltages are being measured, and it may be necessary to carry out a large number of tests before consistent results can be obtained.

The procedure for the measurement of impulse voltages is different: in order to obtain the 50 per cent disruptive discharge voltage, the spacing of the sphere gap or the charging voltage of the impulse generator shall be adjusted in steps corresponding to not more than 2 per cent of the expected disruptive discharge value. Six applications of the impulse should be made at each step. The interval between applications shall not be less than 5 sec. The value giving 50 per cent probability of disruptive discharge is preferably obtained by interpolation between at least two gap or voltage settings, one resulting in two disruptive discharges or less, and the other in four disruptive discharges or more. Another, less accurate, method is to adjust the settings until four to six disruptive discharges are obtained in a series of ten successive applications.

Since in general the actual air density during a measurement differs from the reference conditions, the disruptive voltage of the gap will be given as

$$V_d = k_d V_{d0} \tag{3.1}$$

where V_{d0} corresponds to the table values and k_d is a correction factor related to air density. The actual relative air density (RAD) is given in general terms by

$$\delta = \frac{p}{p_0}\frac{273 + t_0}{273 + t} = \frac{p}{p_0}\frac{T_0}{T} \tag{3.2}$$

where p_0 = air pressure of standard condition, p = air pressure at test conditions, $t_0 = 20°\text{C}$, t = temperature in degrees Centigrade at test conditions.

The correction factor k_d, given in Table 3.5, is a slightly non-linear function of RAD, a result explained by Paschen's law (see Chapter 5).

The influence of humidity is neglected in the recommendations, as its influence (an increase in breakdown voltage with increasing humidity) is unlikely to exceed 2 or 3 per cent over the range of humidity normally encountered in laboratories.

Some factors influencing the gap breakdown such as effects of nearby earthed objects, of humidity, of dust particles, of irradiation and voltage polarity are discussed fully in the previous book[(131)] and will not be dealt with here. The details can be found in references (7 to 24).

Table 3.5 *Air-density correction factor*

Relative air density RAD	*Correction factor* k_d
0.70	0.72
0.75	0.77
0.80	0.82
0.85	0.86
0.90	0.91
0.95	0.95
1.00	1.00
1.05	1.05
1.10	1.09
1.15	1.13

Final remarks

It shall be emphasized that all relevant standards related to the sphere gap are quite old and are essentially based on reference 5, which was submitted to the National Committees for approval in 1958. The publication of IEC 52 in 1960 was then a compromise, accepted from most of the National Committees, as Tables 3.3 and 3.4 are based on calibrations made under conditions which were not always recorded in detail. Also, results from individual researchers have not been in full agreement, especially for impulse voltages. As, however, sphere gaps have been used since then world wide and – apart from the following remarks – no significant errors could be detected during application of this measuring method, the sparking voltages as provided by the tables are obviously within the estimated uncertainties.

IEC Publication 52, since about 1993, has been under revision, which may be finished in about 2000. The main aim of this revision is the inclusion of switching surges and additional hints to the application of irradiation. Although no final decisions have been made up to now, the following information may be valuable:

- *Switching surges.* Some later investigations demonstrated the applicability of the table values for full standard switching impulse voltages, which are identical to those of lightning impulses. This is already considered in reference 6 and in Tables 3.3 and 3.4.
- *Irradiation.* Apart from the requirements as already given in the standards, the special importance of irradiation for the measurement of impulse

voltages will be mentioned. As shown in reference 22, additional irradiation is required if the sphere gap is used in laboratories in which impulse generators with encapsulated gaps are used. Current investigations are also concerned with the influence of irradiation from different kinds of u.v. lamps on breakdown. Only lamps having emission in the far ultraviolet (u.v.-C) are efficient.

- *Influence of humidity*. The systematic influence of humidity to the disruptive voltages, which is about 0.2 per cent per g/m^3, will be mentioned, which is the main source of the uncertainty.[19] In this context, a calculation of all disruptive voltages as provided by Table 3.3 shall be mentioned, see reference 134. These calculations, completely based on the application of the 'streamer breakdown criterion', on the very well-known 'effective ionization coefficients' of dry air, on a very accurate field distribution calculation within the sphere gaps, and on the systematic (see Chapter 5.5) influence of humidity on breakdown, essentially confirmed the validity of the table values with only some exceptions.

3.1.2 Reference measuring systems

Up until the late 1980s the main method for calibration of high-voltage measuring systems for impulse voltages was through the use of sphere gaps in conjunction with step response measurements.[3] The most recent revision of IEC 60-2:1994[53] contains significant differences from the previous version.[3] One of the fundamental changes has been to introduce the application of Reference Measuring Systems in the area of impulse testing. The concept of Reference Measuring Systems in high-voltage impulse testing was introduced to address questions of quality assurance in measurements, an area which has seen a significant increase in attention over the past decade.

The need for better quality assurance in high-voltage impulse measurements was convincingly demonstrated in the 1980s and 1990s through the performance of several round-robin tests designed to quantify the repeatability of measurements between different laboratories. These tests comprised circulating reference divider systems amongst different laboratories and comparing the voltage and time parameters of impulses measured with the reference systems to those derived from the measurement of the same impulses using the regular laboratory dividers. Analysis of the results of these tests showed that while some laboratories were able to make repeatable simultaneous measurements of the voltage and time parameters of impulses using two Measuring Systems with good agreement, others were not.[135,136] For example, reference 135 gives the results of a round-robin test series performed under the sponsorship of the IEEE High Voltage Test Techniques subcommittee. The paper describes the results found when two reference dividers were circulated to a number of laboratories, each having a Measuring System thought to be adequately calibrated in accordance with the previous version of IEC 60-2.

The study revealed significant discrepancies in some laboratories between the results obtained with the Measuring Systems currently in everyday use and the Measuring System using the reference divider which was being circulated. Based on these findings, the concept of Reference Measuring Systems was introduced with the aim of improving the quality of high-voltage impulse measurements.

A Reference Measuring System is defined in IEC Publication 60-2:1994 as a Measuring System having sufficient accuracy and stability for use in the approval of other systems by making simultaneous comparative measurements with specific types of waveforms and ranges of voltage or current. The requirements on a Reference Measuring System for use in high-voltage impulse testing are clearly laid out in IEC Publication 60-2:1994. Reference dividers meeting these requirements are available from several manufacturers or can be constructed by the user.[135] Figure 3.2 shows a photograph of a reference divider which is designed for use in calibrating a.c., d.c., lightning and switching impulse voltages and is referred to as a Universal Reference Divider.

3.1.3 Uniform field gaps

It is often believed that some disadvantages of sphere gaps for peak voltage measurements could be avoided by using properly designed plate electrodes providing a uniform field distribution within a specified volume of air. The procedure to control the electrical field within such an arrangement by appropriately shaped electrodes is discussed in Chapter 4, section 4.2 (Rogowski or Bruce profile). It will also be shown in Chapter 5, section 5.6 that the breakdown voltage of a uniform field gap can be calculated based upon fundamental physical processes and their dependency upon the field strength. According to eqn (5.103) the breakdown voltage V_b can be expressed also by

$$V_b = E_c(\delta S) + B\sqrt{\delta S} \tag{3.3}$$

if the gas pressure p in eqn (5.102) is replaced by the air density δ (see eqn (3.2)) and if the gap distance is designated by S. The values E_c and B in eqn (3.3) are also constants as the values $(E/p)_c$ and $\sqrt{K/C}$ within eqn (5.102). They are, however, dependent upon reference conditions. An equivalent calculation as performed in Chapter 5, section 5.6 shows that

$$E_c = \left(p_0 \frac{T}{T_0}\right) \times \left(\frac{E}{p}\right)_c \tag{3.4}$$

$$B = \sqrt{\frac{K p_0 T}{C T_0}} \tag{3.5}$$

where all values are defined by eqns (5.102) and (3.2). Equation (3.3) would thus simply replace Tables 3.3 and 3.4 which are necessary for sphere gaps.

Figure 3.2 *Universal Reference Voltage Divider for 500 kV lightning and switching impulse, 200 kV a.c. (r.m.s.) and 250 kV d.c. voltage (courtesy Presco AG, Switzerland)*

Apart from this advantage of a uniform field gap, no polarity effect and no influence of nearby earthed objects could be expected if the dimensions are properly designed. All these advantages, however, are compensated by the need for a very accurate mechanical finish of the electrodes, the extremely careful parallel alignment, and – last but not least – the problem arising by unavoidable dust, which cannot be solved for usual air conditions within a laboratory. As the highly stressed electrode areas become much larger than for sphere gaps, erratic disruptive discharges will tend to occur. Therefore, a uniform field gap insulated in atmospheric air is *not* applicable for voltage measurements.

3.1.4 Rod gaps

Rod gaps have earlier been used for the measurement of impulse voltages, but because of the large scatter of the disruptive discharge voltage and the uncertainties of the strong influence of the humidity, they are no longer allowed to be used as measuring devices. A summary of these difficulties may be found in reference 4 of Chapter 2.

Later investigations of Peschke,[14] however, have demonstrated how the simple electrode configuration rod/rod gap may be used for the measurement of d.c. voltages, if the air density and the humidity is taken into account, and if some rules relating to the electrode arrangement are followed. This arrangement must comprise two hemispherically capped rods of about 20 mm diameter as sketched in Fig. 3.3. The earthed rod must be long enough to

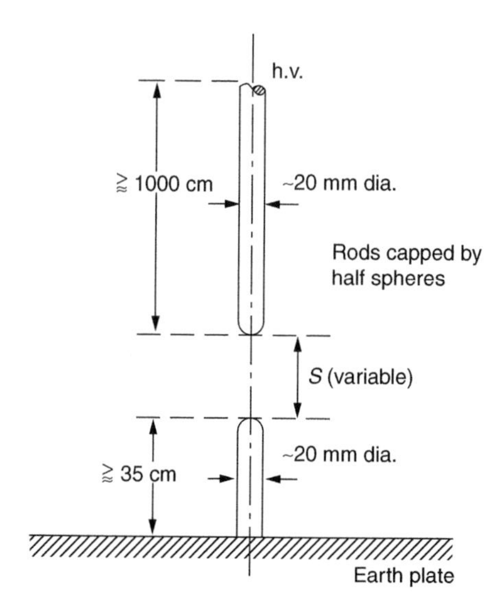

Figure 3.3 *Electrode arrangement for a rod/rod gap to measure high d.c. voltages*

initiate positive breakdown streamers if the h.v. rod is the cathode. Then for both polarities the breakdown will always be initiated by positive streamers giving a very small scatter and being humidity dependent. Apart from too low voltages ($\lessapprox$130 kV), for which the proposed rod/rod gap is not sufficiently inhomogeneous, the breakdown voltage V_b then follows the relationship

$$V_b = \delta(A + BS)\sqrt[4]{5.1 \times 10^{-2}(h + 8.65)} \quad \text{in kV} \tag{3.6}$$

where S = gap distance in cm, δ = relative air density according to eqn (3.2), and h = absolute humidity in g/m^3.

This empirical equation is limited to $4 \leq h \leq 20$ g/m^3 and has been shown to apply in the voltage range up to 1300 kV. V_b shows a linear increase with the gap length S, and the steepness B for the gap configuration shown in Fig. 3.3 is not very dependent on polarity. Also the constant A displays a small polarity effect, and numerical values are

$$A = 20\,\text{kV}; \; B = 5.1\,\text{kV/cm}; \text{for positive polarity}$$

$$A = 15\,\text{kV}; \; B = 5.45\,\text{kV/cm}; \text{for negative polarity}$$

of the h.v. electrode. The estimated uncertainty of eqn (3.6) is lower than ± 2 per cent and therefore smaller than the 'accuracy' provided by sphere gaps.

These investigations of Peschke[(14)] triggered additional work, the results of which are provisionally included within Appendix C of IEC Standard 60-1, 1989, see reference 2 of Chapter 2. The rod/rod gap thus became an approved measuring device for d.c. voltages. The additional investigations showed, that with somewhat different electrode configurations, which are not displayed here, the disruptive voltage U_0 even becomes equal for both voltage polarities, namely

$$U_0 = 2 + 0.534d \tag{3.6a}$$

where U_0 is in kV and d is the gap spacing in millimetres. This equation is valid for gap spacing between 250 and 2500 mm, an air humidity between 1 and 13 g/m^3, and its measurement uncertainty is estimated to be less than ± 3 per cent for these boundary conditions. A disadvantage of the electrode configurations as shown in Figs 19a/b of IEC 60-1 are the much larger dimensions as those displayed in Fig. 3.3.

3.2 Electrostatic voltmeters

Coulomb's law defines the electrical field as a field of forces, and since electrical fields may be produced by voltages, the measurement of voltages can be related to a force measurement. In 1884 Lord Kelvin suggested a design for an electrostatic voltmeter based upon this measuring principle. If the field is produced by the voltage V between a pair of parallel plane disc electrodes, the force F on an area A of the electrode, for which the field gradient E is the same across the area and perpendicular to the surface, can be calculated from the derivative of the stored electrical energy W_{el} taken in the field direction (x). Since each volume element $A\,\mathrm{d}x$ contains the same stored energy $dW_{\mathrm{el}} = (\varepsilon E^2 A\,\mathrm{d}x)/2$, the attracting force $F = -\mathrm{d}W_{\mathrm{el}}/\mathrm{d}x$ becomes

$$|F| = \frac{\varepsilon A E^2}{2} = \frac{\varepsilon A}{2S^2} V^2, \tag{3.7}$$

where ε = permittivity of the insulating medium and S = gap length between the parallel plane electrodes.

The attracting force is always positive independent of the polarity of the voltage. If the voltage is not constant, the force is also time dependent. Then the mean value of the force is used to measure the voltage, thus

$$\frac{1}{T}\int_0^T F(t)\,\mathrm{d}t = \frac{\varepsilon A}{2S^2}\frac{1}{T}\int_0^T v^2(t)\,\mathrm{d}t = \frac{\varepsilon A}{2S^2}(V_{\mathrm{r.m.s}})^2, \tag{3.8}$$

where T is a proper integration time. Thus, electrostatic voltmeters are r.m.s.-indicating instruments!

The design of most of the realized instruments is arranged such that one of the electrodes or a part of it is allowed to move. By this movement, the electrical field will slightly change which in general can be neglected. Besides differences in the construction of the electrode arrangements, the various voltmeters differ in the use of different methods of restoring forces required to balance the electrostatic attraction; these can be a suspension of the moving electrode on one arm of a balance or its suspension on a spring or the use of a pendulous or torsional suspension. The small movement is generally transmitted and amplified by a spotlight and mirror system, but many other systems have also been used. If the movement of the electrode is prevented or minimized and the field distribution can exactly be calculated, the electrostatic measuring device can be used for *absolute* voltage measurements, since the calibration can be made in terms of the fundamental quantities of length and forces.

The paramount advantage is the extremely low loading effect, as only electrical fields have to be built up. The atmospheric air, high-pressure gas or even high vacuum between the electrodes provide very high resistivity, and thus the active power losses are mainly due to the resistance of insulating materials used elsewhere. The measurement of voltages lower than about 50 V is, however, not possible, as the forces become too small.

The measuring principle displays no upper frequency limit. The load inductance and the electrode system capacitance, however, form a series resonant circuit, thus limiting the frequency range. For small voltmeters the upper frequency is generally in the order of some MHz.

Many designs and examples of electrostatic voltmeters have been summarized or described in the books of Schwab,[1] Paasche,[30] Kuffel and Abdullah,[26] Naidu and Kamaraju,[29] and Bowdler.[127] High-precision-type electrostatic voltmeters have been built for very high voltages up to 1000 kV. The construction of such an absolute voltmeter was described by House *et al.*[31]

In spite of the inherent advantages of this kind of instrument, their application for h.v. testing purposes is very limited nowadays. For d.c. voltage measurements, the electrostatic voltmeters compete with resistor voltage dividers or measuring resistors (see next chapter), as the very high input impedance is in general not necessary. For a.c. voltage measurements, the r.m.s. value is either of minor importance for dielectric testing or capacitor voltage dividers (see section 3.6) can be used together with low-voltage electronic r.m.s. instruments, which provide acceptable low uncertainties. Thus the actual use of these instruments is restricted and the number of manufacturers is therefore extremely limited.

3.3 Ammeter in series with high ohmic resistors and high ohmic resistor voltage dividers

The two basic principles

Ohm's law provides a method to reduce high voltages to measurable quantities, i.e. adequate currents or low voltages. The simplest method employs a microammeter in series with a resistor R of sufficiently high value to keep the loading of an h.v. source as small as possible (Fig. 3.4(a)). Thus for a pure resistance R, the measured quantities are related to the unknown high voltage by

$$v(t) = Ri(t) \tag{3.9}$$

or

$$V = RI \tag{3.10}$$

if the voltage drop across the ammeter is neglected, which is usually allowable due to the small terminal impedance of such instruments. For d.c. voltage measurements, average current-indicating instruments such as moving coil or equivalent electronic meters are used giving the arithmetic mean value of V according to eqn (3. 10). Less recommendable is the measurement of r.m.s. values as the polarity of the high voltage would not be shown. Fundamentally

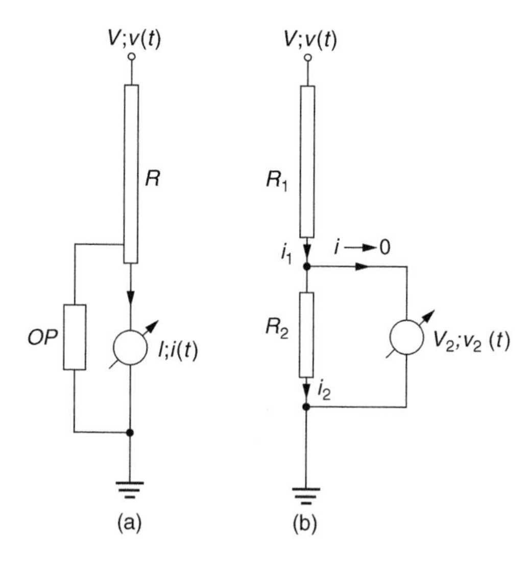

Figure 3.4 *Measurement of high d.c. and a.c. voltages by means of: (a) ammeter in series with resistor R; (b) voltage divider R_1, R_2 and voltmeter of negligible current input. OP, over voltage protection*

the time-dependency $v(t)$ according to eqn (3.9) could also be measured by, for instance, an oscilloscope. The difficulties, however, in treating the resistance R as a pure resistance are limiting this application. This problem will be discussed later on. It is recommended that the instrument be protected against overvoltage with a glow tube or zener diode for safety reasons.

The main difficulties encountered in this method are related to the stability of the resistance R. All types of resistors are more or less temperature dependent and often may show some voltage dependency. Such variations are directly proportional to the voltage to be measured and increase the uncertainty of the measurement result.

Before discussing some details concerning resistor technology, the alternative method shown in Fig. 3.4(b) will be described. If the output voltage of this voltage divider is measured with instruments of negligible current consumption ($i \to 0$ or $i/i_2 \ll 1$), the high voltage is now given by

$$v(t) = v_2(t)\left(1 + \frac{R_1}{R_2}\right) \tag{3.11}$$

$$V = V_2\left(1 + \frac{R_1}{R_2}\right) \tag{3.12}$$

Apart from the uncertainty of the output voltage measurement (V_2 or $v_2(t)$), the magnitude of the high voltage is now only influenced by the ratio R_1/R_2.

As both resistors pass the same current $i_1 = i_2$, the influence of voltage and temperature coefficients of the resistors can be eliminated to a large extent, if both resistors employ equal resistor technology, are subjected to equal voltage stresses, and if provisions are made to prevent accumulation of heat within any section of the resistor column. Thus the uncertainty of the measurement can be greatly reduced. Accurate measurement of V_2 was difficult in earlier times as only electrostatic voltmeters of limited accuracy had been available. Today electronic voltmeters with terminal impedances high enough to keep $i \ll i_2$ and giving high accuracy for d.c. voltage measurements are easy to use.

So far it appears that either method could easily be used for measurement of even very high voltages. The design of the methods starts with dimensioning the h.v. resistor R or R_1 respectively. The current through these resistors is limited by two factors. The first one is set by the heat dissipation and heat transfer to the outside and defines the upper limit of the current. A calculation assuming heat transfer by natural convection only would demonstrate upper limits of 1 to 2 mA. The second factor is due to the loading of the h.v. source; in general, very low currents are desirable. As the resistors predominantly at the input end of the h.v. column are at high potential and thus high field gradients have to be controlled, even with the best insulating materials the leakage along the resistor column or the supporting structure controls the lower limit of the current, which in general shall not be smaller than about 100 μA. This magnitude results in a resistance of $10^{10}\,\Omega$ for a voltage of 1000 kV, and thus the problem of the resistor technology arises.

Comment regarding the resistor technology and design of the h.v. arm

In practice this high ohmic resistor (R, R_1) is composed of a large number of individual elements connected in series, as no commercial types of single unit resistors for very high voltages are available.

Wire-wound metal resistors made from Cu–Mn, Cu–Ni and Ni–Cr alloys or similar compositions have very low temperature coefficients down to about 10^{-5}/K and provide adequate accuracy for the method prescribed in Fig. 3.4(a). As, however, the specific resistivity of these materials is not very large, the length of the wire required becomes very considerable even for currents of 1 mA and even for the finest gauge which can be made. Individual units of about 1 MΩ each then must be small in size as only a voltage drop of 1 kV arises, and thus the manner of winding will enhance self-inductive and self-capacitive components. In addition, the distributed stray capacitance to ground, discussed in more detail in section 3.6 and briefly below, causes a strongly non-linear voltage distribution along a resistor column and overstresses the individual elements during a sudden load drop originated by voltage breakdown of a test object. Wire-wound resistors are thus not only very expensive to produce, but also quite sensitive to sudden voltage drops.

Many constructions have been described in the literature and summaries can be found in references 1, 26, 30 and 127.

Especially for the voltage-dividing system, Fig. 3.4(b), common carbon, carbon composition or metal oxide film resistors are preferably used. They should be carefully selected due to the usually larger temperature coefficients (TC) which may even be different for the same type of such resistors. Nowadays, however, metal oxide products with TC values of about 20 to 30 ppm/K only can be produced. The resistor value of all these resistors may change also with voltage magnitude, and the – in general – negative voltage coefficients may be found in the manufacturer's catalogue. The self-inductance of such resistors is always negligible, as the high values of the individual film resistors are often reached by a bifilar arrangement of the film. Too thin films are generally destroyed by fast voltage breakdown across the resistor column. This effect may well be understood if the stray capacitances to earth are considered, or if high field gradients at the film surfaces are encountered. If the voltage suddenly disappears, high capacitive or displacement currents are injected into the thin film material, which cannot dissipate the heat within a very short time. Thus the temperature rise within the material may be so high that some of the material even explodes. The result is an increase of the original resistance value. Carbon composition resistors have large energy absorption capabilities. Their resistor value may, however, decrease due to short-time overloads, as the individual particles may be additionally sintered. A conditioning performed by prestressing of such resistors with short overloading voltages may decrease the effect. Thus the selection of resistors is not a simple task.

Other problems involved in a skilful design of the h.v. resistor concern the prevention of too high field gradients within the whole arrangement and, related to this, is the effect of stray capacitances upon the frequency-dependent transfer characteristics. To demonstrate these problems the design of a 100-kV standard resistor described by Park[32] will be discussed here. This resistor, shown in Fig. 3.5, is made up of a hundred 1-MΩ wirewound resistors connected in series and arranged to form a vertical helix. Some of these individual resistors are forming resistor elements, as they are placed within small cylindrical housings predominantly made from metal. Figure 3.6 shows a cross-section of such a resistor element; the metal cylinders or 'shields' enclose the individual resistors of small size and thus increase the diameter of the resistors. The metal shield is separated by a gap whose insulation can withstand and insulate the voltage drop $(V_1 - V_2)$ across the element. As the absolute values of the potentials V_1, V_2 can be high, the field gradients at the surface of small wires or small individual resistor units would be too high to withstand the insulation strength of the atmospheric air used for the construction. Therefore, the larger diameter of the shields lowers the field gradients to an acceptable magnitude. A further reduction of these gradients is achieved by the helical

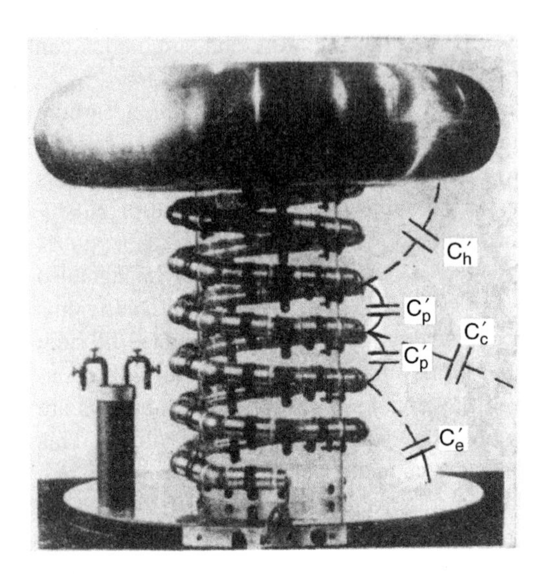

Figure 3.5 *100-MΩ, 100-kV standard resistor according to Park*[32]

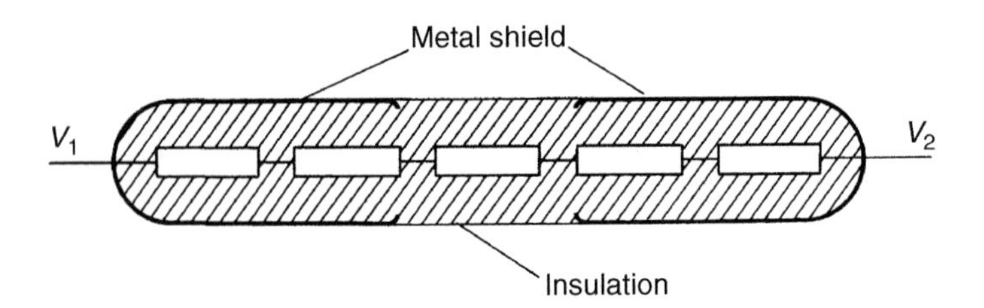

Figure 3.6 *Sketch of cross-section of an h.v. resistor element*

arrangement, as now the helix might be assumed to form a cylinder of much larger diameter, across which the potential continuously decreases from the top to the bottom. These statements could be confirmed by a computation of the very complex field distribution of the three-dimensional structure. The h.v. end of the resistor is fitted with a large 'stress ring' which again prevents concentration of electrical field and thus corona or partial discharge formation. A corona-free design is absolutely necessary to avoid leakage currents, which would decrease the overall resistance value.

For voltages higher than about 100 kV such an air-insulated design becomes difficult. The resistor elements then need improved insulation commonly achieved by mineral oil or highly insulating gases. They have to be placed, therefore, in insulating vessels. Additional oil or gas flow provided by pumps will improve the temperature equalization.

Frequency-dependent transfer characteristics

This problem is closely related to the field distribution phenomena. As charges are the origin and the end of electrostatic field lines, and such field lines will exist between points of differing potentials, the electrostatic field distribution may well be represented by 'stray capacitances'. Such stray capacitances have been included in Fig. 3.5 showing the 100-kV resistor, and three different kinds of capacitances are distinguished: the parallel capacitances C'_p between neighbouring resistor elements within the helix, the stray capacitances to the h.v. electrode C'_h and the stray capacitances C'_e to earth potential. Thus a very complex equivalent network is formed which is shown in Fig. 3.7 by assuming five resistor elements R' only and neglecting any residual inductances of the resistors. For equal values of R', the real values of the different stray capacitances would not be equal as is assumed. Depending upon the magnitude of the individual capacitances the ratio I_1/V will therefore change with frequency. As the number of elements used in Fig. 3.7 is too small in reality, a very large number of results would appear by assuming any combinations of capacitive elements. Thus an ingenious reduction of the circuit parameters is necessary, which can be done by assuming homogeneous ladder networks.

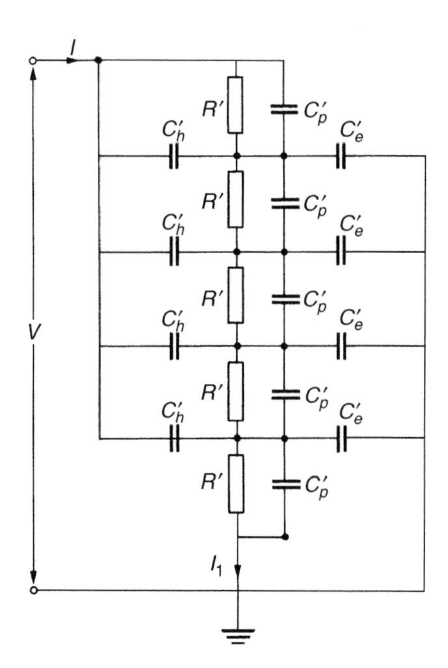

Figure 3.7 *Equivalent network of an h.v. resistor*

Although such ladder networks are treated in more detail in section 3.6, a short calculation is included at this point, originally published by Davis.[33]

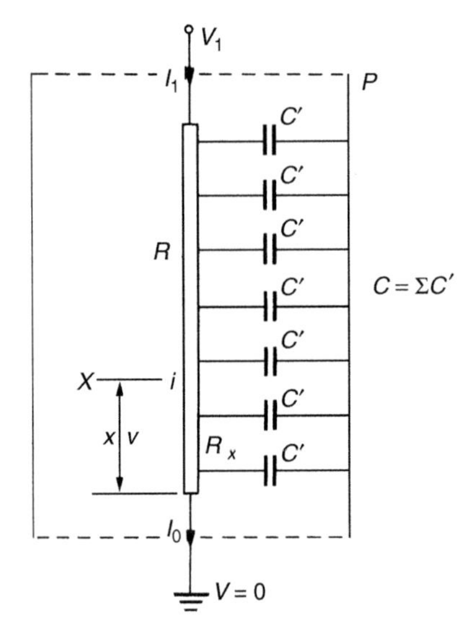

Figure 3.8 *Shielded resistor*

This calculation is based upon a 'shielded resistor' network, shown in Fig. 3.8. Here it is assumed that a resistor R of equally distributed resistance values per unit length $\mathrm{d}x$ is enclosed by a metal shield, whose potential is P. In comparison with Fig. 3.7, the interturn capacitances C'_p are neglected. This metal shield also suppresses the stray capacitances to h.v. electrode C'_h, and this structure leads to only one type of stray capacitance $C = \sum C'$ which is uniformly distributed from the resistance to the shield. Taking a point X at a distance x from the earthed end, the resistance between X and the earthed end is R_x.

Let the ratio $R_x/R = K$, so that $R_x = RK$ and an element of resistance $\mathrm{d}R_x = R\,\mathrm{d}K$. The amount of capacitance associated with $\mathrm{d}R_x$ is then $C' = C\,\mathrm{d}K$. If ϕ is the potential at X and i the current in the resistor at this point,

$$\mathrm{d}i = j\omega C(\phi - P)\,\mathrm{d}K; \quad \mathrm{d}\phi = iR\,\mathrm{d}K.$$

and

$$\frac{\mathrm{d}^2\phi}{\mathrm{d}K^2} = R\frac{\mathrm{d}i}{\mathrm{d}K} = j\omega CR(\phi - P).$$

The general solution of this equation is

$$\phi = A\,\mathrm{e}^{aK} + B\,\mathrm{e}^{-aK} + P,$$

where A and B are constants and $a = \sqrt{j\omega CR}$. The constants are obtained by putting

$$\phi = V_1, \quad \text{where } K = 1,$$
$$\phi = 0, \quad \text{where } K = 0.$$

The equation then becomes

$$\phi = \frac{e^{aK}[V_1 - P(1 - e^{-a})] - e^{-aK}[V_1 - P(1 - e^{a})]}{e^{a} - e^{-a}} + P \tag{3.13}$$

The current i at any point is then

$$i = \frac{1}{R}\frac{d\phi}{dK}$$
$$= \frac{1}{R}\frac{a}{e^{a} - e^{-a}}[e^{aK}\{V_1 - P(1 - e^{-a})\} + e^{-aK}\{V_1 - P(1 - e^{a})\}]. \tag{3.14}$$

Here, the equations for the currents at the earthed end and the h.v. end can be derived by inserting the appropriate values of K.

The current at the earthed end is obtained by putting $K = 0$, and is

$$I_0 = \frac{1}{R}\frac{a}{e^{a} - e^{-a}}[V_1 - P(1 - e^{-a}) + V_1 - P(1 - e^{a})]$$
$$= \frac{a}{R \sinh a}[(V_1 - P + P\cosh a)].$$

By expanding the hyperbolic functions, the result will be:

$$I_0 = \frac{a[V_1 - P + P\{1 + (a^2/2) + (a^4/24) + \ldots\}]}{R\{a + (a^3/6) + (a^5/120) + \ldots\}}$$
$$= \frac{V_1 + (Pa^2/2) + (Pa^4/24)}{R\{1 + (a^2/6) + (a^4/120) + \ldots\}}. \tag{3.15}$$

The current I_1 at the h.v. end is obtained by putting $K = 1$ and by similar treatment

$$I_1 = \frac{V_1 + \{(V_1 - P)(a^2/2)\} + \{(V_1 - P)(a^4/24)\}}{R\{1 + (a^2/6) + (a^4/120)\}} \tag{3.16}$$

The above analysis shows that the current is a function of the shield potential P and it will be of interest to express the currents for the following two special cases:

Case I. When $P = 0$, the uniformly distributed capacitance C is a stray capacitance to earth, C_e (compare with Fig. 3.7), and the current to ground becomes

$$I_0 = \frac{V_1}{R\{1 + (a^2/6) + (a^4/120) + \ldots\}}.$$

The terms containing higher powers of a than a^2 may be neglected, as $a^2 = j\omega RC$ and the following alternating signs as well as decreasing values of the terms do scarcely contribute. Thus

$$I_0 \approx \frac{V_1}{R\left(1 + j\frac{\omega RC_e}{6}\right)} = \frac{V_1}{R\left[1 + \left(j\frac{\omega RC_e}{6}\right)^2\right]}\left(1 - j\frac{\omega RC_e}{6}\right). \quad (3.17)$$

The phase angle between the input voltage V_1 and the current to earth is then $-\omega RC_e/6$. Similarly, the current at the h.v. end is

$$I_1 \approx \frac{V_1(1 + a^2/2)}{R(1 + a^2/6)} = \frac{V_1}{R} \times \frac{\left(1 + \frac{\omega RC_e}{12} + j\frac{\omega RC_e}{3}\right)}{\left[1 + \left(\frac{\omega RC_e}{6}\right)^2\right]}.$$

For not too high frequencies, we may neglect the real frequency terms, and thus

$$I_1 \approx \frac{V_1}{R}\left(1 + j\frac{\omega RC_e}{3}\right) \quad (3.18)$$

The phase angle becomes $+\omega RC_e/3$.

For a.c. voltage measurements only eqn (3.17) is important. Apart from the phase shift the relative change of the current amplitudes with increasing frequency contains the amplitude errors. We thus may define the normalized transfer characteristic

$$H_0(j\omega) = \frac{I_0(\omega)}{I_0(\omega = 0)} = \frac{1}{\left(1 + j\frac{\omega RC_e}{6}\right)}.$$

The amplitude frequency response becomes

$$H_0(\omega) = |H_0(j\omega)| = 1\bigg/\sqrt{1 + \left(\frac{\omega RC_e}{6}\right)^2} \quad (3.19)$$

This equation shows the continuous decrease of the current with frequency. The 3 dB bandwidth f_B, defined by $H_0(\omega) = 1/\sqrt{2}$, is thus

$$f_B = \frac{3}{\pi RC_e} = \frac{0.95}{RC_e}. \tag{3.20}$$

For a decrease of the current amplitude by only 2 per cent, the corresponding frequency is much lower ($\approx 0.095/RC$, or one-tenth of f_B). An h.v. resistor for 100 kV is assumed, with a resistance of 200 MΩ and a stray capacitance C_e of 10 pF. Then eqn (3.20) gives a bandwidth of 475 Hz, demonstrating the limited accuracy for a.c. measurements. As the resistance values cannot be reduced very much due to the heat dissipation, only a decrease of C_e can improve the frequency range.

Case II. One possible way of shielding and thus reducing the stray capacitances to ground would be to raise the potential of the metal shield indicated in Fig. 3.8. When $P = V_1/2$, the expressions for I_0 and I_1 can be obtained in a similar manner as in Case I. Neglecting again in eqns (3.15) and (3.16) powers higher than 2, we obtain for both currents

$$I_0 \approx \frac{V_1}{R}\left(1 + j\frac{\omega RC}{12}\right) \tag{3.21}$$

$$I_1 \approx \frac{V_1}{R}\left(1 + j\frac{\omega RC}{12}\right) \tag{3.22}$$

Thus the expressions for the two currents are the same. In comparison to eqn (3.17) the change in the sign of the phase angle should be emphasized. The output current I_0 thus increases in amplitude also with frequency. Such phenomena are always associated with stray capacitances to h.v. potential C'_h as shown in Fig. 3.7. However, for h.v. resistors or resistor dividers as treated in this chapter, cylindrical metal shields of the type assumed cannot be applied as the external voltage withstand strength would be lowered. But the calculations demonstrated a strategy to enlarge the bandwidth of such systems.

In Fig. 3.9 two suitable methods are therefore sketched, the efficiency of which may well be understood from the results of the above calculation. Figure 3.9(a) shows stress control or grading rings surrounding the resistor. Apart from the toroid fixed to h.v. potential, the other ring potentials would float as long as their potentials are not bound to any voltage-dividing system which is independent of the resistor, i.e. an additional resistor or capacitor voltage divider (see section 3.6). Apart from the additional cost, such voltage dividers are again influenced by stray capacitances and thus it is difficult to control the shield potentials with high accuracy. If the ring potentials are equivalent to the potentials provided by the current of the resistor at the corresponding plane of the toroids, the electrostatic field distribution along

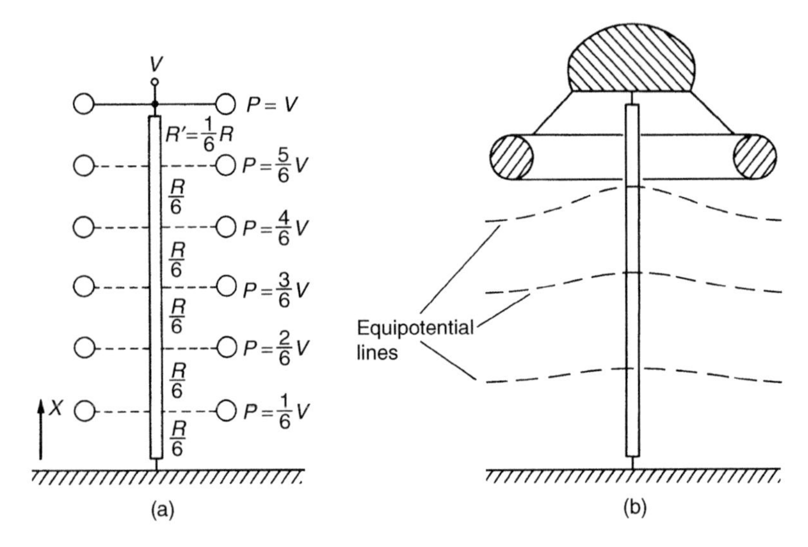

Figure 3.9 *Suitable methods for the shielding of h.v. resistors or resistor dividers. (a) Grading rings. (b) Grading top electrode*

the resistance would have nearly no field component perpendicular to the x-direction. Thus all stray capacitances to ground C'_e or h.v. potential C'_h (Fig. 3.7) are converted to parallel capacitances C'_p, the voltage distribution of which for a.c. voltages equals exactly the voltage distribution along the resistor. With a small number of shielding electrodes equal field distribution can only be approximated.

The top stress ring of the standard resistor in Fig. 3.5 indicates an alternative method of shielding. The comparison of eqns (3.17) and (3.21) shows opposite influences of stray capacitances to ground and to h.v. potentials. Therefore a properly shaped stress control electrode fixed to h.v. potential may also grade the potentials along the resistor, as sketched in Fig. 3.9(b). For a linearly distributed resistor in the x-direction, however, an ideal grading is difficult to achieve. A non-linear resistor distribution originally proposed by Goosens and Provoost[34] for impulse resistor voltage dividers gives an elegant solution to solve the disadvantage. The numerical calculation of the field distribution between h.v. electrode and earthed plane would demonstrate, however, the sensitivity of the distribution to surrounding objects at any potential. Thus the stray capacitance distribution will change with the surroundings, and will influence the frequency-dependent transfer characteristics.

Summarizing the above discussions, the high ohmic resistor in series with an ammeter or the improved method of a voltage dividing system are excellent means for the measurement of high d.c. voltages and, for resistors of smaller size and thus lower amplitudes (about 100–200 kV), also a.c. voltages. A very recent development of a 300 kV d.c. measuring device of very high

Figure 3.10 *300-kV divider for d.c. height 210 cm (PTB, Germany)*[35]

accuracy, described by Peier and Graetsch,[35] takes advantage of all principles discussed before (see Fig. 3.10). Here, 300 equal wire-wound resistors each of about 2 MΩ are series connected, and one of these resistors is used to form the l.v. arm of a divider (ratio ∼300:1). The resistors are aged by a temperature treatment. They form a helix of 50 windings and are installed in a PMMA housing containing insulating oil. The pitch of the helix varies so that the potential distribution of the resistor column equals approximately the electrostatic field potential distribution, although the divider is not provided for the precise measurement of a.c. voltages. Freedom of leakage currents due to corona was confirmed by partial discharge measurements. A very careful investigation of all sources of errors and uncertainties for this device shows a relative uncertainty of $\pm 28 \times 10^{-6}$. The final limit of the uncertainty for d.c. voltage measurement up to 300 kV is now obviously better than 1×10^{-5}, see reference 132.

3.4 Generating voltmeters and field sensors

Similar to electrostatic voltmeters the generating voltmeter, also known as the rotary voltmeter or field mill, provides a lossless measurement of d.c. and, depending upon the construction, a.c. voltages by simple but mainly mechanical means. The physical principle refers to a field strength measurement, and preliminary construction was described by Wilson,[36] who used the principle for the detection of atmospheric fields which are of small magnitude.

The principle of operation is explained by Fig. 3.11. An adequately shaped, corona-free h.v. electrode excites the electrostatic field within a highly insulating medium (gas, vacuum) and ground potential. The earthed electrodes are subdivided into a sensing or pick-up electrode A, a guard electrode G and a movable electrode M, all of which are at same potential. Every field line ending at these electrodes binds free charges, whose density is locally dependent upon the field gradient E acting at every elementary surface area. For measurement purposes, only the elementary surface areas $dA = a$ of the electrode A are of interest. The local charge density is then $\sigma(a) = \varepsilon E(a)$, with ε the permittivity of the dielectric.

If the electrode M is fixed and the voltage V (or field-distribution $E(a)$) is changed, a current $i(t)$ would flow between electrode A and earth. This

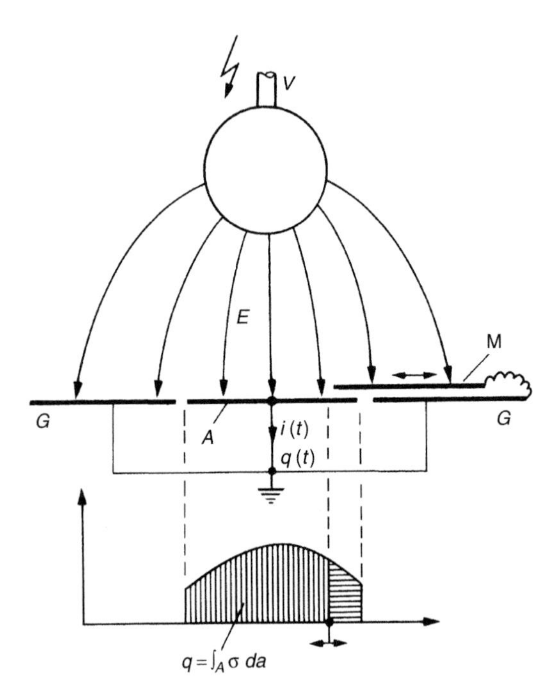

Figure 3.11 *Principle of generating voltmeters and field sensors*

current results then from the time-dependent charge density $\sigma(t, a)$, which is sketched as a one-dimensional distribution only. The amount of charge can be integrated by

$$q(t) = \iint_A \sigma(t, a)\,\mathrm{d}a = \varepsilon \iint_A E(t, a)\,\mathrm{d}a,$$

where A is the area of the sensing electrode exposed to the field. This time-varying charge is used by all kinds of field sensors, which use pick-up electrodes (rods, plates, etc.) only.

If the voltage V is constant, again a current $i(t)$ will flow but only if M is moved, thus steadily altering the surface field strength from full to zero values within the covered areas. Thus the current is

$$i(t) = \frac{\mathrm{d}q}{\mathrm{d}t} = \frac{d}{\mathrm{d}t} \iint_{A(t)} \sigma(a)\,\mathrm{d}a = \varepsilon \frac{d}{\mathrm{d}t} \iint_{A(t)} E(a)\,\mathrm{d}a. \tag{3.23}$$

The integral boundary denotes the time-varying exposed area $A(t)$ and $\sigma(a)$ as well as $E(a)$ are also time dependent if the voltage is not constant.

The field lines between h.v. and sensing electrode comprise a capacitive system. Thus the charge q can be computed by an electrostatic field computation or by calibration of the system. The integration across the time-varying area $A(t)$, however, provides a time-varying capacitance $C(t)$, and also if the voltage changes with time, $q(t) = C(t)V(t)$ and

$$i(t) = \frac{\mathrm{d}}{\mathrm{d}t}[C(t)V(t)]. \tag{3.24}$$

Various kinds of generating voltmeters use these basic equations and the manifold designs differ in the constructional means for providing $C(t)$ and interpreting the current $i(t)$. Such designs and examples can be found in the literature, see, for example, references 1, 29, 131 and 133.

Generating voltmeters are very linear instruments and applicable over a wide range of voltages. The sensitivity may be changed by the area of the sensing electrodes (or iris) as well as by the current instrument or amplification. Their early application for the output voltage measurement of a Van de Graaff's thus may well be understood. Excessive space charge accumulation within the gap between h.v. electrode and generating voltmeter, however, must be avoided. The presence of space charges will be observed if the voltage is switched off.

Vibrating electrometers are also generating voltmeters, but will only be mentioned here as they are not widely used. The principle can well be understood with reference to Fig. 3.11 neglecting the movable disc. If the sensing electrode would oscillate in the direction of the h.v. electrode, again a current $i(t) = \mathrm{d}q/\mathrm{d}t$ is excited with constant voltage V due to a variation of the capacitance $C = C(t)$. This principle was developed by Gahlke and Neubert (see reference 30, p. 77). The sensing electrode may also pick up charges when placed just behind a small aperture drilled in a metal plate. Commercial types of such an instrument are able to measure d.c. voltages down to 10 μV, or currents down to 10^{-17} A, or charges down to 10^{-15} pC, and its terminal resistance is as high as 10^{16} Ω.

3.5 The measurement of peak voltages

Disruptive discharge phenomena within electrical insulation systems or high-quality insulation materials are in general caused by the instantaneous maximum field gradients stressing the materials. Alternating voltages or impulse voltages may produce these high gradients, and even for d.c. voltages with ripple, the maximum amplitude of the instantaneous voltage may initiate the breakdown. The standards for the measurement and application of test voltages therefore limit the ripple factors for d.c. testing voltages, as the peak value of d.c. voltages is usually not measured, and claim for a measurement of the peak values of a.c. and impulse voltages whenever this is adequate.

Up to this point the spark gaps (section 3.1) have been treated to be an adequate means for the measurement of the peak values of all types of voltages. The necessary calibration procedure, however, and the limited accuracy are hindering its daily application and call for more convenient methods. We could already adequately show the disadvantages encountered with high-ohmic resistor voltage dividers (see section 3.3) applied to a.c. voltage measurements, which resulted in limitations within the voltage range of 100–200 kV.

The simplest way to obtain the output peak voltage of a testing transformer is by measuring and recording the primary voltage and then multiplying the value by the transformer ratio. However, the load-dependent magnitude of the ratio as well as unavoidable waveshape variations caused by the transformer impedances which magnify or reduce the higher harmonics render such a method unacceptable. Even simpler would be to calculate the peak value of an impulse voltage from the charging voltage of the impulse voltage generator multiplied by the voltage efficiency factor η (see eqn (2.28), Chapter 2). Here, the unknown voltage drops within the generator and the loading effects by the object under test do not allow, in general, the use of such methods.

The direct measurement of the high voltages across test objects and of their peak values is therefore of great importance. Many of the methods treated in this chapter require voltage dividing systems providing adequate voltage levels for the circuits used to process the peak or crest values. A detailed study and generalized theory of voltage dividing systems will be presented in section 3.6. Therefore, within this chapter the voltage divider's equivalent circuits are simplified and assumed ideal. A treatment of the construction and performance of h.v. capacitors for measuring purposes is, however, added to this chapter, as their application is closely related to the circuits described here.

The measurement of peak voltages by means of oscilloscopes is not treated in detail. Apart from the measurement of impulse crest values their application to a.c. voltages is not convenient and thus unusual. For accurate measurements a very careful adjustment and calibration of the oscilloscope would be necessary. This, however, is beyond the scope of this book.

3.5.1 The Chubb–Fortescue method

This simple but accurate method for the measurement of peak values of a.c. voltages was proposed by Chubb and Fortescue,(37) who as early as 1913 became interested in the use of a sphere gap as a measuring device. The basic

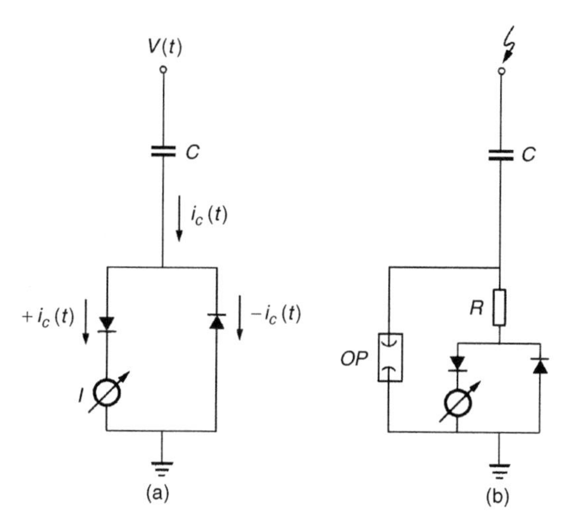

Figure 3.12 *A.C. peak voltage measurement by Chubb and Fortescue. (a) Fundamental circuit. (b) Recommended, actual circuit*

diagram (Fig. 3.12(a)) comprises a standard capacitor, two diodes and a current integrating ammeter (i.e. moving coil or equivalent instrument) only. The displacement current $i_c(t)$ is subdivided into positive and negative components by the back-to-back connected diodes. The voltage drop across these diodes (less than 1 V for Si diodes) may completely be neglected when high voltages are to be measured. The measuring instrument may be included in one of the two branches. In either case it reads a magnitude of charge per cycle, or the mean value of the current $i_c(t) = C\,\mathrm{d}V/\mathrm{d}t$, and thus

$$I = \frac{1}{T}\int_{t_1}^{t_2} i_c(t)\,\mathrm{d}t = \frac{C}{T}\int_{t_1}^{t_2} \mathrm{d}V = \frac{C}{T}(V_{+\mathrm{max}} + |V_{-\mathrm{max}}|)$$

according to Fig. 3.13 which illustrates the integral boundaries and the magnitudes related to Fig. 3.12(a). The difference between the positive and negative peak values may be designated as $V_{\mathrm{p-p}}$, and if both peak values are equal, a

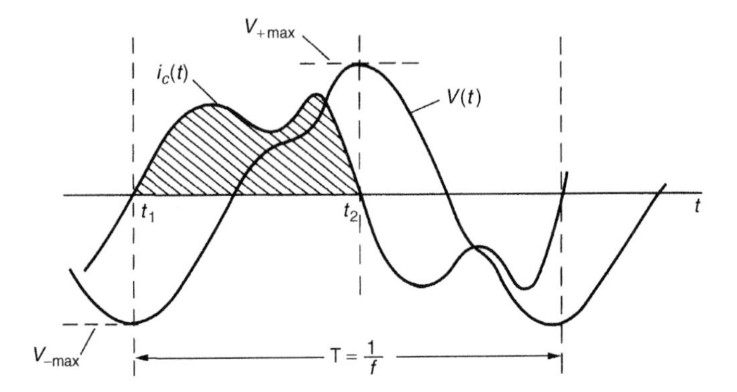

Figure 3.13 *Diagram of voltage $V(t)$ and current $i_c(t)$ from circuit Fig 3.12(a)*

condition which usually applies, we may write

$$I = CfV_{p-p} = 2CfV_{max}. \qquad (3.25)$$

An increased current would be measured if the current reaches zero more than once during one half-cycle. This means the waveshape of the voltage would contain more than one maximum per half-cycle. A.C. testing voltages with such high harmonics contents are, however, not within the limits of standards and therefore only very short and rapid voltage drops caused by heavy predischarges within the test circuit could introduce errors. A filtering of the a.c. voltage by a damping resistor placed between the capacitor C and the object tested will eliminate this problem.

The relationship in eqn (3.25) shows the principal sources of errors. First, the frequency f must be accurately known. In many countries the power frequency often used for testing voltages is very stable and accurately known. The independent measurement of the frequency with extremely high precision (i.e. counters) is possible. The current measurement causes no problem, as these currents are in the mA range. The effective value of the capacitance should also be accurately known, and because of the different constructions available, which will be discussed in section 3.5.4, a very high precision is possible. The main source of error is often introduced by imperfect diodes. These have to subdivide the a.c. current $i_c(t)$ with high precision, this means the charge transferred in the forward direction, which is limited by the capacitance C, must be much higher (10^4–10^5 times) than the charge in the reversed voltage direction. But due to the back-to-back connection of the diodes, the reverse voltages are low. However, the diodes as well as the instrument become highly stressed by short impulse currents during voltage breakdowns. A suitable protection of the rectifying circuit is thus recommended as shown in Fig. 3.12(b). The resistor R introduces a required voltage drop during breakdown to ignite the overvoltage protector *OP* (e.g. a gas discharge tube).

The influence of the frequency on the reading can be eliminated by electronically controlled gates and by sensing the rectified current by analogue-to-digital converters. By this means (see Boeck[38]) and using pressurized standard capacitors, the measurement uncertainty may reach values as low as 0.05 per cent.

3.5.2 Voltage dividers and passive rectifier circuits

Passive circuits are nowadays rarely used in the measurement of peak values of high a.c. or impulse voltages. The rapid development of very cheap integrated operational amplifiers and circuits during the last decades has offered many possibilities to 'sample and hold' such voltages and thus displace passive circuits. Nevertheless, a short treatment of basic passive crest voltmeters will be included because the fundamental problems of such circuits can be shown. The availability of excellent semiconductor diodes has eliminated the earlier difficulties encountered in the application of the circuits to a large extent. Simple, passive circuits can be built cheaply and they are reliable. And, last but not least, they are not sensitive to electromagnetic impact, i.e. their electromagnetic compatibility (EMC) is excellent. In contrast, sophisticated electronic instruments are more expensive and may suffer from EMC problems. Passive as well as active electronic circuits and instruments as used for peak voltage measurements are unable to process high voltages directly and they are always used in conjunction with voltage dividers which are preferably of the capacitive type.

A.C. voltages

The first adequately usable crest voltmeter circuit was described in 1930 by Davis, Bowdler and Standring.[39] This circuit is shown in Fig. 3.14. A

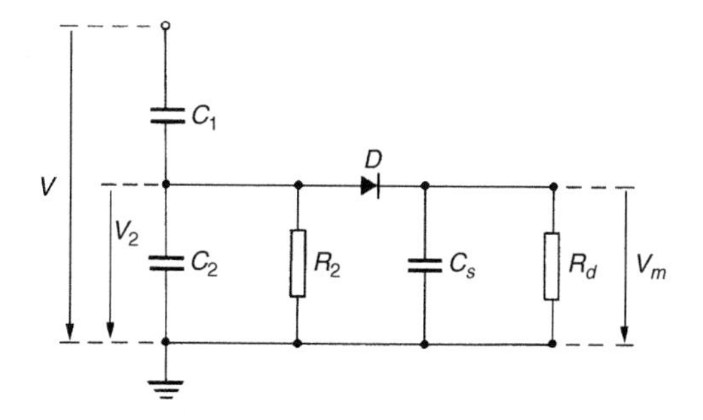

Figure 3.14 *Simple crest voltmeter for a.c. measurements, according to Davis, Bowdler and Standring*

capacitor divider reduces the high voltage V to a low magnitude. If R_2 and R_d are neglected and the voltage V increases, the storage capacitor C_s is charged to the crest value of V_2 neglecting the voltage drop across the diode. Thus the d.c. voltage $V_m \approx +V_{2\,\max}$ could be measured by a suitable instrument of very high input resistance. The capacitor C_s will not significantly discharge during a period, if the reverse current through the diode is very small and the discharge time constant of the storage capacitor very large. If V_2 is now decreased, C_2 will hold the charge and the voltage across it and thus V_m no longer follows the crest value of V_2. Hence, a discharge resistor R_d must be introduced into the circuit. The general rules for the measuring technique require that a measured quantity be indicated within a few seconds. Thus the time constant $R_d C_s$ should be within about 0.5–1 sec. Three new errors, however, are now introduced: an experiment would readily show that the output voltage V_m decreases steadily if a constant high voltage V is switched to the circuit. This effect is caused by a continuous discharge of C_s as well as of C_2. Thus the mean potential of $V_2(t)$ will gain a negative d.c. component, which finally equals to about $+V_{2\,\max}$. Hence a leakage resistor R_2 must be inserted in parallel with C_2 to equalize these unipolar discharge currents. The second error refers to the voltage shape across the storage capacitor. This voltage contains a ripple discussed in Chapter 2, section 2.1. Thus the error, almost independent of the type of instrument used (i.e. mean or r.m.s. value measurement), is due to the ripple and recorded as the difference between peak and mean value of V_m. The error is approximately proportional to the ripple factor (see eqn (2.2)) and thus frequency dependent as the discharge time constant is a fixed value. For $R_d C_s = 1$ sec, this 'discharge error' amounts to ~1 per cent for 50 Hz, ~0.33 per cent for 150 Hz and ~0.17 per cent for 300 Hz. The third source of systematic error is related to this discharge error: during the conduction time of the diode the storage capacitor is recharged to the crest value and thus C_s is in parallel to C_2. If the discharge error is e_d, this 'recharge error' e_r is approximately given by

$$e_r \approx 2e_d \frac{C_s}{C_1 + C_2 + C_s} \tag{3.26}$$

Hence C_s should be small compared to C_2, which for h.v. dividers is the largest capacitance in the circuit. There still remains a negative d.c. component of the mean potential of the voltage V_2, as the equalizing effect of R_2 is not perfect. This 'potential error' e_p is again a negative term, and amounts to $e_p = -(R_2/R_d)$. Hence R_2 should be much smaller than R_d.

This leakage resistor R_2 introduces another error directly related to the now frequency-dependent ratio or attenuation factor of the voltage divider. Apart from a phase shift between V_2 and V, which is not of interest, the relative amplitudes of V_2 decrease with decreasing frequency and the calculation shows the relative error term

$$e_{fd} = \frac{1}{2\{\omega R_2(C_1 + C_2)\}^2} \approx -\frac{1}{2(\omega R_2 C_2)^2} \tag{3.27}$$

Apart from a negligible influence caused by the diode's inherent junction capacitance, we see that many systematic error terms aggravate the exact crest voltage measured.

A numerical example will demonstrate the relative magnitudes of the different errors. Let $C_1 = 100\,\text{pF}$, $C_2 = 100\,\text{nF}$, a realistic measure for a HVAC divider with attenuation or scale factor of 1000. For $R_d C_s = 1\,\text{sec}$, the inherent error term $e_d = -1$ per cent for 50 Hz. Allowing an error of one-half of this value for the recharge error e_r requires a C_s value $C_2/3$ approximately, and thus $C_s = 33\,\text{nF}$. From $R_d C_s = 1\,\text{s}$ the discharge resistor is calculated to be about $30\,\text{M}\Omega$. This value is a measure for the high input resistance of the voltmeter and the diode's reverse resistance necessary. Let the potential error e_p again be 0.5 per cent. Hence $R_2 = R_d/200$ or $150\,\text{k}\Omega$. For a frequency of 50 Hz this leakage resistor gives $e_{fd} \approx 2.25$ percent. Thus the sum of errors *becomes about* -4.25 per cent, still neglecting the voltage drop across the diode.

Hence, for passive rectifying circuits comprising capacitor voltage dividers acting as voltage source, at least too small 'leakage resistors' (R_2) must be avoided. The possible solution to bleed also the h.v. capacitor is too expensive, as it requires an additional h.v. resistor. The addition of an equalizing branch to the l.v. arm of the voltage divider provides an attractive solution. This can be accomplished again using a peak rectifier circuit as already shown in Fig. 3.14 by the addition of a second network comprising D, C_s and R, but for negative polarities. Thus the d.c. currents in both branches are opposite in polarity and compensate each other. All errors related to R_2 are then cancelled.

The most advanced passive circuit to monitor crest values of power frequency voltages was developed in 1950 by Rabus. This 'two-way booster circuit' reduces the sum of systematic error terms to less than 1 per cent even for frequencies down to 20 Hz. More information about this principle is provided in references 1 and 131.

Impulse voltages

The measurement of peak values of impulse voltages with short times to crest (lightning impulses) with passive elements only was impossible up to about 1950. Then the availability of vacuum diodes with relatively low internal resistance and of vacuum tubes to build active d.c. amplifiers offered the opportunity to design circuits for peak impulse voltage measurement but of relatively low accuracy. Now, active highly integrated electronic devices can solve all problems involved with passive circuits, see 3.5.3. The problems, however, shall shortly be indicated by the following explanations.

Impulse voltages are single events and the crest value of an impulse is theoretically available only during an infinitely short time. The actual crest value may less stringently be defined as a crest region in which the voltage amplitude is higher than 99.5 per cent. For a standard 1.2/50 μsec wave the available time is then about 1.1 μsec. Consider now the simple crest voltmeter circuit of Fig. 3.14 discussed earlier, omitting the discharge resistor R_d as well as R_2. The diode D will then conduct for a positive voltage impulse applied to the voltage divider, and the storage capacitor must be charged during the rising front only. But instantaneous charging is only possible if the diode has no forward (dynamic) resistance. The actual forward resistance R_D gives rise to a changing time constant $R_D C_s$ and it will be shown in section 3.6 that a 'response time', which is equal to the time constant $R_D C_s$ for such an RC circuit, of about 0.2 μsec would be necessary to record the crest value with adequate accuracy. For a low C_s value of 1000 pF the required $R_D = 200\,\Omega$. As also the diode's junction capacitance must be very small in comparison to C_s, diodes with adequate values must be properly selected. The more difficult problem, however, is the time required to read the voltage across C_s. The voltage should not decrease significantly, i.e. ≤ 1 per cent for at least about 10 sec. Hence the discharge time constant of C_s must be longer than 10^3 sec, and thus the interaction between the diode's reverse resistance and the input resistance of the instrument necessary to measure the voltage across C_s should provide a resultant leakage resistance of $10^{12}\,\Omega$. A measurement of this voltage with electrostatic or electronic electrometers is essential, but the condition for the diode's reverse resistance can hardly be met. To avoid this problem, a charge exchange circuit shown in Fig. 3.15 was proposed.

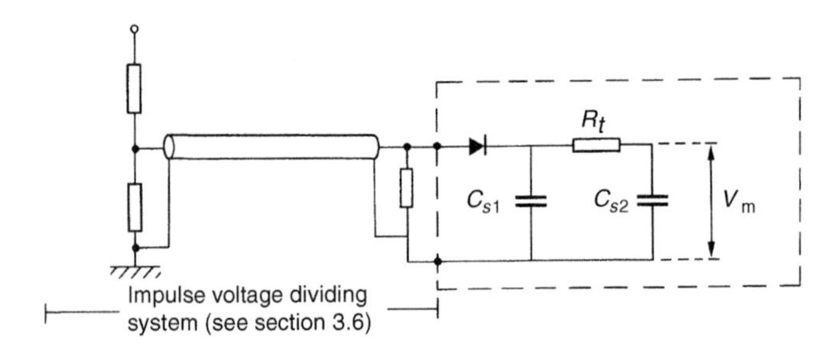

Figure 3.15 *Peak voltmeter within dashed line with continuous charge exchange*

If the capacitor C_{s1} originally charged to crest value transfers most of its charge to a much larger second storage capacitor C_{s2} within a short time, i.e. much shorter than 1 sec, C_{s1} cannot lose much of the charge through the finite reverse resistance of the diode and the discharge time constant after the charge transfer is greatly increased because C_{s1} and C_{s2} are paralleled. As

$C_{s2} \gg C_{s1}$, the output voltage V_m becomes quite low and therefore sensitive (electronic) d.c. voltmeters must be applied. Also, other peak reading devices must always be combined with active electronic circuits and earlier solutions are described elsewhere.[1]

3.5.3 Active peak-reading circuits

Due to the demand within other technical fields, analogue or digital circuits and instruments are now widely commercially available. The main problem encountered with these instruments when applied in h.v. laboratories is in general only related to their electromagnetic compatibility resulting from the transient disturbances following breakdown phenomena. It is not the aim of this section to discuss all possible solutions and instruments as available today. Again, only some hints to basic principles are provided.

The main properties of amplifying circuits may be summarized as follows: a high and linear input impedance ($1-2\,\text{M}\Omega$) is necessary to avoid excessive loading of the h.v. dividers of any kind. Thus the error terms e_r, e_p and e_{fd} discussed in section 3.5.2 can be minimized. In circuits used for continuous measurement of a.c. peak voltages, the reduction of the discharge error e_d is much more difficult. In active analogue circuits this may be achieved by a continuous compensation of the ripple area. To demonstrate the principle only, a simplified circuit is sketched in Fig. 3.16 related to an actual circuit of a specialized manufacturer. A voltage attenuator for low voltages ($<1\,\text{kV}$) reduces and adapts the input voltages to be processed by the first operational amplifier $OP1$, which forms together with the diodes D, the storage capacitor C_s and the discharge resistor R_d an active peak rectifier. $OP1$ charges the storage capacitor C_s and reduces the forward voltage of D to a large extent.

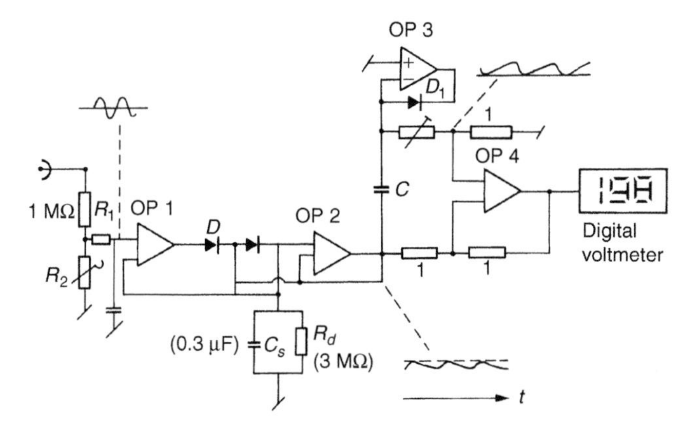

Figure 3.16 *Simplified circuit of a crest voltmeter for a.c. voltages with discharge error compensation (courtesy Haefely & Co.)*

C_s is discharged through R_d with a time constant of 1 sec. The second voltage follower $OP2$ still amplifies the ripple produced by the active peak rectifier. Its feedback to the connection point of the diodes D, however, avoids leakage of C_s by the reversed biased diodes. The ripple is detected by the capacitively coupled operational amplifier $OP3$, which rectifies the voltage by the diode D_1 whose forward conduction voltage is again strongly linearized by the amplifier. By this means, the ripple voltage appears across an adjustable voltage divider. The inversion of the output voltage of $OP2$ and summation with the ripple voltage performed by $OP4$ thus eliminate the ripple and thus also the discharge error to a large extent. The output voltage may then be monitored by an analogue or digital voltmeter.

The principle of the active peak rectifier can also be used for impulse voltage measurements. The discharge of the storage capacitor through R_d is then avoided and replaced by a reset switch. Very high-quality operational amplifiers with high slew rates are necessary, and the long storage time is usually achieved by two or three additional active rectifying circuits. More details may be found in the literature.[39–41,128]

The increasing availability of specialized, fully integrated analogue and digital circuits contributed to the development of a large variety of peak holding circuits. An earlier publication by Schulz[42] describes a mixed circuit for a very precise a.c. peak measurement with a statistical uncertainty of $<2.8 \times 10^{-4}$ which includes a capacitor voltage divider for 200 kV, composed of a pressurized gas capacitor and mica capacitors for the l.v. arm. The peak detecting circuit consists essentially of a special sample-and-hold amplifier (a.c. to d.c converter) and a very precise digital voltmeter, both being controlled by a microprocessor. The main aim of this control is to avoid any discharge error as mentioned earlier. Such precise measuring units are built for calibration purposes only and not for laboratory applications.

For impulse voltage measurements, the transient recorders comprising fast parallel ADCs will increasingly be used. The description of the principle of transient recorders may be found elsewhere[43] and a recent publication by Malewski and Dechamplain[44] demonstrates the necessity of additional shielding of such commercial equipment. Transient recorders are rapidly replacing the CRO technique for impulse voltage measurements.

3.5.4 High-voltage capacitors for measuring circuits

The important influence of the effective capacitance of any h.v. capacitors as used, e.g., in the Chubb–Fortescue circuit of section 3.5.1 or in most of the peak reading circuits for a.c. voltages, makes it necessary to present a short treatment about the technology of h.v. capacitor units widely used in testing and research laboratories.

In comparison to h.v. capacitors used within h.v. transmission and distribution systems for load or series compensation, the requirements for 'measuring capacitors' are different. First, the effective capacitor values are quite low and range between some 10 and 100 pF only. These low values are sufficient to provide the energy or power needed for the measurement and to provide low load for the voltage source. The second requirement is related to the stability of the C values relative to atmospheric conditions (temperature, humidity), external fields and voltage range, including all effects associated with this magnitude, i.e. partial discharges or non-linearity.

An h.v. capacitor may consist of a single capacitance unit, defined basically as a two-electrode arrangement, or of a chain of capacitor units rated for relative 'low' voltages (kV range) electrically connected in series. The technology as well as the electrical behaviour is quite different for the two cases and therefore a separate discussion is appropriate.

Single capacitor units

Ultra high vacuum would provide the ideal dielectric between metal plates forming an arrangement with known and fixed field distribution. Ultra high vacuum has excellent electrical strength although it is limited by well-known, electrode effects. The difficulties and associated costs, however, to place such electrodes in large vessels or tanks providing ultra high vacuum conditions without maintenance are the reasons why vacuum is not used for very high voltages.

According to Paschen's law (Chapter 5) high electric strength can also be achieved with gases at high pressure. Atmospheric pressure may be treated as the lower limit of a high pressure and, dependent upon the type of gas used, the upper limit is set again by predominantly electrode surface effects which place an economic limit given by the decreasing relative dielectric strength of the gas and the increasing cost of pressure vessels. Gases are dielectrics with predominantly electronic polarization only (see section 7.1), providing a very low relative permittivity which is not influenced up to very high frequencies and only by the particle density. Hence a gaseous dielectric is adequate for the construction of h.v. capacitors.

Thus the problem reduces to finding electrode arrangements which provide unchangeable and proper field distributions between two electrodes forming the capacitance. As a certain maximum field strength will limit the insulation strengths of any gas, a uniform field electrode arrangement (see section 4.2) would obviously seem to be most convenient. If the centre part of such an arrangement only would be used to form the effective capacitance, which is easily possible by subdividing the low potential electrode into a 'guard ring' and measuring section, the best field distribution is achieved. The disadvantage of this solution is, however, the very low value of the capacitance for the gap distances necessary for the high voltages. It is also difficult to control exactly the gap distance, if temperature differences and the consequent material movements are considered.

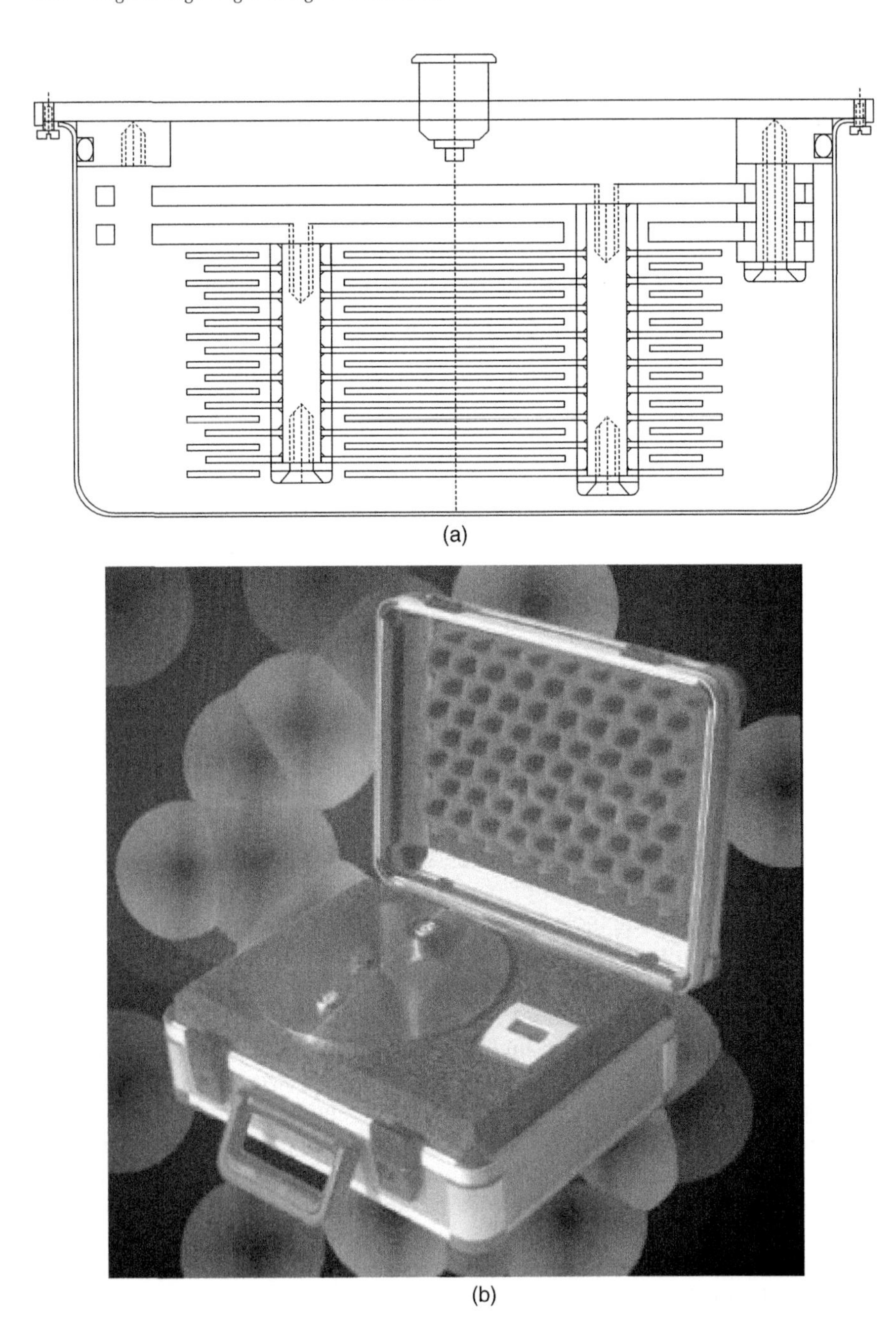

Figure 3.17 *Cross-section (a) and (b) typical view of a 'standard capacitor' for a voltage of 2 kV (r.m.s.) (courtesy Presco AG, Weiningen, Switzerland)*

These disadvantages can be avoided if multiple plate arrangements are used as sketched in Fig. 3.17(a). A larger number of circular metal plates which are insulated by, e.g., dry and very clean air or SF_6 from each other make it possible to realize capacitance values up to some nanofarads for voltages up to some kilovolts. A very careful surface finish of the metal plates is necessary. The plate arrangement is deposited in a grounded metal vessel and forms the guard for field control. Very low temperature coefficients of a few ppm/°K can be reached by a proper selection of the materials used for construction. Such capacitor units are used as 'etalons' or 'capacitance standards' as also the losses are extremely low.

The coaxial cylindrical electrode configuration provides the second opportunity to achieve a fairly good field distribution, if the difference between the two radii of the electrodes is not too large. In Chapter 4, section 4.2, the two-dimensional coaxial field is treated and it is shown that the radii can be optimized to keep the diameter of the outer electrode as small as possible for a given voltage and a limited field strength at the inner electrode. Thus, the radial dimensions do not become very large if the system is pressurized. As the capacitance C per unit axial length l is $C/l = 2\pi\varepsilon/\ln(r_2/r_l)$, where r_1 and r_2 are the radii of the inner and outer cylinders respectively, even with the optimum ratio $r_2/r_1 = e$ this capacitance is about 56 pF/m and thus large enough to achieve adequate capacitance values (30–100 pF) with limited length of the electrodes. A further advantage relates to the possible variation of the capacitance if the inner electrode is not completely centralized. The central position is a position of minimum value of capacitance as shown by a computation of the capacitance varying with eccentricity according to the relevant formula,[45]

$$\frac{C}{l} = \frac{2\pi\varepsilon}{\cosh^{-1}\left(\dfrac{r_1^2 + r_2^2 - D^2}{2r_1r_2}\right)}, \tag{3.28}$$

where D is the distance between the axes of both cylinders. The expression shows that a small eccentricity does not contribute much to a change in capacitance. This is the main reason why most of the 'standard capacitors' used today comprise this coaxial cylinder system. Originally suggested by Schering and Vieweg in 1928,[1] a cross-section of such a compressed gas capacitor is shown in Fig. 3.18. The main h.v. electrode 1 encloses the l.v. electrode with guard ring 2 completely and thus shields the electrode from the influence of all external fields. The pressure vessel 5 is of dielectric material and contributes to minimize the height. The supporting tube 3 is at earth potential. One main insulation problem involved in this construction relates to the tangential field distribution outside the dielectric vessel, as the limited electrical strength of atmospheric air must withstand the strong field concentration in the vicinity of

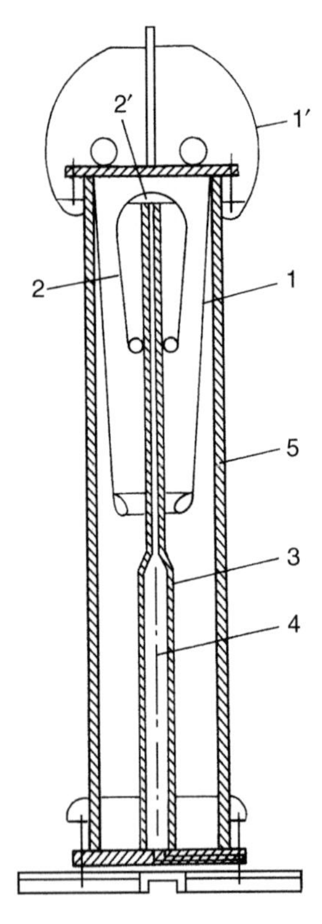

Figure 3.18 *Cross-section of a compressed gas capacitor (standard capacitor). 1. Internal h.v. electrode. 1′. External h.v. electrode. 2. Low-voltage electrode with guarding, 2′. 3. Supporting tube. 4. Coaxial connection to l.v. sensing electrode. 5. Insulating cylinder*

the lower end of the h.v. electrode. Even a rough plot of the equipotential lines surrounding the electrode system may show this field concentration, which is sketched in Fig. 3.19, a result obtained by Keller.[46] The maximum stress which occurs at the end of the h.t. electrode remains approximately the same, and is independent of the length of the dielectric cylinder. The reduction of this external field by simple means is not possible; even the simplest solution to increase the diameter of the vessel and to distribute the equipotential lines within the cylinder is difficult due to the necessary increase in mechanical strength of the vessel construction.

Compressed gas capacitors provide, if well designed and constructed, a h.v. capacitance of highest possible stability; they are, however, expensive if designed for voltages of 100 kV and more. Due to their outstanding performance with regard to the precision of the capacitance value and very low

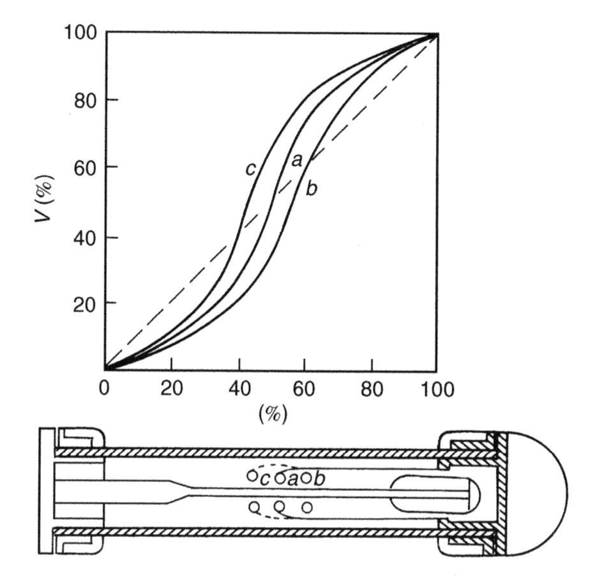

Figure 3.19 *Potential distribution along a compressed capacitor for various lengths of h.t. electrode*

tan δ values they are predominantly used as the standard capacitor within h.v. bridges for C tan δ measurements (see Chapter 7). Variations in the construction are, of course, possible.[47–49] These have been built for a.c. voltages up to 1500 kV. CO_2, N_2 or SF_6 are convenient gases for insulation. SF_6 provides the highest electric strength and thus only pressures up to about 0.4 MPa (in comparison to 1 to 1.5 MPa with other gases) are necessary. The relative influence of the pressure-dependent permittivity upon capacitance value may be calculated taking into account the increase of the relative permittivity ε_r with gas density, given by

$$\varepsilon_r = 1 + \alpha \frac{273}{100} \frac{p}{T}; \quad \begin{array}{l} p \text{ in kPa} \\ T \text{ in °K} \end{array} \tag{3.29}$$

where

$\alpha \approx 0.00232$ for SF_6,

$\alpha \approx 0.00055$ for N_2,

$\alpha \approx 0.00076$ for CO_2.

As the actual gas density in a vessel may also be influenced by the construction, the actual variation with p and T will be specified by the manufacturer. Dissipation factors tan δ are in general about 10^{-5} for power frequency.

Figure 3.20 shows a physical picture of a standard capacitor for a rated voltage of 1000 kV. The increased diameter and thickness of the upper part of the insulating cylinder is made to reduce the electric field stress in the vicinity of the h.v. electrode outside of this cylinder.

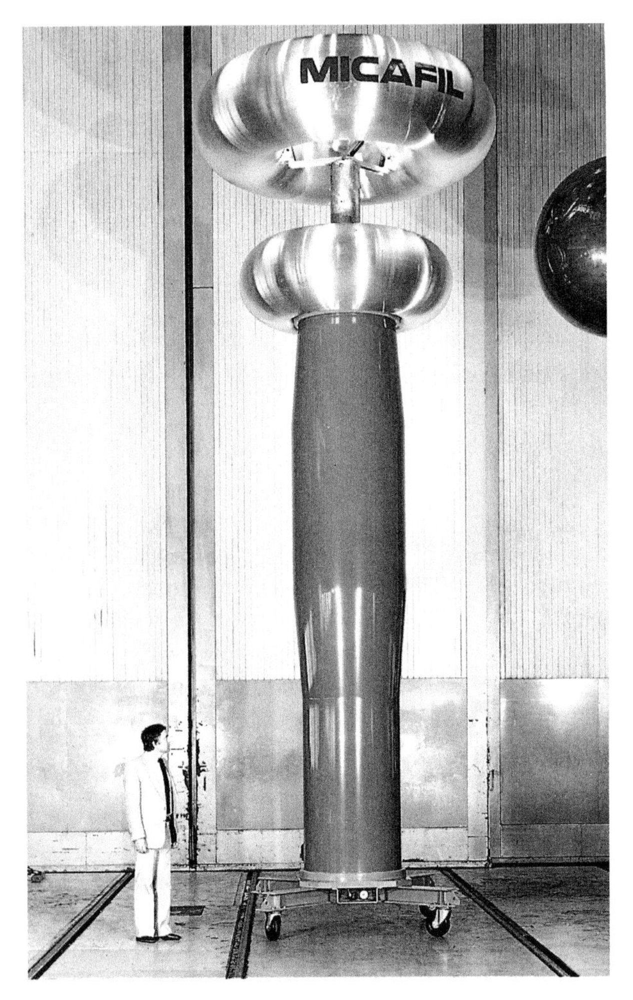

Figure 3.20 *Standard (compressed gas) capacitor for 1000 kV r.m.s. (courtesy Micafil, Switzerland)*

'Stacked' capacitor units

This second type of basic capacitor construction consists of a large number of single capacitor units in series. Single units of conventional capacitors with, e.g., oil-kraft paper or pure solid dielectric insulation cannot be built for voltages higher than about 10 kV, and hence this series connection is necessary. These capacitor units are piled up and thus a stretched stack of large height/diameter ratio is formed.

Whatever the construction of an individual unit, there are always charges located at some parts of the electrodes which do not contribute to the actual series capacitance. These (intermediate) electrodes are at a potential which is essentially given by the terminals of the capacitor elements which form a voltage-dividing system. The 'foreign' charges are thus related to stray capacitances in the same way as discussed in section 3.3.

A realistic equivalent circuit of a stacked capacitor unit established from a certain, usually large number of single capacitors C' is shown in Fig. 3.21. For capacitors within laboratories the stray capacitances to earth may also be influenced by the walls, which are often electromagnetically shielded; this situation is assumed in this figure. The h.v. lead including the top electrode of the whole capacitor may contribute to smaller stray capacitances to h.v. potential V. A calculation of the current I_0 flowing to ground from the last earthed capacitor could be carried out essentially in the same way as the computation performed for the equivalent circuit of resistors in section 3.3, eqns (3.13)–(3.18). Assuming similar approximations within the expressions for the current I_0 we achieve by expansion of the hyperbolic functions the following result:

$$I_0 = V\omega C \frac{1 + \dfrac{C_h}{4C}}{1 + \dfrac{C_e}{6C} + \dfrac{C_h}{12C}} = V\omega C_{\text{eq}} \tag{3.30}$$

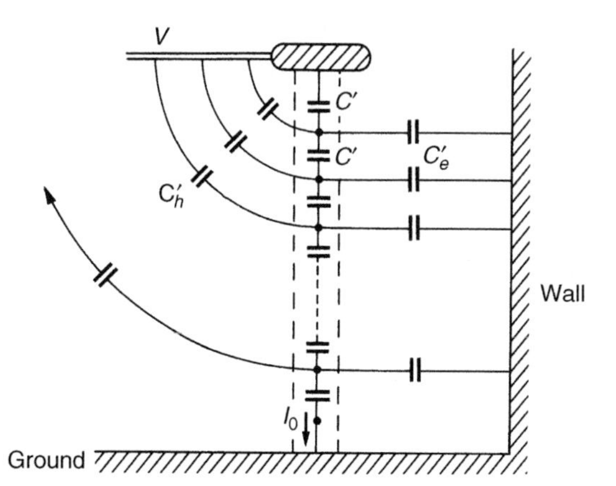

Figure 3.21 *Stray capacitances influencing the equivalent or effective capacitance of h.v. capacitors*

where $C = C'/n$, $C_e = nC'_e$, $C_h = nC'_h$, and n a large (infinite) number of capacitor elements C'. The capacitance $C = C'/n$ is obviously the resultant capacitance of the chain computed from the usual series circuit, i.e. $1/C = 1/C_1 + 1/C_2 + \ldots 1/C_n$. Equation (3.30) demonstrates that the 'equivalent' or 'effective' capacitance C_{eq} of a stacked capacitor cannot be calculated from individual elements, as the stray capacitances C_h and C_e are not exactly known. Therefore, the l.v. capacitor C_2 within the voltage divider of Fig. 3.14 or the diodes within the Chubb–Fortescue circuit (Fig. 3.12) will only 'see' the current I_0.

Many measurements performed with huge stacked capacitor units[(50)] confirmed the fundamental applicability of eqn (3.30). It was also shown that the influence of the stray capacitance to the h.v. side, C'_h, in general can be neglected. Nevertheless, it is necessary to rate the series capacitors so that the term $C_e/6C$ does not exceed 1 to 2 per cent. Thus we may simplify eqn (3.30) to

$$C_{eq} \approx C\left(1 - \frac{C_e}{6C}\right). \tag{3.31}$$

This effect of decreasing capacitance can experimentally be checked by a correct measurement of C_{eq} with an h.v. bridge (Schering or current comparator bridge, see Chapter 7). In such bridge circuits, the unknown capacitor is placed during measurements at its working condition. If the high voltage is applied, the l.v. end of this unknown capacitor remains essentially at earth potential, as the bridge potential is very low; hence, the potential distribution across the test object remains unchanged. A measurement of C_{eq} with a usual two-terminal capacitance bridge should never be made and would indicate wrong results.

The dimensioning of stacked capacitor units for the measurement of high voltages must take this effect into consideration. C_e can approximately be calculated by the assumption that the stacked capacitors are of cylindrical shape, thus forming a metalized vertical cylinder placed upon a horizontal plane, as sketched in Fig. 3.22.

The well-known formula for this arrangement[(51)] is

$$C_e = \frac{2\pi\varepsilon l}{\ln\left[\dfrac{2l}{d}\sqrt{\dfrac{4s+l}{4s+3l}}\right]} \tag{3.32}$$

and for $s \ll 1$:

$$C_e \approx \frac{2\pi\varepsilon l}{\ln\dfrac{1.15l}{d}}. \tag{3.33}$$

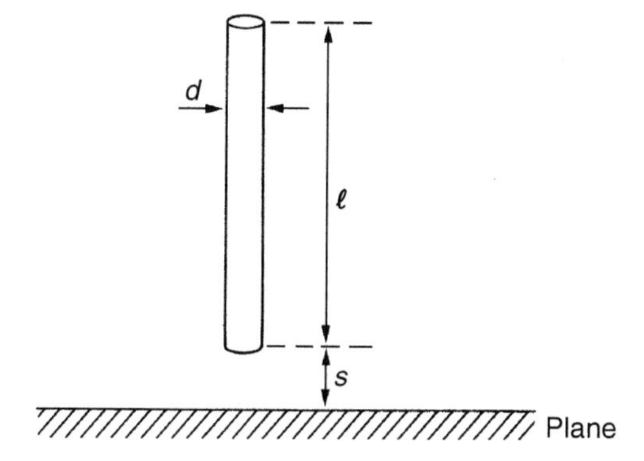

Figure 3.22 *Cylindrical conductor above plane (see eqn (3.32))*

Since even more accurate approximations will not contribute much to the result as shown by Zaengl[51] and Luehrmann,[52] one may evaluate eqn (3.33) only, and one can easily see that the absolute values C_e/l range within about 10 to 15 pF/m.

The effect, that the effective capacitance, eqn (3.31), will change with the dimensions of the capacitor or its surroundings is known as 'proximity effect'. The accurate influence of the surroundings on C_e could only be computed by numerical field calculation (see Chapter 4). As the variations of C_e due to changed surrounding conditions are usually less than 20 per cent, the error term in this equation may be as large as 5 per cent to get variations of C_{eq} smaller than 1 per cent. This condition leads to

$$\frac{C_e}{6C} = \frac{C_e/l}{6C/l} \lessapprox 5\%.$$

With $C_e/l = 10\text{–}15$ pF/m we obtain

$$C/l \approx C_{eq}/l \cong 30\text{–}50\,\text{pF/m}. \tag{3.34}$$

If this condition is not fulfilled, capacitors within measuring circuits should be fixed and placed within a laboratory and no moveable structures or equipment disturbing the potential distribution of the capacitor should be in the vicinity.

Technology of H.V. capacitors

The technology of compressed gas standard capacitors was treated earlier. The following explanations refer thus to the technology of 'stacked', discrete capacitors with special emphasis on those used for measurement purposes. The main requirements for this kind of application are:

- the capacitance C shall be independent of magnitude of voltage level and shall not change with time of application (no ageing effects);
- the temperature coefficient (TC) (in ppm/°K) shall be small or very small, dependent on the kind and temperature range of application, and shall at least be known;
- the effective inductivity of C shall be as small as possible, if used for high-frequency applications, i.e. voltage dividers for impulse voltages.

Foil capacitors. Most of the discrete capacitor units for stacked capacitors are made of long strips of at least two layers of, e.g., 'paper' and one thin layer of Al foil. The condenser is then made by winding up two such strips to form a roll. The rolls are then in general pressed flat to reduce the size, but may also be used in cylindrical form. A suitable number of units are assembled and pressed into an insulating case. After assembly the condenser is vacuum dried and impregnated with a dielectric fluid. Instead of a high-quality paper ('condenser paper', 10 to 30 μm thick) plain plastic films (e.g., polystyrene) or mixtures with plastic films and paper are now in use. Rolled condensers will have high inductance, if the metal electrodes are only joined at their ends or at some intermediate positions. Very low inductivity can only be gained if the technique of 'extended foil construction' is applied. In this method each of the two metal foil electrodes are on one side extended beyond the dielectric foils and after rolling and pressing continuously joined (welded) on each side to form the end caps.

Ceramic capacitors. Only a few manufacturers in the world are able to produce excellent and special types of ceramic capacitors applicable for moderate high voltage. Most types of these 'ceramic radio frequency power capacitors' are made from 'class/type 2' ceramic. The dielectric material used for this kind of ceramic contains mostly alkaline earth titanates giving quite high values of permittivity. This material is prone to ageing caused by a gradual realignment of the crystalline structure, has very high temperature coefficients (TC), is voltage dependent and is not recommended for measurement purposes. There are, however, some kinds of 'class/type 1' ceramic without these disadvantages. Within this class 1 ceramic, mainly manufactured from titanium dioxide or magnesium silicate, the materials with relative permittivities below 1000 show TC values down to some 10 ppm/°K within a temperature range of −25 to +95°C and very low dissipation factors up to the Megahertz range. The main advantage of all types of ceramic capacitors is the extremely low inductance due to their design, which is either of tubular or plate/pot type with conductive noble metal electrodes. The only disadvantage is due to the limitations in rated continuous a.c. voltage (up to about 40 kV peak) and capacitance (up to about 6 nF) per unit.

Both types of capacitors are in general use for voltage dividing systems.

3.6 Voltage dividing systems and impulse voltage measurements

The measurement of impulse voltages even of short duration presents no difficulties, if the amplitudes are low or are in the kilovolt range only. The tremendous developments during the last three decades related to the technique of common CROs, digital scopes or transient recorders provide instruments with very high bandwidth and the possibility to capture nearly every kind of short-duration single phenomena. Although the usual input voltage range of these instruments is low, h.v. probes or attenuators for voltages up to some 10 kV are commercially available.

The problems arise with much higher voltages and it is well known that impulse voltages with magnitudes up to some megavolts are used for testing and research. The voltage dividers necessary to accommodate these voltages are specialized apparatus, and there are only a few manufacturers throughout the world who are ready to produce such dividers with adequate accuracy. Self-provided constructions are often adequate if the problems are known. But also the application of such voltage dividers needs a fundamental understanding of the interactions present in voltage dividing systems. Hence an attempt is made to introduce the reader to the theory as well as to some hints on constructional details on this quite difficult field of h.v. measuring techniques.

We will start with a generalized voltage generation and dividing system and briefly discuss the layout (section 3.6.1). Depending upon the voltage shape to be measured, the requirements related to the whole measuring system must be well defined (section 3.6.2). A generalized analytical treatment of the transfer characteristics of this system involves the complex interactions between the different parts of the circuit (section 3.6.3). The theory of the 'isolated' voltage dividers as the most essential part of the circuit demonstrates the different types of these devices and their possible applications (section 3.6.4). For fast transient voltages the interactions between the dividers and their adherent circuits are briefly discussed and methods for the evaluation of the transfer properties are presented (section 3.6.5). Some advice on a proper design of the l.v. arm of the voltage dividers is given (section 3.6.6). As the transient digital recorder has recently become the most powerful tool for the evaluation of impulse voltages, an up-to-date introduction in this kind of instrument is provided in a separate section (see 3.7), which is partly still related to voltage dividing systems.

3.6.1 Generalized voltage generation and measuring circuit

Figure 3.23 illustrates the common and most adequate layout of any voltage testing circuit within an h.v. testing area. The voltage generator 1 is connected to a test object 3 by a lead 2. These three elements form a voltage generating

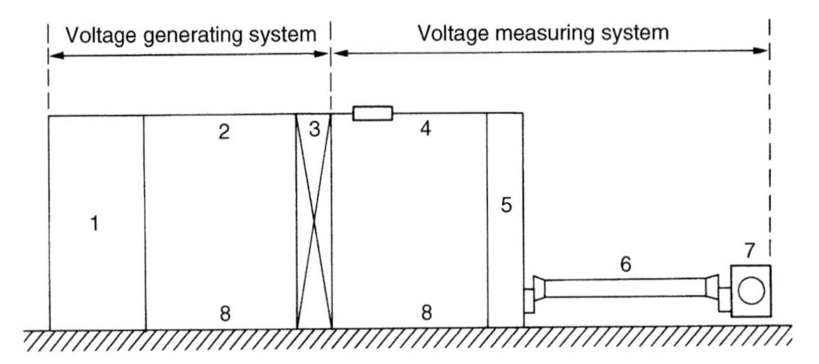

Figure 3.23 *Basic voltage testing system. 1. Voltage supply. 2. Lead to test object. 3. Test object. 4. Lead to voltage divider. 5. Voltage divider. 6. Signal or measuring cable. 7. Recording instrument. 8. Ground return*

system. The lead 2 to the test object may comprise any impedance or resistance to damp oscillations, if necessary, or to limit the short-circuit currents if the test object fails. The measuring system starts at the terminals of the test object and comprises a connecting lead 4 to the voltage divider 5, and a recording instrument 7, whose signal or measuring cable 6 is placed between its input terminals and the bottom or l.v. part of the divider. The appropriate ground return 8 should assure no significant voltage drops for even highly transient phenomena and keep the ground potential to earth as close as possible.

This layout is sometimes altered and there can be acceptable reasons for such a change. For d.c. voltages and small currents drawn by the test object, the voltage divider can be incorporated within the voltage supply, if the lead 2 has no or only a protecting resistance, the voltage drop across which can be neglected. Essentially the same statements are applicable to low-frequency a.c. voltages, but a possible influence of the lead inductance should be checked. In practice, also for impulse voltage testing circuits the voltage divider may form part of the impulse generator. The simple reasons can easily be understood from the impulse generator circuits (see Chapter 2, Fig. 2.25). There, the wave shaping load capacitance C_2 is often combined with an l.v. capacitor connected in series, thus forming an adequate voltage divider. An undamped connection to the object under test then leads to the erroneous assumption that negligible voltage drop can occur across the lead. This assumption may be correct for slowly rising impulse voltages and quite short leads. Connecting leads with lengths of many metres, however, are often used and thus this assumption may become unacceptable. It must be remembered that the test object is a capacitor and thus the circuit formed by the lead and test object is a series resonant circuit. These oscillations are likely to be excited by firing the generator, but will only partly be detected by the voltage divider. Completely wrong is the assumption that such a voltage divider being a part of the generator is measuring the correct voltage across the test object following a voltage collapse or disruptive discharge. The whole generator including voltage divider will be discharged by this short-circuit at the test object and thus the voltage divider is loaded by the voltage drop across lead 2. This lead forms to first approximation an inductance, and hence the oscillatory discharge currents produce heavy (induced) voltage oscillations which are then measured by the capacitor divider. These voltages are often referred to as overvoltages across the test object, but this statement is incorrect. For the measurement of predominantly lightning impulses, therefore, only the layout of the circuit according to Fig. 3.23 shall be used if an accurate measurement of full and chopped voltages is desired.

There is a further reason for placing the voltage dividers away from any energized objects. High-voltage dividers consist of 'open' networks and cannot be shielded against external fields. All objects in the vicinity of the divider which may acquire transient potentials during a test will disturb the field distribution and thus the divider performance. The lead from the voltage divider to the test object 4 is therefore an integral part of the measuring system. The influence of this lead will theoretically be treated in section 3.6.3. There it will be established that a damping resistor at the input end of this lead contributes to improved transfer characteristics of the system.

In order to avoid heavy electromagnetic interactions between the recording instrument and the h.v. test area as well as safety hazards, the length of the signal cable 6 must be adequately chosen. For any type of voltage to be measured, the signal cable should be of a coaxial and shielded type. The shield or outer conductor picks up the transient *electrostatic* fields and thus prevents the penetration of this field to the inner conductor. Although even transient *magnetic* fields can penetrate into the cable, no appreciable voltage (noise) is induced due to the symmetrical arrangement. Ordinary coaxial cables with braided shields may well be used for d.c. and a.c. voltages. For impulse voltage measurements, these cables must provide very low losses to reduce distortion of the voltage pulses to be transmitted. As it is impossible to avoid induced currents within the cable shields which are not related to the transmitted signal, these currents can heavily distort these signals if the so-called *coupling impedance* of the cable is not very low. In the frequency domain, this impedance $Z_c(\omega)$ is defined by

$$Z_c(\omega) = \frac{V_n/l}{I_d}, \tag{3.35}$$

where I_d is the disturbing current flowing in the shield, and V_n/l the voltage drop at the inner surface of the shield per unit length l of the cable. More information about the origin of disturbing cable shield currents may be found in references 1 and 54 and in other publications.[55,56] For a pure d.c. current within the shield, the coupling impedance is given by the voltage drop due to the d.c. resistance of the shield. If the frequency of these currents increases, the coupling impedance will continuously decrease if the shield is of rigid cross-section; then the eddy currents will attenuate the current density at the inner surface of the cylindrical shield. Hence rigid or corrugated shields, i.e. flexwell cables, are best suited for noise reduction. For braided shields, the coupling impedance is in general not a stable quantity, as the current distribution within the shield is likely to be influenced by resistive contacts within the braid. Double-shielded cables with predominantly two insulated braided shields will improve the behaviour. Best conditions are gained by placing the coaxial cable into an additional, non-braided metal tube, which is connected to ground potential at least at the input end of the measuring cable and also at its end.

In Fig. 3.23 there is finally the ground return 8. For h.v. test circuits disruptive discharge must always be taken into account. Large and heavily oscillating short-circuit currents are developed and hence every ground return with simple leads only cannot keep the voltage drops small. The impedance, therefore, must be reduced. Large metal sheets of highly conducting material such as copper or aluminium are best. Many h.v. laboratories provide such ground returns in combination with a Faraday cage for a complete shielding of the laboratory. Expanded metal sheets give similar performance. At least metal tapes of large width should be used to reduce the impedance. A parallel connection of tapes within flat areas will further decrease the inductance and thus approximate the efficiency of huge metal sheets.

Information concerning the layout of testing and measuring circuits is also provided in reference 57. The measuring system thus comprises four main components with quite different electrical behaviour. The simulation of these components will depend upon the necessary frequency range to measure the voltage across the test object. An evaluation of this frequency range shall thus precede this simulation.

3.6.2 Demands upon transfer characteristics of the measuring system

The voltage measuring system defined in Fig. 3.23 is a four-terminal network and can thus be represented as shown in Fig. 3.24. V_i indicates the voltage across the test object (3 in Fig. 3.23), and the output voltage V_0 appears at the recording instrument, i.e. at the screen of a CRO or transient recorder.

The input voltages V_i are either continuous steady state voltages for d.c. and a.c. generating systems, or single events for impulse voltages. In both cases, the instantaneous amplitudes will change with time, even for d.c. voltages with a periodic ripple.

For a sinusoidal input voltage $v_i(t) = V_{mi} \sin(\omega t + \phi_i)$ the magnitude V_{m0} and phase angle ϕ_0 of the output voltage $v_0(t) = V_{m0} \sin(\omega t + \phi_0)$ can be determined either by calculation with known network parameters or by

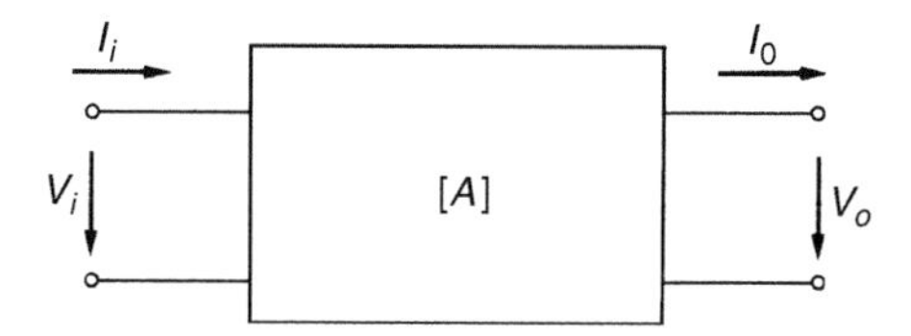

Figure 3.24 *Representation of the measuring system as a four-terminal network*

measurements, although such measurements are difficult to perform for very high ratios of V_{mi}/V_{m0}. The frequency response of the system can then be subdivided into an amplitude (frequency) response $H(\omega) = V_{m0}/V_{mi}$ and a phase (frequency) response $\phi(\omega) = \phi_0(\omega) - \phi_i(\omega)$. It is well known that both quantities are also displayed by assuming complex amplitudes $\mathbf{V}_i = V_{mi}\exp(j\phi_i)$ and $\mathbf{V}_0 = V_{m0}\exp(j\phi_0)$, and the system transfer or network response function

$$\mathbf{H}(j\omega) = \frac{V_0}{V_i} = |\mathbf{H}(j\omega)|\exp\{j[\phi_0(\omega) - \phi_i(\omega)]\} \tag{3.36}$$

where $|\mathbf{H}(j\omega)| = H(\omega)$ as defined above.

Neither d.c. voltages with ripple nor a.c. testing voltages are pure sinusoidal, but periodic in nature. The input voltages may then be described by a – in general – limited number of complex amplitudes $\mathbf{V}_{ik}$ obtained by the application of Fourier series,

$$\begin{aligned}\mathbf{V}_{ik} &= \frac{1}{T}\int_{-T/2}^{T/2} v_i(t)\exp(-jk\omega t)\,\mathrm{d}t \\ &= |\mathbf{V}_{ik}|\exp(j\phi_{ik}),\end{aligned} \tag{3.37}$$

where $\omega = 2\pi/T$, T is the time period and k are discrete numbers. The periodic input quantity is thus analysed into sinusoidal frequency components, and the complex amplitudes are displayed by the amplitude line spectrum $|\mathbf{V}_{ik}|$ and the angular frequency line spectrum. For every component with the frequency $\omega_k = k\omega$, the network response may easily be found with eqn (3.36), and the responses can be summed up using the principle of superposition. Applying again the complex form of the Fourier series, this summation gives:

$$v_0(t) = \sum_{k=-\infty}^{\infty} \mathbf{V}_{ik}\mathbf{H}(j\omega_k)\exp(jk\omega t). \tag{3.38}$$

A direct comparison between $v_0(t)$ and $v_i(t)$ can thus be made and the errors evaluated.

For the single events of impulse voltages, only an infinite number of sinusoidal voltages are able to represent the input voltage $v_i(t)$. This continuous frequency spectrum is defined by the Fourier integral or Fourier transform of $v_i(t)$

$$\mathbf{V}_i(j\omega) = \int_{t=-\infty}^{\infty} v_i(t)\exp(-j\omega\tau)\,\mathrm{d}\tau \tag{3.39}$$

and contains amplitude and phase spectra. The linearity and homogeneity of the time invariant systems assumed enable us again to calculate the time response of the system by a convolution of the continuous frequency spectrum with the network response function and the transition from frequency to time domain by means of the inverse Fourier transform:

$$v_0(t) = \frac{1}{2\pi}\int_{\omega=-\infty}^{\infty} \mathbf{V}_i(j\omega)\mathbf{H}(j\omega)\exp(j\omega t)\,\mathrm{d}\omega. \tag{3.40}$$

In practice, the real input quantity $v_i(t)$ is not known, as only $v_0(t)$ can be measured. This output voltage, however, has suffered from the loss of information contained in $\mathbf{H}(j\omega)$. No appreciable transmission errors could occur, if at least the amplitude frequency response $H(\omega) = |\mathbf{H}(j\omega)|$ would be constant within a frequency range, in which the line or continuous frequency spectra, $\mathbf{V}_{ik}$ or $\mathbf{V}_i(j\omega)$, cannot be neglected. Thus the computation of the spectra of an estimated input quantity is a very efficient tool to judge the necessary frequency range or bandwidth of our measuring system and its individual components.

The highest demands upon the measuring system transfer functions are clearly imposed by impulse voltages. The analysis of the impulse voltage generating circuits (see Chapter 2, section 2.3.1) displayed a waveshape of the generator output voltage, which is a double exponential function. Neglecting the possible interactions between the voltage measuring and generating systems, we thus may assume an input voltage for the measuring system, given by $v_i(t) = A[\exp(-t/\tau_1) - \exp(-t/\tau_2)]$, where A is a constant value and τ_1, τ_2 the time constants according to eqn (2.27). This voltage can be chopped at any instantaneous time T_c as defined in Fig. 2.24 caused by a disruptive discharge of the test object, but the voltage collapse is extremely rapid. The input voltage is then given by

$$v_i(t) = \begin{cases} 0 & \text{for } < 0; t > T_c \\ A[\exp(-t/\tau_1) - \exp(-t/\tau_2)] & \text{for } 0 \le t \le T_c. \end{cases} \tag{3.41}$$

Applying this voltage and its boundary conditions to eqn (3.39) gives $\mathbf{V}_i(j\omega)$. The calculation implies no fundamental difficulties; the result, however, is lengthy and is obtained as

$$V_i(j\omega) = A(Re + jIm) \tag{3.42}$$

where

$$Re = \frac{\tau_1}{1+(\omega\tau_1)^2}\{1 + [\omega\tau_1 \sin(\omega T_c) - \cos(\omega T_c)]\exp(-T_c/\tau_1)\}$$

$$- \ldots - \frac{\tau_2}{1+(\omega\tau_2)^2}\{1 + [\omega\tau_2 \sin(\omega T_c - \cos(\omega T_c)]\exp(-T_c/\tau_2)\};$$

$$Im = \frac{\tau_1}{1+(\omega\tau_1)^2}\{\omega\tau_1 - [\omega\tau_1 \cos(\omega T_c) + \sin(\omega T_c)]\exp(-T_c/\tau_1)\}$$

$$- \ldots - \frac{\tau_2}{1+(\omega\tau_2)^2}\{\omega\tau_2 - [\omega\tau_2 \cos(\omega T_c) + \sin(\omega T_c)]\exp(-T_c/\tau_2)\}.$$

For the special case of a non-chopped voltage ($T_c \to \infty$), the Fourier transform of the input voltage is merely

$$\mathbf{V}_i(j\omega) = A\left[\left(\frac{\tau_1}{1+(\omega\tau_1)^2} - \frac{\tau_2}{1+(\omega\tau_2)^2}\right) - j\left(\frac{\omega\tau_1^2}{1+(\omega\tau_1)^2} - \frac{\omega\tau_2^2}{1+(\omega\tau_2)^2}\right)\right]. \qquad (3.43)$$

The numerical evaluation of eqns (3.42) and (3.43) is shown in Fig. 3.25 for a full lightning impulse of 1.2/50 µsec ($\tau_1 = 68.2$ µsec: $\tau_2 = 0.405$ µsec) and different instants of chopping, T_c. A normalization was made by $\mathbf{v}_i(j\omega) = \mathbf{V}_i(j\omega)/V_i(\omega = 0)$ and only the relative amplitudes $|\mathbf{v}_i(j\omega)| = v_i(\omega)$ are displayed on a linear scale. From the result the following conclusions can be made.

The relative amplitudes for a full lightning impulse ($T_c \to \infty$) become already very small in a frequency range of about 0.5–1 MHz; hence an

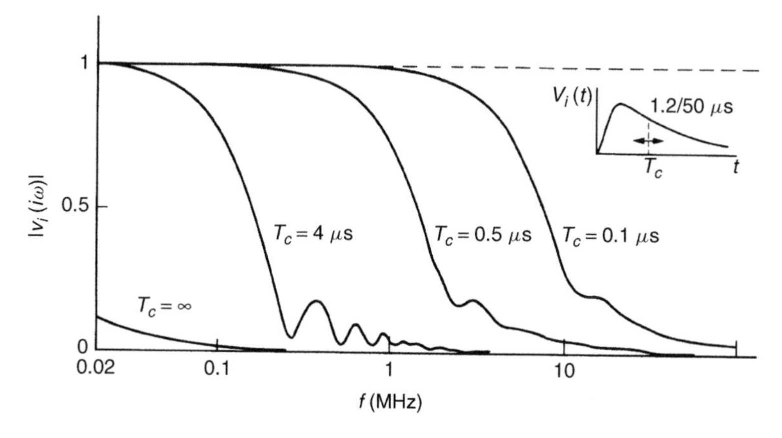

Figure 3.25 *Normalized amplitude frequency spectra (Fourier transform) of a lightning impulse voltage of 1.2/50 µsec, wave full and chopped*

amplitude frequency response of our measuring circuit, which is really flat up to this frequency range, would obviously not provide significant errors. Depending upon the decay of the amplitude frequency response, the bandwidth (−3 dB point) has to be much higher, i.e. about 5–10 MHz.

The chopping of the voltage introduces a heavy increase of the harmonics content. For $T_c = 4\,\mu\text{sec}$, i.e. a chopping at the impulse tail, an accurate measurement of the crest voltage may still be provided by the above-mentioned amplitude response, although appreciable errors might appear during the instant of chopping. The voltages chopped within the front ($T_c = 0.5–0.1\,\mu\text{sec}$), however, will require a very wide bandwidth which must obviously increase with decreasing chopping time. Desirable values of f_B for $T_c = 0.5\,\mu\text{sec}$ only shall obviously reach magnitudes of about 100 MHz, but such large values cannot be achieved with measuring systems for very high voltages.

This frequency domain method described so far for determining a transfer characteristic quantity to estimate measuring errors is difficult to use, as the two quantities, $H(\omega)$ and $\phi(\omega)$, are difficult to measure due to the large 'scale factors' of the measuring systems. For h.v. measuring systems, the transfer characteristic is therefore evaluated by means of a measured (experimental unit) 'step response'.[57,53] This time-domain method is based upon the fact that the Fourier transform (eqn (3.39)) of a single-step function is proportional to $1/j\omega$ and thus all frequencies are contained. Let us, therefore, represent the input voltage of our measuring system by such a step function:

$$v_i(t) = \begin{cases} 0 & \text{for } t < 0 \\ V_{mi} & \text{for } t > 0. \end{cases} \tag{3.44}$$

The output voltage of the measuring system, $v_0(t)$, is then much smaller in amplitude, it may also be time delayed with reference to the voltage input, and it will be distorted mainly at its front. This 'unit step response' (USR) is denoted by the term $G(t)$ and is sketched in Fig. 3.26(a). The time $t = 0$ is defined by eqn (3.44), the time delay described by τ_{de}, and for a good measuring system the final value $V_{mi}(1/N)$ will be reached in a short time. The magnitude N indicates the steady state voltage ratio between input and output voltage, i.e. the scale factor of the system.

In section 3.6.3, $G(t)$ will be calculated based upon equivalent circuits. This quantity is also easy to measure by means of sensitive recorders or CROs. With a known value of the USR, $G(t)$, the output voltage response to any arbitrary input voltage can be calculated from the superposition theorem or Duhamel's integral:

$$\begin{aligned} v_0(t) &= v_i(t)G(+0) + \int_0^t v_i(\tau)G'(t-\tau)\,d\tau \\ &= G(t)v_i(+0) + \int_0^t v_i'(t-\tau)G(\tau)\,d\tau. \end{aligned} \tag{3.45}$$

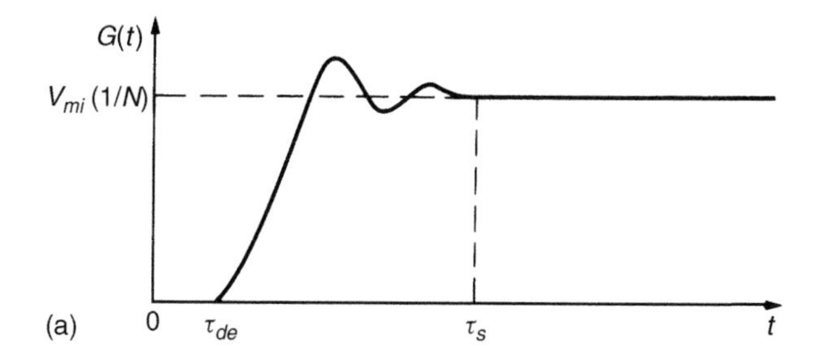

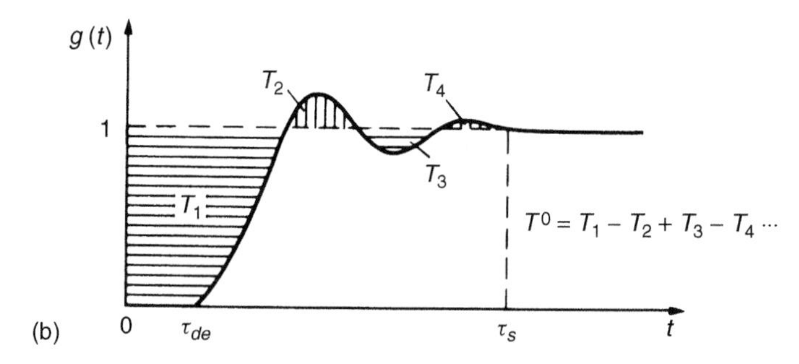

Figure 3.26 *Unit step response and definition of the response time T'. (a) Unit step response as output voltage. (b) Normalized unit step response*

where $G'(t-\tau)$ or $v_i'(t-\tau)$ is the derivative of $G(t)$ or $v_i(t)$ with respect to τ. This integral can always be solved numerically by digital computers, if analytic expressions are not available.(129)

The chopping of a lightning impulse voltage at the front ($T_c \leq 1\,\mu$sec in Fig. 3.25) is sometimes used for h.v. testing and the demands upon the measuring circuits become severe. The chopping on front provides a nearly linearly rising voltage up to T_c. Let us assume an ideally linearly rising voltage,

$$v_i(t) = St, \tag{3.46}$$

where S is the steepness. With eqn (3.45), the output voltage becomes

$$v_0(t) = S\int_0^t G(\tau)\,\mathrm{d}\tau = \frac{S}{N}\int_0^t g(\tau)\cdot \mathrm{d}\tau \tag{3.47}$$

where $g(t)$ is the normalized quantity of the USR voltage $G(t)$, whose final value becomes thus 1 or 100 per cent. Then the term $Nv_0(t)$ represents the high voltage comparable to $v_i(t)$ of eqn (3.46), and we may introduce this term into eqn (3.47) and expand this equation to

$$Nv_0(t) = S\left[t - \int_0^t [1 - g(\tau)]\,\mathrm{d}\tau\right]. \tag{3.48}$$

This expression relates the output to the input voltage as long as (St) increases. The integral term will settle to a final value after a time τ_s indicated in Fig. 3.26. This final value is an interesting quantity, it shows that differences in amplitudes between input (St) and magnified output voltage $N v_0(t)$ remain constant. Hence we may write

$$v_i(t) - N v_0(t) = S \int_0^{t>\tau_s} [1 - g(\tau)]\,d\tau = S \int_0^{\infty} [1 - g(\tau)]\,d\tau = ST^0 \quad (3.49)$$

where

$$T^0 = \int_0^{\infty} [1 - g(\tau)]\,d\tau \quad (3.50)$$

is the 'response time' of the measuring system. This quantity gives the time which can be found by the integration and summation of time areas as shown in Fig. 3.26(b). T^0 includes a real time delay τ_{de} of the output voltage, which is in general not measured, if the time instant of the application of the unit step input is not recorded. The former IEC Recommendations[57] and the newest IEC Standard 60-2[53] therefore neglect this time delay. The justifications for neglecting this delay are shown in Fig. 3.27. There, the linearly rising input voltage is suddenly chopped, and the output voltage multiplied by N is approximately sketched for the USR of Fig. 3.26. Equation (3.48) can be applied up to the instant of chopping, T_c; for later times, eqn (3.45) must be rearranged, and it can easily be seen that a superposition of three terms (response to St, negative USR with amplitude ST_c, and negative response to St for $t > T_c$) will govern this output voltage.

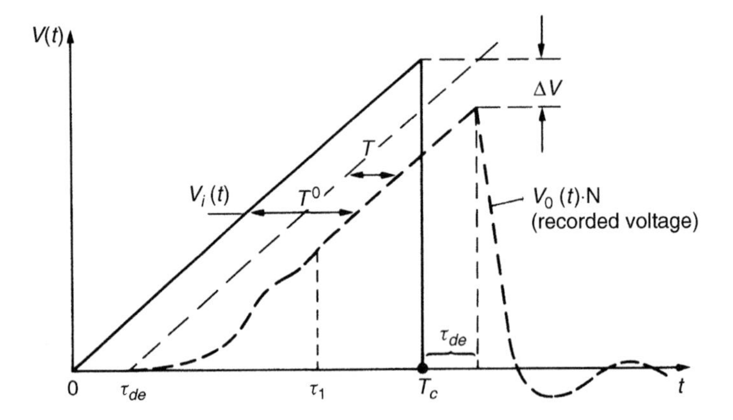

Figure 3.27 *Measuring error ΔV for linearly rising voltages chopped at T_c. Definition of response time T^0 and T*

As the sudden change in the output voltage is also delayed, the amplitude error ΔV is obviously given by

$$\Delta V = v_i(T_c) - N v_0(T_c + \tau_{de}) = S(T^0 - \tau_{de}) = ST$$

if $T_c > \tau_s$. Thus the simple relationship

$$T = T^0 - \tau_{de} \tag{3.51}$$

exists, where T is equal to a response time similar to T_0, but integrated from Fig. 3.26 by

$$T = \int_{\tau_{de}}^{\infty} [1 - g(\tau)]\, \mathrm{d}t. \tag{3.52}$$

The relative amplitude error δ for a chopped linearly rising voltage thus becomes

$$\delta = \frac{\Delta V}{ST_c} = \frac{T}{T_c}. \tag{3.53}$$

For $T = 50\,\text{ns}$, and $T_c = 0.5\,\mu\text{s}$, this error is 10 per cent.

Clearly, this simple qualification criterion for a measuring system has some drawbacks. First, eqn (3.53) can only be used if the assumptions (linearly rising voltage, time to final value or settling time $\tau_s < T_c$, ideal chopping) are fulfilled. Ideal, linearly rising high voltages, however, are difficult to generate and it is even more difficult to confirm this linearity by measurements, as the measured values are not accurate.(50) Due to its definition, the response time T or T^0 can even be negative, see section 3.6.5. Such problems could easily be demonstrated by a systematic evaluation of eqn (3.45) but only one example is shown in Fig. 3.28. This example does not need further explanations and additional information is given in section 3.6.7.

In spite of these disadvantages, the concept of response times cannot be disregarded and it is again used in the latest IEC Standard related to measuring systems.(53) The biggest advantage is related to its computation: it will be demonstrated in the next section that T^0 or T can be calculated analytically even for very sophisticated networks without the knowledge of $g(t)$. Hence, the value of this quantity giving a measure to quantify transfer properties of a measuring system must be acknowledged, although additional response parameters can be used to correlate step responses with other types of measuring errors.(53, 129, 130)

3.6.3 Fundamentals for the computation of the measuring system

Any analysis of a complex and sophisticated network as sketched in Fig. 3.23 either in the frequency or time domain is based on an equivalent circuit,

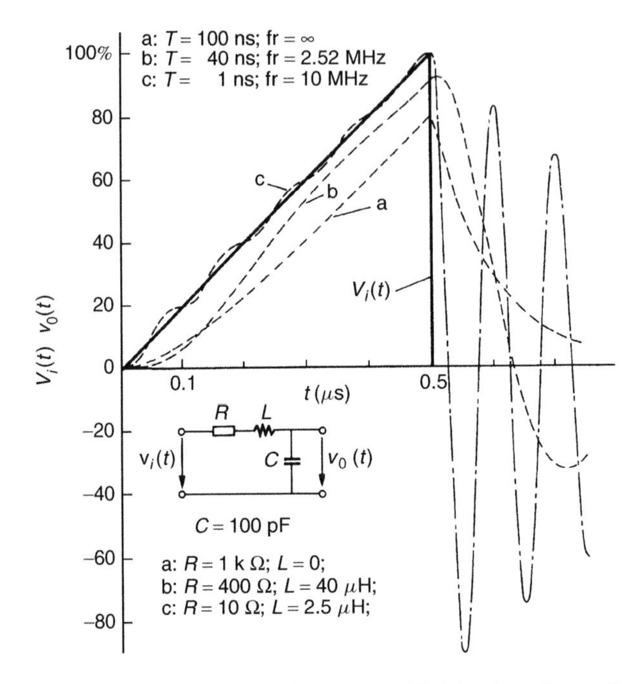

Figure 3.28 *Computed response $V_0(t)$ of an R–L–C circuit with given parameters to a linearly rising input voltage $V_i(t)$ chopped at $T_c = 0.5$ μsec*

which may represent, as closely as possible, the physical reality. Although being aware that the execution of this task is most difficult for h.v. measuring systems due to their dimensions and complex electromagnetic fields involved, we have already simulated the system by a simple 'four-terminal' or 'two-port' network as shown in Fig. 3.24. The analysis or computation of any numerical evaluation of results can be performed nowadays by adequate computer programs in time or frequency domain. The disadvantage of this method is, however, that the influence of individual network parameters is quite difficult to identify. Thus we use as far as possible an analytical treatment of our systems based on the general network theory.

The representation of the actual measuring system (Fig. 3.23) by a four-terminal network imposes, however, certain restrictions. As demonstrated later, the theory of travelling waves or distributed parameters is also used to evaluate the behaviour of the system during transients, and thus it is assumed that the electromagnetic phenomena are quasi-stationary in the two directions perpendicular to the direction of wave propagation. These conditions are somewhat limiting the validity of the calculations when the dimensions of the measuring systems are too large. The limitations are obviously directly correlated with the definition of a voltage as an independent quantity from space within an electromagnetic field, and as the dimensions of our h.v. systems are in the range of metres, the quasi-stationary nature of the electromagnetic phenomena is more or less restricted. For example, the travelling time of a wave at the velocity of light is 20 nsec between two points 6 metres apart. If impulses chopped on the front at $T_c = 200$ nsec are considered, the time is only ten times longer than the field propagation time.

With these restrictions in mind, we nevertheless may start a generalized computation of our four-terminal network, Fig. 3.24, and apply the results to equivalent circuits later on. The Laplace transform will now be used throughout the treatment, with the complex frequency $s = \sigma + j\omega$ being the Laplace operator. Input and output parameters can be described by the matrix equation

$$\begin{bmatrix} V_i(s) \\ I_i(s) \end{bmatrix} = \begin{bmatrix} A_{11}(s); & A_{12}(s) \\ A_{21}(s); & A_{22}(s) \end{bmatrix} \cdot \begin{bmatrix} V_0(s) \\ I_0(s) \end{bmatrix} = [A] \begin{bmatrix} V_0(s) \\ I_0(s) \end{bmatrix} \quad (3.54)$$

where $[A]$ is the network matrix of the system defined by this equation.

The measuring system will load the generating system and thus the input impedance of the measuring system is sometimes necessary. As the output current I_0 for a voltage dividing system with large ratios or scale factors N cannot influence the input, the condition $I_0 = 0$ can always be assumed. From eqn (3.54) the input impedance is

$$Z_i(s) = \frac{V_i(s)}{I_i(s)} = \frac{A_{11}(s)}{A_{21}(s)}. \quad (3.55)$$

The most important quantity is the voltage transfer function. For $I_0 = 0$, this function becomes

$$H(s) = \frac{V_0(s)}{V_i(s)} = \frac{1}{A_{11}(s)}. \quad (3.56)$$

Embedded in this function is the scale factor N of the voltage dividing system. This factor or ratio is a constant quantity for moderate frequencies only and hence we may derive this ratio by

$$N = \lim_{s \to 0} \left[\frac{V_i(s)}{V_0(s)} \right] = \lim_{s \to 0} [A_{11}(s)] = A_{11}(0). \quad (3.57)$$

The voltage transfer function, eqn (3.56), is conveniently normalized by N. Denoting the normalization by $h(s)$, we obtain

$$h(s) = NH(s) = \frac{A_{11}(0)}{A_{11}(s)}. \quad (3.58)$$

The unit step voltage $G(t)$, as described and defined in section 3.6.2, can be found by applying the Laplace inverse integral to the transfer function multiplied by $(1/s)$, the Laplace transform of a unit step. Thus

$$G(t) = L^{-1}\left[\frac{1}{s}H(s)\right] = L^{-1}\left[\frac{1}{sA_{11}(s)}\right]. \tag{3.59}$$

From eqn (3.58), the normalized unit step response is

$$g(t) = NG(t). \tag{3.60}$$

For very complex transfer functions often involved in mixed distributed parameter circuits, the applicability of eqn (3.59) is restricted, as it is too difficult to find solutions in the time domain. Then the response time T^0 cannot be computed by eqn (3.50). Based upon a well-known final value theorem of the Laplace transform, which is

$$\lim_{t\to 0} f(t) = \lim_{s\to 0}[sF(s)],$$

we may compute the response time from the following equation, which can be derived by applying this final value theorem to eqn (3.49):

$$T^0 = \lim_{s\to 0}\left[\frac{1-h(s)}{s}\right].$$

As $\lim_{s\to 0} h(s) \hat{=} 1$ by definition, the rule of Bernoulli–l'Hôpital leads to

$$T^0 = \lim_{s\to 0}\left[-\frac{dh(s)}{ds}\right] = \lim_{s\to 0}[-h'(s)]. \tag{3.61}$$

The final value theorem contains some restraints, i.e. $f(t)$ and df/dt must be Laplace transformable and the product $sF(s)$ must have all its singularities in the left half of the s-plane. Equation (3.61) thus may fail sometimes.

The response time T can be computed from eqn (3.51), if τ_{de} is known. It may be difficult, however, to predict an actual time delay based upon $h(s)$ only. The comparison of experimental and thus actual time delays with computed results may suffer from this disadvantage; for more information about this very specialized question, the reader is referred to the literature.[(50)]

These general results can now be applied to more detailed measuring circuits. Numerous equivalent circuits could be presented. We will, however, follow a representation, developed by Asner,[(58)] Creed *et al.*[(59)] and Zaengl.[(60)] In principle it deals with an adequate simulation of the lead to the voltage divider, i.e. to the simple metal wire or tube used for the connection of test object and divider. It was impossible for a long time to detect the influence of this lead, as no CROs have been available to measure the actual unit step response of the systems. Thus neither the performance of the voltage dividers used nor the performance of the whole measuring circuit could really be checked. Many details within the construction of a voltage divider, however, can completely destroy the fundamentally good layout based upon theoretical investigations.

With this lead, a more detailed representation of our simple four-terminal network, Fig. 3.24, is shown in Fig. 3.29. Three two-port sections are combined, forming a 'three-component system'. System 1 represents a damping impedance Z_d at the input end of the lead 2, connecting this impedance with the voltage dividing system 3, which terminates the lead. Due to their complex structure and frequency-dependent input impedance, the voltage dividers cannot properly match the leads' surge impedance Z_L. The damping impedance Z_d is therefore placed at the input end of the lead, as the travelling wave theory may easily show that only at this place is an efficient damping of oscillations possible. The lead 2 is thus best treated as a lossless transmission line, simulated by its surge impedance Z_L and its travel time τ_L, which implies that the capacitance per unit length is constant. Leads to the voltage dividers consist of metal tubes or metal wires, the diameter of which should be such as to avoid any heavy predischarges. That waves are really travelling with the velocity of light was readily shown in many investigations[50,59,60]. The simple representation of the lead by Z_L and τ_L only was also confirmed by these investigations.

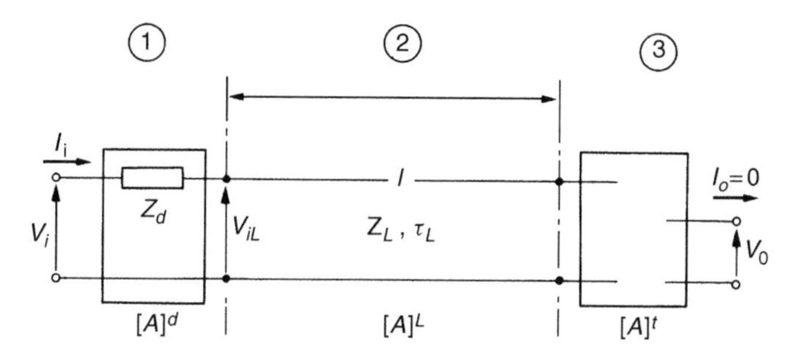

Figure 3.29 *The 'three-component system' comprised of a (1) damping, (2) transmission and (3) terminating system*

Not represented in the circuit of Fig. 3.29 is the signal cable and the recording instrument. It will be shown in section 3.6.6 that a lossless signal cable (see item 6 in Fig. 3.23) can be connected to the different kinds of dividers without appreciably influencing the USR. In this arrangement they form a part of the divider's l.v. arm. As also the recording instruments (item 7, Fig. 3.23) have high input impedances and wide bandwidth, their possible influence on the response is small and can thus be neglected, if the instruments are properly connected to the l.v. arms, see section 3.6.6.

Up to now the terminating or voltage dividing system 3 had not been specified in detail, as its network depends upon the type of divider used. For the computation of the transfer properties, the relevant matrix representation according to eqn (3.54) is used. Thus the matrix $[A]$ of the whole measuring system is

$$[A] = [A]^d[A]^L[A]^t.$$

The matrix $[A]$ can partly be solved by inserting the specific matrix elements for $[A]^d$ and $[A]^L$ defined with the circuit elements of Fig. 3.29. The details of the computation are lengthy and are omitted here. The following results, however, are of general interest.

The normalized transfer function, $h(s) = NV_o(s)/V_i(s)$, is best described by introducing reflection coefficients K for travelling waves, which are reflected either from the terminating system (K_t) or from the damping system (K_d). They are defined by

$$K_t(s) = \frac{\dfrac{A_{11}^t(s)}{A_{21}^t(s)} - Z_L}{\dfrac{A_{11}^t(s)}{A_{21}^t(s)} + Z_L} = \frac{Z_t(s) - Z_L}{Z_t(s) + Z_L} \tag{3.62}$$

$$K_d(s) = \frac{\dfrac{A_{12}^d(s)}{A_{11}^d(s)} - Z_L}{\dfrac{A_{12}^d(s)}{A_{11}^d(s)} + Z_L} = \frac{Z_d(s) - Z_L}{Z_d(s) + Z_L}. \tag{3.63}$$

With these coefficients, the transfer function is:

$$h(s) = e^{-\tau_L s}\frac{Z_d(0) + Z_L}{Z_d(s) + Z_L}\frac{1 + K_t(s)}{1 + K_t(0)} \times \frac{1 - K_t(0)K_d(0)}{1 - K_t(s)K_d(s)\exp(-2\tau_L s)}\frac{A_{11}^t(0)}{A_{11}^t(s)}. \tag{3.64}$$

The inherent time delay caused by the travel time of the lead, τ_L, can well be seen from the first factor; the last factor represents the normalized transfer function of the voltage dividing system. The normalized step response could be calculated using eqn (3.59). A glance at the transfer function indicates the difficulties encountered with its transformation into the time domain. A very simple example, however, will demonstrate the reflection phenomena introduced by the lead.

Let the damping system be a pure resistor, i.e. $Z_d(s) = R_d$, and the terminating system be simulated by a pure resistor divider without any frequency-dependent impedances, i.e. the divider may merely be represented by its input resistance R_t. Thus $K_t(s) = (R_t - Z_L)/(R_t + Z_L) = K_t$, and $K_d(S) = (R_d - Z_L)/(R_D + Z_L) = K_d$, and both are real numbers only. According to eqn (3.58), the normalized transfer function of a pure resistor divider will be equivalent to 1. Then, eqns (3.64) and (3.60) provide the normalized USR:

$$g(t) = L^{-1}\left\{\frac{\exp(-\tau_L s)}{s}\frac{1 - K_t K_d}{1 - K_t K_d \exp(-2\tau_L s)}\right\}.$$

A well-known evaluation of this expression is based upon the expansion of the last factor by a geometric row:

$$g(t) = L^{-1}\left\{\frac{\exp(-\tau_L s)}{s}(1 - K_t K_d)[1 + (K_t K_d)e^{-2\tau_L s} + \ldots \right.$$
$$\left. \ldots + (K_t K_d)^2 e^{-4\tau_L s} + (K_t K_d)^3 e^{-6\tau_L s} + \ldots]\right\} \qquad (3.65)$$

The infinite number of factors thus represents the possible number of reflections. Within the time intervals $(1 + 2n) \leq t/\tau_L < (3 + 2n)$, where $n = 0, 1, 2 \ldots$, the amplitudes of $g(t)$, however, are constant. In Fig. 3.30, eqn (3.65) is evaluated for the most probable case, that when $R_t \gg Z_L$, i.e. $K_t = +1$, and R_d is smaller or larger than Z_L. For $R_d = Z_L$, $K_d = 0$ and any reflection phenomena disappear. This is in fact the reason why the damping resistor is placed at the input end of the lead. This very simplified example shows also that the response time T or T^0 will strongly be influenced by the damping resistor. The magnitude of this influence is related to the length of the lead by τ_L.

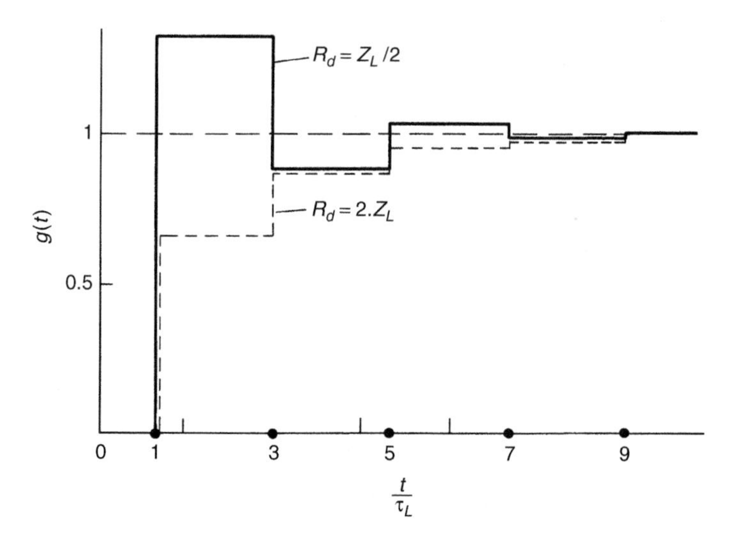

Figure 3.30 *Unit step response according to eqn (3.65)*

An exact evaluation of the response time is based upon eqn (3.61), substituted into eqn (3.64). The result is:

$$T^0 = T_t + \tau_L \left[\frac{Z_t(0)\dfrac{Z_d(0)}{Z_L} + Z_L}{Z_t(0) + Z_d(0)} \right] + \frac{Z_d(0)}{Z_L}$$

$$\times \left[T_{kt} \frac{Z_t(0) + Z_L}{Z_t(0) + Z_d(0)} - T_{kd} \frac{Z_d(0) + Z_L}{Z_t(0) + Z_d(0)} \right] \tag{3.66}$$

In this equation,

$$T_t = \frac{A_{11}^{t'}(0)}{A_{11}^t(0)} \tag{3.67}$$

is the response time of the voltage divider or terminator,

$$T_{kt} = \frac{K_t'(0)}{1 + K_t(0)} \tag{3.68}$$

is the reflection response time of the terminator,

$$T_{kd} = \frac{K_d'(0)}{1 + K_d(0)} \tag{3.69}$$

is the reflection response time of the damping system,

$$Z_t(0) = \frac{A_{11}^t(0)}{A_{21}^t(0)} \tag{3.70}$$

is the d.c. input resistance of the terminator, and finally

$$Z_d(0) = \frac{A_{12}^d(0)}{A_{11}^d(0)} \tag{3.71}$$

is the d.c. resistance of the damping system.

The influence of the dividers' lead is again illustrated by eqn (3.66). The complexity of this result is further discussed in section 3.6.5. In general, the voltage dividing system, mainly represented by its response time T_t, will essentially control the transfer characteristics of the whole system. Thus it is justified to treat the terminating system in advance and isolated from the lead to achieve a general understanding of all h.v. dividing systems and their adequate application.

3.6.4 Voltage dividers

Voltage dividers for d.c., a.c. or impulse voltages consist of resistors or capacitors or convenient combinations of these elements. Inductors are in general not used for voltage dividers for testing purposes, although 'inductance voltage dividers' do exist and are used for the measurement of power frequency voltages,[(139)] independent from inductive voltage transformers as used in power transmission. Inductance voltage transformers consist in the simplest case of a high-quality toroidal core with a tapped winding and some of these elements can be cascaded to form a 'cascade inductance divider'. Measuring uncertainties down to a few ppm can be reached if built for quite low voltages (1 kV or less), but lots of problems arise if they are built for magnitudes of 100 kV or more. Therefore, no further treatment follows here.

The elements of the aforementioned h.v. voltage dividers are usually installed within insulating vessels of cylindrical shape with the ground and h.v. terminals at both ends. The height of a voltage divider depends finally upon the external flashover voltage and this follows from the rated maximum voltage applied; this flashover voltage is also influenced by the potential distribution and is thus influenced by the design of the h.v. electrode, i.e. the top electrode. For voltages in the megavolt region, the height of the dividers becomes large, as one may assume the following relative clearances between top electrode and ground:

2.5 to 3 m/MV for d.c. voltages;
2 to 2.5 m/MV for lightning impulse voltages;
up to or more than 5 m/MV (r.m.s.) for a.c. voltages;
up to and more than 4 m/MV for switching impulse voltages.

Because the breakdown voltages in atmospheric air become strongly non-linear with voltage magnitude for a.c. and lightning impulse voltages, the above suggested clearances may be considered only as guidelines.

The most difficult problems in a simulation of the actual network of voltage dividers is in the inadequate representation of the stray capacitances (see Figs 3.7, 3.8, 3.21). Whereas the location and dimensions of the active parts, i.e. resistor or capacitor units, within a voltage divider are exactly known, the same statements are impossible to achieve for stray capacitances. It would also be too difficult to present equivalent circuits with distributed parameters, which individually account for the physical size of the units, by assuming a too high number of elements of unequal values. Apart from the fundamental difficulties in performing analytical computations of such circuits, the results are then individually related to the high number of parameters.

It has been acknowledged by many investigators that a recurrent or distributed parameter network with *equally* distributed parameters is probably

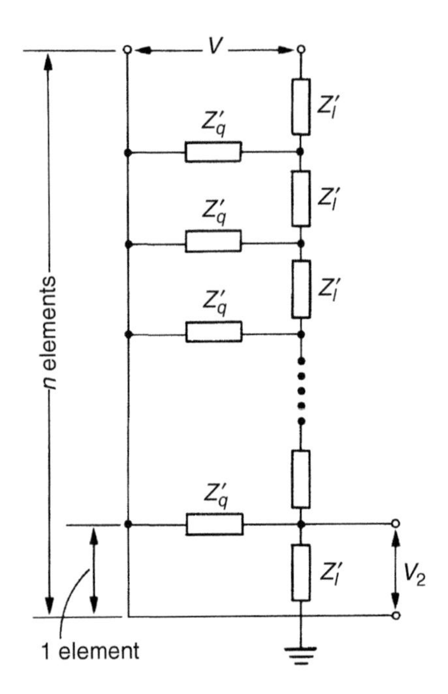

Figure 3.31 *Distributed parameter equivalent circuit of voltage dividers*

the best compromise to simulate transfer characteristics by equivalent circuits. Such a distributed parameter network for a generalized voltage divider is shown in Fig. 3.31. Our 'terminating system' $[A]^t$ of Fig. 3.29 is now simulated by a large number (n) of elements or sections, and the n impedances Z'_l in series are providing the voltage reduction. An equal number of impedances Z'_q to earth are distributed along this column. The input voltage V is thus greatly reduced to the low output voltage V_2. The total impedances are then defined by

$$Z_l = \sum Z'_l = nZ'_l; \quad \text{and} \quad Z_q = \left(\sum \frac{1}{Z'_q}\right)^{-1} = \frac{Z'_q}{n} \tag{3.72}$$

The number n is by this definition equivalent to the voltage ratio or scale factor V/V_2 of the divider; it may differ from N as defined before, as the impedance Z_d of the lead (Fig. 3.29) may change the ratio of the whole voltage measuring system.

The matrix representation of such a network, which is equivalent to a transmission line network, is well known. Applying eqn (3.56) and eqn (3.58) to this network, one may easily find the normalized transfer function (index

$t =$ terminator), which is

$$h_t(s) = \frac{nV_2}{V} = \frac{n \sinh \frac{1}{n} \sqrt{Z_l(s)/Z_q(s)}}{\sinh \sqrt{Z_l(s)/Z_q(s)}}. \tag{3.73}$$

The normalized unit step response is

$$g_t(t) = L^{-1} \left[\frac{1}{s} h_t(s) \right].$$

Both quantities can now be computed and analysed for different equivalent circuits, for which the impedances Z'_l and Z'_q are specified. Z'_q, however, will always be represented by stray capacitances C'_e to earth, as no voltage dividing system is known which would comprise any other passive elements at these locations. This stray capacitance is thus assumed to be equally distributed.

Resistor voltage dividers

The most general representation of such dividers has to assume inductive components L' of the actual resistor R' as well as capacitive elements C'_p in parallel to the resistors (see Fig. 3.32). Inductances are inherent with every flow of current due to the magnetic field, and the parallel capacitors C'_p may be formed by the construction and arrangement of the resistors. The neglecting of any inductance in series to these stray capacitances indicates possible coupling effects and the simulation of electrical fields within insulation media of low permittivity only; the individual values are thus of any small magnitude by the distributed parameter representation.

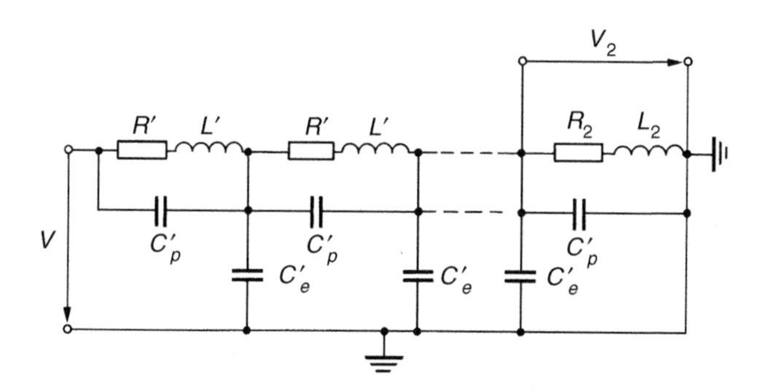

Figure 3.32 *Equivalent circuit for resistor voltage dividers.* $R = nR'$; $L = nL'$; $C_e = nC'_e$; $C_p = C'_p/n$; $R_2 = R'$; $L_2 = L'$; $R_1 = (n-1)R'$

The normalized transfer function is easily found from eqn (3.73) and is

$$h_t(s) = n \frac{\sinh \dfrac{1}{n}\sqrt{\dfrac{(R+sL)sC_e}{1+(R+sL)sC_p}}}{\sinh \sqrt{\dfrac{(R+sL)sC_e}{1+(R+sL)sC_p}}}. \tag{3.74}$$

The computation of $g_t(t)$ for this and all the other circuits presented thereafter can be made with minor approximations justifiable for $n \gg 1$. The details can be found in the literature,(61,62) only the result is presented:

$$g_t(t) = 1 + 2e^{-at}\sum_{k=1}^{\infty}(-1)^k \frac{\cosh(b_k t) + \dfrac{a}{b_k}\sinh(b_k t)}{1 + \dfrac{C_p}{C_e}k^2\pi^2}; \tag{3.75}$$

where

$$a = R/2L;$$

$$b_k = \sqrt{a^2 - \frac{k^2\pi^2}{LC_e[1+(C_p/C_e)k^2\pi^2]}};$$

$$k = 1, 2, 3, \ldots, \infty.$$

Both quantities can be used to demonstrate the limits of applications if representative values for the circuit constants are taken into consideration.

First, it is clear that resistor dividers are ideal for d.c. voltage measurements. The transfer function $h_t(s)$ for high R values and accordingly small values of L/R increase steadily with a decrease of the frequency. For $s \to 0$, $h_t(s) \hat{=} 1$ and therefore

$$V_2 = \frac{V}{n} = V\frac{R_2}{R_1 + R_2}$$

(see Fig. 3.32 for the definition of R_1 and R_2). The advantage of this relationship and its effect upon the accuracy and stability of the divider ratio was already discussed in section 3.3.

The ability to measure a.c. voltages as well as ripple inherent in d.c. voltages depends upon the decrease of $h_t(s)$ with frequency. Since for all constructions of high ohmic resistor dividers the L/R values are lower than about 0.1 μsec, and also $C_p \ll C_e$, the controlling factor of the transfer function is given by the product RC_e. We can thus neglect L and C_p in eqn (3.74) as well as in eqn (3.75) and therefore:

$$h_t(s) \approx n \frac{\sinh \frac{1}{n}\sqrt{sRC_e}}{\sinh \sqrt{sRC_e}} \tag{3.76}$$

$$g_t(t) = 1 + 2\sum_{k=1}^{\infty}(-1)^k \exp\left(-\frac{k^2\pi^2}{RC_e}t\right) \tag{3.77}$$

where again

$$k = 1, 2, 3, \ldots, \infty.$$

Equation (3.76) can be used to calculate the bandwidth f_B from the amplitude frequency response $|g_t(s)|$, if $|g_t(s)| = 1/\sqrt{2}$. The evaluation shows the simple relationship

$$f_B = \frac{1.46}{RC_e}. \tag{3.78}$$

Similarly, the response time T^0 can be computed applying eqn (3.51) to eqn (3.77). The result gives

$$T^0 = \frac{RC_e}{6} \approx T. \tag{3.79}$$

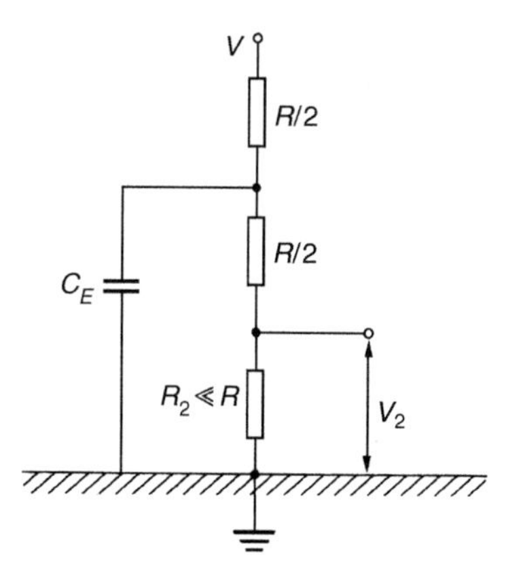

Figure 3.33 *Common equivalent circuit representing approximately the distributed parameter circuit, Fig. 3.32, with $L = C_p = 0$. $C_E = \left(\frac{2}{3}\right) C_e$ for equal response times (eqn (3.80)). $C_E = 0.44\,C_e$ for equal bandwidth (eqn (3.81))*

Although the USR starts continuously, since for $t = 0$; $dg_t/dt = 0$, a very pronounced time delay τ_{de} cannot be defined. Thus $T^0 \approx T$. f_B and T^0 could be used to define much simpler equivalent circuits for the distributed parameter network. Figure 3.33 shows this very common equivalent circuit. For $R_2 \ll R_1$ the USR is obviously

$$g_t(t) = 1 - \exp(-t/\tau);$$

where $\tau = RC_E/4$. Since for this truly exponential rise the response time equals to τ, the not distributed capacitance to ground C_E in this equivalent circuit is

$$T^0 = \frac{RC_e}{6} = \frac{RC_E}{4}; \quad \rightarrow C_E = \frac{2}{3}C_e, \tag{3.80}$$

if equal response times are used for comparison. Comparing, however, the bandwidth of both systems, which is equivalent to $f_B = 1/2\pi\tau$ for the simplified circuit, we obtain

$$\frac{4}{2\pi RC_E} = \frac{1.46}{RC_e}; \quad \rightarrow C_E = 0.44C_e. \tag{3.81}$$

The reasons for these discrepancies can easily be detected if the real unit step response according to eqn (3.77) is compared with a true exponential rise provided by the simplified equivalent circuit (Fig. 3.33). This comparison is shown in Fig. 3.34 for equal response times. The delayed, but faster, increase of $g_t(t)$ for the distributed circuit is the main reason for the discrepancies.

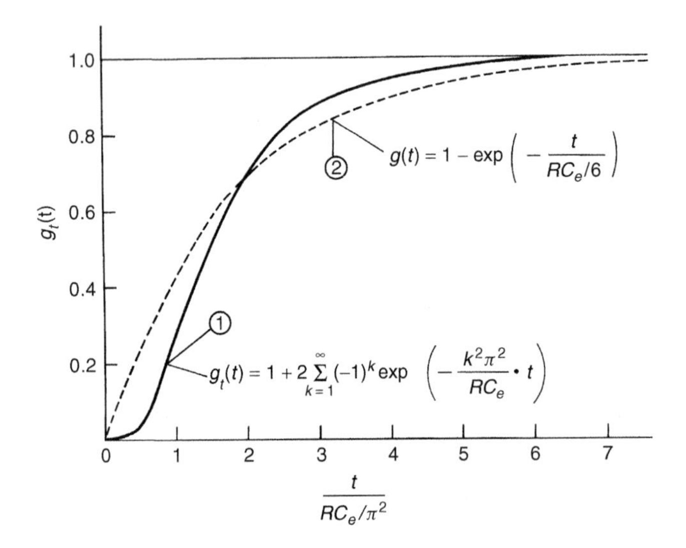

Figure 3.34 *Comparison of the unit step responses with equal response time. (1) For equivalent circuit Fig. 3.32 with* $L = C_p = 0$ *(eqn (3.77)). (2) For equivalent circuit Fig. 3.33 with* $C_E = \left(\frac{2}{3}\right) C_e$

In section 3.5.4 it was shown that the stray capacitances C'_e may approximately be calculated from the dimensions of any structure. In section 3.3 we have also given some guidance for the dimensioning of the resistor values for d.c. or a.c. dividers. Combining both these rules, we may summarize that

$$\frac{C_e}{[\text{pF}]} \approx (10-15)\frac{H}{[\text{m}]}; \quad \frac{R}{[\text{G}\Omega]} \approx (1-2)\frac{V}{[\text{MV}]};$$

where H equals the height of a divider, and V is the rated high voltage. We may introduce these magnitudes into eqn (3.78) and find the following simple relationship:

$$f_B \approx \frac{50\ldots150}{HV} \quad \text{with} \begin{cases} f_B \text{ in Hz} \\ H \text{ in m.} \\ V \text{ in MV} \end{cases} \tag{3.82}$$

Assuming a d.c. voltage divider for $V = 1\,\text{MV}$, which will be about 3 m in height, eqn (3.81) shows a bandwidth of not more than 50 Hz. It is, therefore, impossible to measure d.c. ripple voltages with high-value resistor dividers for voltages higher than some 100 kV. Equation (3.82) also shows the limitations for the application of such dividers without preventive measures: an accurate measurement of power frequency voltages needs f_B values $\gtrapprox 1\,\text{kHz}$, resulting in a product HV of about 100 kV m. This product limits the application of the above to voltages not exceeding 100–200 kV.

The measurement of lightning or even switching impulse voltages demands a much higher bandwidth as already discussed in section 3.6.2. The decrease of C_e by very carefully adapted 'shielding' or potential grading methods is limited, although a reduction by a factor of about 5–10 seems possible. But this is not enough. There is only one practical solution, i.e. to reduce the value of R by some orders of magnitude. Let us assume that we have to build a resistor divider with $T \approx T^0 = 50\,\text{nsec}$, still introducing an amplitude error δ of 10 per cent for linearly rising voltages chopped at $T_c = 0.5\,\mu\text{sec}$ (see eqn (3.53)). Thus the product RC_e becomes 300 nsec according to eqn (3.79). Let the resistance be about 2 m in height, providing a lightning impulse withstand strength of about 1000 kV. Without excessively large top electrodes for forced shielding, C'_e is about 10 pF/m and thus $R \approx 300 \times 10^{-9}/20 \times 10^{-12} = 15\,\text{k}\Omega$. This is indeed the order of magnitude which can be used for voltage dividers applicable for the measurement of lightning impulse voltages. This low value of a resistance will load the impulse generators, but this resistive load is tolerable if the discharge resistors within the generator are adapted. A large increase of the rated voltage is, however, not possible. The reduction of C_e by huge shielding electrodes becomes impractical as the dimensions must increase with the divider's height. Thus the response time with the resistance value unchanged increases proportional to C_e or the product HC'_e. Response times larger than 200 μsec for the measurement of full standard 1.2/50 lightning impulses, also chopped on the crest or the tail, have, however, not been accepted by the former standards[(6)] and the newest IEC Standard[(53)] sets even more stringent requirements, which shall not be discussed here. A further problem is created by the heat dissipation within the resistors. For constant R values and increasing voltage, the energy dissipated in the resistive materials increases proportionally with V^2, and during the short time of voltage application not much heat can be transferred to the surrounding insulation material, the energy must be stored within the resistor. A calculation of the temperature increase within the wire-wound metal resistors would indicate the difficulties of achieving low-inductive resistor units applicable to this h.v. stress. These are the main reasons why resistor voltage dividers for voltages higher than 1.5–2 MV and resistance values of 10–20 kΩ cannot be built.

There are, however, some possibilities to improve the unit step response of such dividers, which will only be treated briefly.

Reduction of resistance value. If only front-chopped impulse voltages of short duration (≤1 μsec) have to be measured, a further reduction of R is possible if the impulse generator has high stored energy and the waveshaping front resistors (R_1 in Fig. 2.26) are of low value. The heat dissipation problem is then solved only by the chopping. It is essential, however, to reduce the inductive time constant L/R of the resistors as far as possible. For assessment, we have to refer to the equivalent circuit, shown in Fig. 3.32, and the relevant transfer properties. The numerical evaluation of eqn (3.75), an example of which is given in Fig. 3.35, shows the appearance of oscillations in the USR with too low resistance values, although L/R was kept constant as well as C_e and C_p. The reasons for this instability can easily be explained using eqn (3.75). Although the damping factor $\exp(-at)$ of the infinite series remains constant, the hyperbolic functions will change to trigonometric ones, depending upon the series number k. The most efficient term within the series is the first one ($k = 1$). For this term, the transition takes place if b_k becomes complex. Hence,

$$R_{\text{crit}} \approx R \leq 2\pi \sqrt{\frac{L}{C_e} \frac{1}{1 + \pi^2 C_p/C_e}} \tag{3.83}$$

This 'critical' resistance R_{crit} is included in the table of Fig. 3.35, and the comparison with the computed responses confirms the validity of the above equation.

Typical examples for such low-resistor voltage dividers are shown by Rohlfs *et al.*[(63)] or Pellinen *et al.*[(64)]

Reduction of C_e. The possibility of reducing the stray capacitance to earth by metal electrodes connected to h.v. potential was theoretically treated in section 3.3. The practical application of field-controlling top electrodes was

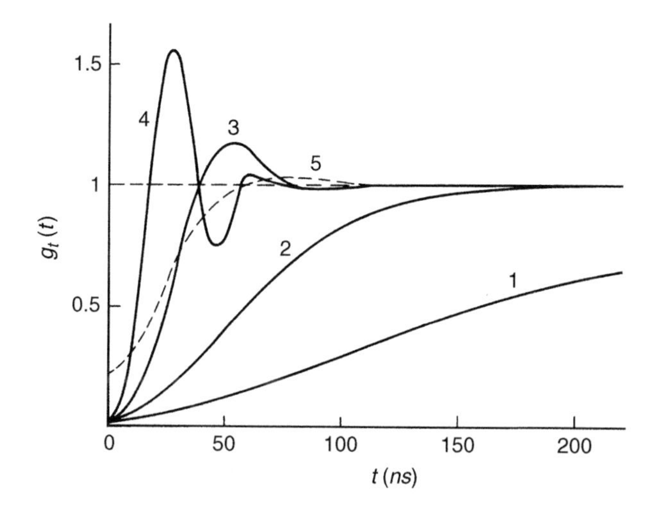

Figure 3.35 *Calculated unit step response for resistor dividers. Equivalent circuit according to Fig. 3.32*

$L/R = 10$ nsec;	$C_e = 40$ pF;	$C_p = 1$ pF;	R_{crit}
(1) $R = 30\,k\Omega$			$15.5\,k\Omega$
(2) $R = 10\,k\Omega$			$8.9\,k\Omega$
(3) $R = 3\,k\Omega$			$4.85\,k\Omega$
(4) $R = 1\,k\Omega$			$2.8\,k\Omega$
$L/R = 10$ nsec;	$C_e = 12$ pF;	$C_p = 1$ pF;	
(5) $R = 10\,k\Omega$			$13.4\,k\Omega$

introduced by Bellaschi,[65] it is a widely used and effective method. The combination of a field-controlling h.v. electrode with a non-linear distribution of the resistance values per unit length was also explained earlier.[34] The inherent disadvantages of all field-controlling methods are twofold. First, the unit step response becomes very sensitive to surrounding objects, as a strong relative change of C_e is likely to be produced by small changes of the external potential distribution. The second disadvantage is related to the interaction between the lead and the divider. Large shielding electrodes introduce a relatively large external parallel capacitance across the divider, which is not equal to C_p in our equivalent circuit. This capacitance loads the lead and enhances travelling wave oscillations, which can only be damped by the impedance Z_d of the lead. Additional explanations are given in section 3.6.5.

C_e can also be reduced by a decrease of the dimensions of the resistor. Harada *et al.*[66] proposed a 1-MV divider with $R = 9.3\,k\Omega$, the resistor of which was only 46 cm in axial length, but placed in a much longer insulating vessel. In this design difficulties arise with the heat dissipation within this small resistor and with the field gradient control in the neighbourhood of the resistor. For further details the reader should refer to the original paper.

Compensation methods. Our equivalent circuits assume an equal distribution of the voltage dividing elements in the resistor column. Also the l.v. arm is assumed to be equal to a resistor unit of the h.v. arm. This is, of course, not true, as the connection of the signal cable with the l.v. arm needs a special construction (see section 3.6.7, Fig. 3.62). For resistor dividers, the voltage USR is about equal to the step response of the current through the l.v. arm. In this way the current also increases in a manner that is similar in shape as is given by the voltage unit step response. As long as R_2 (Fig. 3.32) is not larger than the surge impedance of the signal cable, one may simply increase the inductance L_2 to increase the resistance of the output voltage. The low value of the surge impedance, which is in parallel with R_2, limits the efficiency of this method. In practice, the actual value of L_2 is predominantly determined by the construction of the l.v. arm. The actual USR may, therefore, be quite different from the computed one. Other compensating networks at the input end of the signal cable have been proposed[67] which can be evaluated using the well-known methods of network synthesis.[1] The efficiency of such networks is, however, quite limited.

Parallel-mixed resistor-capacitor dividers

If in the equivalent circuit for resistor dividers of Fig. 3.32 the stray capacitances C'_p are increased, i.e. if real capacitor units are placed in parallel to the resistor R', a 'parallel-mixed resistor–capacitor divider' is formed. This parallel arrangement of resistors and capacitors is a well-known technique used for attenuators within measuring instruments, i.e. CROs, and is often referred to as a compensated resistor voltage divider. The idea to use this circuit for h.v. dividers was introduced by Elsner in 1939,[68] with the goal of reducing the effect of the stray capacitances to earth, C'_e. The efficiency of the C'_p capacitors can actually be seen by comparing unit step responses of Fig. 3.34, curve 1, with those in Fig. 3.35. Neglecting any C_p values within the simplified R-C latter network causes the USR to start continuously with time. Even the small C_p value of 1 pF in Fig. 3.35 excites a small step in the USR, and the value of this step $g_t(+0)$ is obviously dependent upon the capacitance ratio C_p/C_e (compare the curves 1 to 4 with 5). The increase in the ratio of C_p/C_e increases this step and thus the question arises whether it is possible to increase this first step to the final value.

This can be accomplished theoretically only if we assume that the representation of actual capacitor units placed in parallel to the resistors in the equivalent circuit of Fig. 3.32 is correct. It is, however, not correct if this circuit is used to compute high-frequency phenomena or unit step responses in the nanosecond or even microsecond range. The reason for this is simple. The inherent inductance L of every capacitor C causes a series resonance frequency $f_r = 1/2\pi\sqrt{LC}$, which is quite low for capacitance values capable to compensate h.v. dividers (for instance: ($f_r = 10\,\text{MHz}$ for $L = 1\,\mu\text{H}$; $C = 200\,\text{pF}$).

The actual USR of parallel-mixed resistor–capacitor dividers is therefore similar to pure capacitor voltage dividers, which will be treated later. Apart from the fact that this type of divider is still in use for the measurement of impulse voltages, with R values in the 10–100 kΩ range and C_p values in the order of some 100 pF, we shall simulate the transfer properties by a simplified equivalent circuit only, which will *not* cover the high-frequency range. This equivalent circuit is shown in Fig. 3.36.

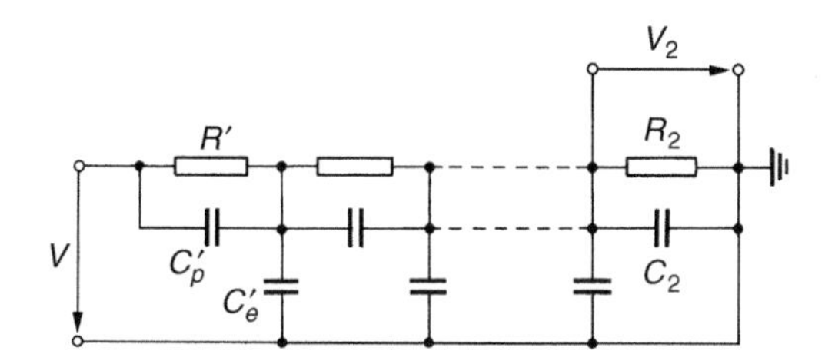

Figure 3.36 *Simplified equivalent circuit for parallel-mixed resistor–capacitor dividers.* $R = nR'$; $C_p = C'_p/n$; $C_e = nC'_e$; $R_2 = R'$; $C_2 = C'_p$

The computation of the normalized transfer function and unit step response yields for equal elements in the h.v. and l.v. arms, i.e. $R_2C_2 = R'C'_p$:

$$h_t(s) = n\frac{\sinh \dfrac{1}{n}\sqrt{\dfrac{sRC_e}{1 + sRC_p}}}{\sinh\sqrt{\dfrac{sRC_e}{1 + sRC_p}}} \tag{3.84}$$

$$g_t(t) = 1 + 2\sum_{k=1}^{\infty}(-1)^k \frac{\exp(-a_k t)}{1 + k^2\pi^2 C_p/C_e} \tag{3.85}$$

where

$$a_k = \frac{k^2\pi^2}{RC_e(1 + k^2\pi^2 C_p/C_e)};$$
$$k = 1, 2, 3, \ldots$$

The peculiar effect of this circuit is detected by the calculation of the limiting values for very high and very low frequencies, or very short and very long times:

$$\lim_{s\to 0}[h_t(s)] = 1; \quad \lim_{t\to\infty}[g_t(t)] = 1.$$

But

$$\lim_{s\to\infty}[h_t(s)] \cong 1 - \frac{C_e}{6C_p}; \quad \lim_{t\to 0}[g_t(t)] \cong 1 - \frac{C_e}{6C_p}.$$

A sketch of the normalized amplitude frequency response and USR in Fig. 3.37 demonstrates the response of this dividing system to different voltage ratios. The difference of these ratios is formed by the relation $C_e/6C_p$, and very high values of C_p would be necessary to reduce this difference to very small values. It is obvious that these differences in scale factors can be reduced by a reduction of C_2 within the l.v. arm to increase the voltage drop across C_2 for high frequencies. A calculation, published by Harada *et al.*,(69) shows the condition

$$R_2C_2 = R_1C_{p1}\left\{\sqrt{\frac{C_e}{C_{p1}}}\frac{1}{\sinh\sqrt{\frac{C_e}{C_{p1}}}}\right\} \approx R_1C_{p1}\left(1 - \frac{C_e}{6C_p}\right) \tag{3.86}$$

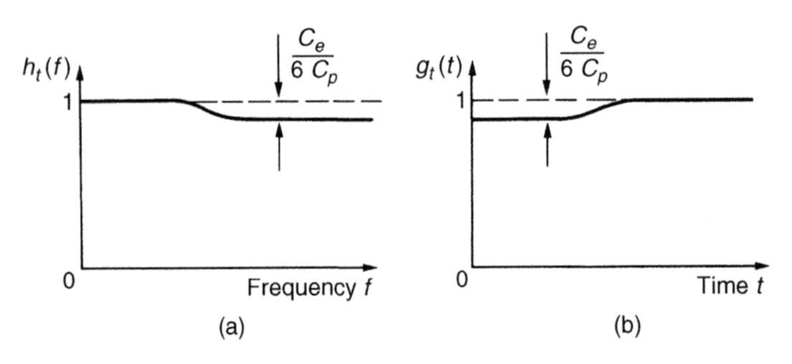

Figure 3.37 *Schematic diagrams for the normalized amplitude frequency response: (a) and unit step response, (b) for voltage dividers according to Fig. 3.36*

where

$$C_{p1} = \frac{C'_p}{(n-1)} \approx \frac{C'_p}{n} = C_p;$$

$$R_1 = (n-1)R' \approx nR' = R.$$

In summary then, it is *not* recommended to compensate resistor dividers for high impulse voltages with parallel capacitor units, as the equivalent circuit of Fig. 3.36 is inadequate to treat short-time phenomena. A compensation of high ohmic dividers commonly used for the measurement of d.c. or a.c. voltages, however, is very attractive to increase the performance in the intermediate frequency range (100 Hz up to some 100 kHz, depending upon the size of the divider).

Capacitor voltage dividers

It was shown in section 3.5.4 that pure capacitor voltage dividers could be made either by using single h.v. capacitance units, i.e. a compressed gas capacitor, in series with a l.v. capacitor, or by applying many stacked and series connected capacitor units to form an h.v. capacitor. The absence of any stray capacitance to earth with compressed gas capacitors provides a very well-defined h.v. capacitance, small in value and small in dimensions, and by this even a pure capacitor voltage divider with quite good high-frequency performance can be built if the l.v. arm or capacitor is constructively integrated in the layout of such a capacitor. This means that this capacitor must be very close to the h.v. capacitance, and this can be provided for instance by inserting a symmetrical arrangement of l.v. capacitors between the l.v. sensory electrode 2 and the guard ring 2′ or supporting tube 3 (see Fig. 3.18). Although such a construction was proposed by Schwab and Pagel,[70] similar systems may well be formed by other coaxial arrangements.[71] The applicability to very high voltages, however, is mainly restricted by the high cost of such constructions and the difficulties involved with the replacement and exchange of l.v. arms to change the scale factors.

A treatment of capacitor voltage dividers with stacked capacitor units is thus justified. The distributed parameter network is able to simulate the transfer properties. Figure 3.38 shows such a network, which may encounter

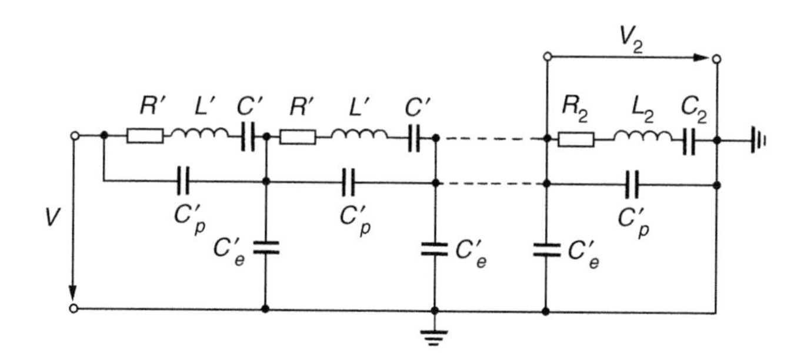

Figure 3.38 *Equivalent circuit for capacitor voltage dividers.* $R = nR'$; $L = nL'$; $C_e = nC'_e$; $C = C'/n$; $C_p = C'_p/n$; $R_2 = R'$; $L_2 = L'$; $C_2 = C'$

all possible passive circuit elements. The actual stacked capacitors are now simulated by the capacitance units C', and L takes into account the inherent inductance. The series resistance R' may be used to simulate either only small losses within the capacitor units C', or even real resistors in series with these units. The small values of stray capacitances in parallel to the stacked columns C'_p and to ground C'_e complete the equivalent circuit.

A glance at the unit step response, which is represented as

$$g_t(t) = 1 - \frac{C_e}{6(C + C_p)} + 2\exp(-at)\sum_{k=1}^{\infty}(-1)^k \frac{\cosh(b_k t) + \dfrac{a}{b_k}\sinh(b_k t)}{AB},$$

where

$$A = \left(1 + \frac{C_p}{C} + \frac{C_e}{Ck^2\pi^2}\right), \quad a = \frac{R}{2L},$$

$$B = \left(1 + \frac{C_p k^2 \pi^2}{C_e}\right), \quad b_k = \sqrt{\frac{k^2\pi^2 \cdot A}{LC_e B}}, \tag{3.87}$$

shows a close similarity to the USR of resistor dividers, eqn (3.75). Both equations are actually the same, if the value C in eqn (3.87) approaches infinite values. With finite values of C, representing capacitor voltage dividers, the main difference is at first related to the negative term $C_e/6(C + C_p) \cong C_e/6C$, which is independent of the time and thus also the frequency. This term was also found in the treatment of the 'equivalent capacitance', see eqn (3.31). It appears again as a result of our procedure of the normalization of the USR. All explanations referring to the proper dimensioning of stacked capacitors, therefore, also apply to this result, which demonstrates the possible variations of the ratio n with C_e.

The time dependency of the USR for 'pure' capacitor dividers, i.e. with $R = 0$ in the equivalent circuit, is obviously very complex. In eqn (3.87), with $R = 0$, the damping term $\exp(-at)$ will be equal to 1, and all hyperbolic functions are converted to trigonometric ones. The numerical evaluation of this equation for this case is impossible due to the infinite number of sinusoidal terms. It is also not realistic to assume no resistance at all, as at least the (frequency-dependent) dissipation factor of the dielectric will cause some damping. For a simple series equivalent of a lossy capacitor, this dissipation factor is $\tan\delta = \omega R'C' = \omega RC$. The relaxation phenomena within the dielectric materials, however, control in reality this dissipation factor for high frequencies. It has been confirmed by measurements[(72)] that an adequate low-resistance value can be assumed to evaluate eqn (3.87) as was done in Fig. 3.39. Here, the oscillations can be related to the travel time $\tau = \sqrt{LC_e}$, as a step voltage applied to the input of such a ladder network can travel along

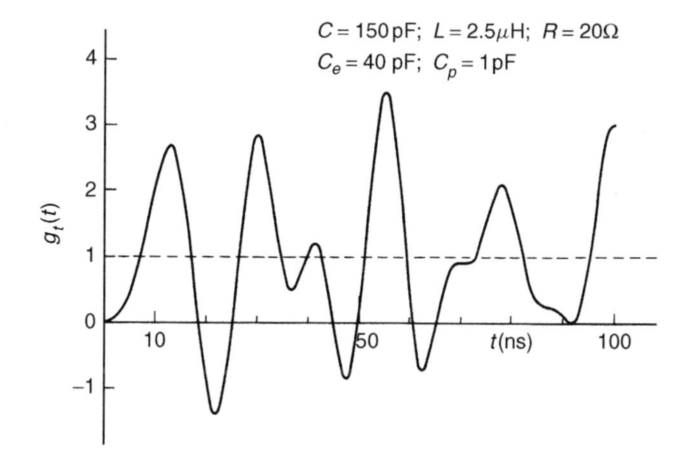

Figure 3.39 *Calculated unit step response for a capacitor voltage divider; the equivalent circuit is Fig. 3.38.* $R = 20\ \Omega$; $L = 2.5\ \mu H$; $C = 150\ pF$; $C_e = 40\ pF$; $C_p = 1\ pF$

the column. If the voltage amplitude is not reduced to a zero value when the wave reaches the earthed l.v. part, it will be reflected and excites oscillations.

Pure capacitor voltage dividers are therefore sensitive to input voltages with short rise times and the output voltage may oscillate with non-oscillating input voltages. In addition, such a capacitance divider within the whole measuring circuit, i.e. with leads connected to its input, will form a series resonant circuit. Thus it is obvious that pure capacitor dividers are not adequate to measure impulse voltages with a steep front (front-chopped lightning impulse voltages) or any highly transient phenomena (voltage during chopping). Crest values of switching impulse or even full lightning impulse voltages, however, can be properly recorded, if the transient phenomena during the front of the impulses have disappeared.

The similarity of the step response equations for resistor voltage dividers to those treated in this part stimulated Zaengl to propose and to realize the possible improvement of pure capacitor dividers by inserting real resistor units in series with the capacitors.[72] If the value of these resistors is not too high, but just sufficient to damp the oscillations, it is likely to achieve an excellent transient performance. A very similar equation to that of eqn (3.83) could be derived by calculating the transition from hyperbolic to trigonometric functions for the argument b_k and $k = 1$ in eqn (3.87), providing again a critical resistance. Such a critical value can also be identified in Fig. 3.40, in which some calculated step responses according to eqn (3.87) are shown. Adequate values for a capacitor voltage divider for a voltage of about 1 MV (height ≈ 3 m) are used for this simulation.

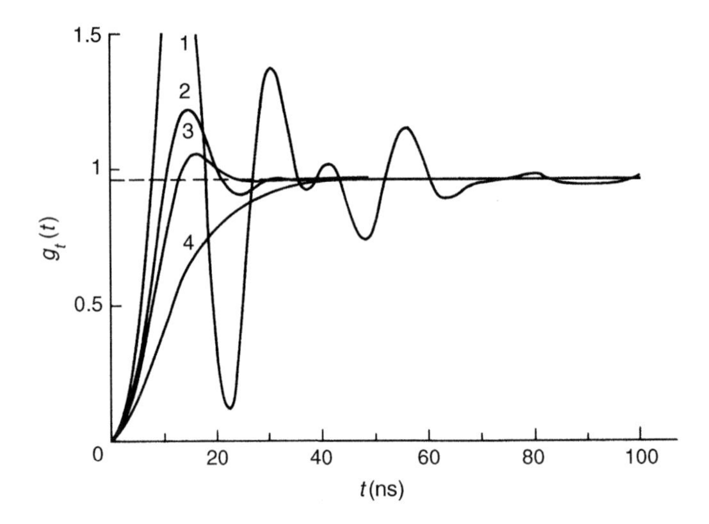

Figure 3.40 *Computed unit step response $G_t(t)$ for damped capacitor dividers according to equivalent circuit, Fig. 3.38*

$C = 150$ pF; $L = 2.5$ μH; $C_e = 40$ pF ; $C_p = 1$ pF

(1) $R = 250\ \Omega$

(2) $R = 750\ \Omega$ $\qquad 4\sqrt{\frac{L}{C_e}} = 1000\ \Omega$

(3) $R = 1000\ \Omega$

(4) $R = 2000\ \Omega$

The influence of the magnitude of the resistance R is obvious. A very well-damped response is reached by a resistance value of about

$$R \approx 4\sqrt{\frac{L}{C_e}} \tag{3.88}$$

although the larger overshoot observed with lower values can still be accepted. The short response time is in accordance with the theory. T^0 or T can be calculated by the transfer function as well as from eqn (3.87). It is equivalent to eqn (3.79), and thus $T^0 = RC_e/6$. The small resistor values as necessary to damp the oscillations are responsible for these low response times, and thus a '*series-damped capacitor divider*' is formed. The input impedance of these dividers increases with decreasing frequencies, and hence the loading effect of the voltage generating system is limited. Their application for a.c., switching or lightning impulse voltages without any restrictions is, therefore, possible.

If a parallel branch of high ohmic resistors is added, d.c. voltages can also be measured as shown before and an 'universal voltage divider' is formed.

These 'series-damped capacitor dividers' are not limited in the voltage range, as a stacking of capacitor units is not limited as well as the insertion of distributed resistors. These resistors are not heavily loaded, as only transient input voltages cause displacement currents. A 'general-purpose' voltage divider is therefore available, and have been in general use since about 1970 up to the highest voltage levels.[54,122]

Figure 3.41(a) shows such a voltage divider for a lightning impulse voltage of 6 MV. The electrodes are not provided to shield the divider, i.e. to reduce C_e, but only to prevent discharges and thus to increase the flashover voltage for switching impulses.

3.6.5 Interaction between voltage divider and its lead

The analytical treatment of our measuring system presented so far is not yet complete. Whereas the USR of the voltage dividers could readily be calculated, similar results are missing for the entire circuit. Now it can be shown that the generalized expression for the response time T and its interaction with the circuit elements, eqn (3.66), can effectively be applied in practice.

As already mentioned in section 3.6.3, it is too difficult to apply an analytical solution to the USR of the whole measuring system, which was represented by the 'three-component system' of Fig. 3.29. Numerical solutions by advanced programming, however, are possible, and many computer programs are available. The results presented here are calculated with the 'transient network program' published by Dommel.[73] Within this program, the lossless transmission line (see 2, Fig. 3.29) is simulated by the exact solution of the partial differential equations of a line and thus does not introduce any errors. The simulation of the terminating system, i.e. the voltage dividers, needs, however, a subdivision of the distributed parameter networks into a finite number of sections. If the number of elements n (for n see Fig. 3.31) is larger than about 5, the results are close to the infinite number solution.

Numerical computations need numerical values for the surge impedance of the lead Z_L to the divider. For the common set-up of a voltage testing system (Fig. 3.22), this lead is more or less horizontal above the ground return, which is assumed to be an extended plane. Many experiments[50] demonstrated that the travel time τ_L is controlled by the velocity of light c_0. As $Z_L = \sqrt{L_L/C_L}$ and $\tau_L = \sqrt{L_L C_L} = l/c_0$, with L_L being the total inductance and C_L the total capacitance of this lead, $Z_L = l/c_0 C_L$, with l being the length of the lead. The capacitance of the lead can be computed assuming that a cylindrical lead of diameter d is at height H above a plane, which is earthed. The well-known

Figure 3.41 *Series-damped capacitor voltage divider for 6-MV impulse voltage (courtesy EdF, Les Renardieres, France)*

capacitance formula

$$C_L = \frac{2\pi\varepsilon_0 l}{A};$$

where

$$A = \ln\left[\frac{2l}{d}\sqrt{\frac{\sqrt{\{1+(2H/l)^2\}}-1}{\sqrt{\{1+(2H/l)^2\}}+1}}\right]$$
$$= \ln\left(\frac{4H}{d}\right) - \ln\frac{1}{2}(1+\sqrt{1+2(H/l)^2})$$

may well be used, although this lead is placed between the test object and the voltage divider. As $c_0 = (\varepsilon_0\mu_0)^{-0.5}$, where ε_0 = permittivity and μ_0 = permeability of free space, the surge impedance becomes

$$(Z_L)_{\text{hor}} = A\frac{1}{2\pi}\sqrt{\frac{\mu_0}{\varepsilon_0}} = 60 \times A(l, d, H) \quad [\Omega] \tag{3.89}$$

for this *horizontal* lead. Sometimes, the horizontal lead is lengthened by a vertical lead to measure the experimental USR of the system. Thus we need Z_L for a vertical lead also. According to Fig. 3.22 and eqn (3.33), this capacitance is known. With the same assumptions as made above, we obtain

$$(Z_L)_{\text{vert}} = \frac{1}{2\pi}\sqrt{\frac{\mu_0}{\varepsilon_0}}\ln\left[\frac{2l}{d}\sqrt{\frac{4s+l}{4s+3l}}\right]$$
$$\approx 60\ln\left(\frac{1.15l}{d}\right) \quad [\Omega] \quad \text{for} \quad s \ll l. \tag{3.90}$$

The differences in the surge impedances are not large if the usual dimensions are taken into account.

In Fig. 3.42(a), a very simplified equivalent circuit represents a 20-kΩ resistor divider with a lead length of 3 m (τ_L = 10 ns). The divider is idealized by the omission of any stray capacitances or inductances, but a parallel capacitance of C_t = 50 pF across the whole divider represents a top electrode which may shield the divider. A pure resistor R_d provides ideal damping conditions for travelling waves. Figure 3.42(b) shows some computed results of the USR. For $R_d = 0$, no noticeable damping effect is observed within the exposed time scale. Although the oscillations are non-sinusoidal, the fundamental frequency can clearly be seen. This frequency is obviously close to the resonance frequency f_r, generated by the lead inductance L_L and the divider's capacitance C_t. As $L_L = Z_L\tau_L$, this inductance is 3 μH, giving f_r = 13 MHz.

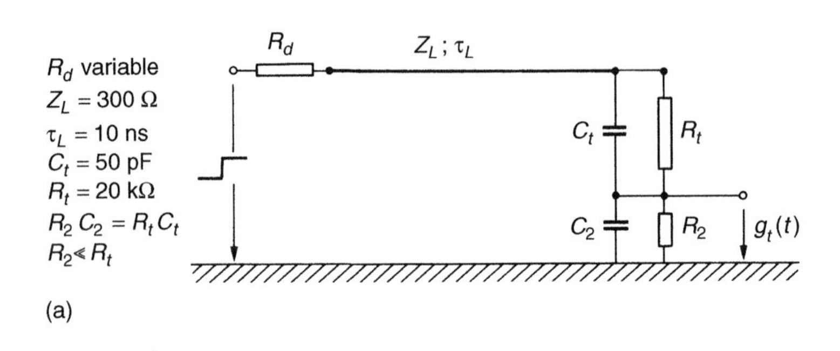

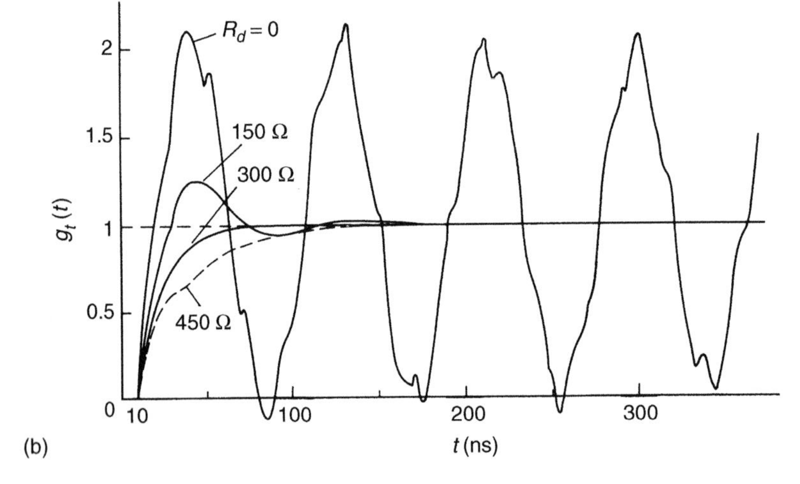

Figure 3.42 *Computed unit step response for idealized resistor or parallel-mixed resistor–capacitor divider with lead. (a) Equivalent circuit. (b) Computed USR*

Thus, this example also implies the typical USR for pure capacitor or parallel-mixed resistor–capacitor voltage dividers, as C_t can well be assumed to represent these types of dividers. Higher values of C_t will decrease the frequency of the oscillations. Acceptable responses are only provided by a damped lead. To prevent any overshoot, R_d must be close to Z_L. The exponential increase in the front and increase of the response time T^0 or $T = T^0 - \tau_L$ is obviously produced by the time constant $R_d C_t$, which equals 15 ns for $R_d = Z_L$ and the specific values assumed. Large capacitor dividers with stacked capacitor units comprise in general much higher capacitance values, and in such cases the large response time of such measuring systems is produced by the necessary damping of the lead.

We may easily compute T^0 or the actual response time $T = T^0 - \tau_L$ from eqns (3.66) to (3.71). It is clear that for this ideal divider $T_t = 0$ and $T_{kd} = 0$ (no frequency dependency of $Z_d(s) = R_d$). With the only frequency-dependent term for the input impedance of the divider $Z_t(s) = R_t/(l + sR_tC_t)$, we may easily find that $T_{kt} = R_tC_t/(1 + Z_L/R_t)$. The final result may be best represented in the form

$$T = T^0 - \tau_L = \frac{1}{(1 + R_d/R_t)}\left[R_dC_t - \tau_L\left(1 - \frac{Z_L}{R_t}\right)\left(1 - \frac{R_d}{Z_L}\right)\right]. \quad (3.91)$$

Some remarkable findings can be observed.

For $R_d = Z_L$, the length of the lead has *no* influence upon the response time. This case corresponds to the 'infinite line response', as the same result would be achieved if a step voltage supplied from an extremely long lead would be applied to the dividing system.

With no damping resistance, or $R_d < Z_L$, the response time taken from the actual beginning of the USR will always decrease proportionally with the lead length $l = \tau_L c_0$. This decrease of T is clearly produced by an overshoot of the USR. As is seen from the computed USR, the determining factor is R_dC_t providing a positive contribution to T. For capacitor dividers, $R_t \to \infty$ and the same equation can be applied.

A second example (Fig. 3.43) simulates a pure resistor divider of low resistance value (2.32 kΩ), which was in reality built from carbon composition resistors to achieve extremely low values of inductances. The stray inductances are therefore neglected in the equivalent circuit (Fig. 3.43(a)), but it comprises distributed stray capacitances to earth, which have been calculated with eqn (3.33). The small input capacitance (5 pF) was estimated as only a very small top electrode was provided. The voltage divider was used for steep-front voltage measurements up to 800 kV. The lead length of 6 m was used for USR measurements only, and the equivalent circuit simulates this lead length. The computed USR (Fig. 3.43(b)) shows again larger oscillations with no damping resistance in the lead. The traces of the oscillations deviate strongly from the USR of a pure resistor network (see Fig. 3.30), due to the stray capacitances involved. Only the 'infinite line response' is smooth. For $R_d = 100\,\Omega$, the computed input voltage of the divider is also plotted to show the distortion introduced by the divider. The small capacitive reflection is mostly suppressed by the divider.

For this equivalent circuit, again the general dependency of the response time from the circuit parameters can be computed by eqn (3.66). The result is

$$T = T^0 - \tau_L = \frac{1}{1 + (R_d/R_t)} \times \left[\frac{R_tC_e}{6} + R_d\left(C_p + \frac{C_e}{2}\right) - \tau_L\left(1 - \frac{Z_L}{R_t}\right)\left(1 - \frac{R_d}{Z_L}\right)\right]. \quad (3.92)$$

Some interesting findings are observed.

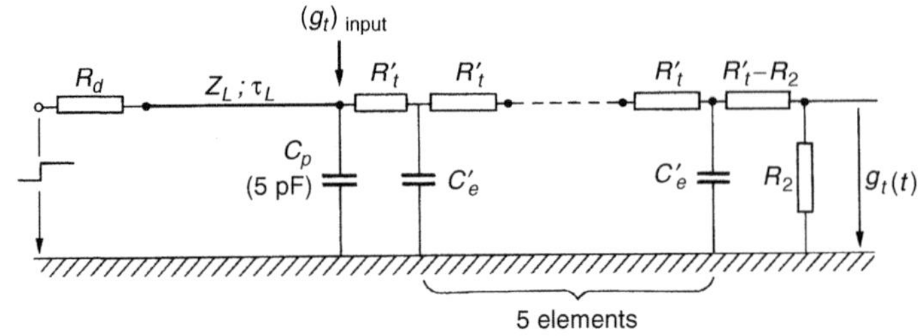

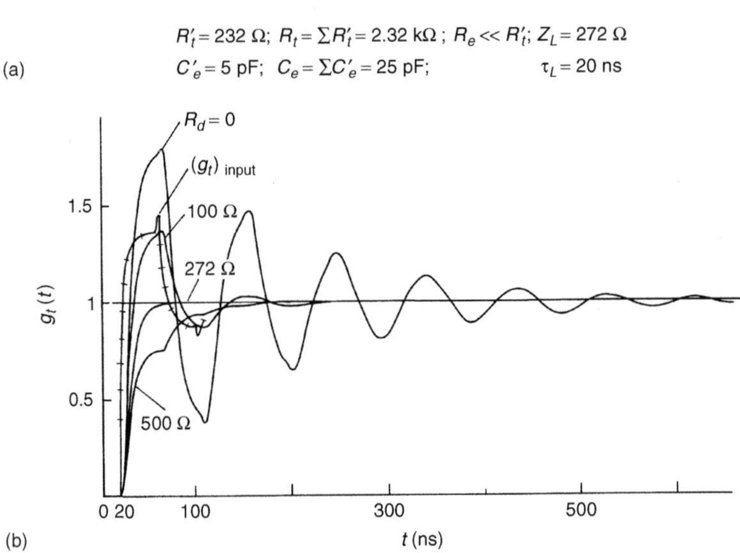

Figure 3.43 *Computed USR for low-value resistor voltage divider. (a) Equivalent circuit. (b) Computed USR (for divider input)*

The influence of the lead on the divider (τ_L) is the same as before. Now the the divider's response time $(R_t C_e/6) = T_t$ appears, as expected. Only a part of the stray capacitance C_e, but the full value of the input capacitance C_p, provides positive response times if the lead is damped.

Figure 3.44 shows oscillograms of measured responses. The lead was placed parallel to the ground and the unit step voltage generator was mounted at the wall of the laboratory, which was shielded by a Faraday cage. There is a very good agreement between the computed and measured values, the USR and the response time.

Finally, the third example (Fig. 3.45) explains the existence of a real time delay between the output and input voltage of a resistor voltage divider. This example is similar to the first one, but the resistors are distributed and comprise a small inductive time constant of $L_t/R_t = 5$ ns. Stray capacitances in parallel to each section, however, are neglected. The USR of the output

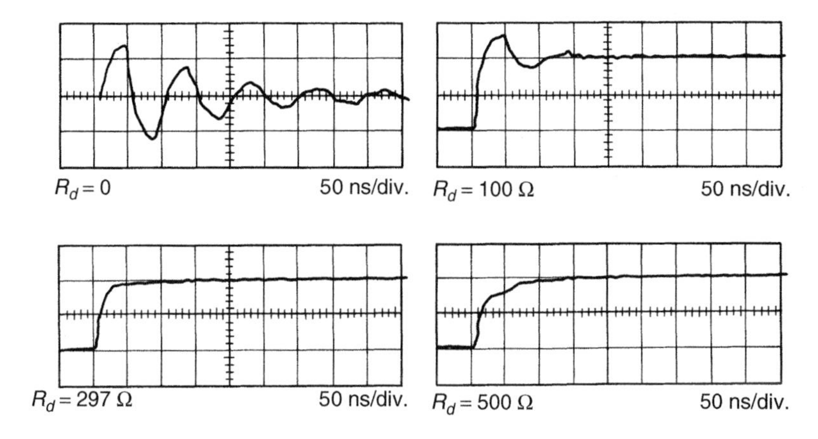

Figure 3.44 *Measured unit step response for the resistor voltage divider. $R = 2320\ \Omega$, with 6-m lead, according to Fig. 3.43*

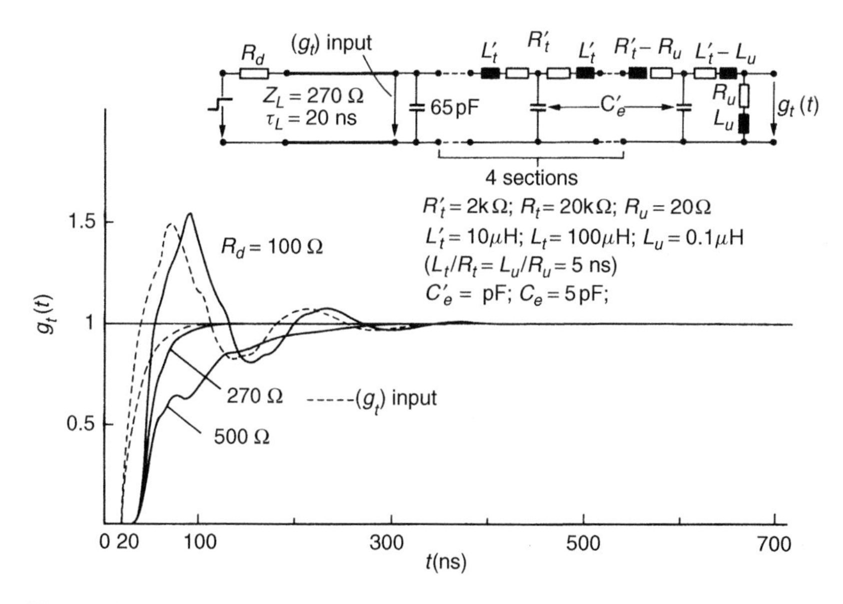

Figure 3.45 *Computed USR of resistor divider with inductance*

voltage now starts with a time delay of about 15–20 ns related to the input voltage. This delay is caused by the travel time of the divider, $\tau_t = \sqrt{L_t C_e}$, which is about 22 ns. A stray capacitance C'_p would only theoretically suppress this travel time, as was shown in reference 50. The very small C_e values assumed in this example should simulate a very good shielded divider. This rise time of the response is accordingly quite short. If the response time is calculated as before, an additional small negative term within the brackets (eqn (3.92)) would appear, being $-R_d L_t/(R_t)^2$. Whereas this term is negligable, the additional time delay decreases the computed response time of the dividers, $R_t C_e/6$. This effect may thus be taken into account by a reduction of this value by a factor of 3/4, and for this some standards[6] recommend a theoretical contribution of the response time for resistor dividers according to $R_t C_e/8$.

Although many other investigations[54] have confirmed the validity of such analytical treatments, a theoretical treatment alone will not satisfy the actual needs. In practice, an experimentally performed step response measurement has to prove the performance of the systems used, as insufficient constructions may show worse results. There are still some unsolved problems inherent to response measurements, but they shall not be treated here in detail. The most difficult problem is related to the fact that the actual 'input terminals' of our voltage measuring system (Fig. 3.22) are separated by a long distance due to the dimensions of the test object. But the existence of a step voltage with a rise time of one nanosecond or less cannot be defined by an electrostatic potential difference between points the distance of which is in the metre-range, due to the basic theory of electromagnetic phenomena. But we still may assume that actual currents charging the test objects within a short but finite time will produce a quasi-stationary field and through it a potential difference for which the expression 'voltage' is justified.

With such restrictions in mind, which are usually neglected, the measurement of the step response belongs now to the routine procedure of an 'acceptance test' for an 'approved measuring system' as used for the measurement of lightning and even switching impulses, see reference 53. The step voltages are usually produced by small generators for some hundred volts, and some different circuit arrangements can nowadays be used to measure the step response. Although some further details can be found in the relevant IEC Standard,[53] some additional information is presented in Fig. 3.46. In this

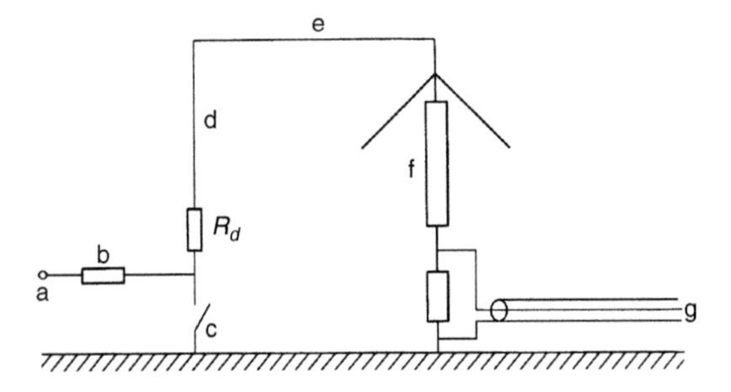

Figure 3.46 *The unit step method. (a) To d.c. supply. (b) Charging resistor. (c) Fast switch. (d) (Added) vertical lead. (e) High-voltage lead. (f) Voltage divider. (g) To recording instrument*

figure, a so-called 'square loop arrangement' is shown as recommended by the earlier standards.[6,57] The step generator a to c must have approximately zero impedance while generating the voltage step and during the subsequent response. Any fast switching device, c, which short-circuits a constant d.c. voltage as used to charge the measuring system before the short-circuit occurs is applicable. Very suitable switches are mercury-wetted relays but also a uniform field gap of about 1 mm spacing at atmospheric air or a uniform gap with a spacing up to some millimetres under increased gas pressure. The gas-insulated gaps can short-circuit after breakdown voltages up to some kilovolts, but only single events can be produced. Whereas these switches fulfil the requirement for 'zero impedance', commercial types of electronic pulse generators are inherent with some 10 ohms of internal impedance which contribute to the damping resistor, see R_d. At least a metallic strip conductor 1 m wide shall serve as the earth return between divider and the step generator. The length of the lead, represented by d and e, shall be equal to the length as used during actual impulse voltage measurement, if the response time is an essential parameter during the measurements. If only part e is used during measurements, it would be possible to calculate the relevant response time due to the theory as shown before and elsewhere.[131] To avoid such effects, the step generator shall be placed at a metallic wall. Then the usual lead length as applied during voltage tests can be used. This 'vertical lead arrangement' was introduced by one of the authors[60] and is the preferred circuit today.

One of the uncertainties of the unit step method is related to the starting point, i.e. the value of τ_{de} in Fig. 3.26, of the response. This starting point on the 'toe' region is influenced by electromagnetic waves radiated from the leads between the step generator and the divider. These phenomena have been thoroughly investigated by an International Research Group;[50] the methods for the computation of these phenomena are based upon Maxwell's equations, which can either be solved in the time domain[76] or in the frequency domain.[77] The solutions are very sophisticated and cannot be treated within this chapter. The new IEC Standard,[53] however, accounts for these effects by defining a 'virtual origin O_1' of the USR together with an 'initial distortion time T_0', the definitions of which may be found in this standard.

3.6.6 The divider's low-voltage arm

We assumed for the theory of dividers, section 3.6.4, that the low-voltage arm is an integral part of the divider and provides an impedance structure which is equivalent to the high-voltage arm. In reality, the structure, i.e. the composition of the circuit elements, is quite different. Therefore, some additional problems may appear concerning adequate construction and layout of the l.v. arm of our measuring system. Many distortions in the response can be related to this part of the system.

For d.c. and a.c. voltage dividers, the design of the l.v. arm is not critical, if only steady state voltages have to be recorded. However, if any fast transients have to be transmitted from the voltage divider to the recording instrument (see Fig. 3.23, items 5 to 7), the l.v. arm of the voltage divider itself may introduce large disturbances to the response. Let us first discuss the adequate impedance matching necessary to transmit impulse voltages from the divider to the recording instrument.

In Fig. 3.47 the somewhat simplified equivalent circuits for the matching procedures for the different types of dividers are sketched. The signal cable is mainly treated as lossless, so that the surge impedance $Z_k = \sqrt{L_k/C_k}$ becomes independent of frequency, and the travel time $\tau_k = \sqrt{L_k C_k}$ is a plain value. For resistor voltage dividers, Fig. 3.47(a), the cable matching is simply done by a pure ohmic resistance $R = Z_k$ at the end of the signal cable. The transmission line theory provides the well-known background for this procedure, the reflection coefficient becomes zero and any unit step voltage appearing across R_2 is undistorted transmitted by the cable. As the input impedance of the signal cable is $R = Z_k$, this resistance is in parallel to R_2 and forms an integral part of the divider's l.v. arm. The low-value of this resistance R, i.e. typically 50 to 75 Ω, should in fact suggest that we consider the losses of the signal cable. These losses are in reality dependent upon frequency due to the skin effect, and the response of such a cable becomes very complex. Whereas the theory of this problem may be found elsewhere,(124,125) the result of this theory shows clearly that the best matching can be achieved with R equal to the surge impedance Z_k defined for high frequencies. For all kinds of signal cables the d.c. resistance for the conductors (inner conductor and shield) will form a voltage dividing system between R_2 and R, which may decrease the voltage across R by an amount of 1 per cent in order of magnitude. As this amount can easily be taken into account by d.c. resistance measurements only, this value should be taken into account. The unit step response from a lossy cable is characterized by a steep increase within a few nanoseconds to values of more than 90 per cent and a slow tripling up to the final value. These effects will introduce larger errors if the impulses to be transmitted are shorter than 0.5 to 1 μsec. At least an experimental test is recommended to check the signal cable with regard to this additional error.

For parallel-mixed resistor-voltage dividers the same procedure for cable matching, Fig. 3.47(a), applies. A matching resistor R, coaxially designed to meet the high-frequency requirements, will not reflect energy. The input impedance of the recording instrument, however, should not comprise appreciable input capacitance, as otherwise too heavy reflections will appear. The l.v. arm for this type of divider reflects heavily due to the parallel capacitance to R_2.

For capacitor voltage dividers, Fig. 3.47(b) or (c), the signal cable cannot be matched at its end. A low ohmic resistor in parallel with C_2 would load the

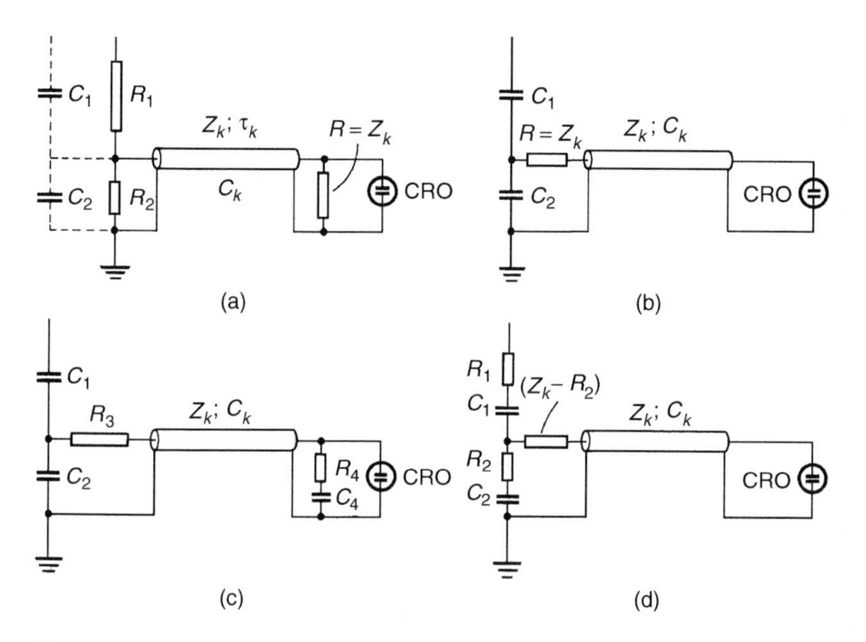

Figure 3.47 *Circuits for signal cable matching. (a) Resistor or parallel-mixed capacitor-dividers. (b) Capacitor dividers, simple matching. (c) Capacitor dividers, compensated matching. (d) Damped capacitor divider, simple matching*

l.v. arm of the divider too heavily and decrease the output voltage with time. To avoid travelling wave oscillations, the cable must then be terminated at its input end. Then, a voltage step of constant amplitude at C_2, i.e. $C_2 \to \infty$, will be halved by $R = Z_k$ at the cable input end, as R and Z_k form a voltage divider. This halved voltage travels to the open end and is doubled by reflection. Thus the original amplitude of the voltage across C_2 appears at the input of the recording instrument. The reflected wave charges the cable to its final voltage amplitude, and is absorbed by R, as the capacitor C_2 forms a short-circuit. In reality, C_2 is of finite value and is therefore discharged during these transient events. The computation shows that the discharge period is very close to twice the travel time. After this time, the cable capacitance is charged to the final voltage, and from this we obtain two ratios of the voltage divider, namely:

$$n_o = \frac{C_1 + C_2}{C_1} \quad \text{for } t = 0;$$

$$n_e = \frac{C_1 + C_2 + C_k}{C_1} \quad \text{for } t \geq 2\tau_k.$$

The signal cable, therefore, introduces an initial 'overshoot' of the voltage of $\Delta V = (n_e/n_o) - 1 = C_k/(C_1 + C_2)$, which may well be neglected for short or medium cable length and high values of C_2, i.e. high ratios of the voltage dividers.

But capacitor dividers are often used for field testing of transient voltages and longer cables thus are often necessary. The response can be improved by transferring a part of the l.v. capacitor C_2 to the cable end and connecting it in series with a resistor, Fig. 3.47(c). This system, first treated by Burch,[120] offers some opportunities to decrease the overshoot effect. Burch proposed to make both matching resistances equal and $R_3 = R_4 = Z_k$. If then the condition $C_1 + C_2 = C_3 + C_k$ is satisfied, the initial and infinite time values of the voltage become the same, and the original overshoot of about $C_k/(C_l + C_2)$ is reduced to about 1/6. There are, however, further opportunities to improve the response as shown by Zaengl.[121]

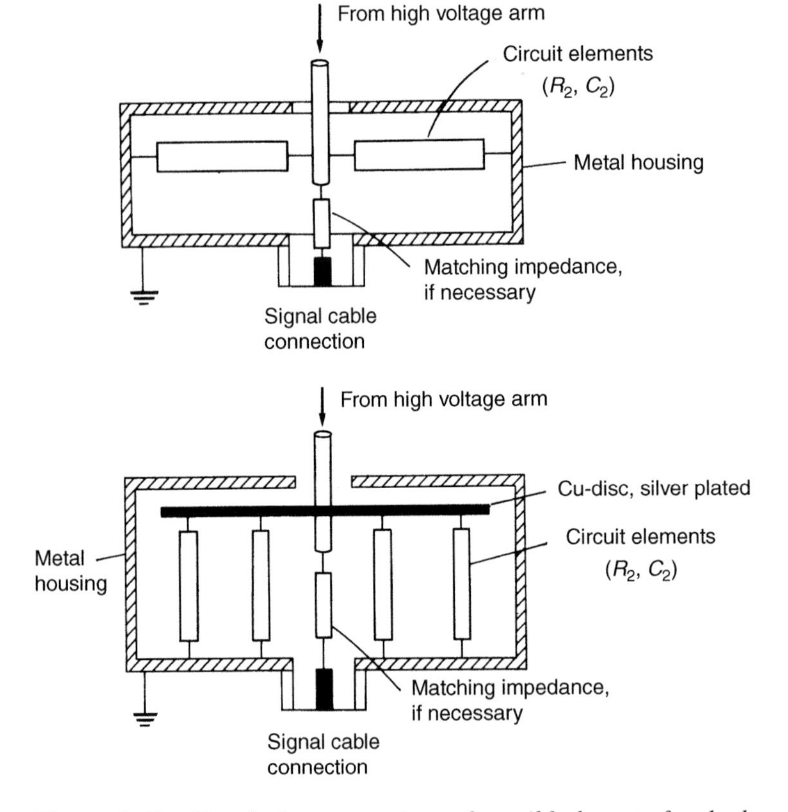

Figure 3.48 *Sketched cross-sections of possible layouts for the l.v. arm of voltage dividers*

For damped capacitor dividers, the resistors R_1 and R_2 necessary within the l.v. arm are for the reflected wave in series to the matching impedance at the l.v. arm, see Fig. 3.47(d). As R_2 is very small in comparison to R_1, the value of this matching resistor must only be reduced by the small value of R_2. The methods of Fig. 3.47(c) can also be applied.

Whereas matching resistors for coaxial cables, i.e. resistors between inner and outer conductors, are commercially available, the series resistors for all capacitor dividers are an integral part of the divider's l.v. arm. It may well be recognized that the path to earth for the reflected wave should not be hindered by too high inductances. This condition dictates the need for every l.v. capacitor to have a very low inductance. The theoretical value of this inductance is given by the ratio of the divider, which divides also the overall inductance of the stacked capacitor column. The physical size of the capacitance C_2 values necessary to reduce the high-voltage to a measurable quantity is, however, relatively large. The coaxial arrangement of any circuit elements used within the l.v. arm is a condition which should be strictly followed. In Fig. 3.48 simple cross-sections of possible layouts of the l.v. arm are sketched. Radially arranged elements tend to give even lower inductance values. The metal housing avoids the impact of electromagnetic fields. By the coaxial connection of the original cable, the input loop of this cable will not 'see' any effective magnetic field, which contributes to reduced voltages by the transient currents. Therefore, the current paths must be evenly distributed within the coaxially arranged elements. A too low inductance may easily be increased by the addition of small wire loops in series with the elements. For more information reference should be made to the literature.[123,126]

3.7 Fast digital transient recorders for impulse measurements

With the advent of high-speed digital recorders the field of high-voltage impulse testing has gained a powerful tool. Recent technological developments have made the use of digital recorders possible in the field of high-voltage impulse measurements.[78–84] Their use has important advantages over the use of traditional analogue oscilloscopes and recorders. Obtaining impulse test records in digital form allows for the introduction of the wide range of digital signal processing techniques into the analysis of high-voltage test data. These techniques enable high-voltage test engineers to correct errors due to non-ideal voltage dividers, to eliminate the effects of slight variations in the form of applied impulses in successive test records, to analyse test object transfer functions rather than merely visually examine the forms of applied stimuli and recorded responses, and to perform statistical analyses on the results of long series of impulse tests.[85–89]

Although digital techniques have been available for many years, it is only over the last decade that their use for measurements in high-voltage impulse tests has become widespread: all the standards covering digital recorders have been published in the last twelve years. Digital recorders for general use are covered by IEEE Std 1057-1994:[(135)] there are no IEC Standards which cover digital recorders for general use. There are presently two standards that deal with digital recorders for measurements in high-voltage impulse tests: IEEE Std 1122-1998[(136)] (revision of IEEE Std 1122-1987) and IEC document 1083-1: 1991[(137)] (a revision is being prepared based on IEEE Std 1122-1998). These standards define terms, identify quantities that affect accuracy, describe tests and set minimum standards to be met by each tested parameter. In many cases, it is not possible to isolate one parameter and test it alone and hence the limits have to allow for other contributions, e.g. noise on the record. Many test engineers make use of these standards to make a preliminary selection of digitizers based on the manufacturer's specifications but the selected digitizer has then to be tested according to the standard. The standards set limits on the sampling rate and the rated resolution. In addition limits are set on measured values of integral non-linearity of the amplitude and the time base, the differential non-linearity of the amplitude, the impulse scale factor, the rise time, the internal noise level, and the effects of interference and ripple.

The following sections of this chapter review the development and fundamental operating principles of digital recorders, outline how these devices differ from their analogue predecessors, cover the sources of static and dynamic errors inherent in digital recorders, and finally provide insight into the test procedures and minimum performance requirements mandated in current standards[(136,137)] related to the use of digital recorders in h.v. impulse testing.

3.7.1 Principles and historical development of transient digital recorders

The first attempts at digital recording of non-repetitive pulses were undertaken in the late 1950s in order to enable on-line processing of recorded transients. Research in nuclear physics and radar signature analysis prompted the design of a hybrid oscilloscope–TV camera system, which employed a conventional high writing-speed CRT coupled to a television vidicon tube scanning the CRT screen. Although very useful in this pioneering period, this hybrid recorder combined the drawbacks of both analogue and digital systems. However, the concept of fast writing on a temporary storage medium and scanning this medium later at a slower rate was a valuable innovation, and the hybrid recorder paved the path to more refined designs. This idea spawned the development of recorders which formed the functional basis of present day scan converters. In the late 1970s electronic circuits utilizing solid state components for high-speed analogue-to-digital conversion came into use. Since then several other mechanisms for realizing the A/D conversion process have been implemented.

The recording errors which characterize a recorder's dynamic accuracy depend to a certain extent on the design and operating principles of the instrument.[90–93] To gain an understanding of the physical principles responsible for these errors it is useful to review briefly the various available designs of recorders. There are four basic A/D conversion schemes utilized in present day high-speed digital recorders. These are: scan conversion, charge coupled device storage, flash conversion, and ribbon beam conversion.

Scan converters

The highest speed A/D conversion technique presently available utilizes scan converters. These consist essentially of an analogue cathode ray tube with the electron beam writing on a matrix of precharged semiconductor elements. The moving electron beam leaves a trace of discharged elements on the target matrix. The target is then read by another beam which scans the matrix at a slower pace. The slowed down replica of the recorded transient is digitized by a conventional ADC and stored for further display and processing. The main errors in such an instrument come from the difference between the electron beam writing speed at the steep and flat portions of the recorded transient. The variation in writing speed results in blooming or thickening of the trace in its slower portions similar to that often encountered in storage oscilloscopes when recording transients with slow and fast portions. However, with a scan converter, data processing can be used to reduce errors caused by blooming and the instruments can achieve a vertical resolution of approximately 1 per cent at a sweep of 5 ns. Further errors are generated by variations in the time base. In addition to these drawbacks, short record length is a limitation inherent in this recording technique. Despite these shortcomings, scan converters are used in many laboratories where high bandwidth and very high equivalent sampling rates are needed.[94]

Charge coupled device converters

Development of the charge coupled device (CCD) formed the basis of another fast digitizer. The CCD input of such a digitizer (often referred to as the bucket-brigade circuit) consists of a number of capacitors and electronic switches integrated on a chip and driven by a clock which can operate at a fast and a slow rate. The first sample of the transient to be recorded is taken by connecting the first capacitor to the instrument's input terminal for a short period. At the next stage, the charge accumulated in the first capacitor is transferred to the second one and the first capacitor is discharged in preparation for the next sample. This procedure is repeated at the high clock rate until the first sample is transferred to the last capacitor. This completes the recording cycle and the recorded transient is stored in the form of charges accumulated in each of the capacitors contained in the CCD. In order to read the signal, the bucket-brigade charge transfer process is repeated in the reverse direction at the slow clock rate. The subsequent samples are then digitized by a conventional ADC at the output terminal of the CCD and stored in the memory for further processing.

By using an advanced control system several MOS integrated circuit registers can be charged sequentially at a high rate thereby increasing the available record length. Although the output signal from the registers can be digitized (at a slower rate) with high resolution, the actual gain accuracy of CCD-based recorders is limited to approximately 1 per cent which is often much less than the high resolution of the actual slow sampling rate A/D converter.[95] This limit is set by the analogue techniques of sampling the recorded transients with CCD elements.

Flash converters

The third type of fast digitizer is based on the conventional flash conversion technique implemented in advanced semiconductor technology.[96,97] In higher resolution instruments two low-resolution monolithic flash converters operating at very high sampling rates are used in a subranging mode. The technique is known as dual rank flash conversion. The input signal is digitized by the first ADC and the digital output is fed into a digital-to-analogue converter (DAC). The DAC output is subtracted from the delayed input signal and the residue is fed into the second ADC to give an overall resolution of double that of each ADC used. The principal limitation of this digitization scheme is the time required for the DAC to settle so that the subtraction yielding the second ADC's input can be performed with sufficient accuracy. This sets the limit on the maximum sampling rate of dual rank flash converters.

A variation of the flash conversion technique can be implemented by using two high-resolution but slower sampling rate ADCs to sample the input alternately, thus giving an effective sampling rate of twice that of each ADC. In this type of digitizer additional errors can arise from differences in the quantization characteristics of the two ADCs and/or from asymmetry in the sampling intervals.

Ribbon beam converters

The last design principle used in digital recorders to achieve high resolution combined with high sampling rate incorporates an electron bombarded semiconductor (EBS) tube.[98] The EBS tube is similar to a conventional cathode ray tube (such as those used in analogue oscilloscopes) except that the luminescent screen is replaced by a target made up of a set of N adjacently positioned strip diodes. The beam is flat in the horizontal plane like a ribbon rather than being focused to a point, and there is only one set of plates used to deflect the electron beam. A metal mask, installed between the electron beam and the diode target, has a pattern of windows cut in it so that as the horizontal beam moves up and down with the input signal it illuminates various combinations of exposed diodes. A maximum of 2^N unique combinations of the N diode states are available. The output of each diode strip is checked at an interval corresponding to the recorder's sampling rate to generate an N-bit binary encoded word which corresponds to the level of the input signal present at the instant that the diode outputs are checked.

Although the different high-speed digital recorders described above may appear similar from the input/output perspective the different conversion schemes can introduce different errors. In fact, the errors obtained using a high-speed recorder are in many instances a direct consequence of the conversion techniques implemented in the particular instrument. In order to assess the actual measuring properties of a digitizer, it is therefore important to examine not only the sampling rate and resolution, but also the type of A/D conversion technique utilized.

3.7.2 Errors inherent in digital recorders

In contrast to an analogue oscilloscope which writes a continuous trace of the measured signal, a digital recorder is able to record and store only instantaneous values of the signal rounded to integer numbers and sampled at a certain rate over a finite period of time. The input signal is reconstructed by positioning in time (according to the sampling rate) and the vector of consecutive values contained in the recorder's memory. This leads to the presence of two types of recording errors which are generally referred to as quantization and discrete time sampling errors. These errors are the only ones present in an ideal recorder. Since they occur even when recording slowly changing or even d.c. input signals they are often referred to as static errors.

Static errors

The quantization error is present because the analogue value of each sample is transformed into a digital word. This A-to-D conversion entails a quantization of the recorder's measuring range into a number of bands or code bins, each represented by its central value which corresponds to a particular digital code or level. The number of bands is given by 2^N, where N is the resolution of the A-to-D converter. The digital output to analogue input relationship of an ideal digitizer is shown diagrammatically in Fig. 3.49. For any input in the range ($i\Delta V_{av} - 0.5 * \Delta V_{av}$ to $i\Delta V_{av} + 0.5 * \Delta V_{av}$), where ΔV_{av} is the voltage corresponding to the width of each code bin, or one least significant bit (LSB), and $i\Delta V_{av}$ is the centre voltage corresponding to the ith code, an

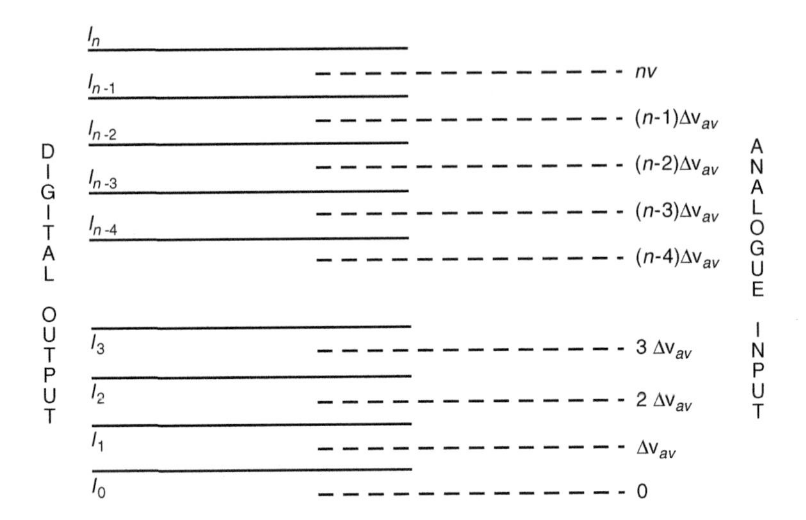

Figure 3.49 *Analogue input to digital output relation of an ideal A/D converter*

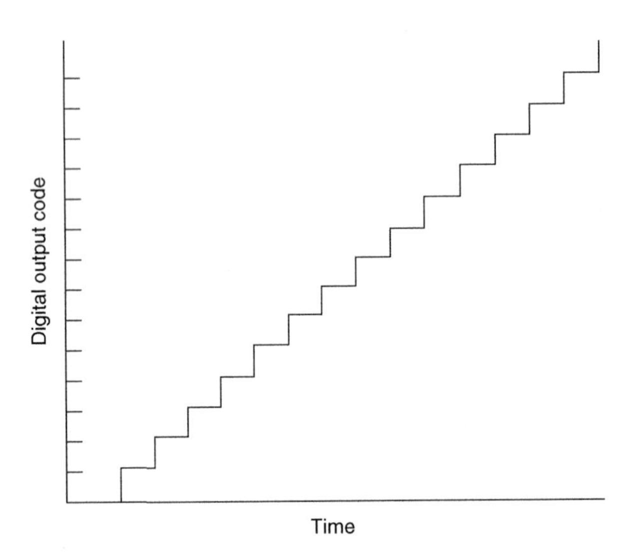

Figure 3.50 *Response of an ideal A/D converter to a slowly rising ramp*

ideal digitizer will return a value of I_i. Therefore, the response of an ideal digitizer to a slowly increasing linear ramp would be a stairway such as that shown in Fig. 3.50. A quick study of these figures reveals the character of the quantization error associated with the ideal A-to-D conversion process. The maximum error possible is equivalent to a voltage corresponding to $\pm(\frac{1}{2})$

of an LSB. For an ideal digital recorder, this quantization would be the only source of error in the recorded samples. For a real digital recorder, this error sets the absolute upper limit on the accuracy of the readings. In the case of an 8-bit machine, this upper limit would be 0.39 per cent of the recorder's full-scale deflection. The corresponding maximum accuracy (lowest uncertainty) of a 10-bit recorder is 0.10 per cent of its full-scale deflection.

The error caused by discrete time sampling is most easily demonstrated with reference to the recording of sinusoidal signals. As an example we can look at the discrete time sampling error introduced in the measurement of a single cycle of a pure sine wave of frequency f, which is sampled at a rate of four times its frequency. When the sinusoid and the sampling clock are in phase, as shown in Fig. 3.51, a sample will fall on the peak value of both positive and negative half-cycles. The next closest samples will lie at $\pi/2$ radians from the peaks. As the phase of the clock is advanced relative to the input sinusoid the sample points which used to lie at the peak values will move to lower amplitude values giving an error (Δ) in the measurement of the amplitude (A) of

$$\Delta = A(1 - \cos\phi)$$

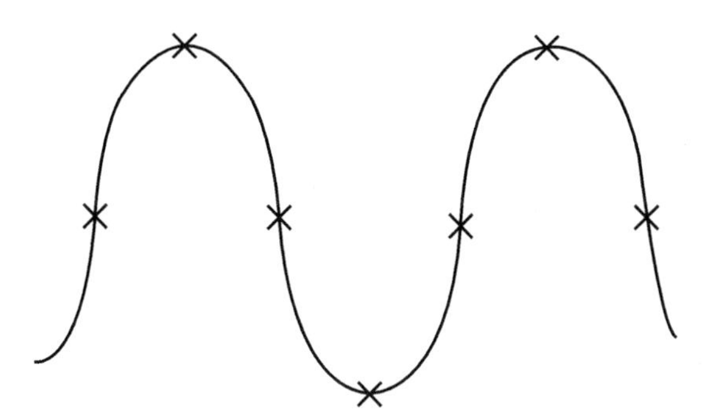

Figure 3.51 *Sample points with sinusoid and sampling clock in phase. (Error in peak amplitude = 0)*

where ϕ is the phase shift in the sample points. This error will increase until $\phi - \pi/4$ (Fig. 3.52). For $\phi > \pi/4$ the point behind the peak value will now be closer to the peak and the error will decrease for a ϕ in the range of $\pi/4$ to $\pi/2$. The maximum per unit value of the discrete time sampling error is given by eqn 3.93,

$$\Delta_{\text{max}} = 1 - \cos(\pi f t_s) \tag{3.93}$$

where t_s is the recorder's sampling interval and f the sinewave frequency.

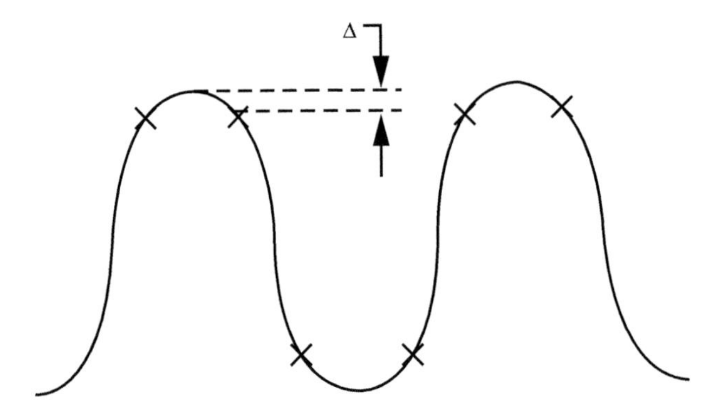

Figure 3.52 *Sample points with sampling clock phase advanced to $\pi/4$ with respect to the sinusoid. Error in peak amplitude (Δ) is at a maximum*

The maximum errors obtained through quantization and sampling when recording a sinusoidal waveform are shown in Fig. 3.53. The plotted quantities were calculated for an 8-bit 200-MHz digitizer.

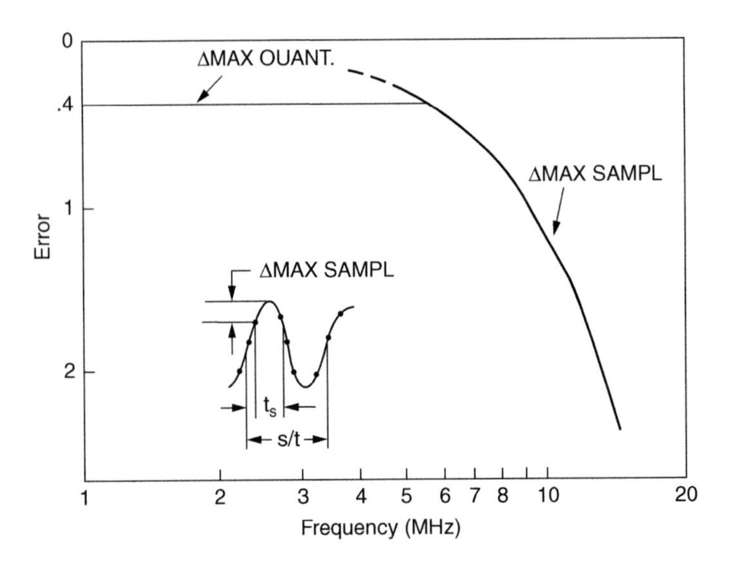

Figure 3.53 *Sampling and quantization errors of an ideal recorder*

In a real digital recorder, an additional two categories of errors are introduced. The first includes the instrument's systematic errors. These are generally due to the digitizer's analogue input circuitry, and are present to some degree in all recording instruments. They include such errors as gain drift, linearity errors, offset errors, etc. They can be compensated by regular calibration without any net loss in accuracy. The second category contains the digitizer's dynamic errors. These become important when recording high-frequency or fast transient signals. The dynamic errors are often random in nature, and cannot be dealt with as simply as their systematic counterparts and are discussed below.

Dynamic errors

In an ideal digitizer each sample of the recorded transient is taken in an infinitely short time window. This precludes any variation in the width of the time window which is often termed aperture uncertainty or jitter. Similarly, there is no uncertainty in the time at which the sample is taken. The widths of all the code bins are equal, and symmetrical about the level representing each bin's centre. When a real digitizer is constructed, the non-ideal nature of all these parameters comes into play. While their effects can be reduced to a negligible level in recorders characterized by slower sampling times, in those digitizers which push today's technology to its limits so as to provide the highest resolution coupled with the fastest sampling rates, these parameters may have a significant effect on the dynamic accuracy of the instruments.[91,93,99–113]

The nature, magnitude and consequences of the random errors encountered in real digital recorders vary depending on the instrument's design.[100,114,115,95] As a result the determination of the suitability of a particular type of recorder to h.v. impulse testing must consist of two phases. The initial phase consists of specifying the sampling rate and resolution if ideal digitizer performance can be assumed. Following this, the dynamic performance of the recorder under consideration must be determined. The former requires that the digitizer's nominal resolution in bits and sampling rate be related to the accuracies required by the standard for instruments used in recording h.v. impulses. The latter entails determination of the nature and limits of the recorder's dynamic errors. This approach was followed in the development of IEEE Std 1122[136] and IEC Pub. 1083,[137] the standards dealing with qualification of digital recorders for h.v. impulse measurements.

3.7.3 Specification of ideal A/D recorder and parameters required for H.V. impulse testing

Standard impulses used in H.V. testing of power apparatus

Requirements on the accuracy of recording instruments used in high-voltage impulse testing vary according to the type of tests, and depend upon the nature of the test object itself, e.g. testing and research into the dielectric strength of gas-insulated substations (GIS) involve generation and measurement of very steep-front high-voltage impulses which may be chopped after a time as short as 0.1–0.2 μs. This type of test does not require a very high measuring accuracy, i.e. an uncertainty of a few per cent on the impulse crest is acceptable. Details regarding impulse testing procedures of testing various types of insulations are described in Chapter 8, section 8.3.1, with appropriate references to national and international standards. Definitions of the prescribed standard voltage waveshape to be used in testing can be found in Chapter 2, Figs 2.23(a) to (c).

Peak measurements

As previously discussed, the maximum ideal quantization error is equal to 1 LSB. This can be restated in terms of per cent of full scale as $100/2^N$, where N is the recorder's resolution in bits. For the error to be less than or equal to 2 per cent of full-scale deflection a simple calculation shows that the recorder's resolution should be not less than 5.7 bits. This must of course be rounded up to the closest integer, 6. Therefore, for a full-scale signal, an ideal 6-bit recorder will meet the amplitude accuracy requirement. In practice it is not possible, or at least not practical, to ensure that all the signals to be recorded will span the instrument's full measuring range. It is more realistic to assume that the input signals will fall within a range of between 50 per cent and 90 per cent of full scale. For a signal which covers 50 per cent of the recorder's full scale, an accuracy limit of 2 per cent would be maintained by using an ideal recorder of 7-bit resolution. The magnitude of the maximum possible discrete time sampling error depends upon the shape of the impulse rather than on its magnitude. Therefore, the sampling-rate requirement must be examined for each different impulse shape. Using the fastest allowable standard lightning impulse as a basis for calculation, the maximum errors in peak voltage measurement as a function of the recorder's sampling rate can be calculated using the same approach as previously illustrated for the case of a sinusoidal input. The maximum possible errors resulting from sampling are given in Table 3.6.

As can be seen from the tabulated results the discrete time sampling error is negligible in comparison to the quantization error when recording the peak value of a full lightning impulse.

Tests often require the use of chopping the standard lightning impulse wave on the front or the tail as shown in Figs 2.23(c) and (b) respectively. With a standard lightning impulse chopped after 2 μs to 5 μs (tail), since the chopping occurs after the peak, the analysis used to derive nominal recorder characteristics required for acceptable recording of the peak value is identical to that outlined above for the case of the standard lightning impulse.

For testing with front-chopped impulse voltages two cases must be considered. The first of these is termed a front-chopped impulse and the second a

Table 3.6 *Maximum error due to sampling when recording the peak of a full standard lightning impulse*

Sampling rate (MHz)	*Error in % FSD*
2	0.195
2.8	0.097
4.0	0.048
5.7	0.024
8.6	0.012
12.0	0.006

linearly rising front-chopped impulse. As far as the measurement of their peak values is concerned these two impulses can be treated under the same analysis.

For impulses chopped on the front, the most demanding situation covered by today's standards specifies that the peak value of an impulse with a time-to-chop of 500 ns be recorded with the recording device contributing an error of less than 2 per cent. In this situation, the discrete time sampling error must be considered as well as the quantization error. For an ideal sawtooth input (the limiting case of a front-chopped impulse) the maximum possible discrete time sampling error is given by the product of the sampling interval and the signal's rate of rise. This is illustrated in Fig. 3.54.

Values of maximum discrete time sampling error (in per cent of the peak value) vs recorder sampling rate are shown in Table 3.7 for the shortest standard time-to-chop of 500 ns.

Table 3.7 *Maximum error due to sampling when recording the peak of an ideal sawtooth waveform with a 500 ns rise time*

Sampling rate (MHz)	*Error in % of peak value*
50	4
100	2
200	1

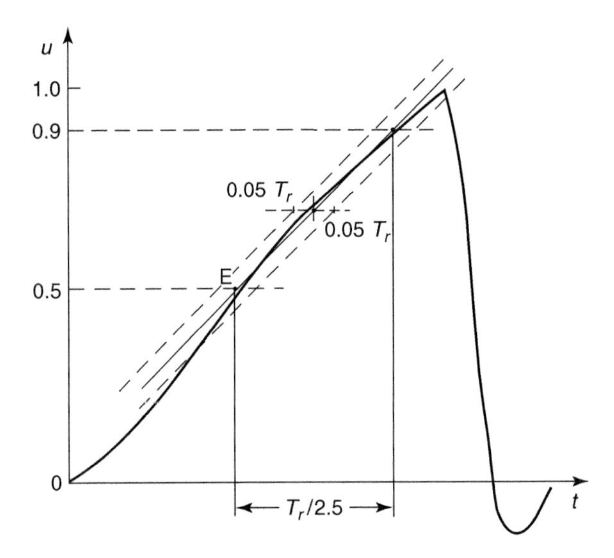

Figure 3.54 *Sampling error for an ideal sawtooth signal*

From a comparison of Tables 3.6 and 3.7 it can be seen that the requirements placed on sampling rate are far more severe when recording front-chopped lightning impulses than when recording standard full lightning impulses or impulses chopped on the tail. As shown in Table 3.7, a minimum sampling rate of 100 MHz is required to ensure a discrete time sampling error of 2 per cent or less in the measurement of the peak.

In reality high-voltage impulses are rounded at the chop. Figure 3.55 shows an example of a linearly rising impulse applied to a 250 mm sphere gap with a gap length of 60 mm. The slope of such a linearly rising front-chopped impulse is, according to the standards, taken as the slope of the best fitting straight line drawn through the portion of the impulse front between 50 per cent and 90 per cent of its peak value. The rise time is defined as being the time interval between the 50 per cent and 90 per cent points multiplied by 2.5. The impulse shown in Fig. 3.55 was measured using a small divider insulated with compressed gas and characterized by an extremely low response time. The measured rate of rise of 10.8 kV/ns and time-to-chop of 36 ns are much more severe than those typifying standard test impulses. However, even in this very severe case, the waveshape shows that the slope close to the chop is very much less than the impulse's rate of rise as calculated between the 50 per cent to 90 per cent points on the impulse front. For higher peak voltages, that is larger breakdown distances and larger (hence slower) impulse measuring systems, the rounding effect just prior to the chop will be more pronounced. This means that in practice the slope at the instant of chopping is significantly less than $(V_{pk}/500)$ kV/ns. Since this was the figure used in calculating the

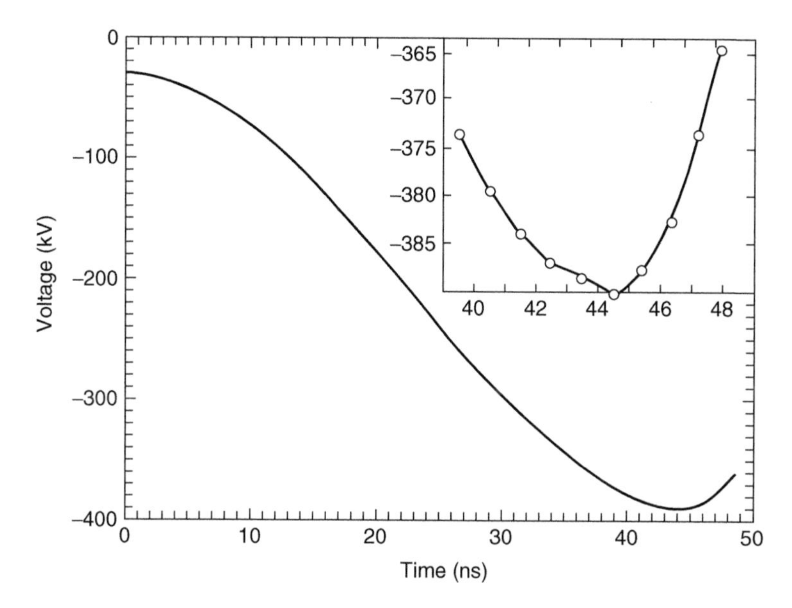

Figure 3.55 *Actual record of a linearly rising front-chopped h.v. impulse. (Inset shows every fifth sample in the vicinity of the chop)*

minimum sampling rates necessary to limit the discrete time sampling errors to acceptable values, the requirements on the sampling rate can be relaxed from those given for the case of the sawtooth wave.

There are two factors responsible for the rounding of the impulse wave near the time of chopping. The first is the mechanism of the breakdown which is responsible for the voltage collapse. Although not in all cases, this mechanism often contributes a significant amount to the rounding. The second, often more important, factor in rounding the impulse peak is the limited bandwidth of high-voltage measuring systems. Because of their physical size, these systems cannot respond quickly to fast changes in their input. The rounding-off introduced in the measurement of an ideal sawtooth waveform with a 500 ns rise time is illustrated in Fig. 3.56. The response to such an input is shown for high-voltage measuring systems of varying response times. The standards require that a measuring system have a minimum response time of 0.05 times the rise time of any linearly rising front-chopped impulse which it is used to record.[6,53] In the worst case of a 500 ns rise time this requirement translates into a response time of 25 ns or less. Looking at the case of the measuring system with 25 ns response time shown in Fig. 3.54 a significant rounding in the area of the peak is evident.

A simple computer analysis can be used to calculate the values of sampling frequency which are necessary to meet the 2 per cent accuracy requirement.

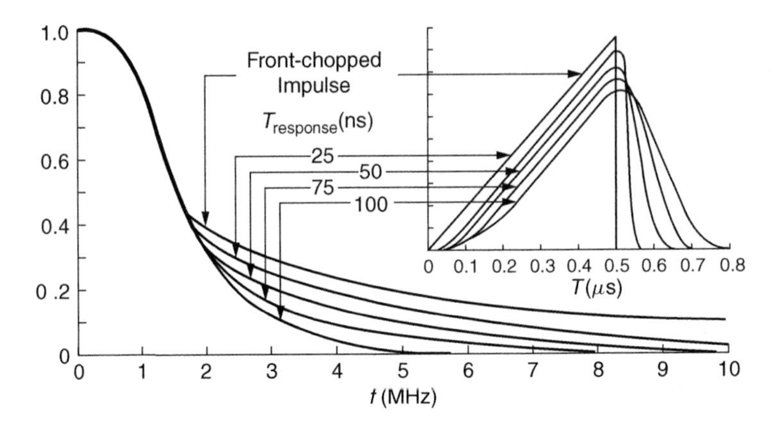

Figure 3.56 *Illustrative example of a front-chopped impulse ($t_{chop} = 0.5$ μs) distorted by an analogue measuring system of a limited bandwidth. Attenuation of higher spectral frequencies of the impulse by the measuring system and corresponding distortion of the impulse form are shown for a few values of the response time of the system*

However, for the practical case of bandwidth limited measuring systems, the minimum requirements on sampling rate necessary to ensure recording of the peak are exceeded by those necessary to meet standard requirements on the measurement of time parameters and front oscillations discussed below.

Measurement of time parameters

Since lightning impulses are defined by time as well as amplitude parameters, the sampling rate required for the evaluation of the time parameters must also be determined. As previously stated, the standards' present requirement on oscilloscopes used to record h.v. impulses is that they must allow for the evaluation of time parameters with an accuracy of better than 4 per cent. As with the determination of minimum sampling rates required to ensure the specified accuracy in the measurement of the impulse peaks, each type of impulse must be examined in turn to derive minimum sampling rates necessary for required accuracy in the measurement of time parameters. A standard full lightning impulse is defined by its front time and its time-to-half value (Fig. 2.23(a)). The front time is defined as 1.67 times the time interval between the instants at which the impulse is between 30 per cent and 90 per cent of its peak value. To determine the minimum sampling rates necessary to evaluate the front time to within 4 per cent the fastest allowable standard lightning impulse must be examined. It is characterized by a front time of 0.84 μs. This means that the shortest time interval which must be measured for such an impulse is 503 ns. The time-to-half value is defined as the time interval between the virtual origin (O_1 in Fig. 2.23(a)) and the instant on the tail at which the voltage has dropped to half of the peak value. The evaluation of the time-to-half value entails recording a much greater time interval than that required to determine the front time. Therefore, the requirement on minimum sampling rate is set by the latter. For a standard impulse chopped on the tail the shortest time interval to be recorded is also the front time and its minimum value is the same as that of a standard full lightning impulse. The second time parameter used to define a chopped impulse is the time-to-chop. However, by definition this time interval is between 2 μs and 5 μs, so the minimum required sampling rate must be based on the front time measurement. For a front-chopped or linearly rising front-chopped lightning impulse the time interval between the virtual origin (O_1 in Fig. 2.23(c)) and the instant of chopping can be much shorter. The minimum time-to-chop allowed for in the standards is 500 ns. This means that the minimum sampling rates required to record the defining time parameters of full, front-chopped, linearly rising front-chopped, or standard chopped impulses are approximately equal. To evaluate a time interval of 500 ns with an accuracy of better than 4 per cent it is necessary to sample at a rate of 500 MHz or more.

Recording of front oscillations

The standard test impulses (Fig. 2.23(a)) are based on pure bi-exponential waves such as would be produced by the resistive and capacitive components present in impulse generating and measuring systems. In practical realizations of large high-voltage systems, there is always some inductance present in the circuit. This inductance results in oscillations being superimposed on the test impulses. A typical record of an impulse with superimposed oscillations is shown in Fig. 3.57. As is expected, the maximum frequency of the oscillations in the circuit decreases as the physical size of the circuit increases.

There are standard requirements placed on impulse test waveforms with regards to the maximum amplitude of superimposed oscillations. In order to

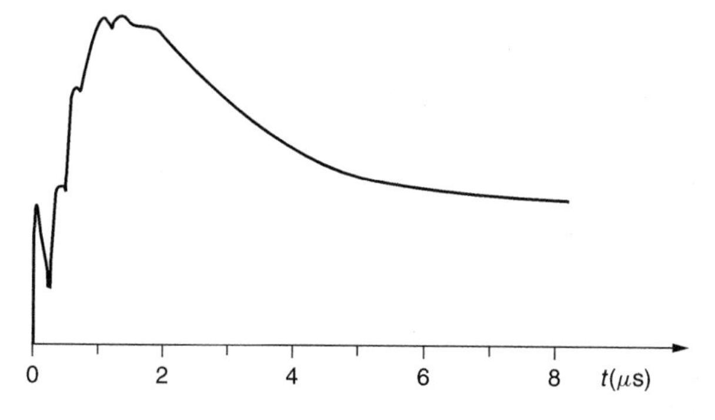

Figure 3.57 *Standard full lightning impulse with superimposed oscillations*

ensure that these criteria are met, it is necessary to be able to evaluate the peaks of the oscillations with a certain precision. The standard requirement intended to ensure sufficiently accurate recording of the oscillations states that the upper 3 dB cut-off frequency for surge oscilloscopes be at least two times f_{max}. The oscillations present on impulses are damped sine waves. Therefore, the formula which gives the maximum discrete sampling time error as a function of signal frequency and sampling rate (eqn 3.93) detailing the errors of ideal recorders can be applied in this instance. Examination of this formula reveals that a maximum error of 30 per cent or 3 dB is obtained when the sampling frequency is four times larger than the frequency of the sine wave being sampled. Therefore, to correspond to the requirement placed on surge oscilloscopes ($f_{3\,dB} > 2f_{max}$) the sampling rate of the recorder must be at least 8 times f_{max}.

Impulse tests on non-self-restoring insulation

H.V. impulse testing of apparatus such as power transformers, which contain non-self-restoring insulation, often requires using non-destructive test techniques. Such tests usually consist of checking the linearity of the test object insulation impedance within a range of test voltages up to the basic insulation level (BIL).[85,116,117,118] The quantities monitored during these tests are the applied voltage and the neutral current which is taken as the response to the voltage application. The analysis of test records comprises a detailed comparison of records taken at different voltage levels. Meaningful deviations between the records indicate that the characteristics of the test object are non-linear, and the test results in a failure. Since even small deviations between records are meaningful and can result in the disqualification of a very expensive piece of equipment, it is imperative that the recorder used provides enough accuracy to resolve such differences. Fortunately, the meaningful frequency content of the records to be examined is band limited to a few MHz (typically less than 2.5 MHz). The requirements on oscilloscopes used for such tests have never been accurately specified as the same impulse oscilloscope has been used for monitoring tests on self-restoring and non-self-restoring insulation. The parameters of an ideal digitizer which meets the requirements are, therefore, not as easily determined as those of one appropriate for testing of self-restoring insulation. The derivation of the necessary parameters can be approached in two ways. The first of these entails matching the accuracy of the analogue h.v. impulse oscilloscopes which have become the *de facto* standard instruments for use in these tests.[119] The second is to individually examine all the test procedures and methods of analysis associated with the testing of objects containing non-self-restoring insulation. A major problem encountered in realizing this second alternative stems from the fact that different apparatus standards have different requirements which must be used in the derivation of the required resolution and accuracy of the digital recorder. This problem causes the latter approach to be less favourable than the former. The high-quality surge oscilloscopes generally used in h.v. impulse testing are characterized by a bandwidth of approximately 35 MHz. The vertical or amplitude resolution of such an oscilloscope cannot be matched by an 8-bit digitizer, but it is most certainly exceeded by the resolution of a 10-bit recorder. The minimum sampling rate required for such tests is not clearly specified, but can be deduced on the basis of test requirements. When monitoring tests on power transformers and reactors, the neutral current is usually recorded by means of a shunt whose bandwidth is generally below 3 MHz. A second aspect of the test which can shed light on the required sampling rate is that in certain design tests involving the use of chopped waveforms, the time taken for the voltage to collapse is specified as being between 200 ns and 500 ns. The accuracy with which this parameter is recorded is a direct function of the sampling rate of the recorder used. However, at present there is no limit specified on the error permissible in the establishment of the duration of voltage collapse. In order to ensure accuracies superior to those attainable with presently used oscilloscopes, it appears that an ideal digital recorder with an amplitude resolution of 10 bits and a sampling rate of 30 MHz is suitable for recording tests on apparatus containing non-self-restoring insulation.

Digitizer dynamic performance

As mentioned earlier, the performance of a real digitizer does not usually match that of its ideal counterpart. This deterioration is principally due to the imperfect performance of the recorder components. The errors caused by these imperfections are usually referred to as dynamic errors, and their magnitude is generally proportional to the slew rate or the rate of change of the input signal. This proportionality results in a deterioration of recorder performance with increasing input signal steepness. The causes and effect of the limited dynamic performance of digital recorders are briefly discussed here. The dynamic errors can be quantified using four parameters: differential non-linearity (DNL), integral non-linearity (INL), aperture uncertainty, and internal noise. Three of these four parameters are assessed directly during the qualification of digital recorders as laid out in references 136 and 137. The effects of the fourth (aperture uncertainty) are included in the assessment of the other three.

Differential non-linearity (DNL)

Differential non-linearity is defined as the variation in the widths of the recorder's code bins. For an ideal recorder all of the code bin widths are equal and given by the recorder's full-scale deflection divided by the number

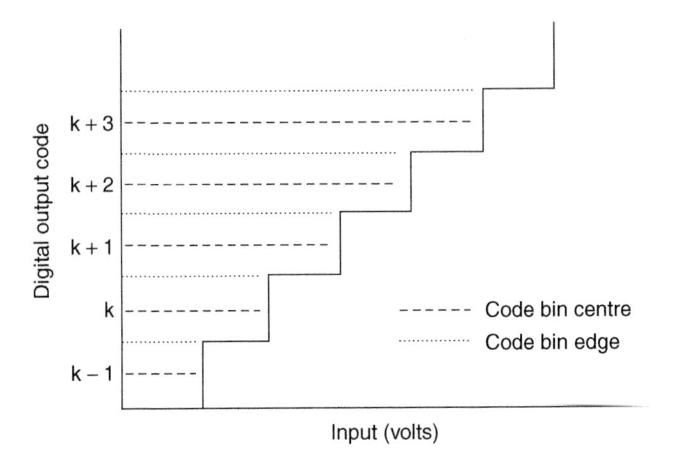

Figure 3.58 *A portion of the quantization characteristic of an ideal digital recorder*

of quantization levels. If the recorder's full-scale deflection is defined in bits, then the width of each ideal quantization band or code bin is equal to 1 least significant bit (LSB). The recorder's measuring range can, therefore, be thought of as being divided into 2^N code bins or quantization bands (where N is the recorder resolution) as shown in Fig. 3.58. In this figure the edges of the quantization bands (or code bins) are basically assigned a zero probability of occurrence. In a real recorder this sort of performance cannot be expected. Since the rate of occurrence of the specific codes can only be determined in statistical terms, the edges of the code bins are defined by a distribution which may resemble the normal bell-shaped curve. These distributions reflect the gradual transition from one state to the next. The areas in which this occurs are referred to as code transition zones. The static profiles of these transition zones can be obtained by repeated testing with incremented d.c. input voltages. The results of this can be used in the establishment of a probability based transition between adjacent code bins. An example of a portion of static quantization characteristic obtained using such a method is shown superimposed on its ideal counterpart in Fig. 3.59. As can be seen from the differences between Figs. 3.58 and 3.59, even the static quantization characteristic obtained using a d.c. input signal deviates from the ideal rectangular characteristic. Under dynamic conditions the limited slew rates and settling times of the recorder's electronic components result in further aberrations of the quantization characteristic. These can take the form of a further sloping out of the code transition zones as well as a non-uniform widening or narrowing of the code bins. These types of deviations are quantified in terms of the recorder's differential non-linearity (DNL), and their magnitudes vary with the signal slope. According to the definition of DNL it can be seen that an ideal digitizer will have a DNL

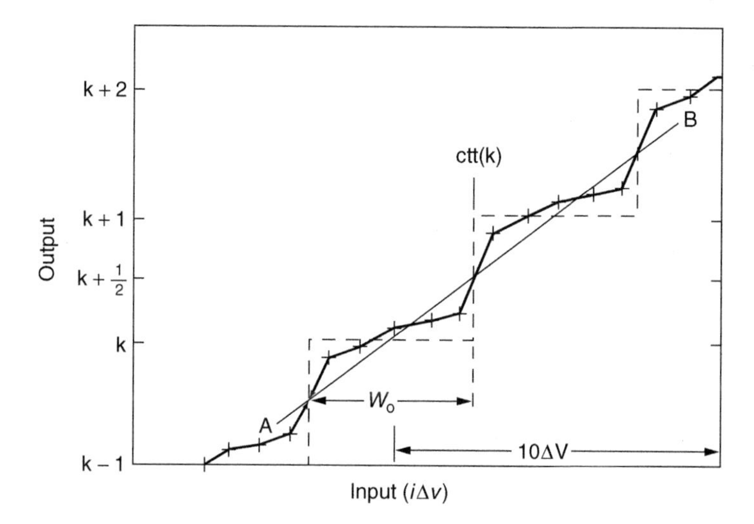

Figure 3.59 *Four codes of a static quantization characteristic. Dashed curve is for an ideal digitizer whose average code bin width is w_0. The code transition threshold from code k to k + 1 is marked (ctt(k)). Measured points are shown as +*

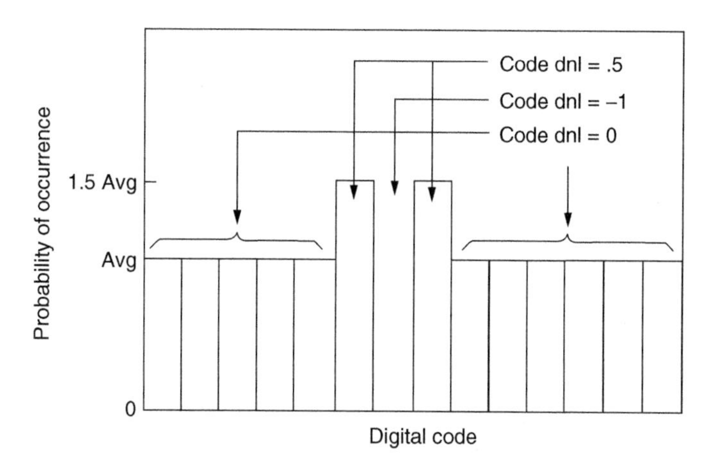

Figure 3.60 *Illustration of DNL values*

of 0 as the width of every code bin is equal (as shown in Fig. 3.58). If a digitizer is ideal except for one code which is missing while the two adjacent codes are 50 per cent wider than the average, then the DNL of the missing code is −1 and the DNL of each of the two adjacent codes is 0.5. An example illustrating this is shown in Fig. 3.60. The effect of this DNL is to reduce the local resolution of the digitizer by 50 per cent at the missing code. The figure of merit used in literature to describe recorder differential non-linearity can be expressed as a vector with each entry representing the DNL as measured for each of the code bins, or alternately as a single figure. When the latter representation is used, the number is taken to represent the maximum DNL of all of those measured across the recorder's range.

The differential non-linearity can also be measured in a dynamic test. This is usually done by repeatedly recording a sinusoidal waveform and relating the observed distribution of code occurrences to the probability density function as expected for a sinusoidal input to an ideal recorder.

Integral non-linearity (INL)

A second parameter used to quantify digitizer performance is the integral non-linearity (INL). Conceptually the INL can be thought of as an assessment of how much the real quantization characteristic of the recorder deviates away from the ideal. This is illustrated in Fig. 3.61.

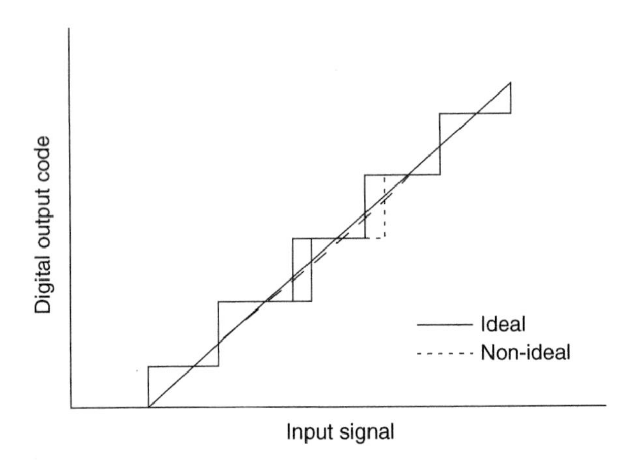

Figure 3.61 *Integral non-linearity as shown by a deviation between the real and ideal quantization characteristics*

The difference between the differential and integral non-linearity lies in the fact that, while the former evaluates the variations in the individual code bin widths, the latter integrates across the DNL of all the code bins and gives a feeling for the deviation of the recorder's quantization characteristic from its ideal counterpart. As with the DNL, the INL is assessed both statistically and dynamically.

Aperture uncertainty

In addition to differential and integral non-linearity related errors a third significant error source associated with real digitizers is the aperture uncertainty or sampling time dispersion. This is described as the variation in timing between successive samples. The recorder will always introduce a delay between the time that it is instructed to sample and the time at which the sample is actually taken. If this delay were constant, then it would not project on the digitizer's measuring properties as it could be corrected for by realigning the samples in time. Unfortunately, as is typical of physical processes, the delay follows a statistical distribution thus precluding any simple corrections. As can be intuitively deduced from the definition of the aperture uncertainty, the magnitude of errors stemming from this cause are dependent on the form of the signal being recorded. For slowly varying inputs the magnitude of the errors is not as prominent as for steeply rising signals. The numerical value of the aperture uncertainty which is often quoted by manufacturers is defined as the standard deviation of the sample instant in time.

Internal noise

Error due to internally generated noise is a fourth type of error associated with real digital recorders. Noise in the digital recorder is equivalent to a smearing of the quantization characteristic. The extent or severity of this effect is of course dependent on the relative magnitude of the noise and the code bin widths thus making high-resolution recorders more vulnerable. When the noise is large, the quantization characteristic degenerates into an approximately straight line. Figure 3.62 illustrates the effect of various relative noise amplitudes on the quantization characteristic. The figure shows a quantization characteristic obtained on a 10-bit recorder. The noise, whose magnitude is in the same range as the quantization step size or code bin width, causes the quantization characteristic to appear as a nearly straight line. The effect of varying degrees of noise magnitude can be illustrated by artificially increasing the quantization step size. This is equivalent to reducing the resolution of the recorder. For an effective resolution of 9 bits, adjacent levels in the recorder were paired together. Similarly 8-bit resolution was simulated by summing every four adjacent levels into one. Looking at the three quantization characteristics shown in Fig. 3.62, the effect of higher and lower noise becomes apparent.

3.7.4 Future trends

The preceding sections have described the development and current state of digital recorders utilized for h.v. impulse testing. Rapid advancements in electronics technology will undoubtedly result in significant developments in this

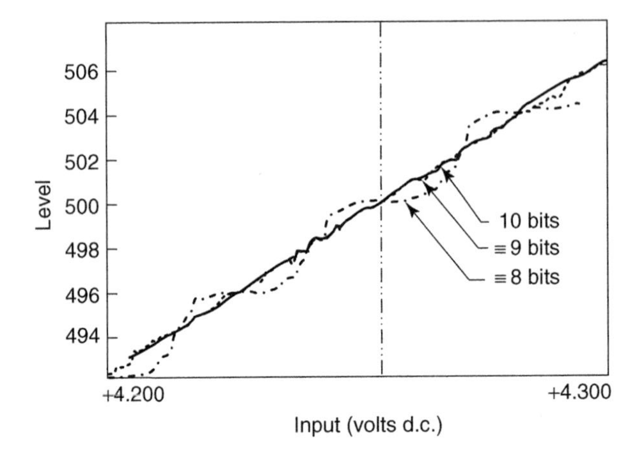

Figure 3.62 *Quantization characteristics obtained by treating a 10-bit digitizer as having 10-, 9-, and 8-bit resolution*

area. Depending on the technologies used in the realization of future generations of these devices, it likely will be necessary to establish new or further refined tests for establishing the ability of digital recorders to faithfully reproduce records of h.v. impulse tests. Evidence of this is reflected in the many improvements and enhancements which have been made to commercially available digitizers over the last decade. Today, various digitizers with rated resolutions corresponding to 8 bits have sampling rates from 1.10^9 samples/s to 1.10^{10} samples/s. These rates are higher than is needed for measurements of lightning impulses and hence it is possible to trade sampling rate for improved amplitude performance. Currently, two modes are of particular interest in h.v. impulse measurements: the 'peak detect' mode and the 'enhanced resolution' mode. In both these modes the digitizer samples the input signal at the maximum sample rate but the sample rate of the record is set at the maximum sampling rate divided by an integer. Examples of the use of these features are given in reference 138.

References

1. A.J. Schwab. *High Voltage Measurement Techniques*. MIT Press, Cambridge, Ma. and London, England, 1972.
2. B.D. Jenkins. *Introduction to Instrument Transformers*. Newnes, London, 1967.
3. Comparative Calibration of Reference Measuring Systems for Lightning Impulse Voltages According to IEC 60-2:1994 Proc. Int. Symp. on H.V. Engineering, Graz 1995, p. 4469.
4. A. Draper. *Electrical Machines*. Longmans, London, 1976.
5. IEC Publication 52, 2nd Edition (1960). Recommendations for Voltage Measurements by Means of Sphere-gaps (one sphere earthed).

6. IEEE Standard Techniques for High Voltage Testing. IEEE Std. 4-1995.
7. F.S. Edwards and J.F. Smee. The Calibration of the Sphere Spark-Gap for Voltage Measurement up to One Million Volts (Effective) at 50 Hz. *JIEE* **82** (1938), pp. 655–657.
8. H.E. Fiegel and W.A. Keen, Jr. Factors influencing the sparkover voltage of asymmetrically connected sphere gaps. *Trans. AIEE, Commun. Electron.* **76** (1957), pp. 307–316.
9. E. Kuffel. The Influence of Nearby Earthed Objects and of the Polarity of the Voltage on the Breakdown of Horizontal Sphere Gaps. *Proc. IEE* **108A** (1961), pp. 302–307.
10. W.G. Standring, D.H. Browning, R.C. Hughes and W.J. Roberts. Effect of humidity on flashover of air gaps and insulators under alternating (50 Hz) and impulse (1/50 μs) voltages. *Proc. IEE* **110** (1963), pp. 1077–1081.
11. S. Guindehi. Influence of humidity on the sparking voltage in air for different kinds of voltages and electrode configurations (in German). *Bull. SEV.* **61** (1970), pp. 97–104.
12. T.E. Allibone and D. Dring. Influence of humidity on the breakdown of sphere and rod gaps under impulse voltages of short and long wavefronts. **Proc. IEE. 119** (1972), pp. 1417–1422.
13. W. Link. PhD Thesis (in German). TU Stuttgart, No. 203 (1975).
14. E. Peschke. Influence of humidity on the breakdown and flashover of air gaps for high d.c. voltages (in German). *ETZ-A* vol. 90 (1969), pp. 7–13.
15. E. Kuffel. The Effect of Irradiation on the Breakdown Voltage of Sphere Gaps in Air under Direct and Alternating Voltages. *Proc. IEE* **106C** (1956), pp. 133–139.
16. F.O. MacMillan and E.G. Starr. *Trans. AIEE* **49** (1930), p. 859.
17. P.L. Bellaschi and P.H. McAuley. Impulse Calibration of Sphere Gaps. *Electric Journal* **31** (1934), pp. 228–232.
18. J.R. Meador. Calibration of the Sphere Gap. *Trans. AIEE* **53** (1934), pp. 942–948.
19. E. Kuffel. Influence of humidity on the breakdown voltage of sphere-gaps and uniform field gaps. *Proc. IEE*, Monograph No. 3322 M, 108C (1961), 295.
20. R. Davis and G.W. Bowdler. The Calibration of Sphere Gaps with Impulse Voltages. *Journal of the IEE* (London) **82** (1938), pp. 645–654.
21. J.M. Meek. The Influence of Irradiation on the Measurement of Impulse Voltages with Sphere Gaps. *Journal of the IEE* (London) **93** (1946), pp. 97–115.
22. A.J. Kachler. Contribution to the problem of impulse voltage measurements by means of sphere gaps. 2nd Int. Symp. on High Voltage Engg., Zurich, 1975, pp. 217–221.
23. W. Schultz. Erratic breakdown in air due to impurities in the presence of direct and alternating voltages. 3rd Int. Symp. on High Voltage Engg., Milan, 1979, Report 52.05.
24. D. Peier and H. Groschopp. *PTB-Mitteilungen* **87** (1977), pp. 396–398.
25. W.O. Schumann. *Elektrische Durchbruchfeldstaerke von Gasen.* Springer, Berlin, 1923.
26. E. Kuffel and M. Abdullah. *High-Voltage Engineering*. Pergamon Press, 1970.
27. G.A. Schroeder. Zeitschr.*f*. Angew. *Physik* **13** (1967), pp. 296–303.
28. H.A. Boyd, F.M. Bruce and D.J. Tedford. *Nature* **210** (1966), pp. 719–720.
29. M.S. Naidu and V. Kamaraju. *High Voltage Engineering*. McGraw-Hill, 1995.
30. P. Paasche. *Hochspannungsmessungen.* VEB Verlag Technik, Berlin, 1957.
31. H. House, F.W. Waterton and J. Chew. 1000 kV standard voltmeter. 3rd Int. Symp. on High Voltage Engg., Milan, 1979, Report 43.05.
32. J.H. Park. *J. Res. Nat. Bur. Stand.* **66C**, 1 (1962), p. 19.
33. R. Davis. *J. Sci. Inst.* **5** (1928), pp. 305–354.
34. R.F. Goosens and P.G. Provoost. *Bull. SEV.* **37** (1946), pp. 175–184.
35. D. Peier and V. Graetsch. 3rd Int. Symp. on High Voltage Engg., Milan, 1979, Report 43.08.
36. C.T.R. Wilson. *Phil. Trans.* **(A)221** (1920), p. 73.
37. L.W. Chubb and C. Fortescue. *Trans. AIEE* **32** (1913), pp. 739–748.
38. W. Boeck. *ETZ-A* **84** (1963), pp. 883–885.
39. R. Davies, G.W. Bowdler and W.G. Standring. The measurement of high voltages with special reference to the measurement of peak voltages. *J. IEE*, London, **68** (1930), 1222.

40. R. Peiser and W. Strauss. Impulse peak voltmeter with extended measuring possibilities. 3rd Int. Symp. on High Voltage Engg. (ISH), Milan, 1979, Report 72.07.
41. J.G. Graeme. *Designing with Operational Amplifiers*. McGraw-Hill, New York, 1977.
42. W. Schulz. High-voltage ac peak measurement with high accuracy. 3rd Int. Symp. on High Voltage Techn., Milan, 1979, Report 43.12.
43. J.C. Whitaker (ed.). *The Electronics Handbook*. IEEE Press, 1996.
44. R. Malewski and A. Dechamplain. Digital impulse recorder for high-voltage laboratories. *Trans. IEEE* **PAS 99** (1980), pp. 636–649.
45. W. Clausnitzer. *Trans. IEEE* **IM 17** (1968), p. 252.
46. A. Keller. Symposium on Precision Electrical Measurements, NPL, London, 1955.
47. H.R. Lucas and D.D. McCarthy. *Trans. IEEE* **PAD 89** (1970), pp. 1513–1521.
48. U. Brand and M. Marckmann. Outdoor high-voltage compressed-gas capacitors using SF_6. 2nd Int. Symp. on High Voltage Engg., Zurich, 1975.
49. D.L. Hillhouse and A.E. Peterson. *Trans. IEEE* **IM 22** (1973), No. 4.
50. IRR-IMS Group. Facing uhv measuring problems. *Electra* No. 35 (1974), pp. 157–254.
51. W. Zaengl. *Arch. Techn. Messen (ATM)*, Blatt Z 130-3 (1969).
52. H. Luehrmann. *ETZ-A* **91** (1971), pp. 332–335.
53. IEC Publication 60: High-voltage test techniques. Part 2: Measuring Systems, 2nd Edition, 1994-11.
54. High voltage measurements, present state and future developments. *Rev. Gen. Electr.*, Special Issue, June 1978.
55. A. Rodewald. Fast transient currents on the shields of auxiliary cables after switching operations in hv substations and hv laboratories. IEEE PES Winter Meeting, New York, 1979, Paper No. A79 086-0.
56. A. Rodewald. *Bull. SEV.* **69** (1978), pp. 171–176.
57. IEC Publication 60-4 (1977). High-voltage test techniques, Part 4: Application Guide for Measuring Devices.
58. A. Asner. *Bull. SEV.* **52** (1961), pp. 192–203.
59. F. Creed, R. Kawamura and G. Newi. *Trans. IEEE* **PAS 86** (1967), pp. 1408–1420.
60. W. Zaengl. *Bull. SEV.* **61** (1970), pp. 1003–1017.
61. W. Zaengl and K. Feser. *Bull. SEV.* **55** (1964), pp. 1250–1256.
62. P.R. Howard. Errors in recording surge voltages. *Proc. IEE* II/99 (1952), pp. 371–383.
63. A.F. Rohlfs, J.F. Kresge and F.A. Fisher. *Trans. AIEE* **76** (1957), Part 1, pp. 634–646.
64. D. Pellinen and M.S. Di Capua. *Rev. Sci. Inst.* **51** (1980), pp. 70–73.
65. P.L. Bellaschi. *Trans. AIEE.* **52** (1933), pp. 544–567.
66. T. Harada, T. Kawamura, Y. Akatsu, K. Kimura and T. Aizawa. *Trans. IEEE* **PAS 90** (1971), pp. 2247–2250.
67. R. Krawczynski. Correction of the high voltage measuring system by means of l.v. transmission line. 2nd Int. Symp. on High Voltage Engg., Zurich, 1975, Report 3.1-07.
68. R. Elsner. *Arch. Elektrot.* **33** (1939), 23–40.
69. T. Harada *et al. Trans. IEEE* **PAS 95** (1976), pp. 595–602.
70. A. Schwab and J. Pagel. *Trans. IEEE* **PAS 91** (1972), pp. 2376–2382.
71. W. Breilmann. Effects of the leads on the transient behavior of coaxial divider for the measurement of high ac and impulse voltage. 3rd Int. Symp. on High Voltage Engg., Milan, 1979, Report 42.12.
72. W. Zaengl. *Bull. SEV.* **56** (1965), pp. 232–240.
73. H. Dommel. *Trans. IEEE* **PAS 88** (1969), pp. 388–399.
74. A. Schwab, H. Bellm and D. Sautter. 3rd Int. Symp. on High Voltage Engg., Milan, 1979, Report 42.13.
75. H. Luehrmann. *Archiv. fur Elektrot.* **57** (1975), pp. 253–264.
76. N. Ari. Electromagnetic phenomena in impulse voltage measuring systems. *Trans. IEEE* **PAS 96** (1977), pp. 1162–1172.

77. K.H. Gonschorek. 3rd Int. Symp. on High Voltage Engg., Milan, 1979, Report 42.02.
78. R. Malewski and A. De Champlain. *IEEE* **PAS 99** (1980), pp. 636–649.
79. H.G. Tempelaar and C.G.A. Koreman. 4th Int. Symp. on High Voltage Engg., Paper 65.07, 1983.
80. R. Malewski. *IEEE* **PAS 101** (1982), pp. 4508–4517.
81. R. Malewski and B. Poulin. *IEEE* **PAS 104** (1982), pp. 3108–3114.
82. B. Beaumont *et al. IEEE* **PAS 96** (1987), pp. 376–382.
83. P.P. Schumacher and M.E. Potter. IEEE Power Engineering Summer Meeting, paper A.78, 1978.
84. T. Miyamoto. *et al. IEEE* **IM 24** (1975), pp. 379–384.
85. K. Schon, W. Gitt. *IEEE* **PAS 101** (1982), pp. 4147–4152.
86. R. Malewski *et al.* Canadian IEEE Communications and Power Conference, 1980.
87. Guo-Xiong, De-Xiang. 4th Int. Symp. on High Voltage Engg., paper 65.09, 1983.
88. S.J. Kiersztyn. *IEEE* **PAS 99** (1980), pp. 1984–1998.
89. R. Malewski and B. Poulin. *IEEE* **PD. 3** (1988), pp. 476–489.
90. K. Schon *et al.* 4th Int. Symp. on High Voltage Engg., paper 65.05, 1983.
91. T.M. Souders and D.R. Flach. National Bureau of Standard Special Publication 634, Proceedings of the Waveform Recorder Seminar, 1982.
92. H.K. Schoenwetter. *IEEE* **IM 33** (1984), pp. 196–200.
93. R.K. Elsley. National Bureau of Standards, Special Publication 596, Ultrasonic Materials Characterization, pp. 311–317, 1980.
94. W.B. Boyer. National Bureau of Standards, Special Publication 634, Proceedings of the Waveform Recorder Seminar, pp. 88–96, 1982.
95. T.R. McComb. *IEEE* **PWRD 2**(3) (1987), pp. 661–670.
96. W. Bucklen. *Electronic Design*, pp. 93–97, Sept. 1980.
97. W. Buchele. *Electrical Design News*, 1983.
98. D.E. Hartman. *Electronic Packaging & Production*, Mar. 1980.
99. R. Maleski *et al. IEEE* **IM 32** (1983), pp. 17–22.
100. R. Maleski *et al.* 4th Int. Symp. on High Voltage Engg., paper 65.06, 1983.
101. National Bureau of Standards, Special Publication 634, 1982.
102. R.A. Lawton. IEEE Instrumentation and Measurements Technology Conference, 1985.
103. M. Neil and A. Muto. *Electronics* **55**(4) (1982), pp. 127–132.
104. B. Peetz. *IEEE* **IM 32**(1) (1983), pp. 12–17m.
105. L. De Witt. *Handshake* (Tektronix Newsletter) **5**(1) (1980), pp. 9–12.
106. Gould. Biomation, Application Note 307.
107. LeCroy. Application Note, 1980.
108. LeCroy. Application Note, #-2004.
109. L. Ocho and P. McClellan. Handshake Application Library, 1977.
110. W.J. Hopkins, M.Sc. Thesis, Arizona State University, 1984.
111. T.R. McComb *et al.* CIGRE WG 33.03, Internal Working Document, No. 33-85, 1985.
112. H. Korff and K. Schon. *IEEE* **IM 36**(2) (1987), pp. 423–427.
113. Monsanto Research Company. Technical Report Commissioned by the US Department of Energy, 1985.
114. R. Maleski and J. Kuffel. CIGRE WG 33.03 Internal Working Document No. 38-84, 1984.
115. J. Kuffel, *et al. IEEE* **IM 35**(4) (1986), pp. 591–596.
116. IEEE Guide for Transformer Impulse Tests. IEEE Std 93, 1968.
117. E.C. Rippon and G.H. Hickling. *AIEE* **96**(2) (1945), pp. 640–644.
118. G.W. Lengnick and S.L. Foster. *AIEE* **76**(3) (1957), pp. 977–980.
119. Products Specifications #1432 and #F143.4, High Voltage Test Systems, Haefely Ltd. Basel, Switzerland.
120. F.G. Burch. On potential dividers for cathode-ray oscillographs. *Phil. Magazine*, Series 7, **13** (1932), pp. 760–774.

121. W. Zaengl. *ETZ-A* **98** (1977), pp. 792–795.
122. K. Feser. *Trans. IEEE* **PAS 93** (1974), pp. 116–127.
123. R. Malewski and N. Hylten-Cavallius. *Trans. IEEE* **PAS 93** (1974), pp. 1797–1804.
124. Ramo, Whinnery and van Duzer. *Fields and Waves in Communication Electronics*. J. Wiley, 1965.
125. H.G. Unger. *Theorie der Leitungen*. Vieweg, 1966.
126. T. Harada *et al.* Development of high-performance low voltage arms for capacitive voltage dividers. 3rd Int. Symp. on High Voltage Engg., Milan, 1979, Report 42.14.
127. G.W. Bowdler. *Measurements in High-voltage Test Circuits*. Pergamon Press, 1973.
128. E. Rinaldi, F. Poletti and A. Zingales. Constructive improvements in impulse peak voltmeters. 4th Int. Symp. on High Voltage Engg., Athens, 1983, Report 61.02.
129. Q.-C. Qi and W. Zaengl. Investigations of errors related to the measured virtual front time T of lightning impulses. *Trans. IEEE* **PAS 102** (1983), pp. 2379–2390.
130. N. Hylten-Cavallius *et al.* A new approach to minimize response errors in the measurement of high voltages. *Trans. IEEE* **PAS 102** (1983), pp. 2077–2091.
131. E. Kuffel and W.S. Zaengl. *High Voltage Engineering Fundamentals*. Pergamon Press, 1984.
132. K. Schon. Zur Rückführbarkeit von Hochspannungsmessungen (To the traceability of high-voltage measurements). PTB-Report No. E-49, Braunschweig/Germany (1994), pp. 41–60.
133. M. Beyer, W. Boeck, K. Moeller and W. Zaengl. *Hochspannungstechnik* (*High Voltage Techniques*). Springer-Verlag, 1986.
134. Komson Petcharaks. Applicability of the Streamer Breakdown Criterion to Inhomogeneous Gas Gaps. Dissertation ETH Zurich/Switzerland No. 11'192, 1995.
135. IEEE Std 1057:1994 (revision of Trial-use IEEE Std 1057-1989), IEEE Standard for digitizing waveform recorders.
136. IEEE Std 1122:1998 (revision of IEEE Std 1122-1987), IEEE Standard for digital recorders for measurements in high-voltage impulse tests.
137. IEC Document 61083:1991 Instruments and software used for measurements in high voltage impulse tests – Part 1: Requirements for instruments and its revision, Committee Draft 42/144/CD.
138. T.R. McComb, J. Dunn and J. Kuffel. Digital Impulse Measurements Meeting Standards While Pushing the Limits. ISH-99, August 1999, London, England.
139. L. Schnell (Editor). *Technology of Electrical Measurements*. John Wiley & Sons Ltd. Chichester, 1993.

Chapter 4

Electrostatic fields and field stress control

In response to an increasing demand for electrical energy, operating transmission level voltages have increased considerably over the last decades. Designers are therefore forced to reduce the size and weight of electrical equipment in order to remain competitive. This, in turn, is possible only through a thorough understanding of the properties of insulating materials and knowledge of electric fields and methods of controlling electric stress.

This chapter is therefore devoted to a discussion of some of the problems encountered when analysing even relatively simple but practical insulating systems. Teaching experience has shown that this is a necessary prerequisite in order to gain a clearer understanding of the behaviour of insulating materials. However, no attempt will be made here to introduce the basic field equations, or to treat systematically the numerous methods available for calculating electrostatic fields as this may be found in many books.[(1–4)]* Rather, this chapter is intended to provide some fundamental understanding of the importance of the interaction between fields and materials involved within an electrical insulation system by discussing some selected examples.

In h.v. engineering most of the problems concerned with the electrical insulation of high direct, alternating and impulse voltages are related to electrostatic and sometimes electrical conduction fields only. It should be emphasized however, that the permissible field strengths in the materials are interlinked with the electrostatic field distributions and thus the problems may become extremely difficult to solve.

4.1 Electrical field distribution and breakdown strength of insulating materials

It is often assumed that a voltage V between two electrodes may be adequately insulated by placing a homogeneous insulating material of breakdown strength E_b which is considered as a characteristic constant of the material, between these electrodes. The necessary separation d may then simply be calculated as $d = V/E_b$. Although the electrodes are usually well defined and are limited

* Superscript numbers are to references at the end of the chapter.

in size, the experienced designer will be able to take care of the entire field distribution between the electrodes and will realize that in many cases only a small portion of the material is stressed to a particular maximum value E_{max}. One may conclude that the condition $E_{max} = E_b$ would provide the optimal solution for the insulation problem, which thus could be solved merely by field analysis. This is true only when E_b has a very specific value directly related to the actual field distribution and can be calculated for very well-known insulating materials, such as gases (see Chapter 5, section 5.8). However, for most solid and liquid dielectrics such values are only approximately known. Hence a special approach is necessary to solve the insulation problem with fair accuracy.

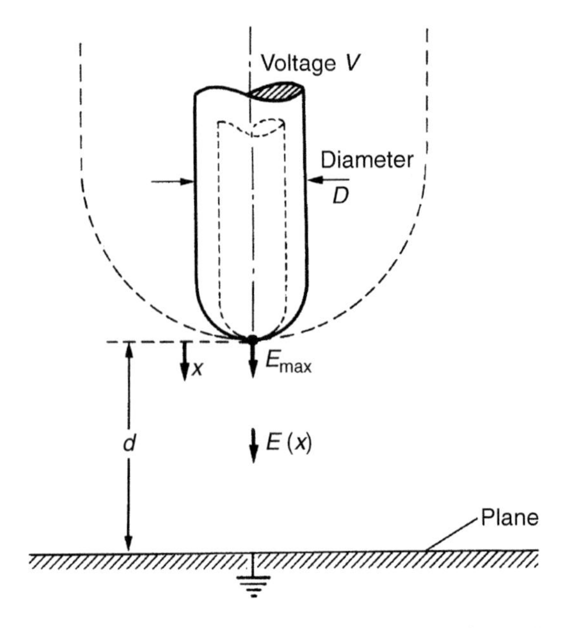

Figure 4.1 *Rod-to-plane electrode configuration (with different field efficiency factor* $\eta = V/(dE_{max})$*)*

These statements will be elucidated and confirmed by considering the simple example of an insulation system shown in Fig. 4.1, which represents a rod–plane electrode configuration insulated by atmospheric air at atmospheric pressure. Whereas the gap length and the air density are assumed to remain constant, the diameter D of the hemispherical-shaped rod will change over a very wide range as indicated by the dashed lines. Two field quantities may be defined for rods of any diameter D. These are the maximum field strength E_{max} at the rod tip and the mean value of the field strength $E_{mean} = V/d$. With these two quantities a 'field efficiency factor' η is defined as

$$\eta = \frac{E_{mean}}{E_{max}} = \frac{V}{dE_{max}} \tag{4.1}$$

originally proposed by Schwaiger.[7] This factor is clearly a pure quantity related to electrostatic field analysis only. In a more complex electrode arrangement E_{max} may appear at any point on an electrode, not necessarily coinciding with the points providing the shortest gap distance, d. η equals unity or 100 per cent for a uniform field, and it approaches zero for an electrode with an edge of zero radius.

If the breakdown of the gap is caused by E_{max} only, then the breakdown voltage V_b is obtained from eqn (4.1) as

$$V_b = E_{max} d\eta = E_b d\eta \quad (\text{with } E_{max} = E_b). \tag{4.2}$$

This equation illustrates the concept of the field efficiency factor. As $1 \geq \eta \geq 0$ for any field distribution, it is obvious that field non-uniformities reduce the breakdown voltage.

Let us now check the validity of eqn (4.2) with experimental results. In Fig. 4.2 the d.c. breakdown voltage V_b is shown for the electrode arrangement of Fig. 4.1 for $d = 10$ cm as function of η. The dashed straight line corresponds to eqn (4.2) with $E_b = 26.6$ kV/cm, a value which agrees well with measured breakdown field intensities in atmospheric air under normal conditions (temperature 20°C; pressure 101.3 kPa; humidity 11 g/m^3) for a uniform field $\eta = 1$. The highest breakdown voltage of the gap $V_b = 26.6 \times 10 = 266$ kV can also be found in Chapter 5, eqn (5.103), or in the calibration tables for measuring sphere gaps discussed in Chapter 3, Table 3.3, for spheres of large diameters, i.e. $D \geq 100$ cm. With small gaps the field distribution is uniform in the highly stressed regions. The measured breakdown voltages, obtained with positive and negative d.c. voltages, are also shown over wide ranges of η or D, the correlation of which can be computed approximately using eqn (4.20), or more accurately by a numerical computation for this special rod-plane system using the charge simulation method.[7] The differences are remarkable. The lowest measured V_b values are polarity dependent; the reason for the dependence of breakdown voltage upon polarities is explained in Chapter 5, section 5.12. Except when $\eta = 100$ per cent, the breakdown voltages are always higher than those predicted by eqn (4.2). For $\eta > 0.3$ for negative and about 0.1 for positive polarity, the breakdown is not preceded by any noticeable predischarge phenomenon (corona, partial discharge; see Chapter 5); thus it is obvious that E_b in eqn (4.2) is not a constant value for a given gap length. A calculation of breakdown field strength in atmospheric air using the streamer breakdown criterion (see eqn (5.90)) and the relevant field distribution within the gap would confirm the dependence of the breakdown

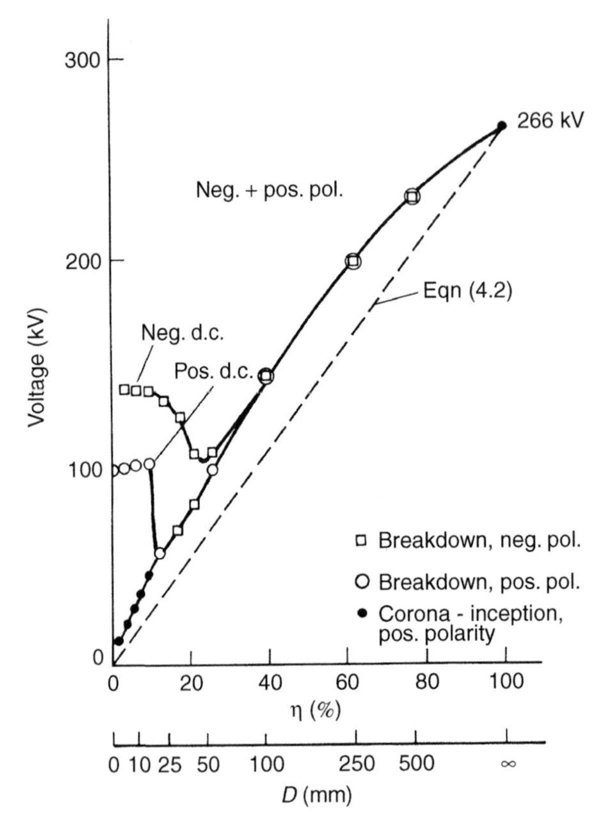

Figure 4.2 *Breakdown and corona inception voltage for the electrode arrangement of Fig. 4.1 in atmospheric air (normal conditions) with $d = 10$ cm, for positive and negative d.c. voltage (η see eqn (4.1))*

strength E_b upon rod or sphere diameter D or – more accurately – upon the actual field distribution. In reality, the lowest breakdown voltage is not reached with the smallest values of η. Below the minimum breakdown voltages, the sparkover of the gap is influenced by predischarges, which, for lower voltages, partially bridge the gap and thus produce charged particles, completely altering the field distribution due to space charges. Computation of the breakdown voltages in this region based upon physical parameters only is inaccurate due to a lack of precise knowledge of the physical data and complications introduced due to the moving space charge.

This example, which is typical for most insulation media, demonstrates the complexity of the problems, i.e. the interaction between the static field distribution, field changes due to discharge development, and parameters related to the insulation materials. Further complications arise from differences in behaviour with direct, alternating and impulse voltages. For any other material, the results would be different, even for the same electrode configuration. The proper design of insulation systems is therefore very difficult. Nevertheless, the maximum field intensity E_{max} within any insulation system may be considered as a significant quantity, even though it only serves as a guide.

In practice, data on the dielectric stresses in the insulation materials used in h.v. equipment obtained by field analysis must be supplemented by extensive tests in which the breakdown stresses are experimentally determined for similar insulation arrangements. Computations of the stresses are most advanced in gaseous dielectrics. Tests necessary for most of the other materials need not, however, involve complete experimental models which precisely simulate the actual equipment. In general, breakdown stresses are dependent upon the field distribution within high field regions, as will be shown in Chapter 5 for gaseous dielectrics. Thus, models representing only those regions in which high stresses occur are, in general, sufficient; this offers definite advantages. Apart from saving time and costs by simplifying the experimental insulation assemblies, the required voltage levels may often also be reduced significantly, as the models can be reduced in size using electrode configurations in which the low field regions are absent.

4.2 Fields in homogeneous, isotropic materials

Many electrical insulation systems contain only one type of dielectric material. Most materials may be considered to be isotropic, i.e. the electric field vector **E** and the displacement **D** are parallel. At least on the macroscopic scale many materials at uniform temperature may also be assumed to be homogeneous. The homogeneity is well confirmed in insulating gases and purified liquids. Solid dielectrics are often composed of large molecular structures forming crystalline and amorphous regions so that the homogeneity of the electrical material properties may not be assured within microscopic structures. The materials will also be assumed to be linear; that means the electric susceptibility – see Chapter 7 – is not a function of electric field strength. On a macroscopic basis, the permittivity ε will then simply be a scalar quantity correlating **D** and **E**, with $\mathbf{D} = \varepsilon\mathbf{E}$.

At this stage it is assumed here that the influence of electrical conductivity σ on the field distribution may be ignored; this is justified for most insulating materials when they are stressed by alternating voltages at frequencies above about 1 Hz. Thus, simple electrostatic field theory may be applied to most of the practical applications concerned with power frequency or impulse voltages. With direct or slowly alternating voltages the use of simple electrostatic field theory is greatly impeded by conduction phenomena. In the limiting case, the field is purely given by conduction and the correlation between field strength

E and current density j is $\mathbf{j} = \sigma\mathbf{E}$, where σ (or the complex permittivity, see Chapter 7) may be highly dependent upon time due to relaxation phenomena, upon temperature and often also upon field intensity. This problem is only mentioned here to emphasize the difficulties encountered with d.c. voltage applications.

The following examples for electrostatic field distributions are typical for h.v. insulation systems.

4.2.1 The uniform field electrode arrangement

The realization of homogeneous fields within a finite volume of insulating material is very difficult. Using parallel metal plates of limited dimensions creates the problem of a proper stress control at the edges of the plates. The field problem becomes thus three dimensional, although a rotational symmetry exists if the parallel plates are circular discs.

Depending upon the material to be tested, the breakdown strength may be very sensitive to local high fields within the whole electrode arrangement. Therefore, the highest stress should only be present in the homogeneous field region, where the plates are in parallel. A certain profile of electrodes is necessary outside the plane region to limit the dimensions, but the field strength at the curved edges should never exceed the value $E = V/d$, if V is the applied voltage and d the distance between the parallel plates. Rogowski[6] proposed electrodes for uniform fields for axially symmetrical systems whose profile follows the analytical function first introduced by Maxwell,

$$z = \frac{a}{\pi}(w + 1 + \mathrm{e}^{w}) \tag{4.3}$$

where z and w represent the complex coordinates in the z- and w-planes. Substitution of the coordinates for the complex values $z = x + iy$ and $w = u + iv$ and separation of the real and imaginary parts gives

$$x = \frac{a}{\pi}(u + 1 + \mathrm{e}^{u}\cos v);$$

$$y = \frac{a}{\pi}(v + \mathrm{e}^{u}\sin v). \tag{4.4}$$

Assuming two infinite, parallel 'plates' in the w-plane, the coordinates of which are given by $v = \pm\pi = \text{const}$, it can be recognized from eqn (4.4) that these plates are transformed into the z-plane to the left half-plane only. All other lines $v = \text{const}$ with $-\pi < v < +\pi$ can be assumed to be other equipotential lines, and all lines $u = \text{const}$ with $-\infty \le u \le +\infty$ can be assumed to be field lines in the w-plane, representing a uniform field distribution. These lines appear in the z-plane as shown in Fig. 4.3, providing the electrical field distribution of parallel plates terminating at $x = 0$. The concentration of the

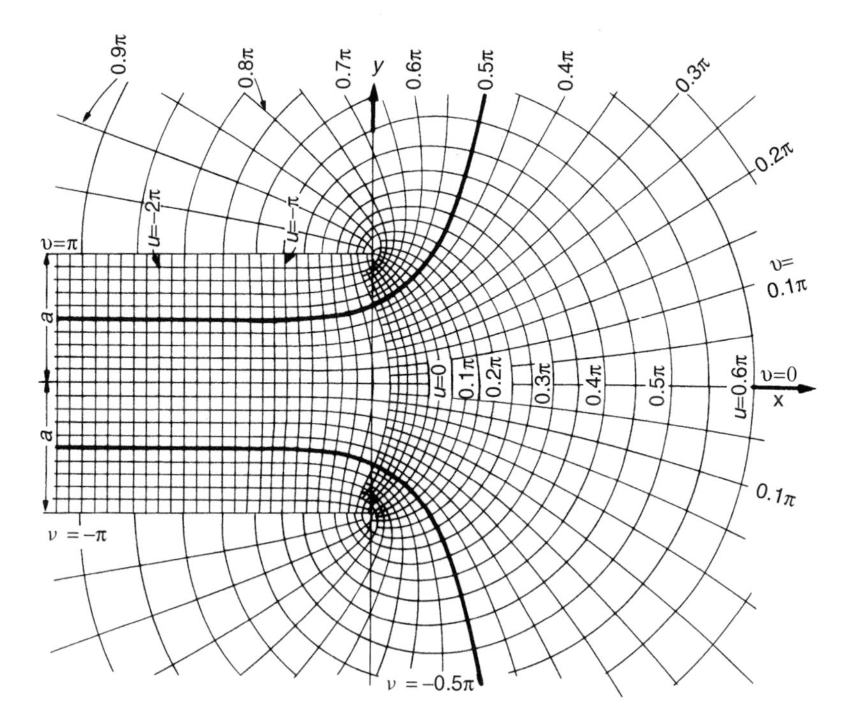

Figure 4.3 *Transformation of a square grid from a w-plane in the displayed z-plane by eqn (4.3): Rogowski's profile* ($\pm\pi 2$)

equipotential lines, $v = \text{const}$, within the z-plane may well be recognized at, or in the vicinity of, the edges of the plates.

The parallel plates, $v = \pm\pi$, are thus inadequate to fulfil the demand for field distribution whose intensity is limited to the field strength within the homogeneous part of the arrangement, i.e. for $u \leq -\pi$. It is obvious that the field strength along equipotential lines for which $-\pi < v < +\pi$ provides better conditions. For quantitative assessment the field strength within the z-plane may be computed in several ways, as follows.

From the conjugate complex field strength in the z-plane

$$E_z^* = E_x - jE_y = j\frac{\mathrm{d}w}{\mathrm{d}z} = j\frac{1}{\mathrm{d}z/\mathrm{d}w} \tag{4.5}$$

the absolute values could be computed by $|E_z^*| = \sqrt{E_x^2 + E_y^2}$.

A second possibility is given by

$$E_z = E_x + jE_y = -\text{grad } v = -\left[\left(\frac{\partial v}{\partial x}\right) + j\left(\frac{\partial v}{\delta y}\right)\right] \tag{4.6}$$

which needs a partial differentiation only.

Finally, the absolute value of E_z may be computed by

$$|E_z| = \frac{1}{\sqrt{\left(\frac{\partial x}{\partial v}\right)^2 + \left(\frac{\partial y}{\partial v}\right)^2}} \tag{4.7}$$

a method which is easiest to apply to our separated analytical function, eqn (4.4). Combining eqns (4.4) and (4.7), we may easily find the field strength as

$$|E_z| = \frac{\pi}{a\sqrt{1 + e^{2u} + 2e^u \cos v}} = f(u; v). \tag{4.8}$$

To quantify this expression with any applied voltage it is necessary to perform a calibration with the field intensity within the original w-plane. If the line $v = \pi$ is at potential $\phi = V$ and the line $v = -\pi$ at potential $\phi = -V$, the magnitude of the field strength in the w-plane is $|E_w| = 2V/2\pi = V/\pi$. Hence, the absolute magnitude in the z-plane becomes $|E_w|\,|E_z|$ or

$$|E_z| = \frac{V}{a\sqrt{1 + e^{2u} + 2e^u \cos v}} \tag{4.9}$$

For $u \lesssim 3 - 5$, $|E_z|$ is practically constant and equals V/a, but for $u = 0$ and $v = \pm\pi$ i.e. at the edges of the plates, $|E_z|$ increases to infinite values. There are, however, many equipotential lines in the z-plane for which $|E_z|$ is always limited to values $\lesssim (V/a)$. The general condition for this behaviour is given by $\cos v \gtrsim 0$ or v within $\pm(\pi/2)$. As the strongest curvature of an equipotential line will provide the smallest possible electrode arrangement, Rogowski has chosen the profile $\cos v = 0$ or $v = \pm\pi/2$, the so-called 90° Rogowski profile, which is marked by a heavier solid line in Fig. 4.3. Along this line the field strength has its maximum values between the plates in the 'homogeneous field region' $u \lesssim -(3 - 5)$ and decreases gradually within the curvature with increasing values u. As for all field lines starting at the curved part, the field strength decreases to a minimum value for $v = 0$, a breakdown should not occur between the curved regions of the electrodes. The actual distance of two metal electrodes shaped in this way would be $d = a$, and eqns (4.4) and (4.9) indicate the necessity of dimensioning the electrodes in accordance to the maximum gap length $d = a$, necessary for breakdown tests. For smaller gap lengths and the same profile, the field strength at the curved profile will decrease relative to the homogeneous field region. Disc-shaped electrodes would have the rotational centre at a field line for $u \cong -5$ or less providing any size or volume of a homogeneous field region desired. The rotation of the profile about the rotational centre converts the field to the third dimension. The additional increase of the field strength components in the x-direction by this additional curvature is, however, in general negligible. Machining of such profiles has to be carried out very carefully. A very efficient test can be made to demonstrate the performance of the electrodes: breakdown tests in pressurized sulphur hexafluoride (SF_6), a gas very sensitive to local field enhancements, must display all sparking events in the plane centre of the electrodes.

The decrease of field intensity at the outer curvature of the Rogowski profile could be prevented by a decrease of the radius of curvature, providing smaller dimensions or diameters of the disc electrodes. Profiles approaching constant field intensities at the electrode surface with magnitudes V/d also outside of the uniform field regions are, for instance, Bruce's profile[5] and Borda's profile.[6] Borda's profiles give a completely constant field intensity along the electrode surface, but as they are also based on a two-dimensional calculation, the uniformity will disappear if this profile is applied to an axisymmetric electrode. Improvements can be made by very accurate numerical, computer-aided field calculations, taking the actual surroundings as additional boundary conditions into account. For Borda's profile, such optimization was already performed by Okubo *et al.*[13]

4.2.2 Coaxial cylindrical and spherical fields

Electrode configurations providing two-dimensional cylindrical or three-dimensional spherical fields are used in h.v. equipment as well as in laboratories for fundamental research or field stress control. In a short treatment of the well-known coaxial arrangements, we shall demonstrate the fundamental differences only; some special cases give useful comparison.

Cross-sections of coaxial cylinders and concentric spheres are sketched in Figs 4.4(a) and (b), and different notations are used to distinguish between the radii of cylinders (r_1, r_2) and spheres (R_1, R_2). The electrical field distribution is symmetrical with reference to the centre of the cylinder axis or the centre point of the sphere. In both cases the lines of force are radial and the field strength E is only a function of the distance x from the centres. The cylinders are then uniformly charged over their surface with a charge per unit length Q/l, and the spheres with a charge Q, if a voltage V is applied to the two electrodes. Using Gauss's law, the field strength $E(x)$ at x is derived from the following:

- Coaxial cylinder:

$$E(x) = \frac{Q/l}{2\pi\varepsilon}\frac{1}{x} = \frac{V}{\ln(r_2/r_1)}\frac{1}{x}, \tag{4.10}$$

- Coaxial spheres:

$$E(x) = \frac{Q}{4\pi\varepsilon}\frac{1}{x^2}$$

$$= \frac{V}{(R_2 - R_1)/R_1R_2}\frac{1}{x^2}, \tag{4.11}$$

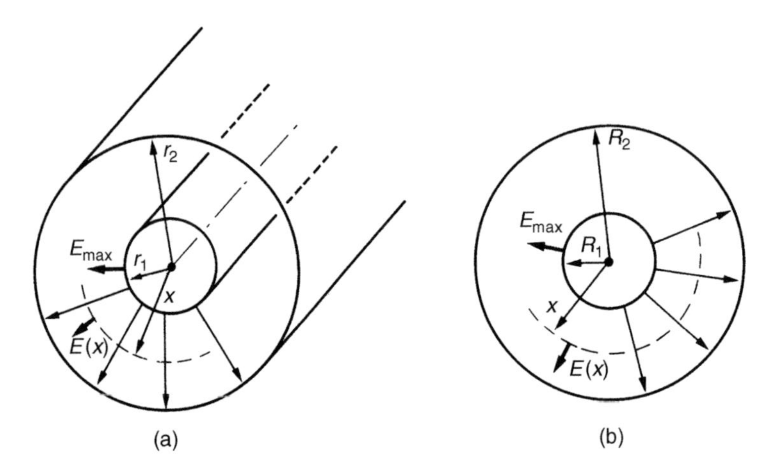

Figure 4.4 *Coaxial cylinders (a) and concentric spheres (b)*

where the subscripts 1 and 2 refer to inner and outer radii respectively. The main difference between the two field distributions is the much faster decrease of the field strength with distance x in the three-dimensional case.

Therefore, for equal geometries ($r_1 = R_1$; $r_2 = R_2$) E_{max} will always be higher in the sphere configuration. As E_{max} is reached for $x = r_1$ or $x = R_1$ respectively, we obtain for:

- Coaxial cylinders:

$$E_{max} = \frac{V}{r_1 \ln(r_2/r_1)}. \tag{4.12}$$

- Coaxial spheres:

$$E_{max} = \frac{V}{R_1(1 - R_1/R_2)}. \tag{4.13}$$

Note that the denominator in eqn (4.12) will always be larger than that in eqn (4.13), confirming the statement made above.

Let us consider a few simple examples. Spheres or sphere-like electrodes are often used as terminating electrodes of h.v. equipment, placed at the top of a bushing or a voltage divider, etc. Neglecting the influence of the structure connected to the sphere, we may roughly estimate its necessary diameter $2R_1$ assuming the ground potential is far away, i.e. $R_2/R_1 \gg 1$ in eqn (4.13). Therefore, $E_{max} \approx V/R_1$. Theoretically, atmospheric air insulation would provide a breakdown strength for large sphere diameters of about 25 kV/cm under normal conditions. Irregularities involved in the production of large electrodes and unavoidable dust particles in the air (see Chapter 3, section 3.1.1) will reduce the permissible breakdown field strength to about $E_b = 12\text{–}15\,\text{kV/cm}$. Therefore, the diameters necessary to avoid discharge inception, or even breakdown, will be

$$(2R_1) \approx 2V_{\text{peak}}/E_b. \tag{4.14}$$

For an a.c. voltage of 1 MV (r.m.s. value) diameters of about 1.9 to 2.4 m are acceptable. In this case, the greatest uncertainty is related to the breakdown strength E_b at the electrode surface used, i.e. the surface irregularities of the electrodes.[15]

A cylindrical conductor used for partial discharge-free connections in h.v. test circuits in laboratories is always limited in length, and no discharges should occur at the end of the cylinder. Obviously, a sphere of larger diameter than that for the cylindrical conductor must be located at the end, as shown in Fig. 4.5(a). The earthed walls of the laboratory will form the outer diameters of the sphere and the cylinder, and we may approximately assume that the field distributions at both electrodes are independent upon each other. Equal maximum values $E_{\max}$ are then achieved by setting eqns (4.12) and (4.13) equal. Thus the

$$\frac{R_1}{r_1} = \frac{\ln(r_2/r_1)}{(1 - R_1/R_2)} \cong \ln(r_2/r_1) \tag{4.15}$$

condition displays the necessary ratio of the diameters. As the 'radius' r_2 or R_2 of the laboratory may well be assumed to be twenty times the radii of the electrodes, this ratio becomes at least 3. For small diameters, the breakdown field strength of gases is not equal for even the same radii, as the increase of E_b is larger with decreasing radii for spherical fields. Exact values, therefore, can only be obtained by exact field computations and taking the properties of the insulation medium into account.

Busbars for SF_6-insulated, metal-enclosed equipment (GIS) are typical coaxial cylindrical arrangements. If the busbar must change the direction, a 'knee' or elbow will be necessary, as shown in Fig. 4.5(b). This problem can approximately be solved by an interconnection of a coaxial sphere with coaxial cylinder configurations, if the edges at the earthed conductors arising at the intersections are adequately rounded. All dimensions are now interlinked, and as a starting point it will first be necessary to demonstrate optimum dimensioning. For every coaxial or concentric system there is an optimum ratio of the radii, as the field stresses $E_{\max}$ reach high values for a given voltage V for small ratios of the radii as well as with too small dimensions. For coaxial cylinders we may rewrite eqn (4.12) as

$$V = E_{\max} r_2 \left(\frac{r_1}{r_2}\right) \ln\left(\frac{r_2}{r_1}\right) \tag{4.16}$$

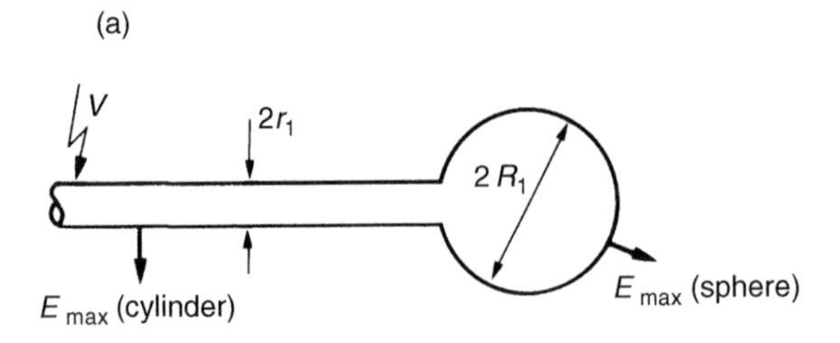

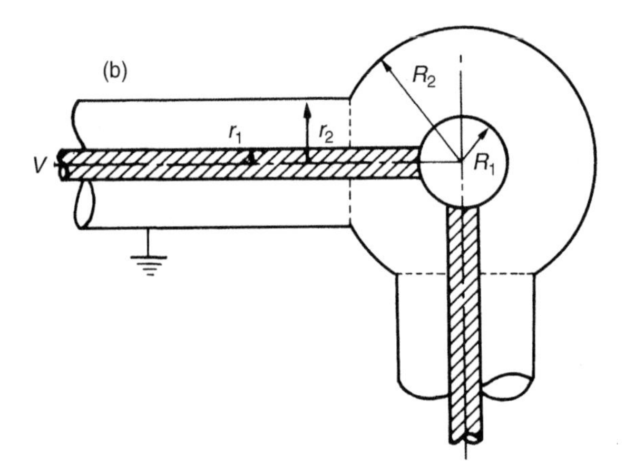

Figure 4.5 *Typical 'coaxial' arrangements. (a) Cylinder ended by a sphere within a laboratory. (b) Busbar arrangement in GIS*

and search for an optimum ratio (r_2/r_1), for which the highest voltage can be insulated with a given breakdown strength $E_b = E_{max}$ of the insulation material. For not too small diameters we may well neglect the fact that E_b depends upon r_1 for all gases or other insulation materials (see Chapter 5, section 5.9, eqn (5.111)). Thus $E_{max} = E_b$ as well as r_2 can be treated to be constant and the differentiation of eqn (4.16) with respect to r_1 gives the condition $\mathrm{d}V/\mathrm{d}r = 0$ for

$$\ln(r_2/r_1) = 1; \quad (r_2/r_1)_{opt} = e \approx 2.72; \qquad (4.17)$$
$$(V_b)_{opt} = E_b r_1.$$

This ratio is obviously a very important one in dimensioning h.v. cables or coaxial conductors insulated by homogeneous materials of any permittivity. The field efficiency factor η defined by eqn (4.1) in coaxial cylindrical system is, according to eqn (4.12),

$$\eta_{\text{cyl}} = \frac{1}{\left(\dfrac{r_2}{r_1} - 1\right)} \ln\left(\frac{r_2}{r_1}\right). \tag{4.18}$$

For $(r_2/r_1) = e$, this efficiency factor becomes 58 per cent, and is therefore quite high. Highest breakdown voltages can actually be reached with ratios of r_2/r_1 very close to the optimum value, which is demonstrated in Fig. 4.6 for SF_6-insulated cylindrical conductors within the most interesting range of r_1/r_2. For small ratios, i.e. for small diameters of the inner conductor, no direct

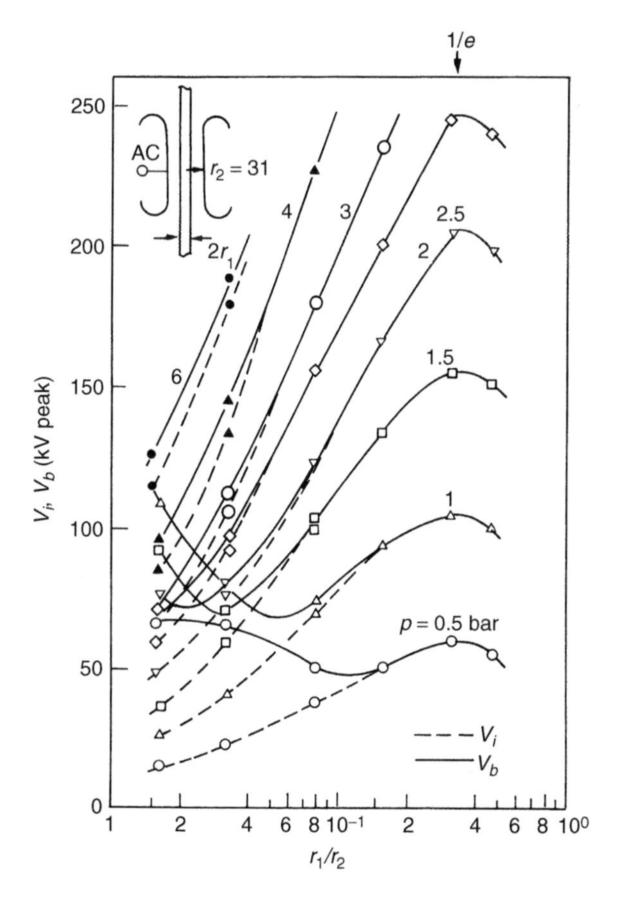

Figure 4.6 *Breakdown (V_b) and discharge inception (V_i) voltages in a coaxial cylindrical system with SF_6 insulation, in dependence of the ratio r_1/r_2. Parameter: gas pressure p. Temperature: 20°C (see reference 16)*

breakdown will occur for $E_{max} = E_b$; similar to Fig. 4.2, the actual breakdown voltage is increased by corona discharges.

For the concentric sphere arrangement, the same statements will be applicable. The optimum values for breakdown can be derived from eqn (4.13), resulting in:

$$(R_2/R_1) = 2; \quad (V_b)_{opt} = E_b R_1/2. \tag{4.19}$$

The field efficiency factor becomes in general terms

$$\eta_{sphere} = (R_1/R_2) \tag{4.20}$$

and thus is only slightly smaller for $(R_2/R_1)_{opt}$ than that found for the coaxial cylinders with optimum conditions.

Now we may solve the example in Fig. 4.5(b). If the busbar is optimally designed, i.e. $r_2 = r_1 e$, and r_1 was calculated by eqn (4.12) for given values of breakdown voltage $V = V_b$ and breakdown field strength $E_{max} = E_b$, one may apply equal breakdown conditions for the concentric sphere arrangement. Equating the values V_b/E_b for the two systems, we obtain

$$R_1(1 - R_1/R_2) = r_1 \ln(r_2/r_1) = r_1,$$

a condition which obviously has many solutions depending upon the magnitude of R_2. We may, however, select the optimum ratio R_2/R_1 for spheres, and thus we obtain the conditions

$$R_1 = 2r_1, \quad R_2 = r_2(4/e)$$

and accordingly the different gap distances related to r_1:

$$R_2 - R_1 = \left(\frac{4}{e}\frac{r_2}{r_1} - 2\right) r_1 = 2r_1,$$

$$r_2 - r_1 = \left(\frac{r_2}{r_1} - 1\right) r_1 \cong 1.72 r_1.$$

These conditions are quite favourable in practice, as the outer sphere diameter is not much bigger than that of the cylindrical system. The gap distance $(R_2 - R_1)$, however, is larger than $(r_2 - r_1)$, which could be expected by the more inhomogeneous field distribution within the three-dimensional field of the sphere arrangement.

4.2.3 Sphere-to-sphere or sphere-to-plane

In practice, the sphere-to-sphere arrangement is used for measuring high voltages with sphere gaps (Chapter 3, section 3.1.1); sphere-to-plane gaps are widely used for fundamental breakdown studies. The field distribution can be computed analytically if the spheres are assumed to become charged to their potential without any connecting leads. The influence of connecting leads upon the field distribution was recently investigated with a charge simulation program by Steinbigler.[7] The analytical results are presented here based upon the method of image charges.[1] Another possible solution based upon bipolar coordinates can be found in the literature.[17]

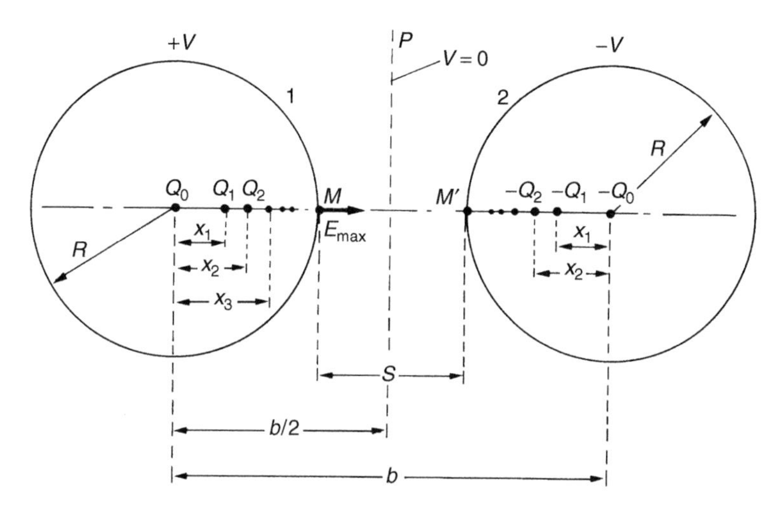

Figure 4.7 *Sphere-to-sphere or sphere-to-plane arrangement*

In Fig. 4.7 two spheres of equal diameter $2R$ separated by distance b between centres are assumed to have the potential $+V$ and $-V$ respectively. Then – and only then – the field distribution is completely symmetrical with reference to an imaginary plane P placed between the two spheres, if the plane has zero potential. Zero potential also exists at distances far away from the spheres. With a point charge $Q_0 = 4\pi\varepsilon_0 RV$ at the centre of the left sphere (1) the surface of this sphere would exactly represent an equipotential surface and could be replaced by a metal conductor, if the right sphere (2) and the plane were not present. A charge Q_0 placed at the centre of sphere (2) will produce a symmetrical field distribution with reference to the plane P, but this charge makes the potentials at the surface of the imaginary spheres non-equipotential. An improvement of these potentials is possible by placing additional image charges $+Q_1$ and $-Q_1$ in both spheres at a certain distance from their centres. This statement is confirmed by a well-known solution of the electrostatic field problem concerning a point charge in the vicinity of a conducting sphere by the image charge technique. A point charge Q and a smaller point charge Q' distant from Q and opposite in polarity are producing a field distribution, for which the zero equipotential surface is formed by a sphere. This sphere does include Q', but not at its centre. The amount of the charge Q' with reference to Q and the distance from the centre of the imaginary sphere may easily be computed by consideration of boundary conditions. Applying this principle to our problem, one may treat the disturbing charge $-Q_0$ of sphere (2) such as the above-mentioned point charge outside of the system (1) and find the necessary image charge $+Q$, within this sphere by

$$|Q_1| = Q_0 \frac{R}{b} = 4\pi\varepsilon_0 RV \frac{R}{b}$$

placed at a distance

$$x_1 = \frac{R^2}{b}$$

from the centre. The charges $+Q_0$ and $+Q_1$ inside of sphere (1) and the charge $-Q_0$ outside would make the surface of sphere (1) precisely equipotential to $+V$; but there is also a charge $-Q_1$ within sphere (2) necessary to gain symmetry with reference to P, and this charge again disturbs the equipotential character of both sphere surfaces. To compensate for these charges, further image charges $+Q_2$ inside sphere (1) and $-Q_2$ inside sphere (2) with magnitudes

$$|Q_2| = Q_1 \frac{R}{(b - x_1)} = 4\pi\varepsilon_0 RV \frac{R}{(b - x_1)} \cdot \frac{R}{b}$$

at distances $x_2 = R^2/(b - x_1)$ from their centres must be added, and this process must be continued indefinitely to reach precisely equipotential sphere surfaces. The potentials or field intensities between the two spheres could now be computed with the knowledge of the charge intensities and their position with reference to the sphere centres. The most interesting quantity is the field strength along a field line of highest field intensity, which is obviously within the shortest distance M–M' of both spheres. As the potentials ϕ at any distance r from a point charge are proportional to $1/r$ and the field strength $E(r) = -\text{grad}\,\phi$ thus proportional to $1/r^2$, the total field intensity is equal to the sum of the single intensities of all image point charges inside of both spheres. The maximum field strength at the points M and M' is, therefore, given by

$$E(R) = E_{\max} = \frac{1}{4\pi\varepsilon_0}\left\{\sum_{n=0}^{\infty}\frac{Q_n}{(R - x_n)^2} + \sum_{n=0}^{\infty}\frac{Q_n}{(b - R - x_n)^2}\right\}$$

$$= RV\sum_{n=0}^{\infty}\left\{\prod_{k=1}^{n}\left(\frac{R}{b - x_{k-1}} \atop k = 1\right)\right\}\left\{\frac{1}{(R - x_n)^2} + \frac{1}{(b - R - x_n)^2}\right\}, \quad (4.21)$$

where

$$Q_n = Q_{n-1}\frac{R}{(b - x_{n-1})} = 4\pi\varepsilon_0 RV \prod_{k=1}^{n}\left(\frac{R}{b - x_{k-1}}\right);$$

$$X_n = \frac{R^2}{(b - x_{n-1})}; \quad \text{with } n = 1, 2, 3 \ldots$$

$$x_0 = 0$$

The same expression can be used to compute the field intensity at any point on the line M–M', if the R values in the expressions $(R - x_n)^2$ and $(b - R - x_n)^2$ are replaced by a distance x measured from the centre of the sphere (1) and the point considered between M and the plane P, i.e. $R \ll x \ll b/2$.

The capacitance between the two spheres can be calculated according to Gauss's law, as the real total charge on metal spheres replacing the imaginary spheres is equal to the sum of all charges Q_n:

$$C = \sum_{n=0}^{\infty}\frac{Q_n}{2V} = 2\pi\varepsilon_0 R \sum_{n=0}^{\infty}\left\{\prod_{k=1}^{n}\left(\frac{R}{b - x_{k-1}}\right)\right\},$$

where again $x_0 = 0$.

Numerical evaluation of eqn (4.21) for different (b/R) ratios displays the following approximation for the maximum field strength E_{max}, if $S > R$:

$$E_{max} \cong 0.9\frac{V}{S/2}\frac{R + S/2}{R} \tag{4.22}$$

where $S = b - 2R$ is equal to the distance M–M', and V equals the potentials as defined in Fig. 4.4, i.e. half the voltage across the two spheres.

For a sphere-to-plane arrangement, the same equation can be used, if $S/2$ is then equal to the gap distance and V identical to the voltage applied.

As mentioned before, eqn (4.21) may be applied to compute the field intensities between oppositely charged metal spheres along a field line of highest field strength, i.e. between the shortest distance M–M'. Numerical examples for the evaluation of this equation are shown in Fig. 4.8 for different values of S/R to demonstrate the increasing non-uniformity of the electrostatic field with increasing S/R ratios. The field strength values are normalized with reference to the mean values E_{mean} according to eqn (4.1); by this the field efficiency factor η may directly be computed from the maximum values of the field intensity. This 'isolated' sphere-to-sphere arrangement is only an approximation of actual electrode arrangements, i.e. sphere gaps for the measurement of the peak values of high voltages (see Chapter 3, section 3.1.1).

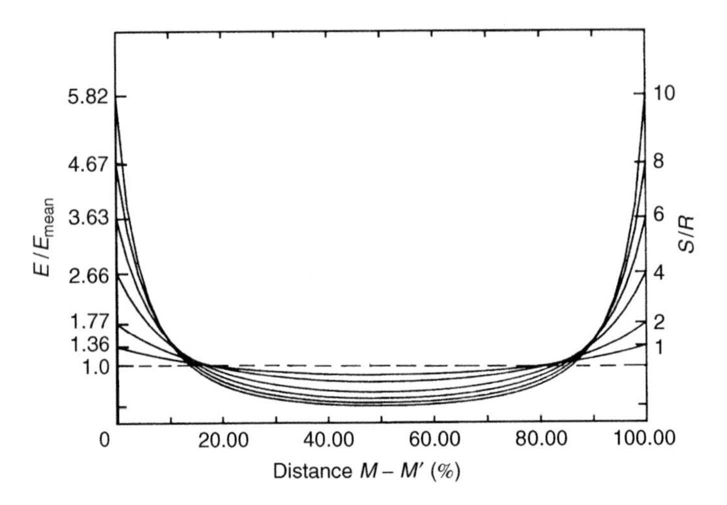

Figure 4.8 *Electric stress distribution along the axis M–M' of the sphere-to-sphere arrangement (Fig. 4.7) for various S/R ratios*

4.2.4 Two cylindrical conductors in parallel

We choose this electrode configuration for comparison with the field distribution between two oppositely charged spheres as treated above. If two or more cylindrical conductors would be at the same potential with reference to predominantly earth potential far away from the parallel conductors, the configuration of so-called 'bundle conductors' is formed, a system extensively applied in h.v. transmission lines. Due to the interaction of the single conductors the maximum field intensity at the conductors is reduced in comparison to a single cylindrical conductor, so that the corona inception voltage can significantly be increased. Solutions of the field distributions for such bundle conductors are possible by the complex variable technique, i.e. conformal mapping.(6)

For our comparison, we have to charge the two cylindrical conductors with opposite polarity to each other. Thus the field distribution can be calculated by assuming only two line charges $\pm Q/1 = \pm\rho_1$ running in parallel and eccentrically placed within the conductors. This statement is confirmed by a short calculation based upon Fig. 4.9, in which the two line charges $\pm\rho_1$ are spaced by a distance b. At any point P within the plane the potential ϕ_p may be found by the principle of superposition. As the field intensity of an individual line charge is $E(r) = \rho/(2\pi\varepsilon r)$ with r being the distance from the charge, the potentials may be found by integration. Superposition leads to

$$\phi_p = \frac{\rho_l}{2\pi\varepsilon} \ln \frac{r''}{r'} + K \tag{4.23}$$

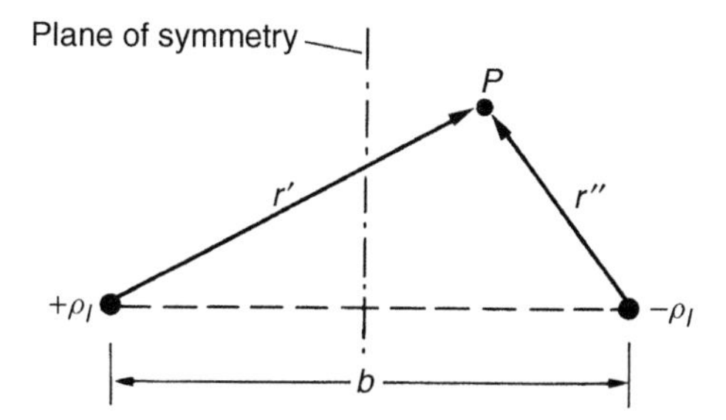

Figure 4.9 *Two line charges* $\pm\rho_l$ *in parallel*

when r' and r'' are defined in Fig. 4.9 and K is a constant found from boundary conditions. For equal line charges of opposite polarity and the potential zero at infinite distances, there is also zero potential, i.e. $\phi_p = 0$ at the plane of symmetry, $r' = r''$. Thus $K = 0$ for this special case of equal charges.

For all other ratios $r''/r' = \text{const}$, also ϕ_p is constant and may lead to any positive or negative potentials. However, all constant ratios of r''/r' generate cylindrical surfaces. These surfaces T may be assumed to be cylindrical conductors of different diameters.

Interested in two conductors of equal diameters, the two line charges will be eccentrically but symmetrically placed within these two conductors as shown in Fig. 4.10. The eccentric position, indicated by the distance c between the

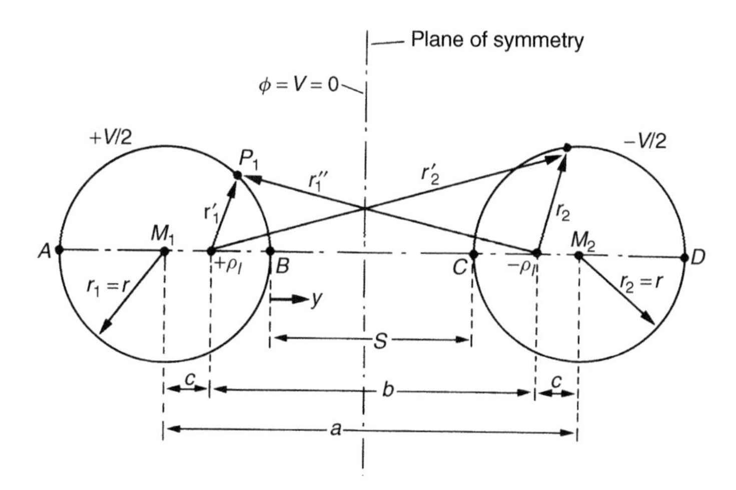

Figure 4.10 *Two equal cylindrical conductors in parallel, symmetrically charged, Fig. 4.7, and the parallel cylindrical conductors, Fig. 4.10, for equal voltages applied*

line charges and the centres M of the conductors, can easily be found for constant ratios r_1''/r_1' and r_2''/r_2' for the points P_1 or P_2 positioned at A,B or C,D. Omitting this simple calculation, we find for equal radii $r_1 = r_2 = r$

$$c = \sqrt{\left(\frac{b}{2}\right)^2 + r^2} - \frac{b}{2} = \frac{a}{2} - \sqrt{\left(\frac{a}{2}\right)^2 - r^2};$$
$$a = \sqrt{b^2 + (2r)^2} \tag{4.24}$$

The distance $c = (a - b)/2$ becomes for $r \ll (a/2)$ very small which demonstrates that for larger gaps the fields in the vicinity of the conductor surface will not be much disturbed in comparison to single conductors. For thinner conductors, we may calculate the field distribution along the flux line for the highest density, i.e. between B and C, where the field strength is highest. The potential $\phi(y)$ along this line starting at $B(y = 0)$ is provided by eqn (4.23), as

$$\phi(y) = A \ln\left(\frac{r''}{r'}\right) = A \ln\left[\frac{\left(\frac{b+S}{2}\right) - y}{\left(\frac{b-S}{2} + y\right)}\right]$$

where A is a constant given by boundary conditions and S is the gap distance. Assuming a total potential difference or voltage of V between the two conductors, A is given by $\phi(y) = +V/2$ for $y = 0$ and thus

$$A = \frac{V/2}{\ln\left(\frac{b+S}{b-S}\right)}$$

The field strength $E(y)$ becomes therefore

$$E(y) = -\frac{d\phi(y)}{dy} = A\left[\frac{1}{\left(\frac{b+S}{2}\right) - y} + \frac{1}{\left(\frac{b-S}{2}\right) + y}\right]$$
$$= \frac{V}{2} \frac{b}{\left[\left(\frac{b}{2}\right)^2 - \left(y - \frac{S}{2}\right)^2\right] \ln\left(\frac{b+S}{b-S}\right)} \tag{4.25}$$

The field distribution is symmetrical to $y = S/2$. For convenience, the distance $b = f(a, r)$ might be expressed by the gap distance also. Then

$$E(y) = \frac{V}{S} \frac{\sqrt{\left(\frac{S}{2r}\right)^2 + \left(\frac{S}{r}\right)}}{\left[1 + \frac{y}{r} - \frac{y^2}{rS}\right] \ln\left(1 + \frac{S}{2r} + \sqrt{\left(\frac{S}{2r}\right)^2 + \frac{S}{r}}\right)}. \tag{4.26}$$

The field distribution between two conductors can easiest be discussed by relating eqn (4.26) with the maximum field intensity E_{max} for $y = 0$. This ratio becomes

$$\frac{E(y)}{E_{max}} = \frac{1}{1 + \frac{y}{r} - \frac{y^2}{rS}} = \frac{r}{r + y\left(1 - \frac{y}{S}\right)}$$

In comparison to a single charged cylindrical conductor, for which this field strength ratio would be given by $r/(r + y)$ only – see eqn (4.10) – it is obvious that for all values $y/S \ll 1$ the parallel conductor is of diminishing influence. As the minimum value of E is reached for $y = S/2$, the ratio E_{min}/E_{max} becomes

$$\frac{E_{min}}{E_{max}} = \frac{1}{1 + (S/4r)}$$

A comparison of the field distributions between the sphere-to-sphere gap and the parallel cylindrical conductors is plotted in Fig. 4.11. Again we can recognize that the cylindrical fields are more uniform for the same ratios of gap distance and radii.

4.2.5 Field distortions by conducting particles

Up to now we have treated 'macroscopic' fields acting between conducting electrodes with dimensions suitable to insulate high voltages by controlling the maximum electrical field strength by large curvatures of the electrodes. In actual insulation systems the real surface of any conductor may not be really plane or shaped as assumed by macroscopic dimensions, or the real homogeneous insulation material may be contaminated by particles of a more or less conducting nature. Although a real surface roughness of an electrode, or the real shape of particles within the insulating material, may be very complex, the local distortion of the electrical field which can be assumed to be 'microscopic' in dimensions can easily lead to partial discharges or even to a breakdown of the whole insulation system.

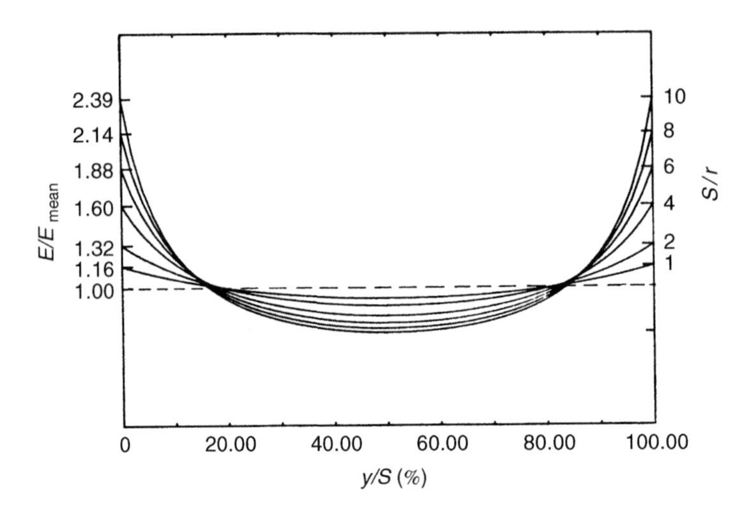

Figure 4.11 *Field strength distribution between two cylindrical conductors of equal radii r, for different ratios S/r, with S = gap distance (see Fig. 4.10). (Compare with Fig. 4.8)*

To account for such phenomena, two results of field distributions produced by spheroidal conducting particles are shown. The results are based upon one of the most powerful methods for solving Laplace's equation, the method of separation of variables, extensively treated in the book of Moon and Spencer.[(3)]

The first example is related to prolate spheroids formed within a prolate spheroidal coordinate system (η, Θ, ψ) shown in Fig. 4.12, which is related to rectangular coordinates by the equations

$$\begin{aligned} x &= a \sinh \eta \sin \Theta \cos \psi; \\ y &= a \sinh \eta \sin \Theta \sin \psi; \\ z &= a \cosh \eta \cos \Theta. \end{aligned} \qquad (4.27)$$

The prolate spheroids are surfaces of constant η values, for which

$$\left(\frac{x}{b}\right)^2 + \left(\frac{y}{b}\right)^2 + \left(\frac{z}{c}\right)^2 = 1, \qquad (4.28)$$

where $b = a \sinh \eta$; $c = a \cosh \eta$. The variable η may be changed from 0 to $+\infty$. For $\eta \to \infty$, $\sinh \eta \cong \cosh \eta$, and thus $b \cong c$, i.e. the spheroid becomes a sphere. For $\eta \to 0$, the spheroid approaches a straight line segment of length $2a$ on the z-axis, as $z = a$ for $\Theta = 0$. Due to the rotational symmetry with reference to the z-axis, the cross-sections of the spheroid for constant z value planes are circles. The surfaces of constant Θ values are hyperboloids, and for the special case of $\Theta = \pi/2$ the hyperboloid becomes the x–y-plane. ψ is the

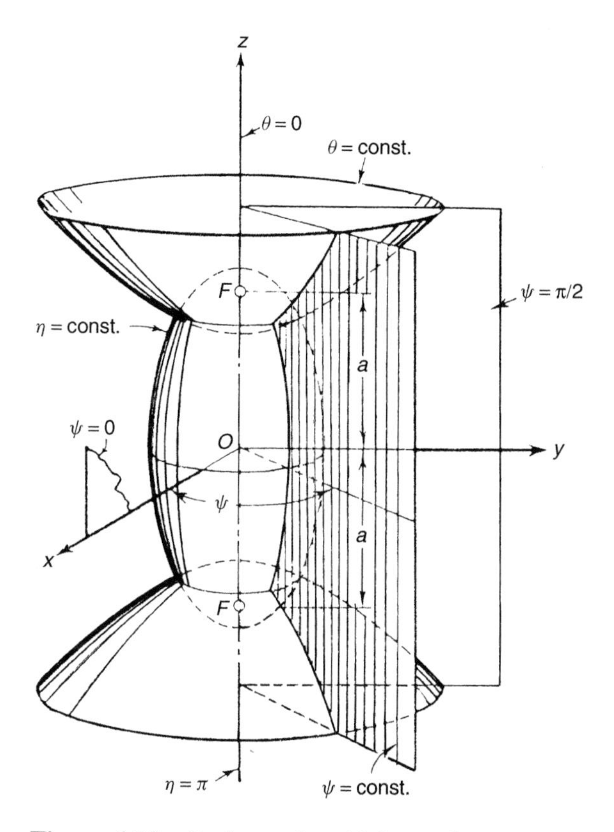

Figure 4.12 *Prolate spheroidal coordinates. The coordinate surfaces are prolate spheroids, $\eta = const$, hyperboloids, $\theta = const$, and meridian planes, $\psi = const$ (see reference 3, p. 237)*

angle measured about the z-axis, and the range of ψ is taken as $0 \leq \psi \leq 2\pi$. Surfaces of constant ψ are half-planes containing the z-axis.

The solution of Laplace's equation for this coordinate system is treated in reference 3. The results depend upon the boundary conditions, i.e. assuming scalar potentials for constant values $\eta = \eta_0$, as well as for distances far away from the centre, $\eta \to \infty$, i.e. for a sphere of infinite large diameter. The lengthy calculations are not shown here, but it may well be recognized that two special cases are of interest. A charged spheroid of potential $\phi = V$ for $\eta = \eta_0$ with a reference potential $\phi = 0$ far away from the spheroid, and a spheroid within an otherwise uniform field $E = E_0 =$ constant. In both cases the field strength $E(z)$ along a flux line in the z-direction ($x = y = 0$) is of main interest; no simple analytical expressions, however, can be achieved as Legendre functions are involved in the solutions. Therefore, only some field distributions and maximum potential gradients are reproduced from computations.[3]

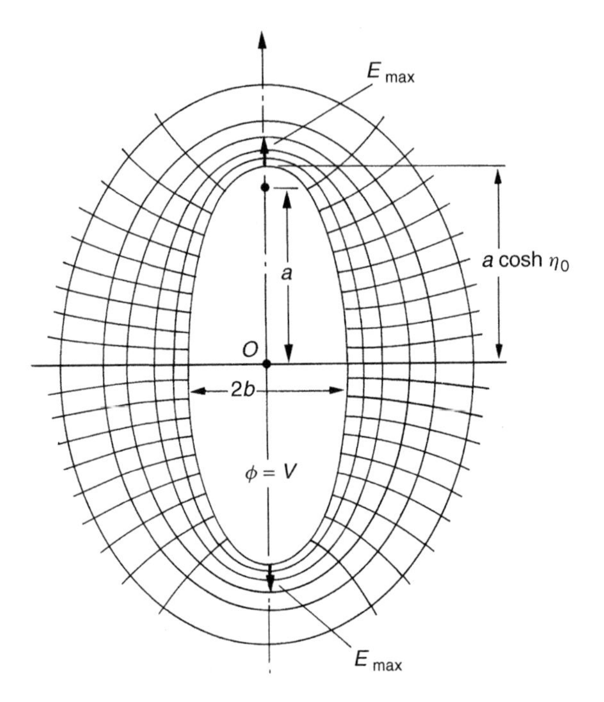

Figure 4.13 *Electrostatic field about the spheroid of Fig. 4.12 with* $\eta = \eta_0$ $\phi = V$ *(see reference 3, p. 245)*

In Fig. 4.13 the electrostatic field about a charged spheroid with potential $\phi = V$ within free space ($\phi = 0$ for $\eta \to \infty$) displays the field enhancement along the z-axis for a ratio of $b/a \cong 0.436$. The maximum field strength E_{max} will heavily increase with decreasing ratios b/a, as the curvature at this point increases. The numerical evaluation of E_{max} is shown in Fig. 4.14. Slim spheroids may be assumed to simulate capped wires whose length is large in comparison to its diameter.

Of more importance is the second case, for which a spheroid of either high permittivity equivalent to a conducting spheroid or a real metal particle placed within a dielectric material in which an originally constant uniform field E_0 was present. A field map is shown in Fig. 4.15. The potential $\phi = 0$ being present not only at the surface of the spheroid, but also for all values $z = 0$, i.e. a plane in the xy-direction simulates also a macroscopic plate-to-plate electrode arrangement, which would produce a uniform field. If a protrusion is present at the plates, whose shape is identical with half of the spheroid, the field is distorted heavily in the vicinity of this protrusion only. The map indicates that the large distortion is limited to dimensions about equivalent to the dimensions of the protrusion only, a region of the field which can be named 'microscopic'.

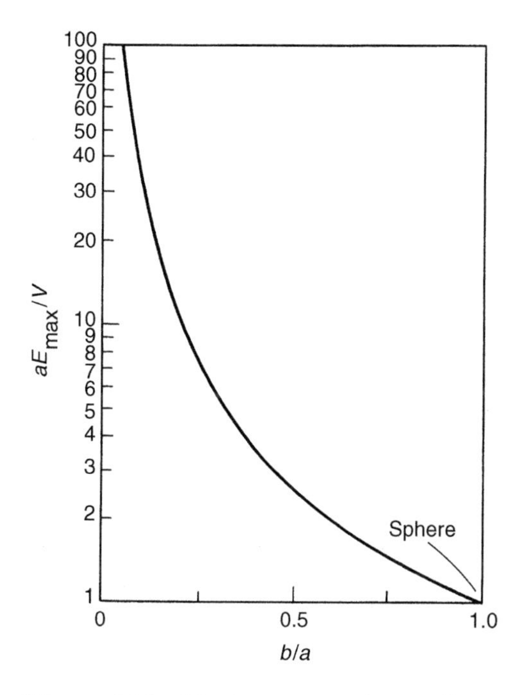

Figure 4.14 *Maximum potential gradient near a charged metal spheroid as affected by the shape of the spheroid. Major semi-axes a, minor semi-axes b, potential of spheroid V. The ordinate (aE_{max}/V) approaches infinity as $b/a \to 0$ and falls to a value of unity for a sphere (see reference 3, p. 246)*

Again, for different shapes of the spheroid the maximum values E_{max} can be calculated with reference to the uniform field strength E_0, the result of which is shown in Fig. 4.16. For $b = a$, i.e. a sphere, E_{max}/E_0 equals to 3, the well-known field enhancement factor for a half-sphere placed upon a plate electrode within a parallel plane-to-plane arrangement. Again, for slender spheroids the E_{max} values will increase to very high values, independently of the absolute size of the spheroids. Such high E_{max} values are responsible for electron emission at metal surfaces. Critical electron avalanches in gases, however, are produced not only by this high value, but also from the field distribution in the vicinity of E_{max}, so that the absolute values of the dimensions a and b become significant also.

4.3 Fields in multidielectric, isotropic materials

Many actual h.v. insulation systems, e.g. a transformer insulation, are composed of various insulation materials, whose permittivities ε are different

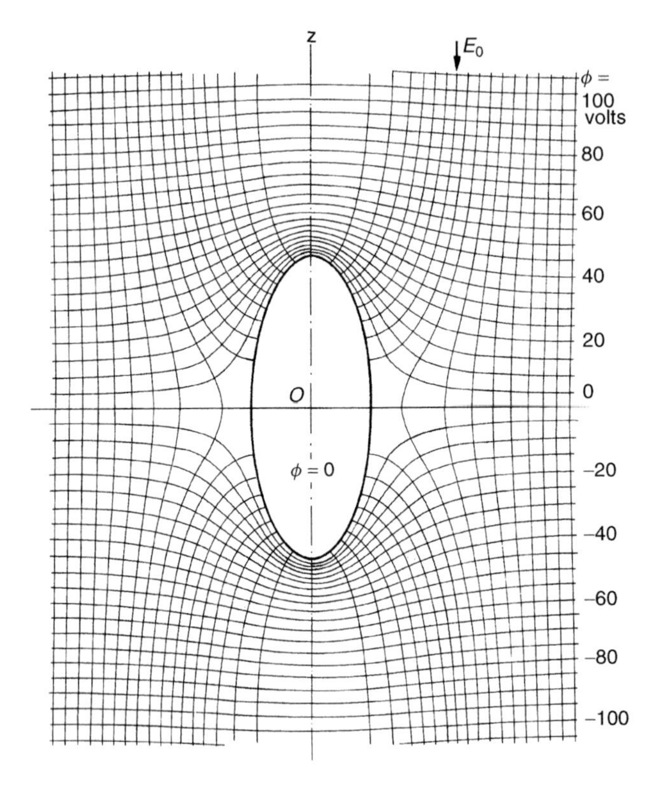

Figure 4.15 *Field distribution produced by a spheroid of high permittivity ($\varepsilon_2/\varepsilon_1 \to \infty$) within a uniform electrostatic field, E_0 (see reference 3, p. 257)*

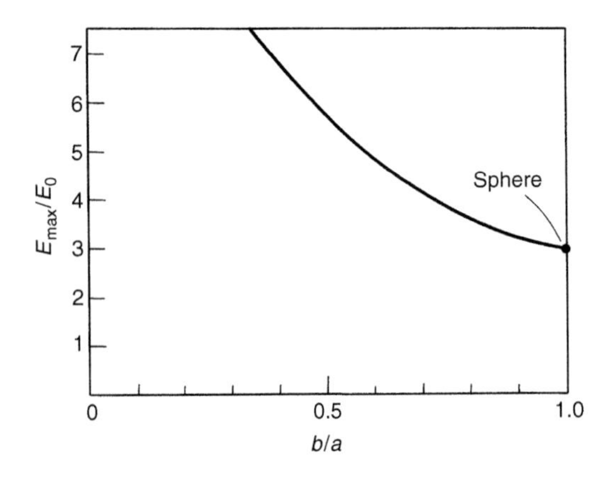

Figure 4.16 *Maximum potential gradient at a metal spheroid introduced into a uniform electric field. Here a and b are semi-axes of the ellipse, Fig. 4.13 (see reference 3, p. 258)*

from each other. The main reasons for the application of such a multidielectric system are often mechanical ones, as gaseous or fluid materials are not able to support or separate the conductors. Layer arrangements may also be applied to control electric stresses. The aim of this section is, therefore, to treat fundamental phenomena for such systems.

Only a few examples have been chosen to demonstrate principally the dangerous effects. Analytical methods for field computations in multidielectric systems containing predetermined shapes of the electrodes as well as the interfaces of the dielectrics are severely restricted. Adequate solutions are in general only possible by numerical computations or experimental field plotting techniques.

4.3.1 Simple configurations

Due to the effect of reduced electrical breakdown at the interface of two different insulation materials, the interfaces in highly stressed field regions should be normal to the field lines. The 'parallel-plate capacitor' containing two layers of different materials represented by their permittivities ε_1 and ε_2 is therefore typical for many applications. Figure 4.17 shows the arrangement and the dimensions assumed. For usual dielectric materials and power-frequency a.c. voltages, the conductivity of the materials can be neglected and hence no free charges are built up at the interface between the two layers. The displacement vectors $\mathbf{D}_1$ and $\mathbf{D}_2$ are then equal, starting from and ending at the equal free charges on the plates only. As $\mathbf{D} = \varepsilon\mathbf{E}$, and identical in both materials, the ratio of the field strength becomes

$$\frac{E_1}{E_2} = \frac{\varepsilon_2}{\varepsilon_1} \tag{4.29}$$

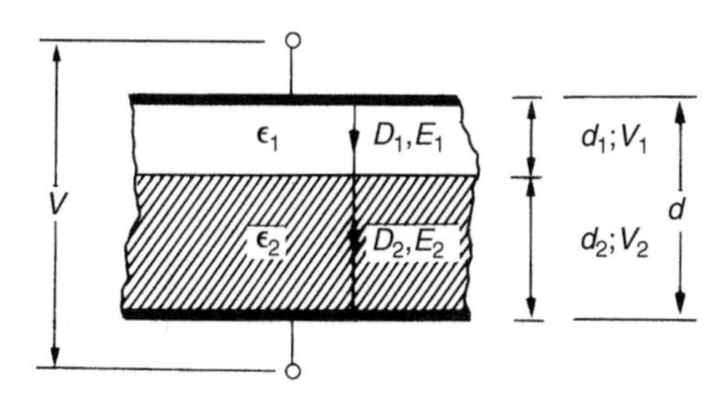

Figure 4.17 *Parallel plate capacitors comprising two layers of different materials*

and as the field remains uniform in each layer, the voltage V or potential difference between the two plates is

$$V = E_1 d_1 + E_2 d_2$$

where d_1, d_2 are the individual values of the thickness of the two dielectrics. Introducing eqn (4.29) into this equation, we obtain the following absolute values of E_1 and E_2 with reference to the voltage applied:

$$E_1 = \frac{V}{\varepsilon_1\left(\dfrac{d_1}{\varepsilon_1} + \dfrac{d_2}{\varepsilon_2}\right)} = \frac{V}{d} \frac{\varepsilon_2/\varepsilon_1}{\dfrac{d_1}{d}\left(\dfrac{\varepsilon_2}{\varepsilon_1} - 1\right) + 1} = \frac{V}{d_1 + d_2\left(\dfrac{\varepsilon_1}{\varepsilon_2}\right)} \tag{4.30}$$

$$E_2 = \frac{V}{\varepsilon_2\left(\dfrac{d_1}{\varepsilon_1} + \dfrac{d_2}{\varepsilon_2}\right)} = \frac{V}{d} \frac{1}{\dfrac{d_1}{d}\left(\dfrac{\varepsilon_2}{\varepsilon_1} - 1\right) + 1} \tag{4.31}$$

This relationship demonstrates some essential effects:

(a) The partial replacement of a given dielectric material of ε_1, for instance a gas within a gap of uniform field, by a material of higher permittivity ε_2 decreases according to eqn (4.30) the 'effective gap distance' $d' = d_1 + d_2(\varepsilon_1/\varepsilon_2)$ defined by the unaltered field strength E_1 in the original gap, as the equivalent thickness of the layer '2' becomes $d_2(\varepsilon_1/\varepsilon_2)$ only. Alternatively, for V_1, d and $(\varepsilon_1/\varepsilon_2)$ remaining constant, the field stress E_1 will always increase if the thickness of the layer '2' with higher permittivity is increased.

Although no distinct relationships exist between the permittivity of an insulation material and its permissible breakdown field strength, gases with the lowest values of ε very close to ε_0, the permittivity of the free space, are in general most sensitive to high field stresses, primarily if the gas pressure is only equal to atmospheric pressure or even lower. Any partial replacement of the gas with solid materials thus does not improve the dielectric strength of an air or gas-insulated system, as the gas will now be even more stressed than in the original system.

(b) The continuous increase of both field intensities E_1 and E_2 in the parallel plate system with increasing thickness d_2 for $\varepsilon_1 < \varepsilon_2$ given by eqns (4.30) and (4.31) can numerically be demonstrated in Fig. 4.18. The worst case is displayed for conditions when $d_1 \to 0$, i.e. for very thin layers of the low permittivity material, as the field strength increases to a value $(\varepsilon_2/\varepsilon_1)$ times the field in a system filled with one type of a material of any permittivity.

'Sandwiched' or multi-dielectric insulation systems can therefore be dangerous if the layers are of very different permittivities. However, it is also very difficult in h.v. insulation technology to avoid such or similar arrangements due to production problems. Examples are the continuous tight contact between metal electrodes and solid insulation materials, or between insulation material interfaces. The remaining voids may then become filled with gases, the breakdown strength of which may be calculated by applying Paschen's law, treated in Chapter 5, if the dimensions and discharge parameters of the gases are known. Only for very thin gaseous layers may the breakdown strength of the gas be high enough to fulfil the requirements. Thus

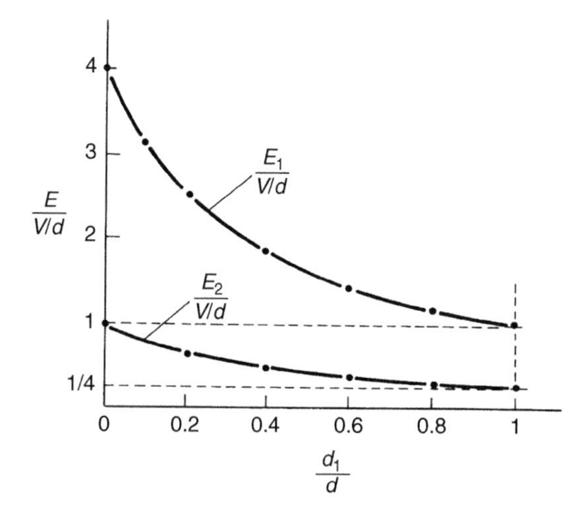

Figure 4.18 *Relative values of field strength E, and E_2 in the system of Fig. 4.17 for constant voltage V and gap distance d by varying d_1 and $d_2 = d - d_1$ for $\varepsilon_2/\varepsilon_1 = 4$*

it is essential to avoid any voids or bubbles within a solid or fluid insulation system, although this was demonstrated by a uniform field configuration only. Actual voids can be more complex in shape, and then the field strength will be more or less reduced (see section 4.3.2).

(c) Either eqn (4.30) or (4.31) may in general be used to calculate the resultant or an 'effective' mean value of the permittivity of any homogeneous mixture of dielectric materials, such as in the case of resin- or oil-impregnated kraft papers which are extensively used in h.v. apparatus. As such layers are usually oriented in parallel to the electrodes, the two-dielectric system can be subdivided into an infinite number of layers with materials designated by their intrinsic properties ε_1 and ε_2 and the resultant permittivity ε_{res} can be defined as

$$D = \varepsilon_{\text{res}} E \tag{4.32}$$

where D and E are macroscopic mean values. As the microscopic values E_1 or E_2 will remain unchanged by multiple layers, we can write

$$D = \varepsilon_{\text{res}} E = \varepsilon_1 E_1 = \varepsilon_2 E_2$$

or after replacement of E_1 or E_2 from eqn (4.30) or (4.31) and rearranging the numbers

$$\varepsilon_{\text{res}} E = \left(\frac{V}{d}\right) \frac{1}{\dfrac{(d_1/d)}{\varepsilon_1} + \dfrac{(d_2/d)}{\varepsilon_2}}$$

As before, V/d represents the mean value of the field strength within the mixture, and the distances can be replaced by relative volumes v_1 and v_2 as the relationships d_1/d and d_2/d represent also the volumes of the two materials. Therefore

$$\varepsilon_{res} = \frac{1}{(v_1/\varepsilon_1) + (v_2/\varepsilon_2)} \tag{4.33}$$

or for a mixture of n materials

$$\varepsilon_{res} = \frac{1}{(v_1/\varepsilon_1) + (v_2/\varepsilon_2) + \ldots + (v_3/\varepsilon_3)} \tag{4.34}$$

with

$$\sum_{i=1}^{n} v_i = 1$$

or 100 per cent.

A kraft paper in which 75 per cent of the volume is filled with cellulose ($\varepsilon_2 \cong 6\varepsilon_0$) should be impregnated with mineral oil ($\varepsilon_1 \cong 2.2\varepsilon_0$). Then $v_1 = 25$ per cent and $\varepsilon_{res} \cong 4.19\varepsilon_0$, which is less than one could expect by a merely linear interpolation.

Please note the assumption which has been made for deriving eqn (4.34): if the materials are not put in layers parallel to the electrodes, but in normal directions, different results will apply which can also easily be calculated.

(d) Multi-dielectric insulation systems provide distinct advantages if made of thin layers making up flexible slabs and which are well impregnated by fluids or even gases of high breakdown strength such as SF_6. Single layers may have weak points of low breakdown strength; overlapping of many layers will provide a statistical distribution of the weak points not spread throughout the insulation. Oil-impregnated h.v. power cables are typical multilayer insulation systems.

(e) The consistency of the electric flux density at interfaces without free charges can in non-uniform electrode arrangements be used to make the field stress more uniform. A typical example is the coaxial cable or coaxial capacitor with sandwiched dielectric materials sketched in Fig. 4.19(a). Applying Gauss's law to each of the individual interfaces forming equipotential areas within the field being symmetrical with reference to the centre of the cylinder axis, one may easily derive the field strength $E(x)$ as

$$E(X) = \frac{V}{\varepsilon_x X \sum_{n=1}^{m} \frac{1}{\varepsilon_n} \ln\left(\frac{r_n + 1}{r_n}\right)} \tag{4.35}$$

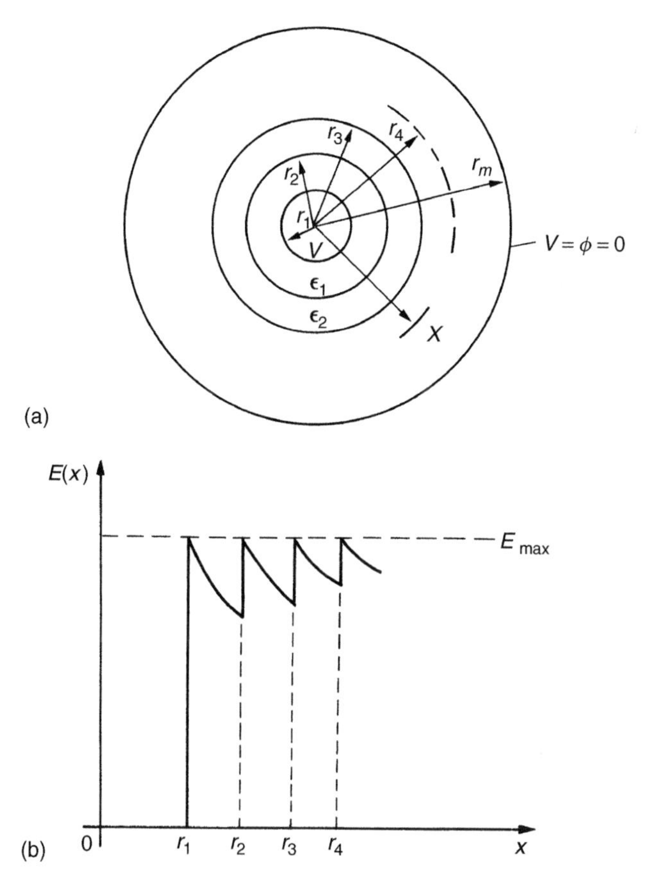

Figure 4.19 *Coaxial cable with layers of different permittivity. (a) Geometry. (b) Field distribution for* $\varepsilon_1 r_1 = \varepsilon_2 r_2 = \ldots \varepsilon_n r_n = const$

with V the voltage applied across all of the m layers, and ε_x the inherent value of permittivity within the layer of distance x from the centre.

For cylindrical conductors within each layer, $E(x)$ is proportional to $1/x$, as this is the case for cylindrical conductors; the discontinuities within the field distributions caused by the interfaces are recognized in eqn (4.35) as for $x \leq r_n$ and $x \geq r_n$ two different values of $E(x)$ will appear in the equation. As the maximum values of $E(x)$ are always at the locations $x \geq r_n$ it is possible to maintain the same values $E_{\max}$ within every layer of the dielectric, if $\varepsilon_x r_n$ remains constant. With $\varepsilon_x = \varepsilon_1, \varepsilon_2, \ldots, \varepsilon_n$ for the individual layers the conditions can be written as

$$\varepsilon_1 r_1 = \varepsilon_2 r_2 = \ldots = \varepsilon_n r_n = \text{const.}$$

The field distribution for this condition is sketched in Fig. 4.19(b). The actual applicability is, however, restricted by the limited availability of dielectric materials capable of taking full advantage of this effect. However, in h.v. oil-filled power cables high-density cellulose papers may be used for the layers close to the inner conductors, whose resultant permittivity ε_{res} is somewhat higher after impregnation than that for a lower density paper used for larger diameters.

4.3.2 Dielectric refraction

In the case when the electrical displacement vector **D** meets the interface between two media of different permittivities at an angle other than 90°, the direction of this vector will change in the second dielectric. In general, it can be assumed that no free charges are present at the interface and only (dipolar) polarization charges define the boundary conditions. Then the angles of incidence and refraction are related as follows:

$$\frac{\tan\alpha_1}{\tan\alpha_2} = \frac{E_{t1}/E_{n1}}{E_{t2}/E_{n2}} = \frac{E_{n2}}{E_{n1}} = \frac{D_{n2}/\varepsilon_2}{D_{n1}/\varepsilon_1} = \frac{\varepsilon_1}{\varepsilon_2}. \quad (4.36)$$

These quantities are illustrated in Fig. 4.20 for the conditions $\varepsilon_1 > \varepsilon_2$. In practical systems stressed with d.c. voltages the accumulation of free surface charges at the interface will take place, caused by the differing conductivities

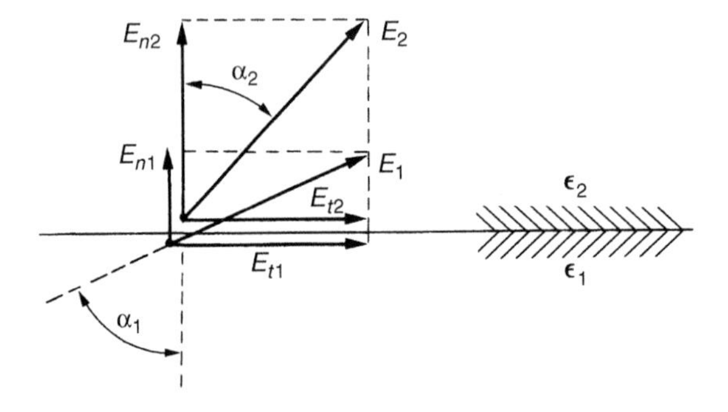

Figure 4.20 *The law of refraction applied to field intensities E for* $\varepsilon_1 > \varepsilon_2$

of the materials ('interfacial polarization', see section 7.1). For a.c. voltage applications eqn (4.36) may be applied.

Figure 4.21 shows the case when two different dielectrics are placed between parallel plane electrodes, the interface of which is not perpendicular to the electrode surface. We observe a compression of equipotential lines at the corner P increasing the field strength at that point.

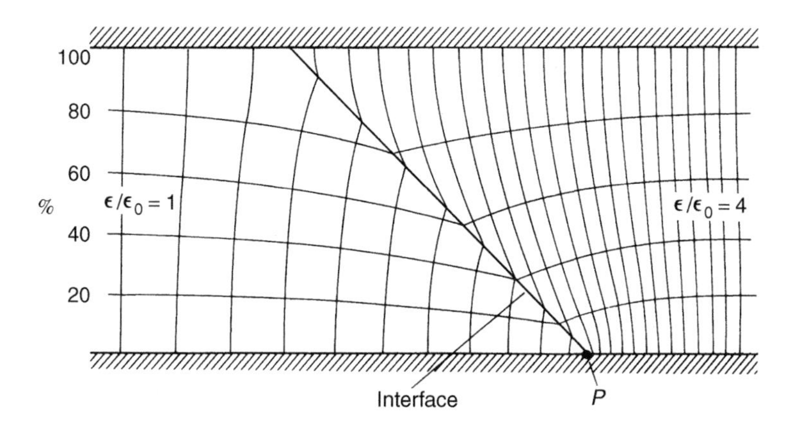

Figure 4.21 *Two different dielectric materials between plane electrodes*

If the angle between interface and electrode in this corner is <90°, the field intensity at point P becomes theoretically infinite.(8,9) This may correspond to the case when a solid dielectric is only partly attached to the electrode, leaving a void filled with dielectric materials of inadequate breakdown strength. A typical example occurs during testing of breakdown strength of solid dielectrics in the form of plates only shown in Fig. 4.22. The metal disc electrodes may be of Rogowski's profile, for which the breakdown could always be achieved within the uniform field region if only one insulation material is present. If plates of solid material with permittivity ε_2 are tested in atmospheric air only, for which the breakdown strength as well as the permittivity $\varepsilon_1 \cong \varepsilon_0$ is much lower than the corresponding values for the solid material, even for voltages much lower than the breakdown voltage, many partial discharges will appear starting from the edges as indicated in the figure. These discharges will spread over the surface of the solid dielectric and will cause breakdown

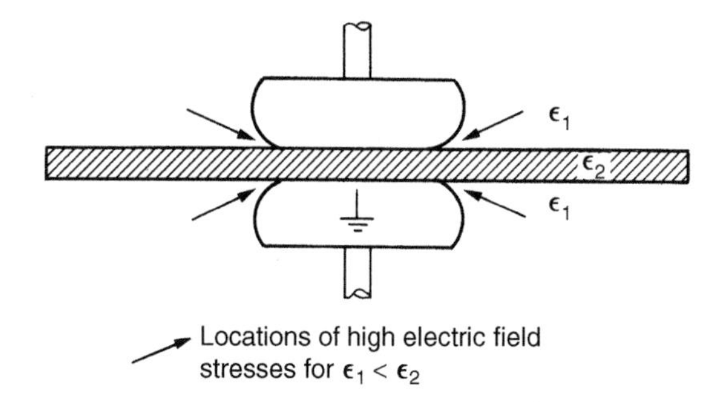

Figure 4.22 *Breakdown tests on solid dielectric plate materials (ε_2). ↑: locations of high electrical field stresses for $\varepsilon_1 < \varepsilon_2$*

outside the uniform field region. To avoid this phenomenon, either compressed gases of very high dielectric strength must be used or insulation fluids, whose permittivity ε_1 should be higher than ε_2 to avoid field enhancement, if the breakdown strength of the fluid is not as high as that of the solid dielectric. Therefore, the testing of the insulation strength of solid materials in which no electrodes can be embedded becomes a troublesome and very difficult task!

However, the law of refraction given by eqn (4.36) can be used to control the electric field, i.e. to improve the dielectric strength of an insulation system. Typical examples include spacers of solid materials used in metal-enclosed gas-insulated substations discussed briefly in section 4.2. The coaxial cylindrical conductors are not only insulated by compressed sulphur hexafluoride (SF_6) but also partly by spacers necessary for mechanical support of the inner conductor. If only a disc of solid material would be used as shown in Fig. 4.23(a), the flux lines would not be refracted or distorted and the field strength $E(x)$ along the interface between gas and solid material would follow eqn (4.10). This means that only tangential components of the electric field, E_t, are stressing the interface and E_t is not constant along the surface. As the permissible E_t values at boundaries are always lower than field magnitudes within the adjacent materials, the spacers can be formed in such a way, that all

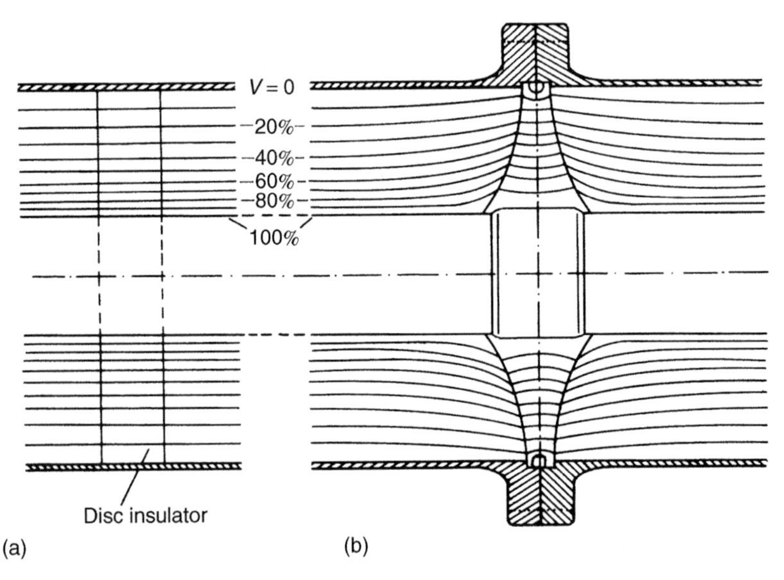

Figure 4.23 *Epoxy disc insulator supporting the inner conductor of a coaxial cylinder system. (a) Simple disc spacer: no refraction of equipotential lines. (b) Shaped spacer for approximate constant tangential field stress at the interface gas insulator*

E_t components along the interface remain nearly constant. One possible solution is shown in Fig. 4.23(b), and the same equipotential lines as in Fig. 4.23(a) are used to demonstrate the change of the field distribution. The field map for this example was computed by numerical methods (see section 4.4).

4.3.3 Stress control by floating screens

The necessity for applying electrostatic stress control in h.v. apparatus was demonstrated up to now for fields in homogeneous materials as well as for multi-dielectric insulation systems. But in all examples only two metal electrodes have been used whose potential was fixed by the applied voltage. For homogeneous or single dielectric materials the field stress control was thus merely possible by providing an adequate shape or contour for these electrodes, and Rogowski's profile (Fig. 4.3) may be considered an example. The insertion of multi-dielectric systems between the main electrodes also provided a means for stress control, as shown for the case of coaxial cables and its two-dimensional field configuration (Fig. 4.19). For this special case, the interface between layers of differing permittivity was equipotential. Dielectric interfaces for general three-dimensional insulation systems, however, are often difficult to shape such as to provide equipotential surfaces, which would avoid any tangential field intensities with its limited breakdown strength.

Equation (4.36) indicates that flux lines penetrating from a dielectric of high permittivity into one of much lower permittivity are forced to leave the material nearly perpendicular to its surface. This means that the equipotential lines or surfaces in the dielectric of smaller permittivity are forced to be nearly parallel to the interface as is found for metal electrodes. A dielectric of very high ε_r values thus behaves similarly to an electrically conducting material, and for $\varepsilon \to \infty$ the boundary conditions for metal surfaces are reached. For this reason insulation systems, including floating screens, whose potential is solely controlled by the field distribution of the dielectric materials attached to the screens, can be treated as a multi-dielectric system.

Field stresses are controlled by means of such screens in many h.v. apparatus such as capacitor-type cable terminations,[14] bushings, potential transformers, etc. The 'capacitor bushing' or 'field stress-controlled bushing' will be treated as a typical example and will demonstrate the complexity of the problems involved. Bushings are used to run a high potential cylindrical conductor H through a grounded wall or barrier W (see Fig. 4.24). The wall may consist of a partially conducting concrete or brick, a grounded metal tank of a transformer or any other metal-enclosed h.v. apparatus. The insulation materials used on both sides of the wall can, therefore, be different. For transformers mineral oil insulation inside the tank is typical and atmospheric air is commonly used outside. For this case, the bushing also provides sealing. The main task, however, is provided by the electrical insulation of the conductor H from the

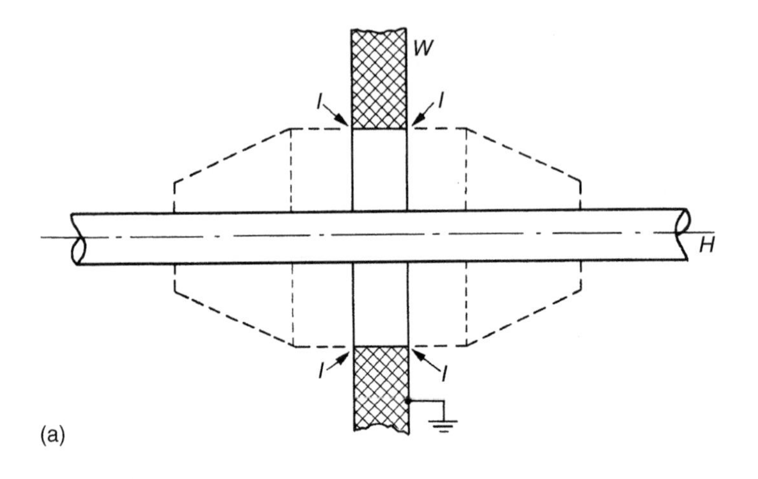

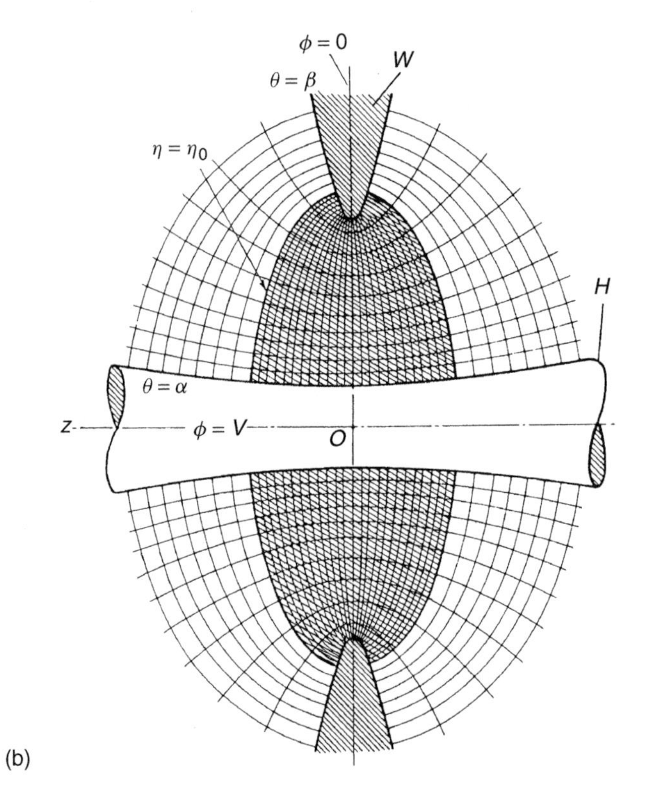

Figure 4.24 *Simple bushing arrangements for introduction into the problems solved by capacitor bushings, Fig. 4.25. (a) The problem. (b) A theoretical solution*[3]

wall and its mechanical support by an insulation system, which is as compact as possible.

To demonstrate the actual complex problems involved in the design of field stress controlled bushings, reference is first made to Fig. 4.24(a). The barrier W perpendicular to the plane of drawing contains a circular opening in which the cylindrical conductor H is centred. Even without taking into account the mechanical support and assuming a homogeneity of the insulation material used, the numerical calculation or graphical field mapping would show that the high field intensity regions are at the conductor surface within the plane of the wall and at the edged contours I of the wall opening. To support the cylindrical conductor and to avoid a breakdown between the wall and the conductor caused by the high field regions, we may add a solid insulation material as shown by dashed lines. The solid dielectric would withstand the high field stresses in the vicinity of the cylindrical conductor; at I at the wall opening, however, the high tangential components of the field intensities at the interface between solid and gaseous (or liquid) dielectrics used on both sides of the wall would cause surface discharges and lead to relatively low flashover voltages.

One solution to the problem is to use special contours conducting electrodes and the solid-type insulator supporting the h.v. conductor. An adequate solution proposed by Moon and Spencer(3) is shown in Fig. 4.24(b), displaying a field map for a three-dimensional arrangement computed with an oblate spheroidal coordinate system. The electrodes W and H are shaped to give equipotential lines. The solid dielectric is shaped such that it prevents refraction of flux lines; the equipotential lines, calculated analytically, remain unchanged. Although the field intensities are still highest at the shaped conductors, the improvement of field distribution in comparison to the simple configurations of Fig. 4.24(a) is clear. As far as we know, bushings of this type have never been used, as it is much too difficult to produce such a device.

In practice, the solution is in the introduction of 'floating' electrodes, as will be shortly demonstrated. Let the cylindrical h.v. conductor H be surrounded by many layers of thin dielectric sheets of permittivity ε where ε is considerably higher than ε_0, the permittivity of vacuum or air used for the 'external' insulation of the bushing. Figure 4.25(a) shows a simplified cross-sectional view of such an arrangement, in which the dielectric sheets of different lengths are interleaved with thin conducting foils providing the floating electrodes; these are shown by the thicker lines. Neglecting now the influence of the dielectric conductivity, i.e. the permittivity ε_0, of the external insulation, which is acceptable for a large number of conducting foils, we may treat this system as an arrangement of coaxial cylindrical capacitor units which are series connected. Thus a 'capacitor bushing' is formed. As indicated, the length $l_0, l_1, \ldots, l_n$ of the sheets is increasing from the wall W to the centre conductor H, and the conditions for the different lengths can be provided by boundary conditions.

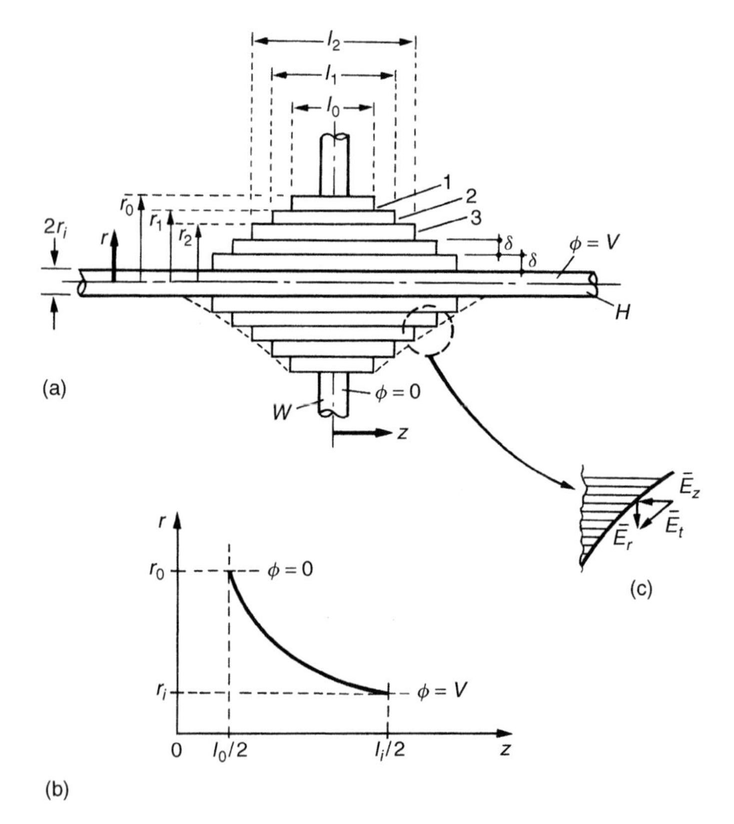

Figure 4.25 *Capacitor bushing. (a) Coaxial capacitor arrangement. (b) Profile of foils for constant radial field intensity $\overline{E_r}$ (mean value). (c) Definition of field intensity components*

Let us assume the simplest boundary condition, for which the mean value of the field intensity $\overline{E}_r$ acting within the sheets remains constant. If every sheet is of equal thickness δ, each of the coaxial capacitor units is stressed by equal voltages $\Delta V = E_r \delta$, if all capacitances are equal. Then $C_1 = C_2 = \ldots C_n$ with

$$C_1 = \frac{2\pi \varepsilon l_0}{\ln(r_0/r_1)},$$

$$C_2 = \frac{2\pi \varepsilon l_1}{\ln(r_1/r_2)},$$

or

$$\frac{l_0}{\ln(r_0/r_1)} = \frac{l_1}{\ln(r_1/r_2)} = \ldots = \frac{l_n}{\ln(r_n/r_{n+1})} \tag{4.37}$$

Apart from this exact solution, an approximation is possible for thin sheets. Then $r_{n+1} = r_n - \delta$ and $(\delta/r_n) \ll 1$ even for the smallest radius r_i of the inner conductor, yielding

$$\ln(r_n/r_{n+1}) = \ln\left[\frac{1}{1-(\delta/r_n)}\right] \cong \delta/r_n$$

With this approximation, eqn (4.37) becomes

$$l_0 r_0 \cong l_1 r_1 \cong \ldots \cong l_n r_n, \tag{4.38}$$

where $0 \leq n \leq N$, with N equal to the total number of sheets. As N is quite high, we may replace the discrete numbers l_n and r_n by the variables $z = 1/2$ (Fig. 4.25(a)) and r. Equation (4.38) then defines a two-dimensional profile or contour of the conducting foil edges as sketched in Fig. 4.25(b). The given boundary condition provides a hyperbolic profile, along which the potential ϕ increases steadily between r_0 and r_i. Neglecting the very local increase of the field intensities produced by the edges of the conducting foils, we can now assume quite constant values between two foils, i.e. a mean radial field strength $\overline{E}_r$ as indicated in Fig. 4.25(c). Whereas $\overline{E}_r$ stresses the insulation material of the sheets only, an even more significant axial component of a field intensity E_z is introduced between the conducting foil edges, as sketched in Fig. 4.25(c). The solid material from the active part of the capacitor bushing also shares a boundary with the surrounding dielectric material, in general atmospheric air, or mineral oil.

Therefore, this interface is in summary stressed by a tangential field intensity E_t, which has the components of $\overline{E}_r$ and $\overline{E}_z$, the latter defined as a mean value of the potential difference $\Delta\phi$ between each adjacent foil and the increase $\Delta l = 2\Delta z$ in sheet length, i.e. $\overline{E}_z = \Delta\phi/\Delta l$. For the small values of $\Delta\phi$, Δl and Δr provided by the large amount of sections we may neglect all discontinuities and write in differential terms

$$\mathrm{d}\phi = -E_r\,\mathrm{d}r = -E_z\,\mathrm{d}z$$

where in addition $\mathrm{d}z = \mathrm{d}l/2$ with the dimensions assumed in Fig. 4.25.

Apart from the sign, the gradient $\mathrm{d}\phi$ can be assumed to be a voltage drop $\mathrm{d}V$ across the capacitor elements as formed by adjacent electrodes. As each capacitor element $C' = (2\pi\varepsilon l r)/\mathrm{d}r$, in which the differential term $\mathrm{d}r$ is used to quantify the quite small distances between the electrodes or thickness of the dielectric sheets, the voltage drop becomes

$$\mathrm{d}V = \frac{i}{\omega C'} = \frac{i}{\omega 2\pi\varepsilon(lr)}\mathrm{d}r = K_0\frac{\mathrm{d}r}{(lr)}$$

In this equation, the product (lr) can be taken from any locus of an electrode as indicated in eqn (4.38). As all capacitor elements are series connected, the displacement current i is always the same and thus K_0 is a constant. Now we are able to indicate the field stresses in general terms as

$$E_r = K_0 \frac{1}{(lr)} \tag{4.39a}$$

$$E_z = 2K_0 \frac{1}{(lr)} \bullet \frac{\mathrm{d}r}{\mathrm{d}l} \tag{4.39b}$$

From both equations, different kinds of boundary conditions may be introduced to find criteria for the field stresses for the tangential components which in general control the external flashover voltages. We use one example only, defined by the assumption that the radial stress E_r shall be constant. Then, according to eqn (4.39a), the product $(lr) = K_0/E_{r0}$ where E_{r0} will be a convenient design criteria. Applying this term to eqn (4.39b) provides the axial component as

$$E_z = -2K_0 \frac{1}{l^2} = -2E_{r0} \frac{r}{l} = -2 \frac{E_{r0}^2}{K_0} r^2 \tag{4.40}$$

This dependency shows the strong increase of the axial field strength with increasing diameter of the dielectric sheets. It contributes to a non-homogeneous potential distribution at the surface of the laminated unit and a highest stress at the grounded flange promoting surface flashover, as the mean value of the tangential field intensity according to Fig. 4.25(c) is

$$\overline{E_t} = \sqrt{(\overline{E_r})^2 + (\overline{E_z})^2}, \tag{4.41}$$

if the surface is very close to the foil edges.

In practice, such a dimensioning of a capacitor bushing due to constant mean values of the radial field intensity is not at all ideal and the calculations performed only indicate the problems. But one could readily see that the conducting foils can be used to control the internal fields E_r as well as the field strength distribution along the boundaries E_t, and that it will not be possible to keep both these values constant. The dimensioning of bushings thus becomes a difficult task, as also other important factors have to be taken into account. First, the surrounding insulation materials cannot be neglected. Secondly, h.v. bushings, in general, are not made from a single dielectric material, which is often provided by oil- or resin-impregnated kraft paper or plastic films. Protection of the active part is provided by porcelain or other solid insulation material housings, having different permittivities and introducing additional field refraction at the interfaces of the differing materials. Due to the heat generated within the h.v. conductor H the permissible radial field intensity may be lower than within the outer regions. Finally, careful attention must be paid to the edges of the conducting foils which form regions of locally high fields, as the equipotential lines will not necessarily leave the foils at the edges, but in its vicinity only. Therefore, foils made from semiconducting materials of still adequate conductivity are sometimes used to adapt the potentials at the foil edges to the field distribution forced by the dielectric materials outside the field-controlled regions. Analytical computations of bushing designs are, therefore, supplemented by numerical computations, which take into account the very different boundary conditions.(18,19)

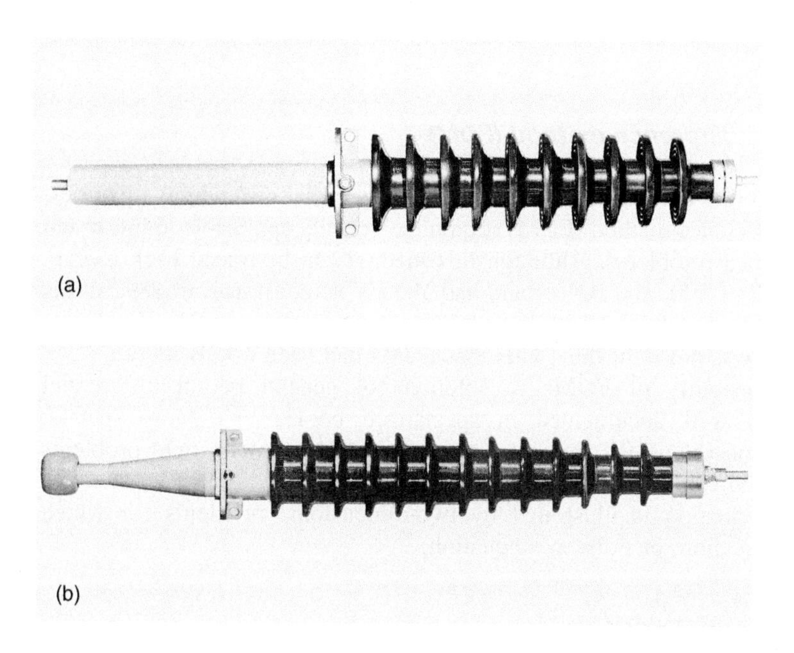

Figure 4.26 *Photographs of bushing (courtesy Micafil, Switzerland). (a) Wall bushing, outdoor–indoor, rated 123 kV/1250 A. (b) Transformer bushing with 'dry' insulation, rated 170 kV a.c./630 A, BIL 750 kV*

Figure 4.26 shows two typical types of bushings, a transformer bushing with its asymmetry due to different external insulation, and a common wall bushing for air-to-air insulation, but one side indoor, the other side (porcelain) outdoor.

4.4 Numerical methods

In recent years several numerical methods for solving partial differential equations and thus also Laplace's and Poisson's equations have become available. There are inherent difficulties in solving partial differential equations and thus in Laplace's or Poisson's equations for general two- or three-dimensional fields with sophisticated boundary conditions, or for insulating materials with different permittivities and/or conductivities. Each of the different numerical methods, however, has inherent advantages or disadvantages, depending upon the actual problem to be solved, and thus the methods are to some extent complementary.[20]

The aim of this chapter is to introduce the most widely used methods in such a way that a fundamental knowledge is provided and to give the user of a computer program an understanding of the limitations of the results and computations.

4.4.1 Finite difference method (FDM)

Apart from other numerical methods for solving partial differential equations, the finite difference method (FDM) is quite universally applicable to linear and even non-linear problems. Although this method can be traced back to C.F. Gauss (1777–1855), and Boltzmann had already demonstrated in 1892 in his lectures in Munich the applicability of difference equations to solve Laplace's equation, it was not until the 1940s that FDMs had been used widely.

The applicability of FDMs to solutions of general partial differential equations is well documented in specialized books.[21,22] More specific references concerning the treatment of electric and magnetic field problems with the FDM can be found in reference 23.

This introduction is illustrated by two-dimensional problems for which Laplace's equation, or Poisson's equation,

$$\nabla^2\phi = \frac{\partial^2\phi}{\partial x^2} + \frac{\partial^2\phi}{\partial y^2} = f(x, y)$$

applies. The field problem is then given within an x–y-plane, the area of which has to be limited by given boundary conditions, i.e. by contours on which some field quantities are known. It is also known that every potential ϕ and its distribution within the area under consideration will be continuous in nature. Therefore, an unlimited number of $\phi(x, y)$ values would be necessary to ascribe the potential distribution. As every numerical computation can provide a limited amount of information, only a discretization of the area will be necessary to exhibit nodes for which the solution may be found. Such nodes are produced by any net or grid laid down upon the area.

As any irregular net, however, would lead to inadequate difference equations replacing the original partial differential equation, and would thus be prohibitive for numerical computations, the FDM is in general applied to regular nets or polygons only. These restrictions will be understood more clearly by the derivation of the differential equations. Regular polygons which can fill a plane are squares, triangles or hexagons, but squares or equilateral triangles are the only regular nets in common use. As also such square or triangular nets will in general not fit into the boundaries, we will derive the difference equations for rectangles, which can at least at given boundaries be formed in such a way that nodes can also be laid down upon the boundary. As squares are a particular case of rectangles, the result applies also for squares, and the inherent difficulties in using irregular nets are better understood.

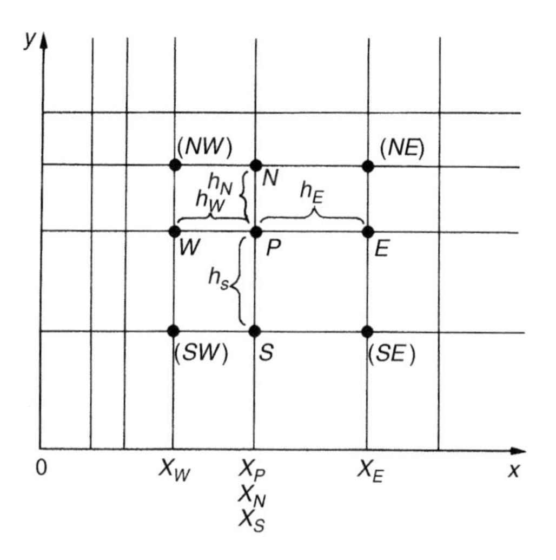

Figure 4.27 *Irregular rectangular net composed of horizontal and vertical lines, with node abbreviations*

In Fig. 4.27 such an irregular net of rectangles is sketched within the $x-y$-plane, with the sides of all rectangles parallel to the x- or y-axis. All points of intersection between the vertical and horizontal lines create nodes, but only five nodes will be of immediate special interest. These are the four neighbouring nodes, N, W, S, and E around a point P, which are given compass notations (N = north, etc.). Let us assume now that the potentials at these nodes, i.e. $\phi(S)$, $\phi(E)$, $\phi(N)$ and $\phi(W)$ are known either from given boundary conditions or other computational results. As the potential within the field region is continuous, it is obviously possible to expand the potential at any point (x, y) by the use of Taylor's series. If this point is identical with node P, the series for the two variables x and y is given by

$$\phi(x, y) = \phi(P) + \frac{1}{1!}[(x - x_P)\phi_x(P) + (y - y_p)\phi_y(P)]$$
$$+ \frac{1}{2!}[(x - x_P)^2\phi_{xx}(P) + 2(x - x_P)(y - y_P)\phi_{xy}(P)$$
$$+ (y - y_P)^2\phi_{yy}(P)]$$

$$+\frac{1}{3!}[(x-x_P)^3\phi_{xxx}(P)+3(x-x_P)^2(y-y_P)\phi_{xxy}(P)$$
$$+3(x-x_P)(y-y_P)^2\phi_{xyy}(P)+(y-y_P)^3\phi_{yyy}(P)]$$
$$+R'_{xy}(P) \qquad (4.42)$$

where the rest of the series $R'_{xy}(P)$ is of still higher order. In eqn (4.42) the derivatives

$$\phi_x(P)=\left[\frac{\partial\phi(x,y)}{\partial x}\right]_P,\ \phi_y(P)=\left[\frac{\partial\phi(x,y)}{\partial y}\right]_P,\ \phi_{xy}(P)=\left[\frac{\partial\phi(x,y)}{\partial x\partial y}\right]_P,\ \text{etc.}$$

are used for abbreviation.

Every potential $\phi(x, y)$ in the close vicinity of node P may be expressed by eqn (4.42) with adequate accuracy, if the Taylor's series is interrupted by ignoring terms containing third derivatives of the potential, as they will be multiplied by small distances h to the power of 3 or more. Thus, the potentials of the nodes E, N, W and S can be expressed by the following equations, in which the small distances $(x - x_P)$ and $(y - y_P)$ are substituted by the proper values h_E, h_N, h_W and h_S:

$$\phi(E)=\phi(P)+h_E\phi_x(P)+\tfrac{1}{2}h_E^2\phi_{xx}(P) \qquad (4.43a)$$

$$\phi(N)=\phi(P)+h_N\phi_y(P)+\tfrac{1}{2}h_N^2\phi_{yy}(P) \qquad (4.43b)$$

$$\phi(W)=\phi(P)+h_W\phi_x(P)+\tfrac{1}{2}h_W^2\phi_{xx}(P) \qquad (4.43c)$$

$$\phi(S)=\phi(P)+h_S\phi_y(P)+\tfrac{1}{2}h_S^2\phi_{yy}(P) \qquad (4.43d)$$

The sums of eqns (4.43a) and (4.43c), and eqns (4.43b) and (4.43d) respectively yield the following two equations:

$$\phi(E)+\phi(W)-2\phi(P)=(h_E-h_W)\phi_x(P)+\tfrac{1}{2}(h_E^2+h_W^2)\phi_{xx}(P), \qquad (4.44a)$$

$$\phi(N)+\phi(S)-2\phi(P)=(h_N-h_S)\phi_y(P)+\tfrac{1}{2}(h_N^2+h_S^2)\phi_{yy}(P), \qquad (4.44b)$$

The derivatives $\phi_x(P)$ and $\phi_y(P)$ may be expressed by the well-known first order approximations

$$\phi_x(P)\cong\frac{\dfrac{h_W}{h_E}[\phi(E)-\phi(P)]+\dfrac{h_E}{h_w}[\phi(P)-\phi(W)]}{(h_E+h_W)}$$

$$=\frac{h_W}{h_E(h_E+h_W)}\phi(E)+\frac{(h_E-h_W)}{h_Eh_W}\phi(P)-\frac{h_E}{h_W(h_E+h_W)}\phi(W) \qquad (4.45a)$$

$$\phi_y(P)\cong\frac{h_S}{h_N(h_N+h_S)}\phi(N)+\frac{(h_N-h_S)}{h_Nh_S}\phi(P)-\frac{h_N}{h_S(h_N+h_S)}\phi(S) \qquad (4.45b)$$

Introducing eqn (4.45a) into (4.44a) and eqn (4.45b) into (4.44b) will result in

$$\phi_{xx}(P) = \frac{2\phi(E)}{h_E(h_E + h_W)} + \frac{2\phi(W)}{h_W(h_E + h_W)} - \frac{2\phi(P)}{h_E h_W}, \tag{4.46a}$$

$$\phi_{yy}(P) = \frac{2\phi(N)}{h_N(h_N + h_S)} + \frac{2\phi(S)}{h_S(h_N + h_S)} - \frac{2\phi(P)}{h_N h_S}, \tag{4.46b}$$

With these approximations for the second derivatives of the potential functions in the x- and y-direction at node P it is now possible to solve Laplace's or Poisson's equation,

$$\nabla^2\phi = \phi_{xx} + \phi_{yy} = \begin{cases} 0 & \text{(Laplacian region)} \\ -F(x, y) & \text{(Poissonian region)} \end{cases} \tag{4.47}$$

where $F(x, y) = \rho/\varepsilon$ for electrostatic fields within a medium of permittivity ε and containing distributed charges of density $\rho(x, y)$. The solution may then be written as

$$D_{EP}\phi(E) + D_{NP}\phi(N) + D_{WP}\phi(W) + D_{SP}\phi(S) + D_{PP}\phi(P) + \tfrac{1}{2}F(P) = 0 \tag{4.48}$$

with

$$D_{EP} = \frac{1}{h_E(h_E + h_W)}, \quad D_{NP} = \frac{1}{h_N(h_N + h_S)},$$

$$D_{WP} = \frac{1}{h_W(h_E + h_W)}, \quad D_{SP} = \frac{1}{h_S(h_N + h_S)},$$

$$D_{PP} = -\left(\frac{1}{h_E h_W} + \frac{1}{h_N h_S}\right).$$

This difference equation is a valid approximation of the original differential equation (4.47), but it should be recalled that the validity is restricted to the individual point P under consideration. The same form is, however, valid for every node within a net.

Before further considerations we shall discuss briefly the common simplifications. For every two-dimensional problem most of the field regions can be subdivided by a regular square net. Then $h_E = h_N = h_W = h_S = h$, and eqn (4.48) is reduced to

$$\phi(E) + \phi(N) + \phi(W) + \phi(S) - 4\phi(P) + h^2F(P) = 0 \tag{4.49}$$

It may well be understood now that difference equations similar to eqn (4.48) can be derived for other nets or other neighbouring nodes to P within our rectangular net shown in Fig. 4.27 if the proper derivations are performed. In this figure, for instance, one could involve the nodes NE, NW, SW and SE

either neglecting the nodes N, W, S and E or including the nodes. In all cases, the unknown potential $\phi(P)$ can be expressed by the surrounding potentials which are assumed to be known for the single difference equation.

All difference equations, however, are approximations to the field equation due to the omission of higher order terms in eqns (4.42) and (4.45). The error due to these approximations is known as truncation error, and it is important to investigate this error carefully if the values h are not chosen properly. The treatment of the truncation error is beyond the scope of this chapter and may be found elsewhere.[21,23]

The numerical evaluation of the difference equation (4.49) is obviously simple, but time consuming and therefore FDM is now seldom used.

4.4.2 Finite element method (FEM)

By reviewing the theory of the FDM it was readily demonstrated that the partial derivatives of the basic field equations (4.47) have been replaced by their algebraic difference form, eqns (4.46a, b), resulting in a system of algebraic equations which have to be solved. Due to the approximations made during this derivation the algorithm was linear of the first order ('first order FDM algorithm').

Although there are different approaches to arrive mathematically at finite element approximations[24] and the most general approach is traced back to the variational problem of extremization of a specific functional, the most common basis is related to a very well-known physical property of fields. The FEM concerns itself with minimizing the energy in the whole field region of interest, when the field may be electric or magnetic, of Laplacian or Poissonian type.

In this section a specific rather than general treatment of the method will be presented. To reduce the size of equations, we will restrict ourselves to two-dimensional electric fields of Laplacian type. Convenient applications even for complicated Poissonian electric fields as, for instance, present around coronating h.v. lines are documented in references 28 and 29, as well as the practical application to magnetic fields.[25–27]

Let us consider a steady state electrostatic field within a dielectric material whose conductivity may be neglected and whose permittivity may be dependent upon the direction of the field strength $\mathbf{E}$ (anisotropic material) or not (isotropic dielectric). Then as no space charge should be present or accumulated, the potentials would be excited from boundaries (metal electrodes) between which the dielectric material is placed. Assuming a Cartesian coordinate system, for such a Laplacian field, the electrical energy W stored within the whole volume R of the region under consideration is

$$W = \iiint_V \left[\frac{1}{2} \left\{ \varepsilon_x \left(\frac{\partial \phi}{\partial x} \right)^2 + \varepsilon_y \left(\frac{\partial \phi}{\partial y} \right)^2 + \varepsilon_z \left(\frac{\partial \phi}{\partial z} \right)^2 \right\} \right] \mathrm{d}x\,\mathrm{d}y\,\mathrm{d}z. \quad (4.50)$$

ε_x, ε_y and ε_z would be anisotropic permittivity coefficients, and it should be noted that even in an isotropic material with $\varepsilon_x = \varepsilon_y = \varepsilon_z = \varepsilon$, the absolute values of ε may change at boundaries between different dielectric materials. The reader may easily verify from any small volume element $\mathrm{d}V = (\mathrm{d}x\,\mathrm{d}y\,\mathrm{d}z)$ that the expressions $(\varepsilon\nabla^2\phi/2)$ within eqn (4.50) are energy densities per unit volumes $\mathrm{d}V$.

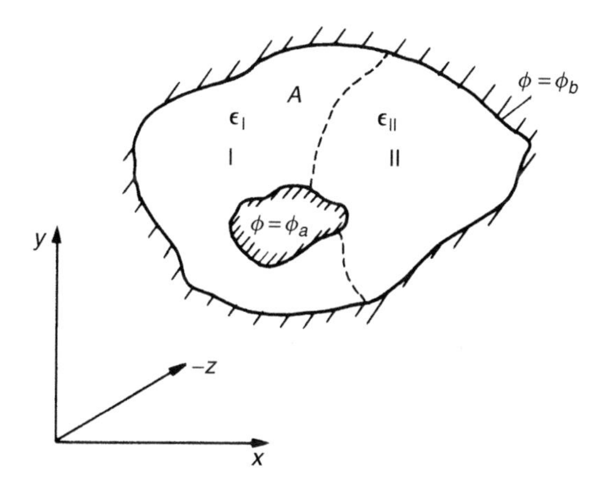

Figure 4.28 *Limited field area A within x–y-plane representing a two-dimensional field within space (x, y, z-coordinates). Dielectric material subdivided by dashed line into regions I and II*

Furthermore, it is assumed that the potential distribution does not change in the z-direction, i.e. a two-dimensional case. Figure 4.28 displays the situation for which the field space is reduced from the volume R to the area A limited by boundaries with given potentials ϕ_a and ϕ_b (Dirichlet boundaries). The dielectric may be subdivided into two parts, I and II, indicated by the dashed interface, for which the boundary condition is well known (see section 4.3), if no free charges are built up at the interface. The total stored energy within this area-limited system is now given according to eqn (4.50) by

$$W = z \iint_A \left[\frac{1}{2} \left\{ \varepsilon_x \left(\frac{\partial\phi}{\partial x} \right)^2 + \varepsilon_y \left(\frac{\partial\phi}{\partial y} \right)^2 \right\} \right] \mathrm{d}x\,\mathrm{d}y \tag{4.51}$$

where z is a constant. W/z is thus an energy density per elementary area $\mathrm{d}A$.

Before any minimization criteria based upon eqn (4.51) can be applied, appropriate assumptions about the potential distribution $\phi(x, y)$ must be made. It should be emphasized that this function is continuous and a finite number of derivatives may exist. As it will be impossible to find a continuous function for the whole area A, an adequate discretization must be made.

For our two-dimensional problem it is possible to use rectangular or square elements, as was done for the FDM (see Fig. 4.27), or multiple node composite elements for three-dimensional regions. There are, however, definite advantages in using simple, irregularly distributed elements with an arbitrary triangular shape (or tetrahedrons for three-dimensional problems). Such triangles can easily be fitted to coincide with boundary shapes, i.e. the nodes of a triangular element system can be placed upon curved boundaries, a situation often met in h.v. insulation systems.

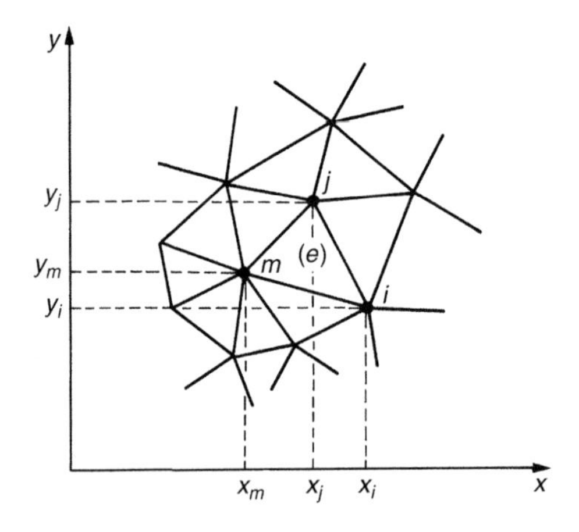

Figure 4.29 *A section of area A (Fig. 4.28) subdivided into irregular triangular elements. Notation of nodes i, j, m for element e*

Figure 4.29 shows such a subdivision of a part of a two-dimensional region A (in Fig. 4.28) into triangular elements. Let us consider one of these elements (indicated by e) and the nodes i, j and m, and formulate basic functions for the potential distributions $\phi(x, y)$ within this element. In the development of the FEM no *a priori* restrictions are placed on this basic function. However, for triangular elements, or a 'triangular element family',(29) polynomials can be of higher order, such as

$$\phi(x, y) = \alpha_1 + \alpha_2 x + \alpha_3 y + \alpha_4 x^2 + \alpha_5 xy + \alpha_6 y^2 + \ldots, \tag{4.52}$$

for which the inter-element compatibility can be improved. The increase in accuracy by applying higher order functions is compensated, however, by an increase in computation time and computation complexity, and thus most of the algorithms used are based upon a first order approximation, i.e. a linear dependency of ϕ on x and y in eqn (4.52). Following this simple basic function, this equation is reduced to

$$\phi(x, y) = \phi = \alpha_1 + \alpha_2 x + \alpha_3 y. \tag{4.53}$$

This means that the potentials within each element are linearly distributed and the field intensity, whose components in the x- and y-directions can be computed for eqn (4.53) by simple derivation, is constant. In this respect, the FEM and the FDM coincide.

For such a first order approximation, the three coefficients α_1, α_2 and α_3 for element e are easily computed by means of the three *a priori* unknown potentials at the respective nodes i, j and m, given by the equations

$$\left.\begin{aligned} \phi_i &= \alpha_1 + \alpha_2 x_i + \alpha_3 y_i \\ \phi_j &= \alpha_1 + \alpha_2 x_j + \alpha_3 y_j \\ \phi_m &= \alpha_1 + \alpha_2 x_m + \alpha_3 y_m \end{aligned}\right\} \tag{4.54}$$

The coefficients may be computed applying Cramer's rule, the result being

$$\alpha_1 = \frac{1}{2\Delta_e}(a_i\phi_i + a_j\phi_j + a_m\phi_m); \tag{4.55a}$$

$$\alpha_2 = \frac{1}{2\Delta_e}(b_i\phi_i + b_j\phi_j + b_m\phi_m); \tag{4.55b}$$

$$\alpha_3 = \frac{1}{2\Delta_e}(c_i\phi_i + c_j\phi_j + c_m\phi_m); \tag{4.55c}$$

where

$$\left.\begin{aligned} a_i &= x_j y_m - x_m y_j \\ a_j &= x_m y_i - x_i y_m \\ a_m &= x_i y_j - x_j y_i \end{aligned}\right\} \tag{4.55d}$$

$$\left.\begin{aligned} b_i &= y_j - y_m \\ b_j &= y_m - y_i \\ b_m &= y_i - y_j \end{aligned}\right\} \tag{4.55e}$$

$$\left.\begin{aligned} c_i &= x_m - x_j \\ c_j &= x_i - x_m \\ c_m &= x_j - x_i \end{aligned}\right\} \tag{4.55f}$$

and

$$\left.\begin{aligned} 2\Delta_e &= a_i + a_j + a_m \\ &= b_i c_j - b_j c_i \end{aligned}\right\} \tag{4.56}$$

From eqn (4.56) and Fig. 4.29, one may easily see that the symbol Δ_e is used to describe the area of the triangular element i, j, m.

With eqns (4.53), (4.54) and (4.55), the potential distribution of the element can thus be related to the potentials of the adjoining nodes, and simple numbers a_i, b_i, etc. for each element can be computed once division of the two-dimensional region into triangular elements has been performed. Introducing these values into eqn (4.53) the result is (index e used for 'element'):

$$\phi_e(x, y) = \frac{1}{2\Delta_e}[(a_i + b_i + c_i)\phi_i + \dots + (a_j + b_j + c_j)\phi_j + (a_m + b_m + c_m)\phi_m] \quad (4.57)$$

This equation may also be written as

$$\phi_e = [N_i, N_j, N_m] \begin{Bmatrix} \phi_i \\ \phi_j \\ \phi_m \end{Bmatrix} \quad (4.58)$$

in which the functions N are the 'shape functions', as they will depend upon the shape of the finite elements used. Such shape functions can be derived for many kinds and shapes of elements including the rectangles used for the FDM.[(24)]

With eqn (4.57) or eqn (4.53), the energy noted within the element is easily computed. According to eqn (4.51), the partial derivatives for each element are:

$$\frac{\partial \phi}{\partial x} = \alpha_2 = f(\phi_i, \phi_j, \phi_m)$$
$$\frac{\partial \phi}{\partial y} = \alpha_3 = f(\phi_i, \phi_j, \phi_m) \quad (4.59)$$

However, as we are not interested in the absolute values of these energies, the components of the electric field intensities should not be introduced into eqn (4.51) at this stage. The FEM is based upon the minimization of the energy within the whole system, and thus only derivatives of the energies with respect to the potential distribution are of interest. According to eqn (4.51), the energy functional, i.e. the energy per unit length in the z-direction for our specific case, is for the element under consideration

$$\chi^e = \frac{W_e}{z} = \frac{1}{2}\Delta_e \left\{ \varepsilon_x \left(\frac{\partial \phi}{\partial x}\right)^2 + \varepsilon_y \left(\frac{\partial \phi}{\partial y}\right)^2 \right\}_e \quad (4.60)$$

as $\iint dx\,dy$ provides the area of the element, Δ_e. For further consideration only, isotropic dielectric material is assumed within each individual element, i.e $\varepsilon_x = \varepsilon_y = \varepsilon_e$.

Whereas the functional χ^e in eqn (4.60) is only dependent upon the node potentials of the individual element (eqn (4.59)), an equivalent functional χ for the whole system (area A, Fig. 4.28) will exist. The formulation regarding the minimization of the energy within the complete system may thus be written as

$$\frac{\partial \chi}{\partial \{\phi\}} = 0 \quad (4.61)$$

where $\{\phi\}$ is the potential vector for all nodes within this system. For our specific element, the minimizing equations can easily be derived by differentiating eqn (4.60) partially with respect to ϕ_i, ϕ_j and ϕ_m. Taking also eqns (4.55) and (4.59) into account, the differentiation with respect to ϕ_i yields

$$\begin{aligned}\frac{\partial \chi^e}{\partial \phi_i} &= \frac{1}{2}\varepsilon_e \Delta_e \left(2\alpha_2 \frac{\partial \alpha_2}{\partial \phi_i} + 2\alpha_3 \frac{\partial \alpha_3}{\partial \phi_i}\right) \\ &= \frac{1}{2}\varepsilon_e(\alpha_2 b_i + \alpha_3 c_i) \\ &= \frac{\varepsilon_e}{4\Delta_e}[(b_i^2 + c_i^2)\phi_i + (b_i b_j + c_i c_j)\phi_j + (b_i b_m + c_i c_m)\phi_m]\end{aligned} \tag{4.62}$$

The set of all three equations may best be expressed in matrix form as

$$\begin{aligned}\frac{\partial \chi^e}{\partial \{\phi\}^e} &= \frac{\varepsilon_e}{4\Delta_e}\begin{bmatrix} (b_i^2 + c_i^2) & (b_i b_j + c_i c_j) & (b_i b_m + c_i c_m) \\ & (b_j^2 + c_j^2) & (b_j b_m + c_j c_m) \\ \text{sym} & & (b_m^2 + c_m^2) \end{bmatrix} \begin{Bmatrix} \phi_i \\ \phi_j \\ \phi_m \end{Bmatrix} \\ &= [h]^e \{\phi^e\}\end{aligned} \tag{4.63}$$

The matrix $[h]^e$ is well known as the 'stiffness matrix' for the individual element, as it contains the sensitivity of the functional with respect to the potentials. (Within a mechanical, elastic system, this matrix relates mechanical nodal forces to displacements.) It contains well-known geometric quantities (eqns (4.55), (4.56)) and the material's permittivity ε_e.

It is now possible to establish a set of algebraic equations with which the still unknown potentials can be computed. No assumptions have been made so far concerning the *a priori* known potentials at the boundaries, and Fig. 4.28 displayed only the finite field regions with a Dirichlet boundary. The triangular element (e) within Fig. 4.29 is surrounded by other triangular elements and it is seen that any node potential within such a system will depend upon the potentials of the surrounding nodes. The number of these nodes is dependent upon the triangular network, but that number is always small. Thus it is sufficient to demonstrate the last step with a set of only four triangular elements as shown in Fig. 4.30. The elements are numbered from 1 to 4, and the nodes by 1 to 5.

Application of eqn (4.61) to this set of elements yields

$$\frac{\partial \chi}{\partial \phi_5} = 0, \tag{4.64}$$

where χ is the energy functional of the system with the four elements. Before this equation is evaluated, it is convenient to write the stiffness matrix,

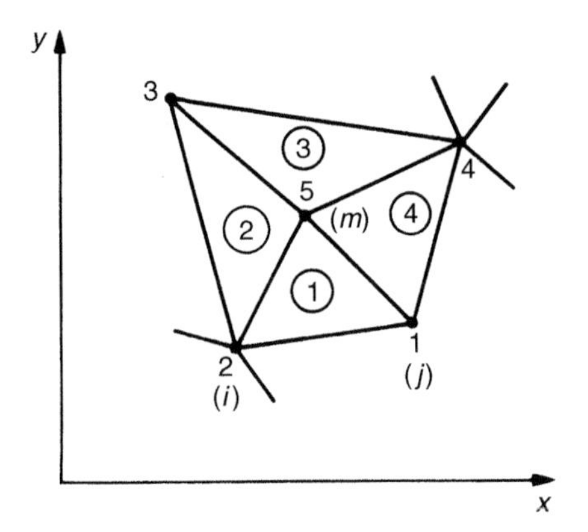

Figure 4.30 *Node 5 connected to four triangular elements (i, j, m identify element l)*

eqn (4.63), as

$$[h]^e = \begin{bmatrix} (h_{ii})_e & (h_{ij})_e & (h_{im})_e \\ & (h_{jj})_e & (h_{jm})_e \\ \text{sym} & & (h_{mm})_e \end{bmatrix} \tag{4.65}$$

where

$$(h_{ii})_e = \frac{\varepsilon_e}{4\Delta_e}(b_i^2 + c_i^2);$$

$$(h_{ij})_e = \frac{\varepsilon_e}{4\Delta_e}(b_i b_j + c_i c_j);$$

$$\vdots$$

etc.

Replacing the index e by the individual numbers of the elements of Fig. 4.30 results in

$$\frac{\partial \chi}{\partial \phi_5} = 0 = \tag{4.66}$$

$$\text{(from element 1)} = [(h_{im})_1\phi_2 + (h_{jm})_1\phi_1 + (h_{mm})_1\phi_5 + \ldots$$

$$\text{(from element 2)} = [(h_{im})_2\phi_3 + (h_{jm})_2\phi_2 + (h_{mm})_2\phi_5 + \ldots$$

$$\text{(from element 3)} = [(h_{im})_3\phi_4 + (h_{jm})_3\phi_3 + (h_{mm})_3\phi_5 + \ldots$$

$$\text{(from element 4)} = [(h_{im})_4\phi_1 + (h_{jm})_4\phi_4 + (h_{mm})_4\phi_5].$$

This equation may be written as

$$H_{15}\phi_1 + H_{25}\phi_2 + H_{35}\phi_3 + H_{45}\phi_4 + H_{55}\phi_5 = 0 \tag{4.67}$$

where

$$H_{15} = [(h_{im})_4 + (h_{jm})_1]$$

$$H_{25} = [(h_{im})_1 + (h_{jm})_2]$$

$$\vdots$$

$$H_{55} = [(h_{mm})_1 + (h_{mm})_2 + (h_{mm})_3 + (h_{mm})_4] = \sum_{r=1}^{4}(h_{mm})_r$$

If the potentials ϕ_1 to ϕ_4 were known, ϕ_5 could immediately be calculated from this equation. As, however, the potentials of the nodes 1 to 4 might still be embedded in a larger triangular network, for every unknown potential a corresponding equation has to be set up. For our system with Laplace conditions the FEM solution may thus be written as

$$\frac{\partial \chi}{\partial \{\phi\}} = 0 = [H]\{\phi\} \tag{4.68}$$

indicating the assembly of the whole set of minimizing equations, which can be solved following the usual rules (see section 4.4.1).

This short, detailed introduction of the FEM cannot demonstrate all the advantages and disadvantages of the method. In the application to electric field problems within insulation systems, the advantages may be summarized as follows:

(a) It is readily applicable to non-homogeneous systems (i.e. with materials of different permittivities) as well as to anisotropic systems (refer to eqn (4.51)).
(b) The shapes and sizes of the elements may be chosen to fit arbitrary boundaries and the grid size may easily be adapted to the gradient of the potentials, i.e. small elements can be placed into regions with high gradients and vice versa.
(c) Accuracy may also be improved using higher order elements (compare with eqn (4.52)), without complicating boundary conditions.
(d) Dielectric materials may also be treated as the case where conduction currents contribute to the potential distribution. This can be done by assuming complex permittivity with real and imaginary parts (i.e. $\varepsilon = \varepsilon' - j\varepsilon''$, where $\tan\delta = \varepsilon''/\varepsilon'$).[30]

For the calculation of electric field intensities within electric insulation systems, the only disadvantage of the FEM is still related to the limited and *a priori* unknown accuracy which can be achieved. Even for two-dimensional problems and highly divergent fields, a very large number of triangular elements or nodes would be necessary to obtain an adequate accuracy within the highly divergent field regions, which are responsible for the breakdown of the whole system. It should be remembered that the often used first order algorithm (see eqn (4.53)) does result in a constant field strength within each element (see eqn (4.59)), which is only approximately correct for the case of continuous field distribution within homogeneous materials. Although the size of the elements can well be adapted to the divergence of the field distribution, too large a number of elements or nodes would be required for high accuracy. Efficient computation algorithms are necessary to solve eqn (4.68), as the stiffness matrix $[H]$, although highly sparse and symmetric, will become very large.

Finally, Fig. 4.31 shows an example of a field computation using the FEM. Figure 4.31(a) displays the original triangular grid used for computation of a coaxial section of a GIS comprising a conical space $r(\varepsilon_r = 6.5)$ within the gaseous insulation system ($\varepsilon_r \approx 1$). The result of this grid displayed by 5 per cent equipotential lines (Fig. 4.31(b)) still shows some discontinuities, although much smaller triangular elements have been used in regions with high field non-uniformity. Figures 4.31(c) and (d) are sections of the figures shown before. The same sections computed by a much higher grid density (see Figs 4.31(e), (f)) confirm the large improvement displayed by the new equipotential lines, the discontinuity of which disappeared.

4.4.3 Charge simulation method (CSM)

A third numerical method widely and successfully used today to calculate electric fields is known as the charge simulation method (CSM). Though the fundamentals of this method may be familiar to most electrical engineers, as it is based upon frequently used analytical field computation methods, it may be useful to review some fundamentals of Maxwell's equations.

The Poisson's equation may be written as

$$\mathrm{div}(-\varepsilon \ \mathrm{grad}\ \phi) = \mathrm{div}\, D = \rho \tag{4.69}$$

(which is a differential form of Maxwell's equation) where D is the electric flux density and p is the volume charge density. Equation (4.69) is independent of any particular coordinate system. This equation may be integrated by means of a volume integral, resulting in

$$\int_V \mathrm{div} D\, \mathrm{d}V = \oint_A D\, \mathrm{d}A = \int_V \rho\, \mathrm{d}V \tag{4.70}$$

Here the volume integral of divergence is transferred to a surface integral with the closed surface A, and the volume integral applied to the charge density can easily be identified with the total charge enclosed by the surface A. Equation (4.70) represents the well-known Gauss's law. This law gives an exact solution of Poisson's equation, and many direct methods for field computations are based upon this law.

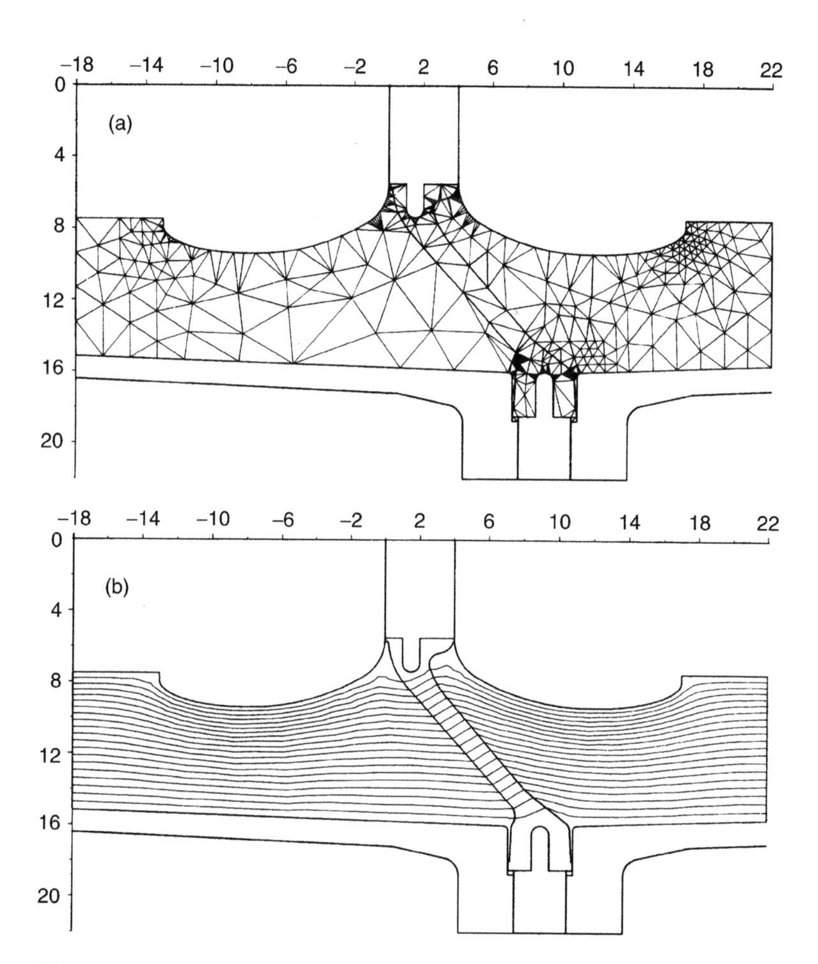

Figure 4.31 *Field computation by FEM. Coaxial section of GIS. (a) Triangular elements for the computation of a limited axial section comprising the spacer. (b) Result of the computation (5 per cent equipotential lines) based upon the grid displayed in (a). (c), (d) A section of (a) and (b) enlarged. Same axial and radial notations. (e) A section of the improved grid with a much higher number of elements in comparison to (a) or (c) respectively. (f) Equipotential lines due to (e) (courtesy of BBC, Baden, Switzerland)*

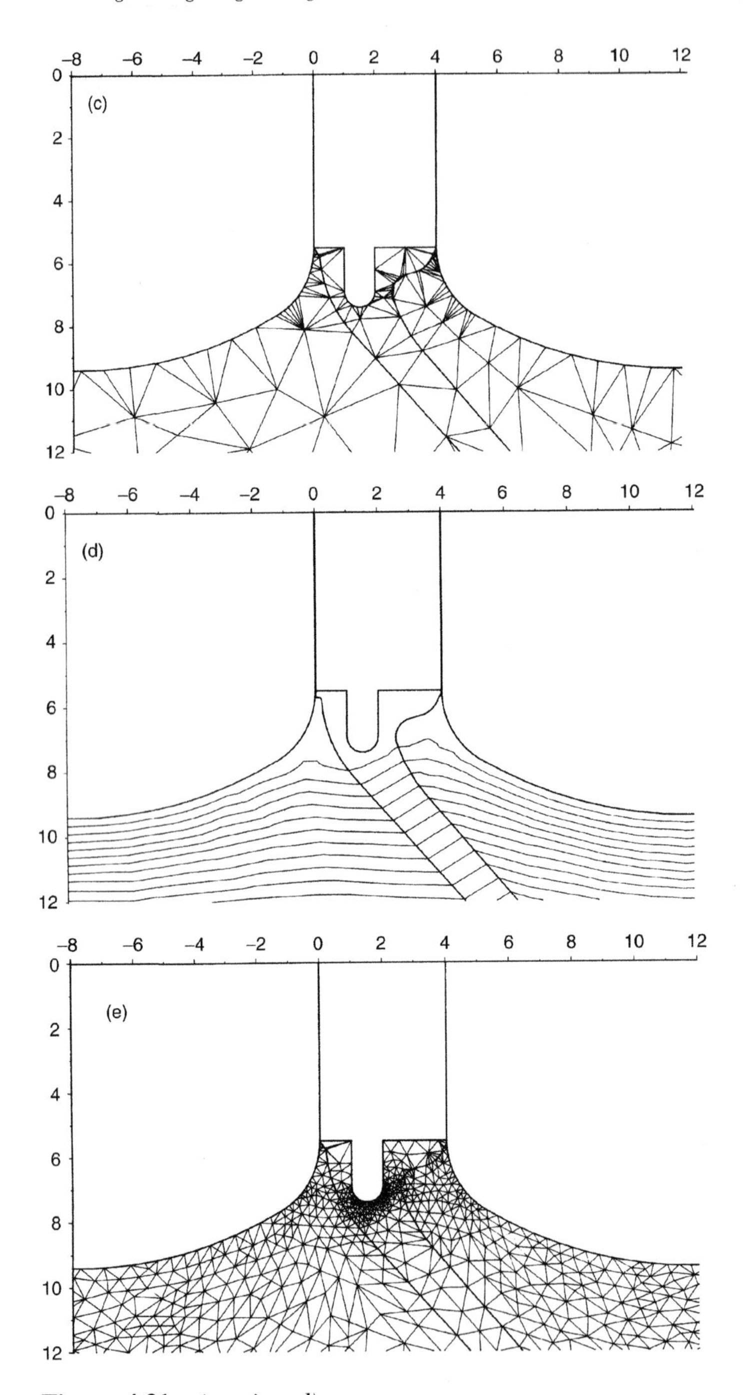

Figure 4.31 *(continued)*

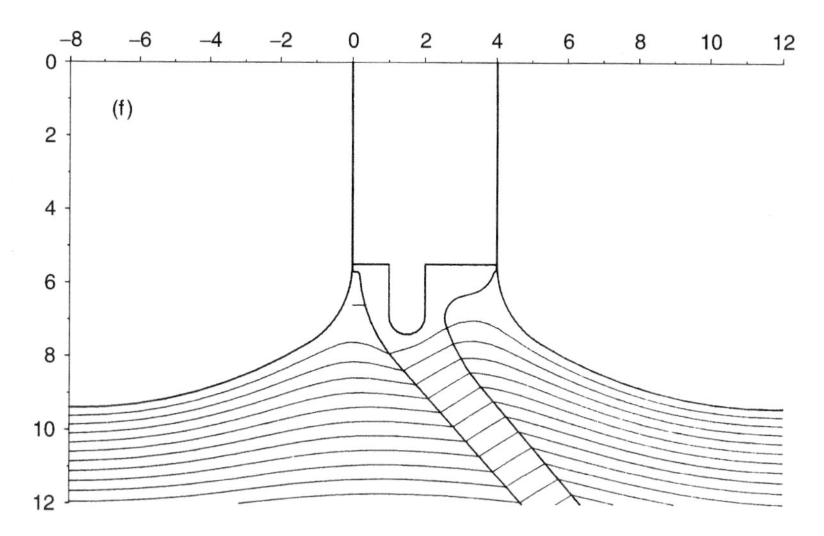

Figure 4.31 *(continued)*

We have seen that both the FDM (section 4.4.1) and the FEM (section 4.4.2) are directly based upon the differential form of a Maxwell equation (see eqn (4.69)). We noted also, in previous sections, that solving the differential equations, either analytically or numerically, involves difficulties inherent in the formulation of boundary conditions as well as due to inaccuracies arising within numerical procedures. On the other hand, Gauss's law is much easier to apply, at least for the cases where some symmetry boundary conditions are apparent. This advantage was used by applying eqn (4.70) to calculate analytically some simple field configurations, e.g. coaxial cylindrical or spherical fields (see section 4.2), for which the integrals of the left-hand side of eqn (4.70) could easily be solved due to symmetry conditions, arising from a concentration of the charge distribution (right-hand side of this equation) within line or point charges respectively.

Directly related to the application of Gauss's law is the method of images (or image charges), which could be used to compute analytically some important problems by means of ready-made solutions, thus eliminating the need for formal solutions of Laplace's or Poisson's equations in differential form. This method, which can be traced to Lord Kelvin[31] and Maxwell,[32] was also used for field computation of a sphere-to-sphere arrangement.

Steinbigler[7] introduced this technique as an efficient method for digital computation of electric fields. Since its publication in English[34] this method (CSM) has been recognized to be very competitive and often superior to FEM or FDM, at least for treating two- or three-dimensional fields within h.v. insulation systems, particularly where high accuracies within highly divergent field areas are demanded. Although the efficiency and applicability of the CSM may not have been fully developed up to now, many recent publications have shown interest in this technique.(20,35,42)

The basic principle of CSM is very easy to formulate. Using the superposition principle, the potential functions of the fields of individual charges of any type (point, line or ring charges, for instance) can be found by a summation of the potentials (scalars) resulting from the individual charges. Let Q_j be a number n of individual charges, and Q_j be the potential at any point within the space (independent of the coordinate system used). The superposition principle results in

$$\phi_i = \sum_{j=1}^{n} p_{ij} Q_j \tag{4.71}$$

where p_{ij} are the potential coefficients, which are known for many types of individual charges by particular solutions of Laplace's or Poisson's equations mentioned earlier. Figure 4.32 displays a point charge Q_P and a line charge Q_l placed at the x- and y-axis respectively and an arbitrary point P_i at which the potential ϕ_i would apply.

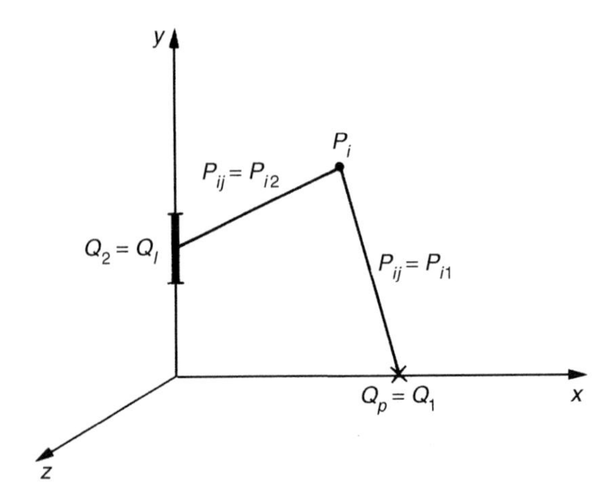

Figure 4.32 *A finite line charge Q_l and a point charge Q_P related to a field point P_i*

Whereas the potential coefficients, $p_{ij}, \ldots$, are known, only additional boundary conditions enable us to relate ϕ_i with Q_j quantitatively. If the individual charges are placed outside the space in which the field is to be computed (or inside a closed metal electrode, whose surface is an quipotential area), the magnitudes of these charges are related to the distributed surface charges which are physically bonded by the electric flux leaving or entering the surface of any electrode or conductor surrounding these charges. If n charges Q_j are assumed, we require also at least n known potentials to solve eqn (4.71) for the *a priori* unknown charge magnitudes. This can easily be done by identifying the potentials ϕ_i with n potentials on the surface of the conductors ('contour points'), which are adequately placed at a given electrode configuration. If this potential is $\phi_\iota = \phi_X$, we may rewrite eqn (4.71) as

$$\sum_{j=1}^{n} p_{ij} Q_j = \phi_c. \tag{4.72}$$

This equation leads to a system of n linear equations for the n unknown charges

$$\begin{bmatrix} p_{11} & p_{12} & \cdots & p_{1n} \\ p_{21} & p_{22} & \cdots & p_{2n} \\ \vdots & & & \\ p_{n1} & p_{n2} & \cdots & p_{nn} \end{bmatrix} \begin{Bmatrix} Q_1 \\ Q_2 \\ \vdots \\ Q_n \end{Bmatrix} = \begin{Bmatrix} \phi_1 \\ \phi_2 \\ \vdots \\ \phi_n \end{Bmatrix} \tag{4.72a}$$

or

$$[p]\{Q\} = \{\phi\}.$$

After this system has been solved, it is necessary to check whether the set of calculated charges fits the actual boundary conditions. It must be emphasized that only n discrete contour points of the real electrode system have been used to solve eqn (4.72), and thus the potentials at any other contour points considered in this calculation might still be different from ϕ_X. Therefore, eqn (4.71) must be additionally used to compute the potentials at a number of 'check points' located on the electrode boundary (with known potential). The difference between these potentials and the given boundary potential is then a measure of the accuracy and applicability of the simulation. The development and introduction of special objective functions is thus an important procedure within the optimization of the CSM.[34–36]

As soon as an adequate charge system has been adopted, the potentials and the field strength within the space can be computed. Whereas the potentials are found by superposition, i.e. by eqn (4.71) or the corresponding set of linear equations (compare with eqn (4.72)), the field stresses are calculated by superposition of magnitudes and directional components. For a Cartesian coordinate system, for instance, the x-coordinate E_x would then be for a number of n charges.

$$E_x = \sum_{j=1}^{n} \frac{\partial p_{ij}}{\partial x} Q_j = \sum_{j=1}^{n} (f_{ij})_x Q_j \tag{4.73}$$

where f_{ij} are 'field intensity coefficients' in the x-direction. Before further considerations, the computation algorithm may be applied to a simple example. In Fig. 4.33(a), a symmetrical sphere-to-sphere electrode system is sketched symmetrically charged to $\pm V$. This condition implies zero potential for the plane $z = 0$ as well as for the dielectric space at a distance from the spheres (unlimited dielectric space). Thus the field configuration is axisymmetric with the rotation centre being the z-axis. This simple example would be difficult to compute by FDM or FEM, as the space is unlimited.

Let us consider the case of two point charges $\pm Q_1$ and $\pm Q_2$ symmetrically placed along the axis at $r = 0$; $z = \pm 0.75/1.25D$ and only two contour points P_1, P_2 at $r = 0$ as shown in Fig. 4.33(a). The symmetric arrangement of the charges (imaging) gives $V = 0$ at $z = 0$. Thus also a sphere-to-plane geometry is computed.

To solve eqn (4.72), the potential coefficients for a point charge are necessary. The potential related to a point charge Q at distant d is given by:

$$\phi = \frac{Q}{4\pi\varepsilon}\frac{1}{d} = pQ \tag{4.74}$$

or

$$p = \frac{1}{4\pi\varepsilon d}$$

Thus the potential coefficients p_{ij} are dependent upon the distance d between the charges Q_j and the contour points P_i. For our r–z-coordinate system, the coefficients may be expressed by

$$p_{ij} = \frac{1}{4\pi\varepsilon\sqrt{r_i^2 + (z_i - z_j)^2}} \tag{4.74a}$$

from simple geometric considerations. Let the computer now solve the four simple equations using eqn (4.72), in order to obtain the magnitudes of Q_1 and Q_2, and to compute a sufficient set of other potentials within the r–z-coordinate system. These potentials can be used to draw equipotential lines; such lines are shown in Fig. 4.33(b) for part of the positive z-axis. The result may appear disappointing, since the equipotential line $+V$ deviates grossly from the circle, representing the cross-section of the sphere. An agreement of the computed and given potential is only found for the contour points P_1, P_2, but for other contour points a disagreement extending up to about 39 per cent can be observed. This suggests a very poor simulation and bad assumption of point charges. Therefore, we may add a third point charge Q_3 and contour point P_3, as also indicated in Fig. 4.33(a), and repeat the calculations. The result is now shown in Fig. 4.33(c). The disagreement between the real contour of the

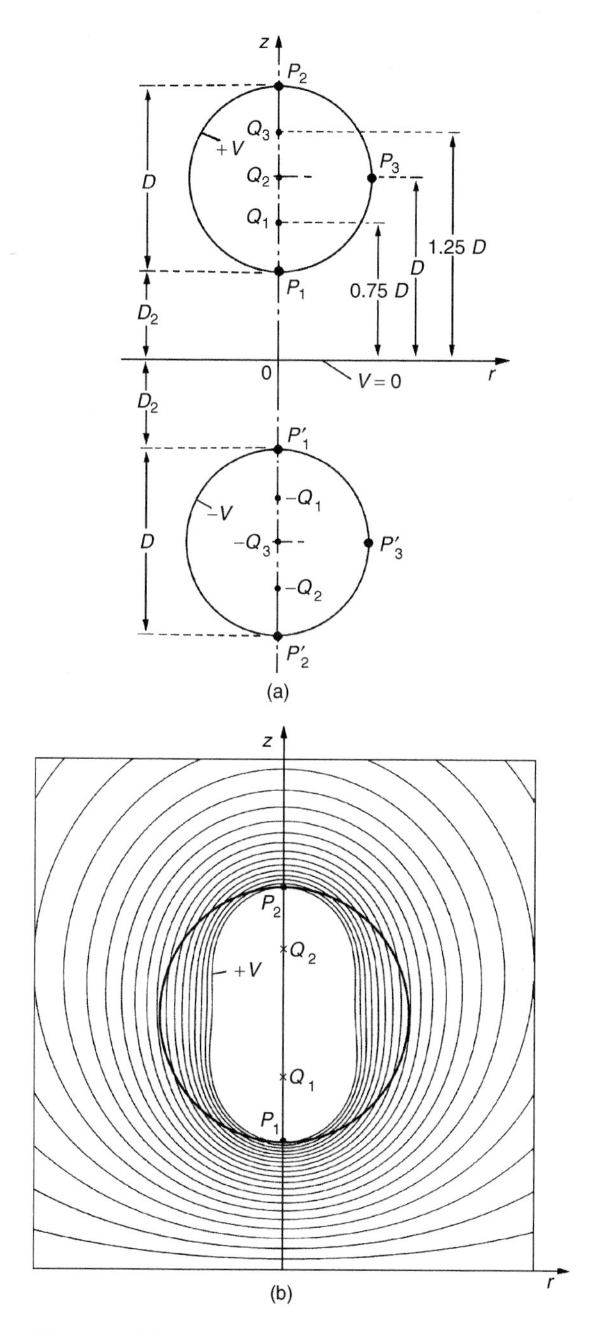

Figure 4.33 *Example for CSM. (a) Sphere-to-sphere electrode arrangement. (b) to (d) Computed results. (For more information see text)*

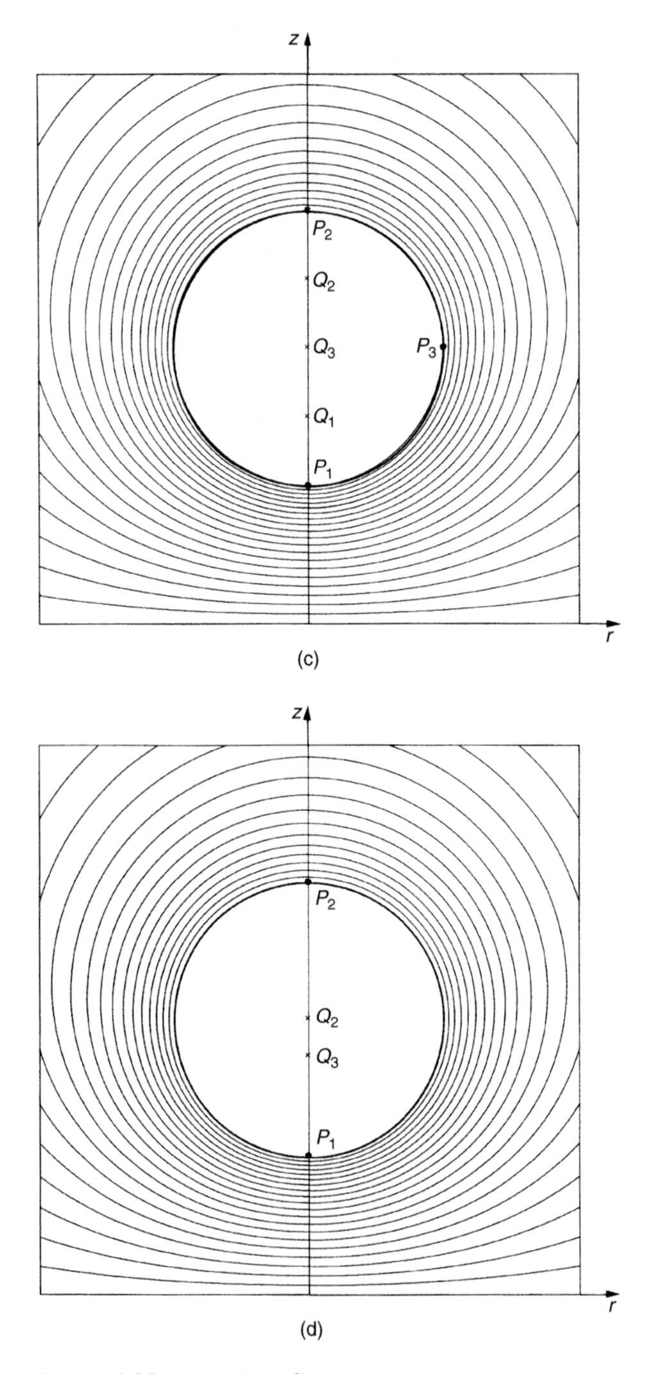

Figure 4.33 *(continued)*

electrode and the computed equipotential line $\phi = +V$ is now very small, not exceeding 1.98 per cent, an error difficult to establish within the figure. This means that the simulation was greatly improved and it is easy to recognize that more charges improve the computation. We can, however, also find excellent solutions using only two simulation charges placed at proper positions. This was done in Fig. 4.33(d), where again only two charges and two contour points were used to solve the problem. The largest deviation for the computed potential $\phi = V$ from the sphere is now less than 0.2 per cent.

This simple example demonstrates two essential features concerning an effective application of the CSM. The first relates to the proper selection of the types of simulation charges, and the second to a suitable arrangement of the charges and contour points.

Various other charge types are available for which the potential coefficients are known from analytical solutions. For our example, the application of toroidal line charges (ring charges) of constant charge density and centred on the axis of symmetry would have been an effective method of discretization. One could also use infinite or finite line charges, or even plane or curved surface charges. The complexity of computation, however, in general increases with the complexity of the simulation charges used, as the potential coefficients become more difficult to compute numerically. As an exercise, only the coefficients for finite line charges and toroidal line charges are reproduced here. With the notations of Fig. 4.34, the potential coefficients are for:

Finite straight line charges (Fig. 4.34(a)):

$$p_{ij} = \frac{1}{4\pi\varepsilon(z_{j2} - z_{j1})} \ln \frac{(z_{j2} - z_i + \gamma_1)(z_{j1} + z_i + \gamma_2)}{(z_{j1} - z_i + \delta_1)(z_{j2} + z_i + \delta_2)} \tag{4.75}$$

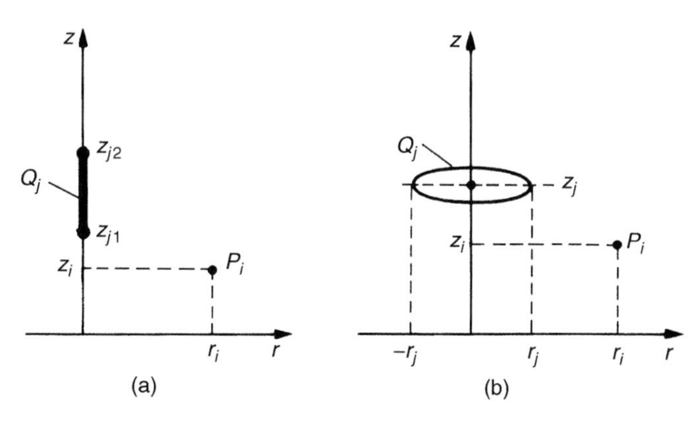

Figure 4.34 *Finite straight line charges (a) and toroidal line (ring) charges (b), with constant charge densities. Total charge: Q_j*

where

$$\gamma_1 = \sqrt{r_i^2 + (z_{j2} - z_i)^2}; \quad \gamma_2 = \sqrt{r_i^2 + (z_{j1} + z_i)^2},$$

$$\delta_1 = \sqrt{r_i^2 + (z_{j1} - z_i)^2}; \quad \delta_2 = \sqrt{r_i^2 + (z_{j2} + z_i)^2}.$$

A suitable application of eqn (4.73) leads also to an analytical expression for the field strength components in the r- and z-directions (Q_j individual line charges):

$$E_r = \sum_{j=1}^{n} \frac{-Q_j}{4\pi\varepsilon(z_{j2} - z_{j1})} \times \left[\frac{z_{j2} - z_i}{r_i\gamma_1} - \frac{z_{j1} - z_i}{r_i\delta_1} + \frac{z_{j1} + z_i}{r_i\gamma_2} - \frac{z_{j2} + z_i}{r_i\delta_2}\right], \tag{4.76a}$$

$$E_z = \sum_{j=1}^{n} \frac{Q_j}{4\pi\varepsilon(z_{j2} - z_{j1})} \left[\frac{1}{\gamma_1} - \frac{1}{\delta_1} - \frac{1}{\gamma_2} + \frac{1}{\delta_2}\right]. \tag{4.76b}$$

Ring charges (Fig. 4.34(b)):

$$p_{ij} = \frac{1}{4\pi\varepsilon}\frac{2}{\pi}\left[\frac{K(k_1)}{\alpha_1} - \frac{K(k_2)}{\alpha_2}\right], \tag{4.77}$$

where

$$\alpha_1 = \sqrt{(r_i + r_j)^2 + (z_i - z_j)^2}, \quad \alpha_2 = \sqrt{(r_i + r_j)^2 + (z_i + z_j)^2},$$

$$\beta_1 = \sqrt{(r_i - r_j)^2 + (z_i - z_j)^2}, \quad \beta_2 = \sqrt{(r_i - r_j)^2 + (z_i + z_j)^2},$$

and

$$k_1 = \frac{2\sqrt{r_j r_i}}{\alpha_1}, \quad k_2 = \frac{2\sqrt{r_j r_i}}{\alpha_2}$$

with the complete elliptic integrals of the first kind $K(k)$ and the second kind $E(k)$.

The field stress components become

$$E_r = \sum_{j=1}^{n} \frac{Q_j}{4\pi\varepsilon}\frac{1}{\pi r_i}\left\{\frac{\left[r_j^2 - r_i^2 + (z_i + z_j)^2\right]E(k_1) - \beta_1^2 K(k_1)}{\alpha_1\beta_1^2} - \frac{\left[r_j^2 - r_i^2 + (z_i + z_j)^2\right]E(k_2) - \beta_2^2 K(K_2)}{\alpha_2\beta_2^2}\right\} \tag{4.78a}$$

$$E_z = \sum_{j=1}^{n} \frac{-Q_j}{4\pi\varepsilon}\frac{2}{\pi r_i}\left\{\frac{(z_i - z_j)\,E(k_1)}{\alpha_1\beta_1^2} + \frac{(z_i + z_j)\,E(k_2)}{\alpha_2\beta_2^2}\right\}. \tag{4.78b}$$

As far as the most suitable arrangement of discrete charges within an electrode is concerned, these may either be found by optimization techniques based upon objective functions[35] or a more practical approach is by the definition of an assignment factor,[34] which relates the successive distances of the contour points with the distances between a contour point and the adjoining corresponding charge. Details of this method may be found in the literature.

For a field space containing only one type of dielectric material ($\varepsilon =$ constant), the application of the CSM to real three-dimensional problems does not present fundamental difficulties. Even sophisticated electrode configurations can be treated by means of discrete charges and images, at least if types of charges with variable charge densities are used (ring charges with periodically variable charge distribution,[34] multipoles,[39] elliptic cylinder charges,[37] axispheroidal charges.[38] Even electric fields with even moving space charges can be treated.[34]

In contrast to the simple solutions within the FDM or FEM for treating multi-dielectric materials, the CSM when used for field calculations in systems composed of two or more materials increases the expenditure. This may be understood by considering the fundamental mathematical solutions and the physical mechanisms involved. The CSM is directly based upon physical charges and in every dielectric material polarization processes take place. Whereas in a homogeneous material placed between electrodes the absolute value of its permittivity does not contribute to the field strength (or potentials), but only the flux density D, the field distribution at the boundaries of different materials is heavily distorted due to the dipole charges at the boundaries which do not have counterparts at the adjacent medium. The law of dielectric refraction (section 4.3.2, eqn (4.36)) results from this physical effect and is associated with an infinitely thin layer of bonded charges located in the two media. The free surface charges physically present due to electrical conduction of the interface surface also contribute to field distortions, but the common dielectric refraction is not related to such additional charges.

This realignment of dipoles within different dielectric materials must thus be considered within the CSM. An exact solution with CSM must be based upon the physical dipole surface charge density as has been shown recently.[47] But continuous surface charges can also be simulated by discrete charges by replacing the surface charge density at metal electrodes, whose potential is a fixed value, by discrete charges within this electrode. This method, originally presented by P. Weiss,[34] will be presented briefly through a simple example.

Figure 4.35 displays a cross-section of a part of an insulation system, in which a metal electrode with fixed potential, $\phi = \phi_c$, meets two adjoining dielectric materials I and II. The actual shapes of the two-dimensional surfaces of the three different boundaries (electrode–dielectric I, electrode–dielectric II, dielectric I–dielectric II) determine the optimal types of discrete charges simulating the problem. Thus, the localized charges 1 to 7 will represent point

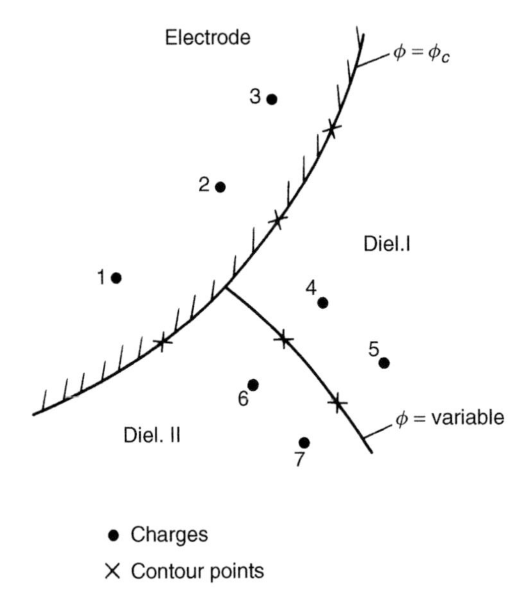

Figure 4.35 *Simulation of a dielectric boundary by discrete charges*

charges as well as intersections with line or ring charges. From earlier considerations it is obvious that a part of the charges (nos 1–3, denoted as n_E) have been placed inside the electrode, i.e. behind the metal surface. However, the same is correct for the charges placed on both sides of the dielectric interface (nos 4–7), as the influence of the dipolar charges within dielectric I upon the field in dielectric II can be simulated by the discrete charges nos 4 and 5 within dielectric I and vice versa. It was also shown earlier that a limited number of contour points placed at a '$\phi =$ constant' boundary is necessary, which is equal to the number of simulated charges within an electrode, and thus a number of $n_E = 3$ contour points (nos 1–3) is adequate. For the dielectric interface, however, it will be sufficient for our example to place only two contour points corresponding to the two pairs of simulation charges (nos 4 and 6, nos 5 and 7), as each contour point belongs to dielectric I as well as to dielectric II. Equal numbers of charges, designated by n_B, on both sides of the dielectric interface are thus convenient and they should be placed at positions equally distributed between the mutual contour points and adjacent charges respectively. For our example, n_B is only 2. Now it is possible to set up a system of equations for our unknown charges based upon well-known boundary conditions. These boundary conditions can be subdivided into three parts:

(1) The electrode–dielectric interface is a boundary with known potential, $\phi = \phi_X$. The absolute magnitude of the surface charge density at this electrode is, due to the polarization mechanisms in both dielectric materials,

dependent upon the relative permittivity ε_r of the dielectric materials, as $D = \varepsilon E = \varepsilon_r \varepsilon_0 E$, where ε_0 is the permittivity of vacuum. Also the absolute magnitudes of our simulation charges would depend upon these materials' constants. However, it is not necessary to take these physical effects into account, which are indeed included within our potential coefficients (see eqns (4.74), (4.74a), etc.). For any homogeneous dielectric material, the electric field may be computed independent of any relative permittivity ε_r, and the potential coefficients, for which eqn (4.74) is one example only, are in general always computed by assuming $\varepsilon = \varepsilon_0$, or any other number as long as the simulated charges are not used to derive capacitance values from the results, which is also possible. It is easy to understand, however, that the absolute magnitudes of the discrete charges used within our system are based upon a superposition of potentials. And thus we can use the known potential at the electrode interface to derive two sets of equations due to the two dielectrics used. The first set of three equations based upon the three or n_E contour points takes only dielectric I into account, for which the charges within dielectric II can be neglected:

$$\underset{(1-3)}{\sum_{j=1}^{n_E} Q_j p_{ij}} + \underset{(4-5)}{\sum_{j=n_E+1}^{n_E+n_B} Q_j p_{ij}} = \Phi_c \tag{4.79a}$$

Using eqns (4.80) and (4.81) subject to two new boundary conditions, the electric field within dielectric II could be computed, as all Q_j charges within eqn (4.87a) which are not yet known define the potentials within this material.

For the computation of the field distribution within dielectric I, the same considerations apply. But now we neglect the charges within dielectric I, which results in an equal set of three or n_E equations, as

$$\underset{(1-3)}{\sum_{j=1}^{n_E} Q_j p_{ij}} + \underset{(6-7)}{\sum_{j=n_E+1}^{n_E+2n_B} Q_j p_{ij}} = \phi_c \tag{4.79b}$$

(2) The potential at the dielectric interface is unknown. We know, however, that due to the continuity of the potential at either side of the interface, the potentials must be equal at each contour point. As the charges within the electrode (nos 1–3) will not disturb the continuity condition, the potentials due to the charges within the dielectric materials must satisfy the condition

$$\underset{(4-5)}{\sum_{j=n_E+1}^{n_E+n_B} Q_j p_{ij}} = \underset{(6-7)}{\sum_{j=n_E+n_B+1}^{n_E+2n_B} Q_j p_{ij}} \tag{4.80}$$

This equation refers to a number n_b contour points giving an equal number of new equations, in which those charges Q_j are involved, which have not yet been used within eqn (4.79a) or eqn (4.79b) respectively. It should be noticed that this potential continuity condition implies that the field stress components tangential to the interface are equal.

(3) Finally, the third boundary condition refers to the continuity of the normal component of the electric flux density crossing the dielectric interface, or the discontinuity of the normal components of the field intensity (see eqn (4.29)). To include this condition, the 'field intensity coefficient' f_{ij} must be considered (see eqn (4.73)), being the contribution of the charge j to that component of the field vector, which is normal to the dielectric boundary at a contour point i. These factors are in general also known from analytical computations, as this applies to the potential coefficients p_{ij}, and specific f_{ij} values can be taken directly from the earlier equations ((4.76) or (4.78)) for line or ring charges. Then for any normal component $(E_n)_i = Q_j f_{ij}$, this condition may be written as

$$\varepsilon_I \left[\underbrace{\sum_{j=1}^{n_E} Q_j f_{ij}}_{(1-3)} + \underbrace{\sum_{j=n_E+n_B+1}^{n_E+2n_B} Q_j f_{ij}}_{(6-7)} \right] = \varepsilon_{II} \left[\underbrace{\sum_{j=1}^{n_E} Q_j f_{ij}}_{(1-3)} + \underbrace{\sum_{j=n_E+1}^{n_E+n_B} Q_j f_{ij}}_{(4-5)} \right] \quad (4.81)$$

where ε, $\varepsilon_{\mathrm{II}}$ are the permittivities of the two dielectrics.

This equation refers again to a number of n_B contour points, and thus a total number of $(n_E + 2n_B)$ linear equations are given for the calculations of the same number of unknown charges. This procedure demonstrates the difficulties involved with the implementation of dielectric boundaries, as a significant number of additional charges increase the computational efforts.

This section will be concluded with an example of a numerical field computation based upon the CSM with surface simulation charges.[45] Figure 4.36 shows the computed arrangement as well as the essential sections of the fields computed. A cylindrical epoxy spacer ($\varepsilon_r = 3.75$) with recessments is placed between parallel electrodes shaped at the outer parts similar to Bruce's profile, but also recessed to reduce the field intensity at the triple point (gas–solid–electrode interface). Outside the spacer, gas insulation ($\varepsilon_r = 1$) is provided, and the whole system is placed within a cylindrical metal pressure vessel with zero potential, the vessel not being shown. As the diameter of the vessel is large compared with the diameter of the electrode system, for the field calculation zero potential is assumed also in infinite space.

The result of the computation is displayed by a number of field stress arrows starting at the points at which the field intensity is computed. These sites are located at the electrode contours as well as at the interface between gas and dielectric, for which the normal and tangential components of the field

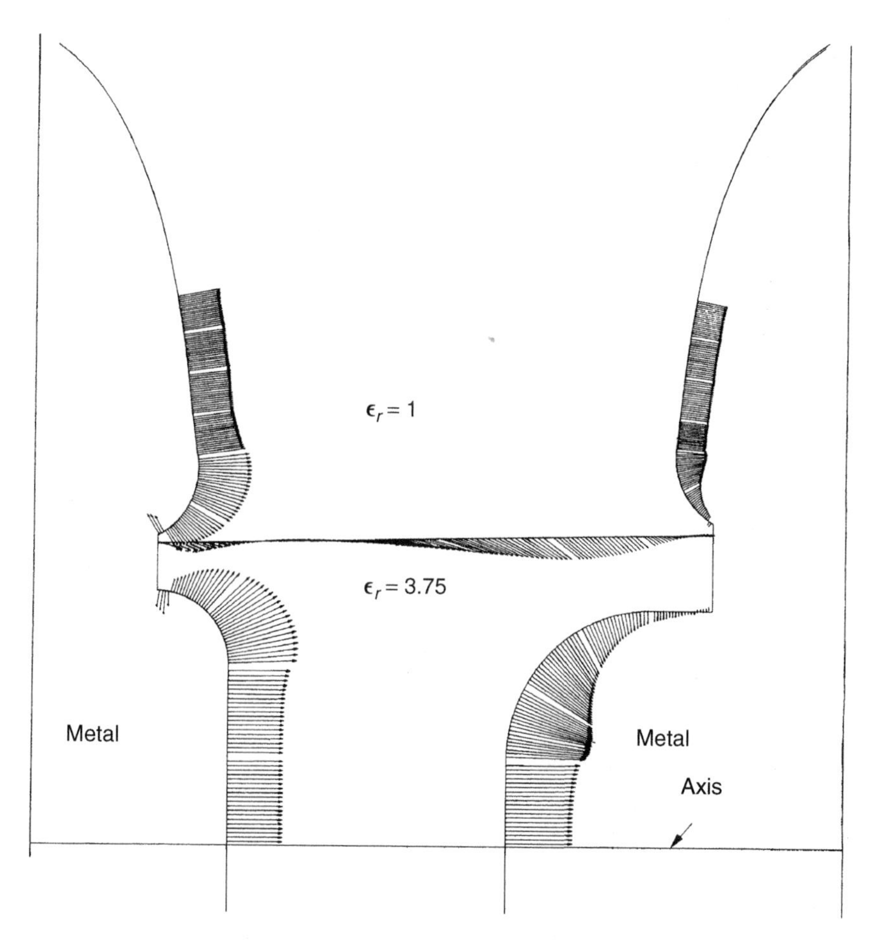

Figure 4.36 *Field computation by CSM with surface charges.*[45] *Epoxy spacer between parallel plate electrodes*

intensities are of utmost interest. The length of each arrow is proportional to the absolute value of the field strength, and the direction of an arrow displays the field direction at each site. No tangential components act at the electrode–dielectric interfaces, and normal components can barely be noted in the upper part of the spacer. This example is taken from an investigation concerning surface charge accumulation at the gas–dielectric interface under a d.c. voltage application. The experimental investigations showed a high accumulation of positive or negative surface charges after applying high d.c. voltages to the electrode for a long time, i.e. up to hours, but the polarity and magnitude of these charges are directly related to this original electrostatic field, i.e. the field before charge is deposited and fixed to the interface.

4.4.4 Boundary element method

The boundary element method (BEM) is a relatively new technique for solving Laplace's and Poisson's equations (and other partial differential equations).[33,47,48,49,50] The unknown function u is first solved on the boundary of the domain, the value of u and its partial derivatives are then calculated by integration of the number of elements on the boundary. In this way, the number of elements and thus the number of unknowns of the resulting linear equations is greatly reduced compared with domain approaches such as finite element and final difference methods. The boundary approach also makes it possible to handle problems with infinite domain.

The formulation of boundary integral equations[48]

The problem is to solve Poisson's equation (or Laplace's equation if $\rho(x, y) = 0$):

$$\nabla^2 u(x, y) = -\rho(x, y) \qquad (x, y) \ \text{in } \Omega \tag{4.82}$$

$$u(x, y) = f(x, y) \qquad (x, y) \ \text{in } \Gamma_1 \tag{4.83}$$

$$q(x, y) = \partial u/\partial n = g(x, y) \qquad (x, y) \ \text{in } \Gamma_2 \tag{4.84}$$

$$\Gamma = \Gamma_1 + \Gamma_2 \tag{4.85}$$

where Γ is the boundary of the domain Ω, and q is the directional derivative of $u(x, y)$ with respect to the outwards normal n of the boundary.

If the solution is approximated, the residual $\text{Res}(x, y)$ is not zero,

$$\text{Res}(x, y) = \sigma^2 u(x, y) + \rho(x, y) \neq 0 \tag{4.86}$$

Therefore we force it to zero by a weighting function $w(x, y)$,

$$\int_\Omega (\nabla^2 u(x, y) + \rho(x, y) w(x, y)\, dx\, dy = 0 \tag{4.87}$$

or

$$\int_\Omega \nabla^2 u(x, y) w(x, y)\, dx\, dy = -\int_\Omega \rho(x, y) w(x, y)\, dx\, dy \tag{4.88}$$

Integrating the left-hand side by parts twice yields

$$\oint_\Gamma \frac{\partial u}{\partial n} w\, ds - \int_\Omega \left(\frac{\partial u}{\partial x}\frac{\partial w}{\partial x} + \frac{\partial u}{\partial y}\frac{\partial w}{\partial y} \right) dx\, dy = -\int_\Omega \rho(x, y) w(x, y)\, dx\, dy \tag{4.89}$$

$$\oint_\Gamma \left(\frac{\partial u}{\partial n} w - u \frac{\partial w}{\partial n} \right) ds + \int_\Omega u \nabla^2 w\, dx\, dy = -\int_\Omega \rho(x, y) w(x, y)\, dx\, dy \tag{4.90}$$

or

$$\int_{\Omega} u\nabla^2 w \, \mathrm{d}x \, \mathrm{d}y = \oint_{\Gamma} \left(u\frac{\partial w}{\partial n} - \frac{\partial u}{\partial n} w \right) \mathrm{d}s - \int_{\Omega} \rho(x, y) w(x, y) \, \mathrm{d}x \, \mathrm{d}y \qquad (4.91)$$

Choosing

$$w(x, y) = \ln r$$

$$r = \sqrt{(x - x_0)^2 + (y - y_0)^2} \qquad (4.92)$$

where r is the distance between points $P(x, y)$ and $P_0(x_{0,y} y_0)$, and P_0 are either in the domain Ω or on the boundary Γ.

It is easy to verify that $\ln r$ is a solution to Laplace's equation in the domain with the singular point P_0 being excluded:

$$\nabla^2(\ln r) = 0 \quad (\text{for } P \neq P_0) \qquad (4.93)$$

The function $\ln r$ is called the fundamental solution of Laplace's equation in an infinite region.

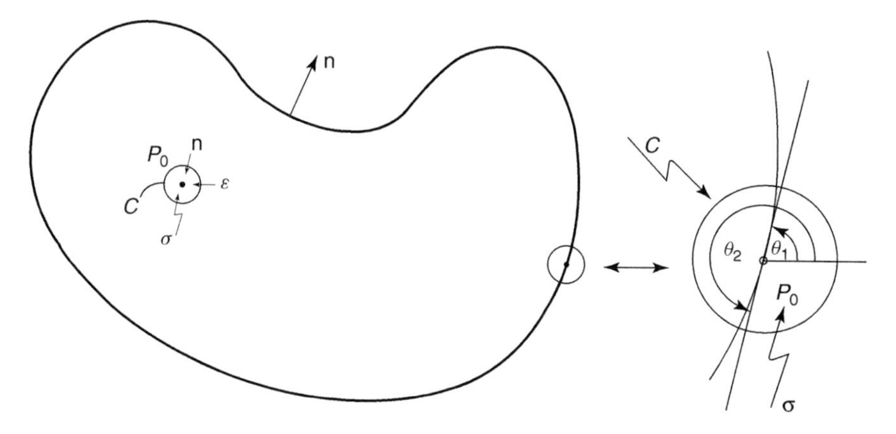

Figure 4.37 *Boundary integral equation (courtesy of Ming Yu)*

If P_0 is located in Ω (as shown in Fig. 4.37), to exclude it from the domain, P_0 is encircled with a circle of very small radius ε centred at P_0. The domain and the boundary of the circle are denoted by σ and c respectively.

$$\oint_{\Gamma+c} \left(u\frac{\partial \ln r}{\partial n} - q \ln r \right) \mathrm{d}s - \int_{\Omega-\sigma} Q(x, y) \ln r \, \mathrm{d}s \, \mathrm{d}y$$

$$= \int_{\Omega-\sigma} u\nabla^2(\ln r) \, \mathrm{d}x \, \mathrm{d}y = 0 \qquad (4.94)$$

Note that the normal vector of c is inward,

$$\oint_{\Gamma+c}\left(u\frac{\partial \ln r}{\partial n} - q\ln r\right)\mathrm{d}s = \oint_{\Gamma}\left(u\frac{\partial \ln r}{\partial n} - q\ln r\right)\mathrm{d}s + \oint_{c}\left(u\frac{\partial \ln r}{\partial n} - q\ln r\right)\mathrm{d}s \tag{4.95}$$

$$\frac{\partial \ln r}{\partial n} = \frac{\partial \ln r}{\partial r}\frac{\partial r}{\partial n} = \frac{1}{r}(-1) = -\frac{1}{r} \tag{4.96}$$

Since it is a first type line integration, it does not matter whether it is integrated from 0 to 2π or from 2π to 0, so long as we keep $\mathrm{d}s > 0$. Thus if integrating from 0 to 2π (note that if it is integrated from 2π to 0, then $\mathrm{d}\theta < 0$, $\mathrm{d}s = r(-\mathrm{d}\theta) = -r\mathrm{d}\theta$, which will give the same result),

$$\mathrm{d}s = r\mathrm{d}\theta \tag{4.97}$$

$$\oint_{c}\left(u\frac{\partial \ln r}{\partial n} - q\ln r\right)\mathrm{d}s = \int_0^{2\pi}\left(u\frac{-1}{\varepsilon} - \frac{\partial u}{\partial n}\ln\varepsilon\right)\varepsilon\,\mathrm{d}\theta \tag{4.98}$$

$$= \int_0^{2\pi}\left(u + \frac{\partial u}{\partial n}\varepsilon\ln\varepsilon\right)\mathrm{d}\theta \tag{4.99}$$

$$= -\left(u(\varepsilon,\xi) + \frac{\partial u(\varepsilon,\xi)}{\partial n}\varepsilon\ln\varepsilon\right)2\pi \tag{4.100}$$

Note that

$$u(\varepsilon,\xi) = u(\varepsilon\cos\xi + x_0, \varepsilon\sin\xi + y_0) \tag{4.101}$$

$$0 \le \xi \le 2\pi \tag{4.102}$$

Let $\varepsilon \to 0$, $\Gamma + c$ becomes Γ while $\Omega - \sigma$ approaches Ω, and

$$\lim_{\varepsilon\to 0}\varepsilon\ln\varepsilon = 0 \tag{4.103}$$

$$\lim_{\varepsilon\to 0}u(\varepsilon,\xi) = u(x_0, y_0) \tag{4.104}$$

which leads to

$$u(x_0, y_0) = \frac{1}{2\pi}\left[\oint_{\Gamma}\left(u\frac{\partial \ln r}{\partial n} - q\ln r\right)\mathrm{d}s - \int_{\Omega}Q(x,y)\ln r\,\mathrm{d}x\,\mathrm{d}y\right] \tag{4.105}$$

This is the boundary integral equation which links the values of the unknown function in the domain with the line integral along the boundary. If u and q are known on the boundary, $u(x, y)$ in the domain can be calculated by eqn (4.105). Because on the boundary either u or q is known but not both, it is necessary to find out another half of the u and q on the boundary.

Let us now locate the P_0 on the boundary. It can be done similarly as in the case when P_0 was located in Ω (as in Fig. 4.37), except that the integral is now taken from θ_1 to θ_2, and when $\varepsilon \to 0$, $\theta_2 - \theta_1 = \pi$ (if the boundary is smooth at P_0), therefore the boundary integral equation for P_0 on the boundary is

$$u(x_0, y_0) = \frac{1}{\pi}\left[\oint_\Gamma \left(u\frac{\partial \ln r}{\partial nn} - q\ln r\right)\mathrm{d}s - \int_\Omega Q(X, Y)\ln r\,\mathrm{d}x\,\mathrm{d}y\right] \tag{4.106}$$

Another approach to obtain (4.91) is to use Green's identity:

$$\int_\Omega (u\nabla^2 w - w\nabla^2 u)\,\mathrm{d}x\,\mathrm{d}y = \oint_\Gamma \left(u\frac{\partial w}{\partial n} - w\frac{\partial u}{\partial n}\right)\mathrm{d}s \tag{4.107}$$

Since $\nabla^2 u = -Q$, the above equation yields (4.91) naturally. From this approach it is clear that if u and q are known exactly on the boundary, theoretically u in the domain can be calculated exactly, not approximately (the fundamental solution $\ln r$ is actually the Green's function for an infinite region).

In formulating the boundary element equations consider the case shown in Fig. 4.37 with the boundary Γ discretized into elements which can be modelled by curves or straight lines as shown in Fig. 4.38. On each element, u and q are approximated by constant, linear, quadratic or other basis functions. For simplicity, we will use constant elements. The boundary Γ is discretized into

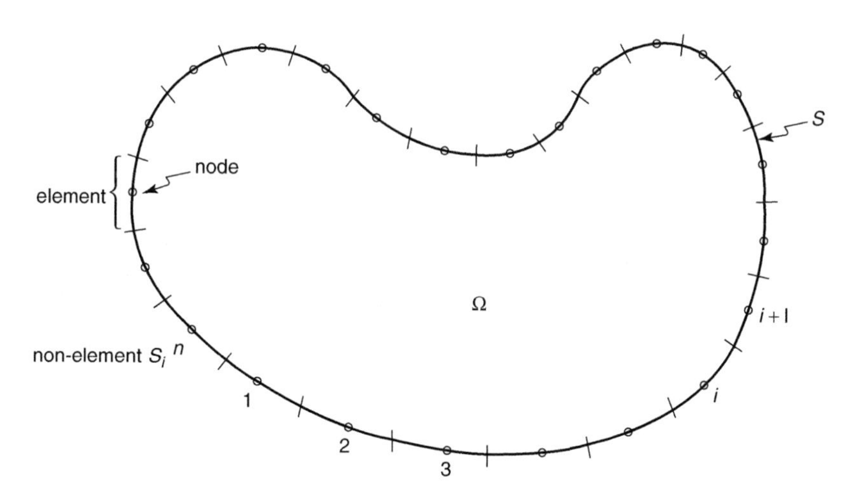

Figure 4.38 *Discretization of the boundary constant element*

n straight line elements S_i $(i = 1, \ldots, n)$,

$$S = \sum_{i=1}^{n} S_i \tag{4.108}$$

where S is the discretized boundary (refer to Fig. 4.38). That is, S is an approximation of boundary Γ.

On each element a node is located at the centre. The values of u and q on the boundary are approximated as follows:

$$u(x, y) = \sum_{i=1}^{n} U_i \phi_i(x, y) \tag{4.110}$$

$$q(x, y) = \sum_{i=1}^{n} Q_i \phi_i(x, y) \tag{4.111}$$

$$\phi_i(x, y) = \begin{cases} 1 & (x, y) \quad \text{in} \quad S_i \\ 0 & \text{otherwise} \end{cases} \tag{4.112}$$

where U_i and Q_i are nodal values of u and q. For node i, a weighting function is defined,

$$w_i(x, y) = \ln r_i \tag{4.113}$$

$$r_i = \sqrt{(x - x_i)^2 + (y - y_i)^2} \quad (i = 1, 2, \ldots, n) \tag{4.114}$$

where n is the number of boundary elements. For every element we have the following equation,

$$\pi u(si, yi) = \oint_{\Gamma} \left(u \frac{\partial \ln r}{\partial n} - q \ln r \right) \mathrm{d}s \tag{4.115}$$

$$\approx \oint_{s} \left(u \frac{\partial \ln r}{\partial n} - q \ln r \right) \mathrm{d}s \tag{4.116}$$

$$= \sum_{j=1}^{n} \int_{s_j} \left(U_j \frac{\partial \ln r_i}{\partial n} - Q_j \ln r_i \right) \mathrm{d}s \tag{4.117}$$

$$\pi u(x_i, y_i) \approx \sum_{j=1}^{n} U_j \int_{s_j} \frac{\partial \ln r_i}{\partial n} \mathrm{d}s - \sum_{j=1}^{n} Q_j \int_{s_j} \ln r_i \, \mathrm{d}s \quad (i = 1, 2, \ldots, n) \tag{4.118}$$

or

$$\sum_{j=1}^{n} H_{ij} U_j = \sum_{j=1}^{n} GijQ_j \quad (i = 1, 2, \ldots, n) \tag{4.119}$$

where

$$H_{ij} = \int_{s_j} \frac{\partial \ln r_i}{\partial n} \mathrm{d}s \quad (i = 1, 2, \ldots, n) \tag{4.120}$$

$$H_{ij} = \int_{s_j} \frac{\partial \ln r_i}{\partial n} \mathrm{d}s - \pi \quad (i = 1, 2, \ldots, n) \tag{4.121}$$

$$G_{ij} = \int_{s_j} \ln r_i \,\mathrm{d}s \quad (i = 1, 2, \ldots, n) \tag{4.122}$$

Now there are n equations. Because half of the values of U_i and Q_i are known, there are only n unknowns. Therefore the above equations can be rearranged to obtain a set of algebraic equations in matrix form,

$$A_n z = b \tag{4.123}$$

After solving this equation, all the values of U_i and Q_i on the boundary are known. Thus the values of $u(x_0, y_0)$ in the domain Ω can be calculated by (refer to eqn 4.105)

$$u(x_0, y_0) = \frac{1}{2\pi}\left(\sum_{j=1}^{n} U_j \int_{s_j} \frac{\partial \ln r}{\partial n} \mathrm{d}s - \sum_{j=1}^{n} Q_j \int_{s_j} \ln r \,\mathrm{d}s\right) - \int_{\Omega} Q(x, y) \ln r \,\mathrm{d}x\,\mathrm{d}y \tag{4.124}$$

$$r = \sqrt{(x - x_0)^2 + (y - y_0)^2} \tag{4.125}$$

The partial derivatives of u can also be calculated,

$$\frac{\partial u(x, y)}{\partial x_0} = \frac{1}{2\pi}\left\{\sum_{j=1}^{n} \int_{S_j} \frac{\partial}{\partial x_0}\left(U_j \frac{\partial \ln r}{\partial n} - Q_j \ln r\right) \mathrm{d}s - \int_{\Omega} Q(x, y) \frac{\partial \ln r}{\partial x_0} \mathrm{d}s\right\} \tag{4.126}$$

$$\frac{\partial u(x, y)}{\partial y_0} = \frac{1}{2\pi}\left\{\sum_{j=1}^{n} \int_{S_j} \frac{\partial}{\partial y_0}\left(U_j \frac{\partial \ln r}{\partial n} - Q_j \ln r\right) \mathrm{d}s - \int_{\Omega} Q(x, y) \frac{\partial \ln r}{\partial y_0} \mathrm{d}s\right\} \tag{4.127}$$

Note that in the above equations x and y are integral variables, and the partial derivatives are taken with respect to x_0 and y_0.

The third type boundary condition $au + bq = h$ can be treated as follows,

$$a_i U_i + b_i Q_i = h_i \quad (i = 1, 2, \ldots, k) \tag{4.128}$$

where k is the number of nodes with third type boundary condition. Combining eqn (4.128) with eqn (4.119), there are $(n + k)$ unknowns and $(n + k)$ equations which are sufficient for finding the unique solutions.

Numerical examples

In the following illustration, the symbols below are used:

N : number of points being checked on an equipotential line

N_e : number of boundary elements on a closed boundary

u : exact solution

$\hat{u}$: approximates solution

$\hat{u}_{\text{avg}}$: average $\hat{u}$ on a equipotential line

$\hat{u}_{\text{max}}$: *max*imum $\hat{u}$ on a equipotential line

$\hat{u}_{\text{min}}$: *min*imum $\hat{u}$ on a equipotential line

E : exact solution of field strengh

$\hat{E}_{\text{avg}}$: average $\hat{E}$ on a boundary

$\hat{E}_{\text{max}}$: *max*imum $\hat{E}$ on a boundary

$\hat{E}_{\text{min}}$: *min*imum $\hat{E}$ on a boundary

Note that the reference direction of E is opposite to n on the conductor surface, thus on the conductor surface $E = q$. The numbers in parentheses are percentage errors defined by,

$$\text{error} = \frac{\text{approximation} - \text{exact}}{\text{exact}} 100\% \tag{4.129}$$

Example 1 A conductor of infinite length above the ground

A conductor with $R = 1$ is centred at (0.5), i.e. its height $h = 5$. Boundary condition:

$$U_1 = 100 \tag{4.130}$$

The exact solution can be obtained by the method of image (conductor 2 is the image),

$$u(x, y) = u_1 \frac{\ln(r_1/r_2)}{\ln[(2(h-d)+a)/a]} \tag{4.131}$$

where

$$d = \sqrt{h^2 - R^2}, \quad a = R - (h - d).$$

The equipotential line with potential U_0 is given by,

$$x^2 + (y - y_0)^2 = b^2 \tag{4.132}$$

$$y_0 = d\frac{1 + K_0^2}{1 - K_0^2} \tag{4.133}$$

$$b = \frac{2dk_0}{1 - K_0^2} \tag{4.134}$$

$$K_0 = \exp\left(-\frac{U_0}{u_1}\ln\frac{2(h-d)+a}{a}\right) \tag{4.135}$$

The boundary elements used is are 12. Comparison of results is given in Table 4.1.

Table 4.1 *Comparison for Example 1 ($N = 10$)*[48]

u(exact)	*0*	*30*	*50*
$\hat{u}_{avg}$ (error %)	0.0010 (N/A)	29.9128 (−0.29%)	49.8845 (−0.23%)
$\hat{u}_{max}$ (error %)	0.0017 (N/A)	29.9179 (−0.27%)	49.9003 (−0.20%)
$\hat{u}_{min}$ (error %)	0.0007 (N/A)	29.9016 (−0.33%)	49.8838 (−0.23%)

N/A: not applicable.

Example 2 Coaxial cylinders

Boundary condition (subscripts 1 and 2 denote the inner and outer cylinders respectively),

$$E_0 = 144.2695041 \quad u_2 = 10 \tag{4.136}$$

Exact solution,

$$u(x, y) = E_1R_1\ln\frac{R_2}{r} + u_2 \tag{4.137}$$

$$E(x, y) = E_1\frac{R_1}{r} \tag{4.138}$$

$$r = \sqrt{x^2 + y^2} \tag{4.139}$$

The boundary element used is also 12. Comparison of results is given in Table 4.2

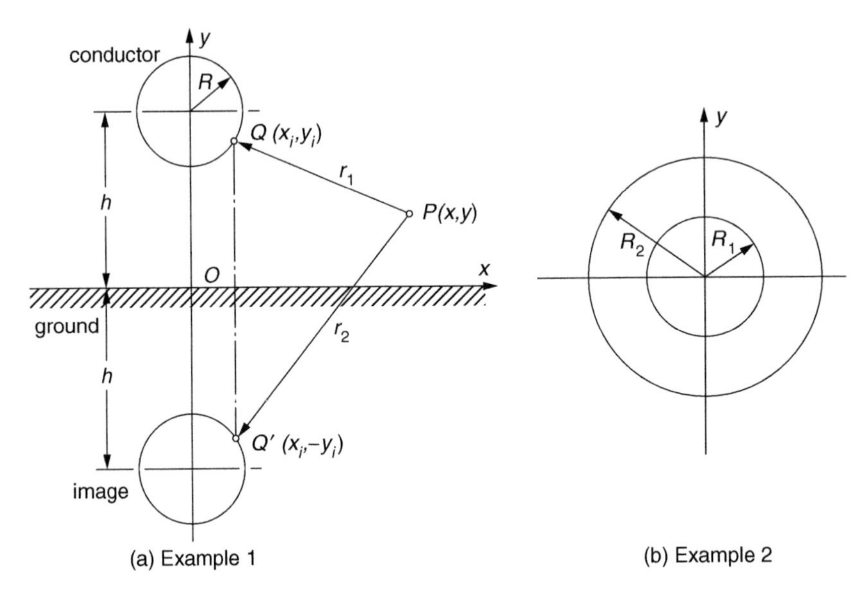

Figure 4.39 *Conductor models for the numerical examples. (a) Conductor above ground. (b) Two coaxial cylinders*[48]

Table 4.2 *Comparison for Example 2*[48]

	10	*N*	*32*
u (exact)	51.5037	E (exact)	72.1348
$\hat{u}_{avg}$ (error %)	50.8005(−1.37%)	$\hat{E}_{avg}$ (error %)	71.9526(−0.25%)
$\hat{u}_{max}$ (error %)	50.8016(−1.36%)	$\hat{E}_{max}$ (error %)	72.0011(−0.19%)
$\hat{u}_{min}$ (error %)	50.7995(−1.37%)	$\hat{E}_{min}$ (error %)	71.8905(−0.34%)

It can be observed that the numerical solution is very accurate even with constant elements. The errors of potentials in the second example are greater than that of the first example. This is due to the use of the Neumann boundary condition.

With a higher order interpolation function on the boundary, more accurate numerical results can be achieved.[48]

References

1. B.D. Popovic. *Introductory Engineering Electromagnetics*. Addison-Wesley, 1971.
2. D. Vitkovitch. *Field Analysis: Experimental and Computational Methods*. D. Van Nostrand, 1966.

3. P. Moon and D.E. Spencer. *Field Theory for Engineers*. D. Van Nostrand, 1961; *Field Theory Handbook*. Springer, 1961.
4. J.D. Kraus and K.R. Carver. *Electromagnetics* (2nd edn). McGraw-Hill, New York, 1973.
5. F.M. Bruce. Calibration of uniform field spark gaps for high voltage measurements at power frequencies. *Proc. IEE* **94**, Part II (1947), p. 138.
6. H. Prinz. *Hochspannungsfelder*. Oldenburg, Munich, 1969.
7. H. Steinbigler. Anfangsfeldstaerken und Ausnutzungsfaktoren rotations-symmetrischer Elektrodenanordnungen in Luft. Dr.-Thesis, TH Munich, 1969.
8. T. Takuma, T. Kouno and H. Matsuba. Field behaviour near singular points in composite dielectric arrangements. *Trans. IEEE* **EI 13** (1978), pp. 426–435.
9. P. Weiss. Fictious peaks and edges in electric fields. 3rd Int. Symp. on High Voltage Engg., Milan, 1979, Report 11.21.
10. K. Wechsler and M. Riccitiello. Electric breakdown of a parallel solid and liquid dielectric system. *Trans. AIEE* **80**, Part III (1961), pp. 365–369.
11. E.A. Cherney. High voltage flashover along a solid–liquid interface. Conf. on El. Ins. and Diel. Phenomena, Annual Report 1970, pp. 187–190.
12. H. Kaerner and H.-J. Voss. The particle influenced breakdown of insulating surfaces in SF_6 under oscillating switching impulse voltage. 3rd Int. Symp. on High Voltage Engg., Milan, 1979, Report 32.04.
13. H. Okubo, T. Amemiya and M. Honda. Borda's profile and electrical field optimization by using charge simulation method. 3rd Int. Symp. on High Voltage Engg., Milan, 1979, Report 11.16.
14. B.M. Weedy. *Underground Transmission of Electric Power*. J. Wiley, 1980.
15. K. Feser. Bemessung von Elektroden im UHV-Bereich am Beispiel von Toroidelektroden für Spannungsteiler. *ETZ-A* **96** (1975), pp. 207–210.
16. S. Sangkasaad. Dielectric strength of compressed SF_6 in nonuniform fields. Dr.-Thesis, ETH Zurich, No. 5738 (1976).
17. G.W. Carter and S.C. Loh. The calculation of the electric field in a sphere-gap by means of bipolar coordinates. *Proc. IEE* **106C** (1959), pp. 108–111.
18. A. Roth. *Hochspannungstechnik* (5th edn). Springer-Verlag, Vienna/New York, 1965.
19. H. Singer. Ein Rechenverfahren von Steuerbelaegen in Durchfuehrungen und Kabelendverschluessen. 2nd Int. Symp. on High Voltage Engg., Zurich, 1975, Report 1.1–12.
20. M.D.R. Beasley *et al.* Comparative study of three methods for computing electric fields. *Proc. IEE* **126** (1979), pp. 126–134.
21. G.E. Forsythe and W.R. Wasow. *Finite-Difference Methods for Partial Differential Equations*. John Wiley, 1960.
22. H. Rutishauser. *Vorlesungen über numerische Mathematik*. Bd. 2: *Differentialgleichungen und Eigenwertprobleme*. Birkhäuser Verlag, 1976.
23. K.J. Binns and P.J. Lawrenson. Analysis and Computation of Electric and Magnetic Field Problems, (2nd edn). Pergamon Press, 1973.
24. O.C. Zienkiewicz. *The Finite Element Method in Engineering Science*. McGraw-Hill, London, 1971 (3rd edn 1977) (in German: *Methode der finiten Elemente*, Carl Hanser,1975).
25. N.A. Demerdash and T.W. Nehl. An evaluation of the methods of finite elements and finite differences in the solution of nonlinear electromagnetic fields in electrical machines. *Trans. IEEE* **PAS 98** (1979), pp. 74–87.
26. P. Unterweger. Computation of magnetic fields in electrical apparatus. *Trans. IEEE* **PAS 93** (1974), pp. 991–1002.
27. E.F. Fuchs and G.A. McNaughton. Comparison of first-order finite difference and finite element algorithms for the analysis of magnetic fields. Part 1: Theoretical analysis, Part II: Numerical results. *Trans. IEEE* **PAS 101** (1982), pp. 1170–1201.
28. W. Janischewskyj and G. Gela. Finite element solution for electric fields of coronating dc transmission lines. *Trans. IEEE* **PAS 98** (1979), pp. 1000–1012.

29. W. Janischewskyj, P. Sarma Maruvada and G. Gela. Corona losses and ionized fields of HVDC transmission lines. CIGRE-Session 1982, Report 3609.
30. O.W. Andersen. Finite element solution of complex potential electric fields. *Trans. IEEE* **PAS 96** (1977), pp. 1156–1160.
31. W. Thomson (Lord Kelvin). *Reprint of Papers on Electrostatics and Magnetism.* Macmillan, London, 1872.
32. J.C. Maxwell. A Treatise on Electricity and Magnetism. (3rd edn). Clarendon Press, Oxford, 1891.
33. L. Lapidus and G.F. Pinder. *Numerical Solutions of Differential Equations in Science and Engineering*. J. Wiley, New York, 1982.
34. H. Singer, H. Steinbigler and P. Weiss. A charge simulation method for the calculation of high voltage fields. *Trans. IEEE* **PAS 93** (1974), pp. 1660–1668.
35. A. Yializis, E. Kuffel and P.H. Alexander. An optimized charge simulation method for the calculation of high voltage fields. *Trans. IEEE* **PAS 97** (1978), pp. 2434–2440.
36. F. Yousef. Ein Verfahren zur genauen Nachbildung und Feldoptimierung von Elektrodensystemen auf der Basis von Ersatzladungen. Dr.-Thesis, Techn. Univ. Aachen, 1982.
37. S. Sato *et al.* Electric field calculation in two-dimensional multiple dielectric by the use of elliptic cylinder charge. 3rd Int. Symp. on High Voltage Engg., Milan, 1979, Report 11.03.
38. S. Sato *et al.* Electric field calculation by charge simulation method using axi-spheroidal charge. 3rd Int. Symp. on High Voltage Engg., Milan, 1979, Report 11.07.
39. H. Singer. Numerische Feldberechnung mit Hilfe von Multipolen. *Arch. für Elektrotechnik* **59** (1977), pp. 191–195.
40. H. Singer. Feldberechnung mit Oberflaechenleitschichten und Volumenleitfaehigkeit. *ETZ Archiv* **3** (1981), pp. 265–267.
41. H. Singer and P. Grafoner. Optimization of electrode and insulator contours. 2nd Int. Symp. on High Voltage Engg., Zurich, 1975, Report 1.3–03, pp. 111–116.
42. D. Metz. Optimization of high voltage fields. 3rd Int. Symp. on High Voltage Engg., Milan, 1979, Report 11.12.
43. H. Singer. Flaechenladungen zur Feldberechnung von Hochspannungs-systemen. *Bull. SEV.* **65** (1974), pp. 739–746.
44. J.H. McWhirter and J.J. Oravec. Three-dimensional electrostatic field solutions in a rod gap by a Fredholm integral equation. 3rd Int. Symp. on High Voltage Engg., Milan, 1979, Report 11.14.
45. S. Sato *et al.* High speed charge simulation method. *Trans. IEE Japan*, **101** A (1981), pp. 1–8 (in Japanese).
46. A. Knecht. Development of surface charges on epoxy resin spacers stressed with direct applied voltages. Proc. 3rd Int. Symp. on Gaseous Diel., Knoxville, 1982. (Gaseous Dielectrics III, 1982, pp. 356–364.)
47. C.A. Brebbia, J.F.C. Telles and L.C. Wrobel. Boundary Element Techniques; Applications Theory and Applications in Engineering. Berlin, New York, Springer, Verlag, 1984.
48. Ming Yu. The Studies of Corona and Ion Flow Fields Associated with HVDC Power Transmission Lines. Ph.D. Thesis, Elec. Eng. Dpt. University of Manitoba, 1993.
49. Ming Yu and E. Kuffel. The Integration Functions with Irregularities in Boundary Element Method for Poisson's Equations. Pro. 6th Int. Symp. on Electromagnetic Fields, Sept. 1993. Warsaw, Poland.
50. Ming Yu and E. Kuffel. Spline Element for Boundary Element Method. *COMPUMAG* **93**, Miami Fl. USA, Nov. 1993.

Chapter 5

Electrical breakdown in gases

Before proceeding to discuss breakdown in gases a brief review of the fundamental principles of kinetic theory of gases, which are pertinent to the study of gaseous ionization and breakdown, will be presented. The review will include the classical gas laws, followed by the ionization and decay processes which lead to conduction of current through a gas and ultimately to a complete breakdown or spark formation.

5.1 Classical gas laws

In the absence of electric or magnetic fields charged particles in weakly ionized gases participate in molecular collisions. Their motions follow closely the classical kinetic gas theory.

The oldest gas law established experimentally by Boyle and Mariotte states that for a given amount of enclosed gas at a constant temperature the product of pressure (p) and volume (V) is constant or

$$pV = C = \text{const.} \tag{5.1}$$

In the same system, if the pressure is kept constant, then the volumes V and V_0 are related to their absolute temperatures T and T_0 (in K) by Gay–Lussac's law:

$$\frac{V}{V_0} = \frac{T}{T_0}. \tag{5.2}$$

When temperatures are expressed in degrees Celsius, eqn (5.2) becomes

$$\frac{V}{V_0} = \frac{273 + \theta}{273} \tag{5.3}$$

Equation (5.3) suggests that as we approach $\theta = -273°\text{C}$ the volume of gas shrinks to zero. In reality, all gases liquefy before reaching this value.

According to eqn (5.2) the constant C in eqn (5.1) is related to a given temperature T_0 for the volume V_0:

$$pV_0 = C_0. \tag{5.4}$$

Substituting V_0 from eqn (5.2) gives

$$pV = \left(\frac{C_0}{T_0}\right) T. \tag{5.5}$$

The ratio (C_0/T_0) is called the universal gas constant and is denoted by R. Equation (5.5) then becomes

$$pV = RT = C. \tag{5.6}$$

Numerically R is equal to 8.314 joules/°K mol. If we take n as the number of moles, i.e. the mass m of the gas divided by its mol-mass, then for the general case eqn (5.1) takes the form

$$pV = nC = nRT, \tag{5.7}$$

Equation (5.7) then describes the state of an ideal gas, since we assumed that R is a constant independent of the nature of the gas. Equation (5.7) may be written in terms of gas density N in volume V containing N_1 molecules. Putting $N = N_A$ where $N_A = 6.02 \times 10^{23}$ molecules/mole, N_A is known as the Avogadro's number. Then eqn (5.7) becomes

$$\frac{N_1}{V} = N = \frac{N_A}{R}\frac{p}{T}$$

or

$$pV = \frac{N_1}{N_A}RT = N_1 kT \text{ or } p = NkT \tag{5.8}$$

The constant $k = R/N_A$ is the universal Boltzmann's constant (=1.3804 × 10^{-23} joules/°K) and N is the number of molecules in the gas.

If two gases with initial volumes V_1 and V_2 are combined at the same temperature and pressure, then the new volume will be given by

$$V = V_1 + V_2$$

or in general

$$V = V_1 + V_2 + V_2 + \ldots + V_n. \tag{5.9}$$

Combining eqns (5.7) and (5.9) gives

$$V = \frac{n_1 RT}{p} + \frac{n_2 RT}{p} + \ldots + \frac{n_n RT}{p}$$

rearranging

$$p = \frac{n_1 RT}{V} + \frac{n_2 RT}{V} + \ldots + \frac{n_n RT}{V}$$

or

$$p = p_1 + p_2 + \ldots p_n. \tag{5.10}$$

where $p_1, p_2, \ldots, p_n$ denote the partial pressures of gases $1, 2, \ldots, n$. Equation (5.10) is generally referred to as the law of partial pressures.

Equations (5.1) to (5.10) can be derived directly from the kinetic theory of gases developed by Maxwell in the middle of the nineteenth century. A brief derivation will be presented.

The fundamental equation for the kinetic theory of gas is derived with the following assumed conditions:

1. Gas consists of molecules of the same mass which are assumed spheres.
2. Molecules are in continuous random motion.
3. Collisions are elastic – simple mechanical.
4. Mean distance between molecules is much greater than their diameter.
5. Forces between molecules and the walls of the container are negligible.

Consider a cubical container of side $l = 1\,\text{m}$ as shown in Fig. 5.1 with N_1 molecules, each of mass m and r.m.s. velocity u. Let us resolve the velocity into components, u_x, u_y, u_z where $u^2 = u_x^2 + u_y^2 + u_z^2$. Suppose a molecule of mass m is moving in the x-direction with velocity u_x. As it strikes the wall of container plane YZ it rebounds with the velocity $-u_x$. The change in momentum, therefore, is

$$\Delta m = mu_x - (-mu_x) = 2mu_x.$$

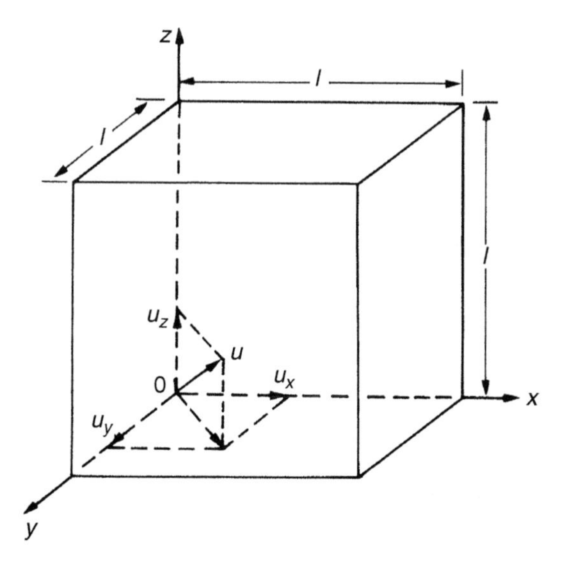

Figure 5.1 *Resolution of molecular forces*

For the cube of side l the number of collisions per second with the right-hand wall is $u_x/2l$, therefore

$$\Delta m/\text{sec/molecule} = \frac{2mu_x u_x}{2l} = \frac{mu_x^2}{l},$$

but the same molecule will experience the same change in momentum at the opposite wall. Hence Δm/sec/molecule in the x-direction $= 2mu_x^2/l$. For the three-dimensional cube with total change in momentum per second per molecule (which is the force) we obtain the force per particle as

$$F = \frac{2m}{l}(u_x^2 + u_y^2 + u_z^2) = \frac{2mu^2}{l}. \tag{5.11}$$

As kinetic energy for a particle $W = \frac{1}{2}mu^2$, therefore,

$$F = 4\frac{W}{l}.$$

For N_1 particles the energy due to different velocities u of particles will become the mean energy, and therefore

$$F = 4\frac{N_1\overline{W}}{l}.$$

Force leads to pressure p, taking into account the total area of the cube ($A = 6l^2$)

$$p = \frac{F}{A} = \frac{4N_1\overline{W}}{6l \cdot l^2} = \frac{2}{3}\frac{N_1\overline{W}}{l^3} \tag{5.12}$$

with $l^3 = V =$ volume. Comparing eqns (5.8) and (5.12) leads to:

$$pV = \tfrac{2}{3}N_1\overline{W}.$$

Comparing eqn (5.12) with eqn (5.1) we note that these equations are identical for constant temperature. Using eqn (5.8) gives

$$p = \frac{2}{3}\frac{N_1}{V}\overline{W} = \frac{2}{3}N\overline{W} = NkT$$

which leads to the expression for mean energy per molecule:

$$\overline{W} = \tfrac{3}{2}kT. \tag{5.13}$$

5.1.1 Velocity distribution of a swarm of molecules

It has been shown using probability consideration that the distribution of molecular velocities depends on both the temperature and the molecular weight of the gas. The mathematical analysis shows the most probable velocity is neither the average nor the r.m.s. velocity of all the molecules.

The velocity u of gas molecules or particles has a statistical distribution and follows the Boltzmann–Maxwell distribution given by the expression[(1)*]

$$f(u)\,\mathrm{d}u = \frac{\mathrm{d}N_u}{N} = \frac{4}{\sqrt{\pi}}\left(\frac{u}{u_P}\right)^2\left[\mathrm{e}^{-(u/u_P)^2}\right]\frac{\mathrm{d}u}{u_P} \tag{5.14}$$

where u_p is the most probable velocity and $\mathrm{d}N_u/N$ the relative number of particles whose instantaneous velocities lie in the range u/u_p and $(u+\mathrm{d}u)/u_p$.

Let

$$f\left(\frac{u}{u_p}\right) = \frac{\mathrm{d}N_u}{N}\bigg/\frac{\mathrm{d}u}{u_p}$$

and

$$u_r = \frac{u}{u_p} \text{ (relative velocity).}$$

Introducing this dimensionless variable into eqn (5.14) gives the function representing velocity distribution

$$f(u_r) = \frac{4}{\sqrt{\pi}}u_r^2\mathrm{e}^{-u_r^2} \tag{5.14a}$$

with

$$\frac{\mathrm{d}N_u}{N} = f(u_r)\,\mathrm{d}u_r.$$

The distribution function corresponding to eqn (5.14a) is shown in Fig. 5.2. It should be noted that the function is asymmetrical about the most probable velocity u_p. A greater number of particles has a velocity higher than u_p. The average velocity $\bar{u}$ is obtained from integrating u_r from 0 to ∞.

$$\bar{u}_r = \int_{u_r=0}^{\infty} u_r f(u_r)\,\mathrm{d}u_r = \frac{4}{\sqrt{\pi}}\underbrace{\int_0^{\infty} u_r^3\mathrm{e}^{-u_r^2}\,\mathrm{d}u_r}_{1/2} = \frac{2}{\sqrt{\pi}};$$

or

$$\bar{u} = \bar{u}_r u_p = 1.128 u_p \tag{5.15}$$

* Superscript numbers are to references at the end of the chapter.

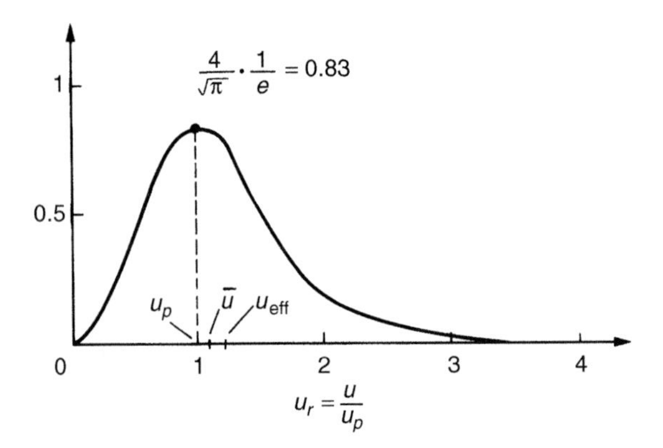

Figure 5.2 *Distribution of velocities (u_p most probable; $\bar{u}$ average; u_{eff} effective or r.m.s.)*

The r.m.s. or effective value of velocity is obtained by squaring u_r and obtaining the average square value

$$(u_r)^2_{\text{eff}} = \int_{u_r=0}^{\infty} u_r^2 f(u_r)\,\mathrm{d}u_r = \frac{4}{\sqrt{\pi}} \underbrace{\int_0^{\infty} u_r^4 \mathrm{e}^{-u_r^2}\,\mathrm{d}u_r}_{3/8\sqrt{\pi}} = \frac{3}{2}$$

$$u_{\text{eff}} = u_{r\text{eff}} u_p = \sqrt{\frac{3}{2}} u_p = 1.224 u_p. \tag{5.16}$$

The mean kinetic energy of the particle given by eqn (5.13) relates its effective velocity to the temperature ($\frac{1}{2}mu^2_{\text{eff}} = \frac{3}{2}kT$) and we obtain

$$u_{\text{eff}} = \sqrt{\frac{3kT}{m}}; \quad \bar{u} = \sqrt{\frac{8kT}{\pi m}}; \quad u_p = \sqrt{\frac{2KT}{m}}. \tag{5.17}$$

Hence the respective velocities remain in the ratio $u_p : \bar{u} : u_{\text{eff}} = 1 : 1.128 : 1.224$.

It should be noted that the foregoing considerations apply only when the molecules or particles remain in thermal equilibrium, and in the absence of particle acceleration by external fields, diffusion, etc. If the gas contains electrons or ions or other atoms that are at the same temperature, the average particle energy of such mixture is

$$\tfrac{1}{2}mu^2_{\text{eff}} = \tfrac{1}{2}m_e u^2_{\text{eeff}} = \tfrac{1}{2}m_i u^2_{\text{ieff}} = \ldots = \tfrac{3}{2}kT \tag{5.18}$$

where m, m_i, m_e are the respective masses of gas molecules, ions, electron, and u_{eff}, u_{ieff}, u_{eeff} are their corresponding velocities.

Table 5.1 *Mean molecular velocities at 20°C and 760 Torr*[2]

Gas	*Electron*	H_2	O_2	N_2	*Air*	CO_2	H_2O *(vapour)*	SF_6
$\bar{u}$ (m/sec)	100×10^3	1760	441	470	465	375	556	199

The values of the mean molecular velocities calculated for 20°C and 760 Torr for several of the common gases are included in Table 5.1.

5.1.2 The free path λ of molecules and electrons

Knowledge of dependency and distribution of free paths (λ) may explain (with restrictions) the dependency of $\alpha = f(E, N)$ discussed later, even assuming a simple 'ballistic' model. For this reason a short treatment of free paths will be presented. The free path (λ) is defined as the distance molecules or particles travel between collisions. The free path is a random quantity and as we shall see its mean value depends upon the concentration of particles or the density of the gas.

To derive the mean free path ($\bar{\lambda}$) assume an assembly of stationary molecules of radius r_1, and a moving layer of smaller particles of radius r_2 as particles move, their density will decrease as shown in Fig. 5.3. As the smaller particles move, their density will decrease due to scattering caused by collisions with gas molecules. If we assume that the moving particles and molecules behave as solid spheres, then a collision will occur every time the centres of two particles come within a distance $r_1 + r_2$. The area for collision presented by a

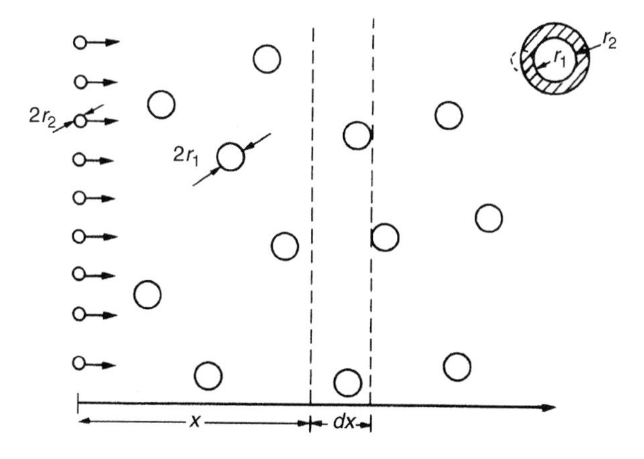

Figure 5.3 *Model for determining free paths*

molecule is then $\pi(r_1 + r_2)^2$ and in a unit volume it is $N\pi(r_1 + r_2)^2$. This is often called the effective area for interception where N = number of particles per unit volume of gas.

If we consider a layer of thickness dx, distant x from the origin (Fig. 5.3) and $n(x)$ the number of particles that survived the distance x, then the decrease in the moving particles due to scattering in layer dx is

$$dn = -n(x)N\pi(r_1 + r_2)^2\, dx.$$

Assuming the number of particles entering (at $x = 0$) is n_0, integration gives

$$n(x) = N_0\, e^{-N\pi(r_1+r_2)^2 x}. \tag{5.19}$$

The probability of free path of length x is equal to the probability of collisions between x and $x + dx$. The mean free path $\bar{\lambda} = \bar{x}$ is obtained as follows. Differentiating eqn (5.19) we obtain

$$f(x) = \frac{dn}{n_0} = N\pi(r_1 + r_2)^2 e^{-N\pi(r_1+r_2)^2 x}\, dx.$$

For the mean free path

$$\begin{aligned}\bar{x} = \bar{\lambda} &= \int_{x=0}^{\infty} x f(x)\, dx \\ &= N\pi(r_1 + r_2)^2 \int_{x=0}^{\infty} x\, e^{-N\pi(r_1+r_2)^2 x}\, dx \\ &= \frac{1}{N\pi(r_1 + r_2)^2}\end{aligned} \tag{5.20}$$

The denominator in eqn (5.20) has the dimensions of area and the value $\pi(r_1 + r_2)^2$ is usually called the cross-section for interception or simply collision cross-section and is denoted by σ:

$$\sigma = \frac{1}{N\bar{\lambda}} \tag{5.21}$$

We shall see later that the collisions between the incoming particles and the stationary molecules may lead to processes such as ionization, excitation, attachment, etc.

If we put in eqn (5.21) $Q = N\sigma$, then Q will represent the effective cross-section presented by molecules or particles in unit volume of gas for all collisions for density of N molecules/volume. If, for example, only a fraction P_i of collisions between the incoming particles and the gas particles leads to ionization then P_i is the probability of ionization. Thus if only ionizing collisions are counted, the molecules present an effective area of only $P_i Q = Q_i$;

Q_i is the effective cross-section for ionization. Similarly for other processes, excitation (Q_e), photoionization Q_{ph} attachment (Q_a), etc., including elastic collisions can be taken into account

$$Q = Q_{\text{elastic}} + Q_i + Q_e + Q_a + \dots \tag{5.22}$$

Atomic cross-sections (σ) for different processes vary over a wide range. For ionization they can rise to some $2 \times 10^{-16}\,\text{cm}^2$, but for collisions resulting in nuclear reactions they may be $10^{-24}\,\text{cm}^2$ or less.

In deriving the expression (5.20) it was assumed that the struck molecules were stationary, i.e. the molecules of gas 2 had no thermal velocity. In reality this is not true. It can be shown that the expression giving the collisional cross-section must be still multiplied by a factor

$$\eta = \sqrt{1 + \frac{m_1}{m_2}}$$

with m_1 and m_2 the mass of each gas component. In a gas mixture the collisional cross-section of particles of type 1 of gas (m_1, r_1, N_1) becomes equal to the sum of all collisional cross-sections of the other particles of types of gas $(m_2, m_3, \dots, r_2, r_3, \dots, N_2, N_3, \dots)$. Thus the mean free path of particles of type 1 is

$$\bar{\lambda}_1 = \frac{1}{\pi \sum_{i=1}^{n} N_i (r_1 + r_i)^2 \sqrt{1 + \frac{m_1}{m_i}}} \tag{5.23}$$

For an atom in its own gas $r_1 = r_2 = r$; $u_1 = u_2$. Then

$$\bar{\lambda}_a = \frac{1}{4\sqrt{2}\pi r^2 N}. \tag{5.24}$$

For an electron in a gas $r_1 \ll r_2$ and $m_1 \ll m_2$ eqn (5.23) gives

$$\bar{\lambda}_e = \frac{1}{\pi r_2^2 N}$$

or

$$\bar{\lambda}_e = 4\sqrt{2}\bar{\lambda}_a = 5.66\bar{\lambda}_a. \tag{5.25}$$

Table 5.2 shows examples of mean free path (gas) for gases of different molecular weight.

From eqn (5.8), $N = p/kT$, it follows that the mean free path is directly proportional to temperature and inversely as the gas pressure

$$\lambda(p, T) = \lambda_0 \frac{p_0}{p} \frac{T}{T_0}. \tag{5.26}$$

Table 5.2 *Mean free paths measured at 15°C and 760 Torr*[2]

Type of gas	H_2	O_2	N_2	CO_2	H_2O	*Dimensions*
λ	11.77	6.79	6.28	4.19	4.18	10^{-8} m
Molecular weight	2.016	32.00	28.020	44.00	18.00	

Considering a typical practical case with values for average velocity of gas $\overline{u} \approx 500$ m/sec and the mean free path $\overline{\lambda} \approx 10^{-7}$ m we obtain the number of collisions per second:

$$\nu = \frac{\overline{u}}{\overline{\lambda}} = 5 \times 10^9 \frac{1}{\text{sec}} \approx 5 \quad \text{collisions/nsec.}$$

The average time between two collisions

$$\Delta t = \frac{1}{\nu} = \frac{1}{5 \times 10^9} = 0.2\,\text{nsec.}$$

5.1.3 Distribution of free paths

In the earlier sections it was shown that molecular collisions are random events and these determine free paths. Hence, free path is a random quantity and will have a distribution about a mean value. For the system in Fig. 5.3 the mean free path is given by eqn (5.20)

$$\overline{\lambda} = \frac{1}{N\pi(r_1 + r_2)^2},$$

N being the gas density and r_1 and r_2 the radii of the two types of particles. The distribution function of free paths is obtained from eqn (5.19)

$$\int_{n_0}^{n} \mathrm{d}n = -\int_{x=0}^{x} \ln \frac{\mathrm{d}x}{\overline{\lambda}}$$

or

$$n(x) = n_0\,\mathrm{e}^{-x/\overline{\lambda}} \tag{5.27}$$

where $n(x)$ = number of molecules reaching a distance x without collision, $\mathrm{d}n$ = number of molecules colliding thereafter within a distance $\mathrm{d}x$, n_0 = total number of molecules at $x = 0$. Equation (5.27) is plotted in Fig. 5.4.

It is seen that the percentage of molecules that survive collisions is only 37 per cent. The exponent in eqn (5.27) may also be written in terms of

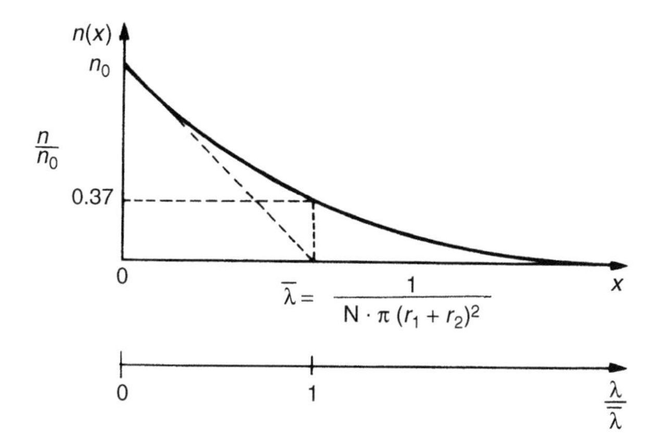

Figure 5.4 *Distribution of free paths*

collision cross-sections defined by eqn (5.21), to represent absorption or decay of particles along the path x or

$$n = n_0 e^{-N\sigma x} \tag{5.28}$$

where σ may include photoabsorption, attachment, etc.

5.1.4 Collision-energy transfer

The collisions between gas particles are of two types: (i) elastic or simple mechanical collisions in which the energy exchange is always kinetic, and (ii) inelastic in which some of the kinetic energy of the colliding particles is transferred into potential energy of the struck particle or vice versa. Examples of the second type of collisions include excitation, ionization, attachment, etc., which will be discussed later.

To derive an expression for energy transfer between two colliding particles, let us consider first the case of an elastic collision between two particles[(2)] of masses m and M. Assume that before collision the particle of large mass M was at rest and the velocity of the smaller particle was u_0 in the direction shown in Fig. 5.5. After collision let the corresponding velocities be u_1 and V, the latter along line of centres as shown. θ is the incidence angle and ψ is the scattering angle.

The fractional energy loss by the incoming particle during a collision at an angle θ is then given by

$$\Delta(\theta) = \left(\frac{u_0^2 - u_1^2}{u_0^2} \right). \tag{5.29}$$

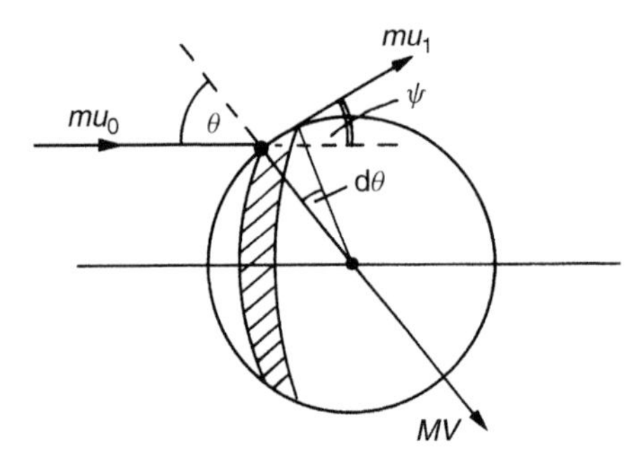

Figure 5.5 *Energy transfer during elastic collision*

Since the collision is assumed to be kinetic, the equations for conservation of momentum and energy are

$$mu_0 - mu_1 \cos\psi = MV\cos\theta \tag{5.30}$$

$$mu_1 \sin\psi = MV\sin\theta \tag{5.31}$$

$$\tfrac{1}{2}mu_0^2 - \tfrac{1}{2}mu_1^2 = \tfrac{1}{2}MV^2. \tag{5.32}$$

Squaring eqns (5.30) and (5.31) and adding and combining with eqn (5.32) we obtain

$$V = \frac{2mu_0\cos\theta}{M+m}.$$

Rearranging eqn (5.32) and combining with eqn (5.29) gives

$$\Delta\theta = \frac{MV^2}{mu_0^2} = \frac{4mM\cos^2\theta}{(m+M)^2}. \tag{5.33}$$

To obtain the mean fractional energy loss per collision, let $P(\theta)$ be the probability of a collision at an angle of incidence between θ and $\theta + d\theta$. The total area presented for collision is $\pi(r_1 + r_2)^2$. The probability of a collision taking place between θ and $\theta + d\theta$ is the ratio of the projected area (Fig. 5.5) to the whole area or

$$P(\theta)\,d\theta = \left\{\frac{2\pi(r_1+r_2)\sin\theta\cos\theta\,d\theta}{\pi(r_1+r_2)^2} = \sin 2\theta\,d\theta \left\{\begin{matrix}\text{for} & 0 \le \theta \\ \text{for} & \pi/2 \le \theta \le \pi\end{matrix}\right.\right. .$$

The mean fractional loss of energy per collision allowing for collisions at all angles is

$$\overline{\Delta(\theta)} = \int_0^{\pi/2} P(\theta)\Delta(\theta)\,d\theta \Big/ \int_0^{\pi/2} P(\theta)\,d\theta. \tag{5.34}$$

Using eqns (5.33) and (5.34), we obtain

$$\overline{\Delta(\theta)} = \frac{2mM}{(m+M)^2}. \tag{5.35}$$

If we consider the case when the incoming particle is an ion of the same mass as the struck particle, then $m = M$ and eqn (5.35) gives $\overline{\Delta\theta} = \frac{1}{2}$ which indicates a high rate of energy loss in each elastic collision. On the other hand, if the incoming particle is an electron, then $m \ll M$ and eqn (5.35) gives $\overline{\Delta\theta} = 2m/M$. The average fraction of energy lost by an electron in an elastic collision is therefore very small. For example, if we consider the case of electrons colliding with He gas atoms, the average fractional energy loss per collision $\overline{\Delta\theta}$ is 2.7×10^{-4} and in argon it is 2.7×10^{-5}. Thus electrons will not readily lose energy in elastic collisions whereas ions will.

Let us now consider the case when part of the kinetic energy of the incoming particle is converted into potential energy of the struck particle. Then applying the laws of energy and momentum conservation we obtain

$$\tfrac{1}{2}mu_0^2 = \tfrac{1}{2}mu_1^2 + \tfrac{1}{2}MV^2 + W_p \tag{5.36}$$

$$mu_0 = mu_1 + MV \tag{5.37}$$

where W_p is the increase in potential energy of the particle of mass M initially at rest. Substituting eqn (5.37) into eqn (5.36) and rearranging we obtain

$$W_p = \frac{1}{2}\left[m(u_0^2 - u_1^2) - \frac{m^2}{M}(u_0 - u_1)^2\right]. \tag{5.38}$$

For the conditions of constant kinetic energy of the incoming particles, differentiation of eqn (5.38) with respect to u_1 gives the maximum energy transfer when

$$\frac{dW_{p\max}}{du} = 0$$

or

$$\frac{u_1}{u_0} = \frac{m}{m+M}. \tag{5.39}$$

Equation (5.39) shows that the potential energy gained from the incident particle reaches a maximum value when the ratio of its final to initial velocity equals the ratio of its mass to the sum of masses of the individual particles.

When the colliding particles are identical, the maximum kinetic to potential energy transfer occurs when $u_1 = u_0/2$. On the other hand, if the colliding particle is an electron of mass $m \ll M$ the maximum energy transfer corresponds to $u_1 - (m/M)u_0$ which means that the new velocity u_1 becomes only a small fraction of the original velocity.

For the case when the target particle was initially at rest, the maximum amount of potential energy gained will be given by the expression obtained by inserting the value of velocity u_1 from eqn (5.39) into eqn (5.38) or

$$W_{p\max} = \frac{M}{m+M} \frac{mu_0^2}{2}. \tag{5.40}$$

For an electron $m \ll M$, eqn (5.40) becomes

$$W_{p\max} \cong \frac{1}{2} \frac{mu_0^2}{2} \tag{5.41}$$

or almost all its kinetic energy is converted into potential energy. Thus we shall see later that electrons are good ionizers of gas, while ions are not. To cause ionization the incoming electron must have a kinetic energy of at least $\frac{1}{2}mu_0^2 \geq eV_i$, where V_i is the ionization potential of the atom or molecule.

5.2 Ionization and decay processes

At normal temperature and pressure gases are excellent insulators. The conduction in air at low field is in the region $10^{-16} - 10^{-17}$ A/cm^2.* This current results from cosmic radiations and radioactive substances present in earth and the atmosphere. At higher fields charged particles may gain sufficient energy between collisions to cause ionization on impact with neutral molecules. It was shown in the previous section that electrons on average lose little energy in elastic collisions and readily build up their kinetic energy which may be supplied by an external source, e.g. an applied field. On the other hand, during inelastic collisions a large fraction of their kinetic energy is transferred into potential energy, causing, for example, ionization of the struck molecule. Ionization by electron impact is for higher field strength the most important process leading to breakdown of gases. The effectiveness of ionization by electron impact depends upon the energy that an electron can gain along the mean free path in the direction of the field.

If $\bar{\lambda}_e$ is the mean free path in the field direction of strength E then the average energy gained over a distance $\bar{\lambda}$ is $\Delta W = eE\bar{\lambda}_e$. This quantity is proportional to E/p since $\bar{\lambda}_e \propto 1/p$ (eqn (5.26)). To cause ionization on impact the energy ΔW must be at least equal to the ionization energy of the molecule (eV_i). Electrons with lower energy than eV_i may excite particles and the excited particles

* The figure $10^{-16} - 10^{-17}$ A/cm^2 correlates with the current flowing to the whole surface of earth (due to natural electric field). This current is 1000–1200 A. With earth surface of about 5×10^{19} cm^2, we get

$$j = \frac{I}{\text{surface}} = \frac{1000}{5 \times 10^{19}} = 0.2 \times 10^{-16}\ \text{A/cm}^2.$$

on collision with electrons of low energy may become ionized. Furthermore, not all electrons having gained energy $\Delta W \geq eV_i$ upon collision will cause ionization. This simple model is not applicable for quantitative calculations, because ionization by collision, as are all other processes in gas discharges, is a probability phenomenon, and is generally expressed in terms of cross-section for ionization defined as the product $P_i\sigma = \sigma_i$ where P_i is the probability of ionization on impact and σ is the molecular or atomic cross-sectional area for interception defined earlier. The cross-section σ_i is measured using monoenergetic electron beams of different energy. The variation of ionization cross-sections for H_2, O_2, and N_2 with electron energy is shown in Fig. 5.6.[3] It is seen that the cross-section is strongly dependent upon the electron energy. At energies below ionization potential the collision may lead to excitation of the struck atom or molecule which on collision with another slow moving electron may become ionized. This process becomes significant only when densities of electrons are high. Very fast moving electrons may pass near an atom without ejecting an electron from it. For every gas there exists an optimum electron energy range which gives a maximum ionization probability.

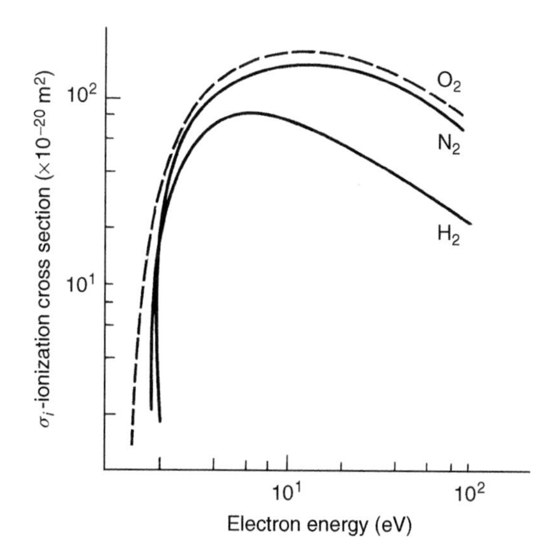

Figure 5.6 *Variation of ionization cross-sections for O_2, N_2, H_2 with electron energy*

5.2.1 Townsend first ionization coefficient

In the absence of electric field the rate of electron and positive ion generation in an ordinary gas is counterbalanced by decay processes and a state of equilibrium exists. This state of equilibrium will be upset upon the application of a sufficiently high field. The variation of the gas current measured between two parallel plate electrodes was first studied as a function of the applied voltage by Townsend.[4]

Townsend found that the current at first increased proportionately with the applied voltage and then remained nearly constant at a value i_0 which corresponded to the background current (saturation current), or if the cathode was irradiated with a u.v. light, i_0 gave the emitted photocurrent. At still higher voltage the current increased above the value i_0 at an exponential rate. The general pattern of the current–voltage relationship is shown schematically in Fig. 5.7.

The increase in current beyond V_2 Townsend ascribed to ionization of the gas by electron collision. As the field increases, electrons leaving the cathode are accelerated more and more between collisions until they gain enough energy to cause ionization on collision with gas molecules or atoms.

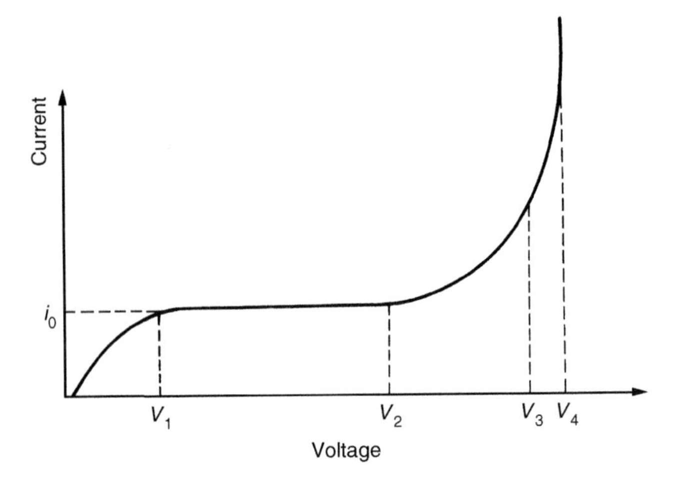

Figure 5.7 *Current–voltage relationship in prespark region*

To explain this current increase Townsend introduced a quantity α, known as Townsend's first ionization coefficient, defined as the number of electrons produced by an electron per unit length of path in the direction of the field.

Thus if we assume that n is the number of electrons at a distance x from the cathode in field direction (Fig. 5.8) the increase in electrons $\mathrm{d}n$ in additional distance $\mathrm{d}x$ is given by

$$\mathrm{d}n = \alpha n \mathrm{d}x.$$

Integration over the distance (d) from cathode to anode gives

$$n = n_0 \mathrm{e}^{\alpha d} \tag{5.42}$$

where n_0 is the number of primary electrons generated at the cathode. In terms of current, with I_0 the current leaving the cathode, eqn (5.42) becomes

$$I = I_0 e^{\alpha d}. \tag{5.43}$$

The term $e^{\alpha d}$ in eqn (5.42) is called the electron avalanche and it represents the number of electrons produced by one electron in travelling from cathode to anode. The electron multiplication within the avalanche is shown diagrammatically in Fig. 5.8.

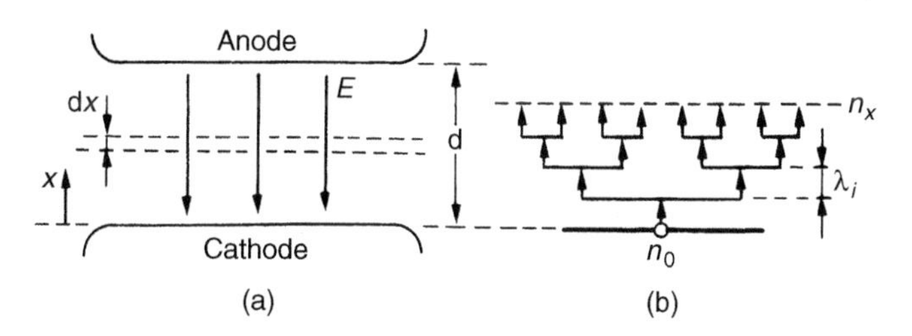

Figure 5.8 *Schematic representation of electron multiplication (a) gap arrangement, (b) electron avalanche*

The increase of current (avalanche growth) shown in the diagram (Fig. 5.8(b)) would be $I = I_0 e^k$, with k = number of ionizing steps ($k = x/\lambda_i$). The transition for infinitely small values of

$$dx \begin{pmatrix} \lim \lambda_i \\ \lambda_i \to dx \end{pmatrix}$$

leads to the expression $e^{\alpha x}$.

The quantity α, although a basic quantity describing the rate of ionization by electron collision, cannot be readily calculated from the measured cross-section for ionization (σ_i). The latter is determined for monoenergetic electrons and calculation of α from value of σ_i is only possible when the electron energy distribution in the gas is known. For 'swarm' conditions Raether(5) derived a relationship between α and σ_i, which is of the form

$$\frac{\alpha}{N} = \frac{1}{u_e} \int_0^\infty v\sigma_i(v) f(v)\, dv$$

with N the concentration, molecules/atoms, $f(v_{th})$ the distribution of velocities of electrons, and u_e the drift velocity of electrons in the field direction. A simple derivation is possible for simple gases (non-attaching) using the Clausius distribution of free paths (Fig. 5.4) and applying it to electrons.

We have seen that at a constant temperature for a given gas the energy distribution ΔW depends only on the value E/p. Also for a given energy distribution the probability of an ionization occurring will depend on the gas density or pressure.

Therefore, we can write

$$\alpha = pf\left(\frac{E}{p}\right)$$

or

$$\frac{\alpha}{p} = f\left(\frac{E}{p}\right). \tag{5.44}$$

Equation (5.44) describes a general dependence of α/p upon E/p which has been confirmed experimentally.

A derivation of expression for this dependence is possible for simple gases, using the Clausius distribution (eqn (5.27)) for free paths applied to electrons. This means that we assume that this distribution will not be altered by the additional velocity of electrons in field direction. Then all electrons which acquire energy $\Delta W \geq eV_i$, where V_i is the ionization potential, will ionize the gas. These electrons have travelled a distance x, and using eqn (5.27) the fraction of electrons with paths exceeding a given value x is

$$f'(x) = \mathrm{e}^{-x/\bar{\lambda}}.$$

Therefore, only with a very small probability electrons can gain high energies if they reach long distances.

The number of successful collisions – the ionization coefficient α – is clearly related to this distribution, and is certainly directly proportional to the decay of collisions in the intervals between x and $(x + \mathrm{d}x)$, or

$$\alpha = -\frac{\mathrm{d}f'(x)}{\mathrm{d}x} = \frac{1}{\bar{\lambda}}\mathrm{e}^{-(\lambda_i/\bar{\lambda})} \tag{5.44a}$$

where $\lambda_i = x$ is the ionizing free path. The above treatment assumes $\bar{\lambda}_E = \bar{\lambda}$, i.e. the velocity distribution is not altered by the additional velocity of electrons in the field direction. In reality there is a difference between $\bar{\lambda}$ and $\bar{\lambda}_E$ as shown below.

Hence

$$\frac{\bar{v}}{\bar{\lambda}} = \frac{v}{\bar{\lambda}_E} \quad \text{and} \quad \bar{\lambda}_E = \frac{1}{n_E}$$

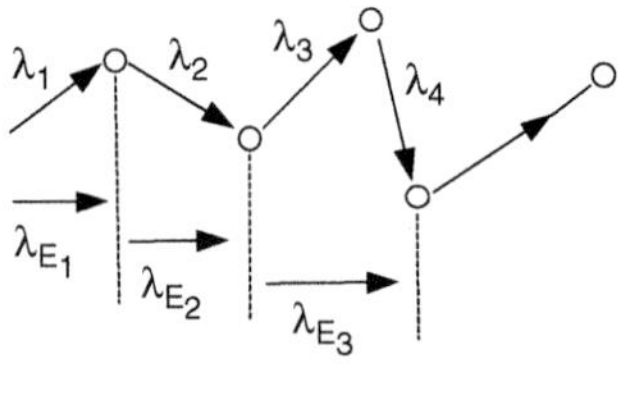

v being the electron drift velocity.

Then eqn (5.44) when corrected for field drift velocity becomes

$$\alpha = n_E\,\mathrm{e}^{-\lambda_i/\bar{\lambda}_E} = \frac{1}{\bar{\lambda}_E}\mathrm{e}^{-\lambda_i/\bar{\lambda}_E} \ldots. \tag{5.45}$$

Using eqn (5.21), with σ_i as true cross-section for ionization and N the gas density, we obtain

$$\bar{\lambda} = \frac{1}{N\sigma_i}.$$

Introducing from eqn (5.8) $N = p/kT$, for a gas pressure p the mean free path becomes

$$\bar{\lambda} = \frac{kT}{p\sigma_i}.$$

If in addition we put $\lambda_i = V_i/E$, then

$$\alpha = -\frac{\mathrm{d}f'(x)}{\mathrm{d}x} = \frac{p\sigma_i}{kT}\mathrm{e}^{-(V_i/E)(p\sigma_i/kT)}, \tag{5.46}$$

or

$$\frac{\alpha}{p} = \frac{\sigma_i}{kT}\mathrm{e}^{-(\sigma_i/kT)[V_i/(E/p)]} = A_{(T)}\,\mathrm{e}^{-[B_{(T)}/(E/p)]} \tag{5.47}$$

where

$$A_{(T)} = \frac{\sigma_i}{kT}; \quad B_{(T)} = \frac{V_i\sigma_i}{kT}. \tag{5.47a}$$

It cannot be expected that the real dependence of α/p upon E/p agrees with measured values within the whole range of E/p, because phenomena which have not been taken into account are influencing the ionization rate. However, even with constant values of A and B, eqn (5.47) determines the ionization process within certain ranges of E/p. Therefore, for various gases the 'constants' A and B have been determined experimentally and can be found in the literature.[6]

Some of these experimental values for several of the more common gases are listed in Table 5.3.

Table 5.3 *Ionization constants A and B ($T = 20°C$)*

Gas	*A* *ion pairs* $cm^{-1}\,Torr^{-1}$	*B* $V\,cm^{-1}$ $Torr^{-1}$	E/p *range* $V\,cm^{-1}\,Torr^{-1}$	V_i *volts*
H_2	5	130	150–600	15.4
N_2	12	342	100–600	15.5
air	15	365	100–800	–
CO_2	20	466	500–1000	12.6
He	3	34	20–150	24.5
Hg	20	370	200–600	–

The constants A and B in eqn (5.47a), as derived from kinetic theory, rarely agree with the experimentally determined values. The reasons for this disagreement lies in the assumptions made in our derivations. We assumed that every electron whose energy exceeds eV_i will automatically lead to ionization. In reality the probability of ionization for electrons with energy just above the ionization threshold is small and it rises slowly to a maximum value of about 0.5 at 4 to 6 times the ionization energy. Beyond that it decreases. We have also assumed that the mean free path is independent of electron energy which is not necessarily true. A rigorous treatment would require taking account of the dependence of the ionization cross-section upon the electron energy.

Using the experimental values for the constants A and B for N_2 and H_2 in eqn (5.47), the graphical relationship between the parameters α/p and E/p has been plotted in Fig. 5.9. The values have been corrected to $T = 0°C$.

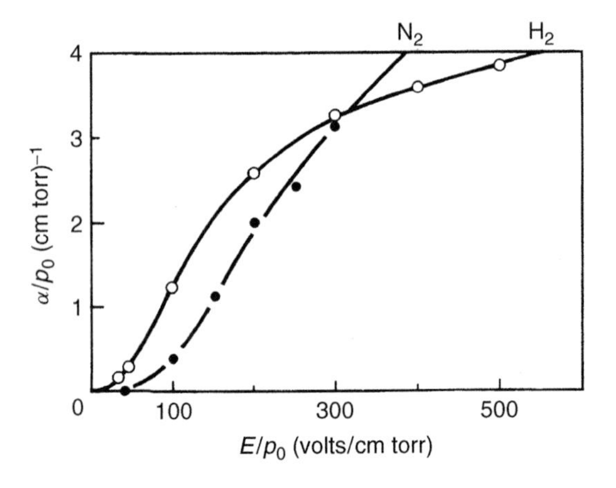

Figure 5.9 *Dependence of α/p on E/p in N_2 and H_2, reduced to 0°C*

It should be noted that theoretically α/p begins at zero value of E/p, which follows from the distribution of free paths which have values from 0 to ∞. In practice in many gases attachment q will also be present, and at low values of E/p it is difficult to obtain the values for 'real' α and for 'real' η. Experimental measurements yield the 'effective' ionization coefficient ($\overline{\alpha} = \alpha - \eta$). In this case $\overline{\alpha}p$ begins at a finite value of E/p corresponding to the lowest breakdown strength.

Numerous measurements of α in various gases have been made by Townsend[(4)] and subsequent workers and the data can be found in the literature.[(7–9)] The Geballe and Harrison's data are included in Table 5.5.

5.2.2 Photoionization

Electrons of lower energy than the ionization energy eV_i may on collision excite the gas atoms to higher states. The reaction may be symbolically represented as $A + e + K$ energy $\rightarrow A^* + e$; $A^* \rightarrow A + hv$; A^* represents the atom in an excited state. On recovering from the excited state in some 10^{-7}–10^{-10} sec, the atom radiates a quantum of energy of photon (hv) which in turn may ionize another atom whose ionization potential energy is equal to or less than the photon energy. The process is known as photoionization and may be represented as $A + hv \rightarrow A^+ + e$, where A represents a neutral atom or molecule in the gas and hv the photon energy. For ionization to occur $hv \geq eV_i$ or the photon wavelength $\lambda \leq c_0 h/eV_i$, c_0 being the velocity of light and h Planck's constant. Therefore, only very short wavelength light quanta can cause photoionization of gas. For example, the shortest wavelength radiated from a u.v. light with quartz envelope is 145 nm, which corresponds to $eV_i = 8.5\,\text{eV}$, lower than the ionization potential of most gases.

The probability of photon ionizing a gas or molecule is maximum when $(hv - eV_i)$ is small (0.1–1 eV). Photoionization is a secondary ionization process and may be acting in the Townsend breakdown mechanism and is essential in the streamer breakdown mechanism and in some corona discharges. If the photon energy is less than eV_i it may still be absorbed by the atom and raise the atom to a higher energy level. This process is known as photoexcitation.

5.2.3 Ionization by interaction of metastables with atoms

In certain elements the lifetime in some of the excited electronic states extends to seconds. These states are known as metastable states and the atoms in these states are simply referred to as metastables represented by A^m. Metastables have a relatively high potential energy and are therefore able to ionize neutral particles. If V^m, the energy of a metastable A^m, exceeds V_i, the ionization of another atom B, then on collision ionization may result according to the reaction

$$A^m + B \rightarrow A^+ + B + e.$$

For V^m of an atom $A^m < V_i$ of an atom B the reaction may lead to the exciting of the atom B which may be represented by $A^m + B \rightarrow A + B^*$.

Another possibility for ionization by metastables is when $2V^m$ for A^m is greater than V_i for A. Then the reaction may proceed as

$$A^m + A^m \rightarrow A^+ + A + e + \text{K.E.}$$

This last reaction is important only when the density of metastables is high.

Another reaction may follow as

$$A^m + 2A \rightarrow A_2^* + A;$$

$$A_2^* - A + A + h\upsilon.$$

The photon released in the last reaction is of too low energy to cause ionization in pure gas, but it may release electrons from the cathode.

Ionization by metastable interactions comes into operation long after excitation, and it has been shown that these reactions are responsible for long time lags observed in some gases.(10) It is effective in gas mixtures.

5.2.4 Thermal ionization

The term thermal ionization, in general, applies to the ionizing actions of molecular collisions, radiation and electron collisions occurring in gases at high temperature. If a gas is heated to sufficiently high temperature many of the gas atoms or molecules acquire sufficiently high velocity to cause ionization on collision with other atoms or molecules. Thermal ionization is the principal source of ionization in flames and high-pressure arcs.

In analysing the process of thermal ionization, the recombination between positive ions and electrons must be taken into account. Under thermodynamic equilibrium conditions the rate of new ion formation must be equal to the rate of recombination. Using this assumption Saha(11) derived an expression for the degree of ionization θ in terms of the gas pressure and absolute temperature as follows:

$$\frac{\theta^2}{1-\theta^2} = \frac{1}{p}\frac{(2\pi m_e)^{3/2}}{h}(kT)^{5/2}\,\mathrm{e}^{-W_i/kT}$$

or

$$\frac{\theta^2}{1-\theta^2} = \frac{2.4\times 10^{-4}}{p}T^{5/2}\,\mathrm{e}^{-W_i/kT} \qquad (5.48)$$

where p is the pressure in torr, W_i the ionization energy of the gas, k Boltzmann's constant, θ the ratio of n_i/n, and n_i the number of ionized particles of total n particles. The strong dependence of θ on temperature in eqn (5.48) shows that the degree of ionization is negligible at room temperature. On substitution of values W_i, kT, p and T in eqn (5.48) we find that thermal ionization becomes significant for temperatures above 1000 K.

5.2.5 Deionization by recombination

Whenever there are positively and negatively charged particles present, recombination takes place. The potential energy and the relative kinetic energy of the recombining electron–ion is released as quantum of radiation. Symbolically the reaction may be represented as

$$\left.\begin{array}{l} A^+ + e \rightarrow A + hv \\ A^+ + e \rightarrow A^m + hv \end{array}\right\} \begin{array}{l}\text{radiation}\\ \text{recombination}\end{array}$$

or

Alternatively a third body C may be involved and may absorb the excess energy released in the recombination. The third body C may be another heavy particle or electron. Symbolically

$$A^+ + C + e \rightarrow A^* + C \rightarrow A + C + hv$$

or

$$A^+ + e + e \rightarrow A^* + e \rightarrow A + e + hv.$$

At high pressures, ion–ion recombination takes place. The rate of recombination in either case is directly proportional to the concentration of both positive ions and negative ions. For equal concentrations of positive ions n_+ and negative ions n_- the rate of recombination

$$\frac{dn_+}{dt} = \frac{dn_-}{dt} = -\beta n_+ n_- \tag{5.49}$$

where β is a constant known as the recombination rate coefficient.

Since $n_+ \approx n_- = n_i$ and if we assume at time $t = 0$: $n_i = n_{i0}$ and at time t: $n_i = n_i(t)$, then eqn (5.49) becomes

$$\frac{dn_i}{dt} = -\beta_i^2.$$

Integration gives

$$\int_{n_{i0}}^{n_i} \frac{dn_i}{n_2^i} = -\beta \int_0^t dt$$

or

$$n_i(t) = \frac{n_{i0}}{1 + n_{i0}\beta t}. \tag{5.50}$$

The half-time duration, during which time the concentration of ions has decreased to half its original value, is given by

$$t_n = \frac{1}{n_{i0}\beta}. \tag{5.51}$$

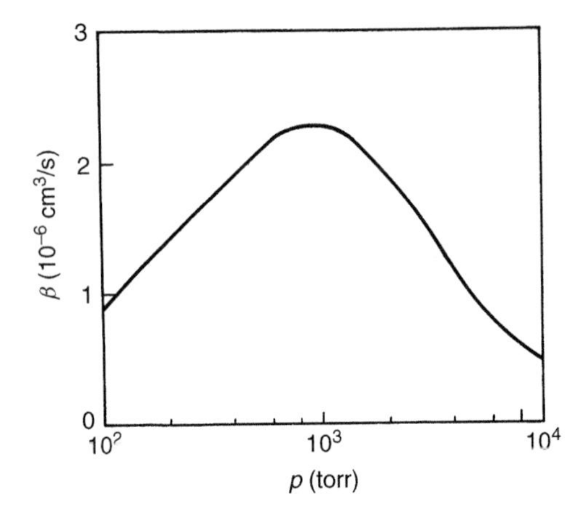

Figure 5.10 *Recombination coefficient (ion–ion) in air at 20°C*

The variation of the recombination rate coefficient β with pressure in air is shown in Fig. 5.10. The recombination process is particularly important at high pressures for which diffusion is relatively unimportant.

5.2.6 Deionization by attachment – negative ion formation

Electron affinity

Certain atoms or molecules in their gaseous state can readily acquire a free electron to form a stable negative ion. Gases, whether atomic or molecular, that have this tendency are those that are lacking one or two electrons in their outer shell and are known as electronegative gases. Examples include the halogens (F, Cl, Br, I and At) with one electron missing in their outer shell, O, S, Se with two electrons deficient in the outer shell.

For a negative ion to remain stable for some time, the total energy must be lower than that of an atom in the ground state. The change in energy that occurs when an electron is added to a gaseous atom or molecule is called the electron affinity of the atom and is designated by W_a. This energy is released as a quantum or kinetic energy upon attachment. Table 5.4 shows electron affinities of some elements.

There are several processes of negative ion formation:

(1) The simplest mechanism is one in which the excess energy upon attachment is released as quantum known as radiative attachment. This process is reversible, that is the captured electron can be released by absorption of a photon known as photodetachment. Symbolically the process is represented as:

$$A + e \Leftrightarrow A^- + hv \quad (W_a = hv).$$

Table 5.4 *Electron affinities of some elements*

Element	*Ion formed*	W_a *(kJ/mole)*
H	H^-	−72
O	O^-	−135
F	F^-	−330
Cl	Cl^-	−350
Br	Br^-	−325
I	I^-	−295

(2) The excess energy upon attachment can be acquired as kinetic energy of a third body upon collision and is known as a third body collision attachment, represented symbolically as:

$$e + A + B \rightarrow A^- + (B + W_k) \quad (W_a = W_k).$$

(3) A third process is known as dissociative attachment which is predominant in molecular gases. Here the excess energy is used to separate the molecule into a neutral particle and an atomic negative ion, symbolically expressed as:

$$e + AB \Leftrightarrow (AB^-)* \Leftrightarrow A^- + B.$$

(4) In process (3) in the intermediate stage the molecular ion is at a higher potential level and upon collision with a different particle this excitation energy may be lost to the colliding particle as potential and/or kinetic energy. The two stages of the process here are:

$$e + AB \Leftrightarrow (AB^-)*$$

$$(AB^-) * +A \Leftrightarrow (AB)^- + A + W_k + W_p.$$

Other processes of negative ion formation include splitting of a molecule into positive and negative ions upon impact of an electron without attaching the electron:

$$e + AB \Leftrightarrow A^+ + B^- + e$$

and a charge transfer following heavy particle collision, yielding an ion pair according to:

$$A + B \rightarrow A^+ + B^-.$$

All the above electron attachment processes are reversible, leading to electron detachment.

The process of electron attachment may be expressed by cross-section for negative ion formation σ_A in an analogous way to ionization by electron impact. Typical examples of the variation of attachment cross-section with electron energy for processes (2) and (3) measured in SF_6 and CO_2 are shown in Figs 5.11 and 5.12 respectively.

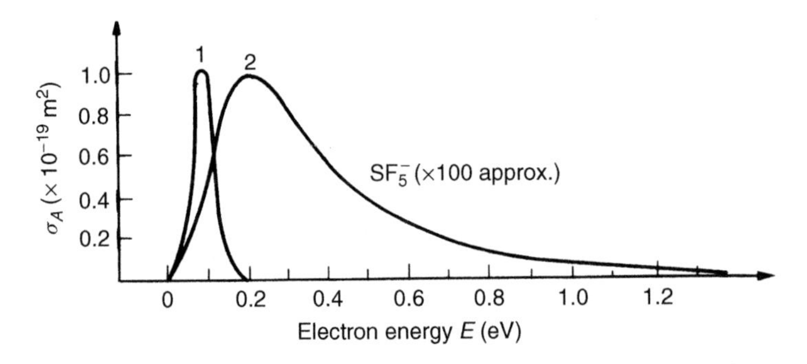

Figure 5.11 *Variation of attachment cross-section with electron energy in SF_6. 1. Radiative attachment. 2. Dissociative attachment*

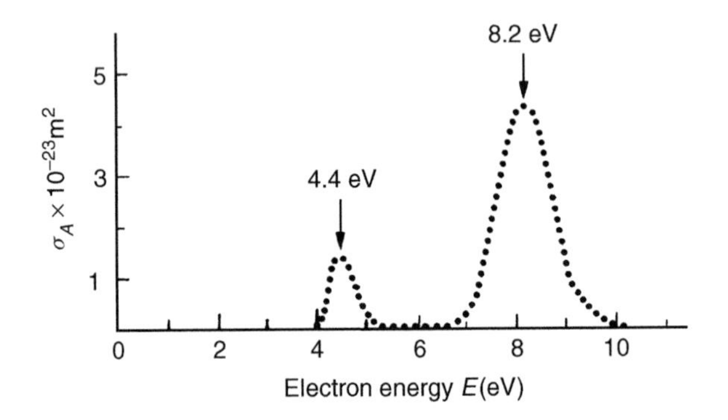

Figure 5.12 *Variation of electron attachment cross-section with electron energy in CO_2 (both peaks O^-)*

Cumulatively the process of electron attachment describing the removal of electrons by attachment from ionized gas by any of the above processes may be expressed by a relation analogous to the expression (5.43) which defines electron multiplication in a gas. If η is the attachment coefficient defined by analogy with the first Townsend ionization coefficient α, as the number of attachments produced in a path of a single electron travelling a distance of 1 cm in the direction of field, then the loss of electron current in a distance dx due to this cause is

$$dI = -\eta I \, dx$$

or for a gap of length d with electron current I_0 starting at cathode

$$I = I_0 e^{-\eta d} \quad (5.52)$$

Several methods for the measurements of the attachment coefficient have been described in the literature.[12] Methods for the determination of attachment coefficient utilizing eqn (5.52) rely on the measurement of the surviving electronic current[13] at grids distance d apart inserted at two points along the path of the current between the electrodes. Such methods are applicable only at relatively low values of E/p when ionization by electron collision can be neglected. At higher values of E/p it becomes necessary to measure both the ionization coefficient α and the attachment coefficient η simultaneously.

If the processes of electron multiplication by electron, collision and electron loss by attachment are considered to operate simultaneously, then neglecting other processes the number of electrons produced by collision in distance dx is

$$dn_i = n\alpha \, dx$$

where x is the distance from the cathode. At the same time the number of electrons lost in dx by attachment is

$$dn_A = -n\eta \, dx$$

so that the number of electrons still free is

$$dn = dn_i + dn_A = n(\alpha - \eta) \, dx.$$

Integration from $x = 0$ to x with n_0 electrons starting from the cathode gives the number of electrons at any point in the gap as

$$n = n_0 e^{(\alpha - \eta)X} \quad (5.53)$$

The steady state current under such conditions will have two components, one resulting from the flow of electrons and the other from negative ions. To determine the total current we must find the negative ion current component. We note that the increase in negative ions in distance dx is

$$dn_- = n\eta \, dx = n_0 \eta e^{(\alpha - \eta)} \, dx.$$

Integration from 0 to x gives

$$n_- = \frac{\eta_0 \eta}{\alpha - \eta} [e^{(\alpha - \eta)x} - 1]$$

The total current equals the sum of the two components or

$$\frac{n + n_-}{n_0} = \frac{\alpha}{\alpha - \eta} e^{(\alpha-\eta)d} - \frac{\eta}{\alpha - \eta} \tag{5.54}$$

and the expression for current becomes

$$I = I_0 \left[\frac{\alpha}{\alpha - \eta} e^{(\alpha-\eta)d} - \frac{\eta}{\alpha - \eta} \right]. \tag{5.55}$$

In the absence of attachment when η is zero the expression (5.55) reduces to the form $i = i_0 e^{\alpha d}$ and the $\log i - d$ plot of eqn (5.55) gives a straight line, with α representing the slope. When the value of η is appreciable, there may be a decrease in currents, especially at large values of d, such that the $\log i$ against d curve drops below the straight line relation. The departure from linearity in plotting $\log i$ against d gives a measure of the attachment coefficient. Several workers(9) have used this method for determining the α and η coefficients. The results obtained by this method by Geballe and Harrison for ionization α and attachment η in oxygen and in air are included in Table 5.5. It is convenient to represent the observed ionization coefficient by a single coefficient $\overline{\alpha} = \alpha - \eta$ defined as the effective ionization coefficient.

As electron attachment reduces electron amplification in a gas, gases with a high attachment coefficient such as sulphur hexafluoride or freon have much higher dielectric strength than air or nitrogen. The measured data for ionization and attachment coefficients for SF_6 are included in Table 5.6. These gases are technically important and are widely used as insulating medium in compact h.v. apparatus including totally enclosed substations and h.v. cables as will be discussed later in this chapter.

5.2.7 Mobility of gaseous ions and deionization by diffusion

Mobility

In the presence of an electric field charged particles in a gas will experience a force causing them to drift with a velocity that varies directly with the field and inversely with the density of the gas through which it moves. The drift velocity component in the field direction of unit strength is defined as the mobility (K) or symbolically

$$K = \frac{u}{E} (\mathrm{m^2/V\,sec}),$$

where u is the average drift velocity in field direction and E is the electric field strength. The mobility (K) is mainly a characteristic of the gas through which the ion moves and is independent of E/p over a wide range

Table 5.5 *Geballe and Harrison's values for α/p and η/p in oxygen and air*

E/p V/cm. torr	Oxygen			Air		
	α/p	η/p	$\alpha/p - \eta/p$	α/p	η/p	$\alpha/p - \eta/p$
25.0	0.0215	0.0945	−0.0730	0.00120	0.00495	−0.00375
27.5	0.0293	0.0900	−0.0607	0.00205	0.00473	−0.00268
30.0	0.0400	0.0851	−0.0451	0.00340	0.00460	−0.00120
32.5	0.0532	0.0795	−0.0263	0.00560	0.00460	+0.00100
35.0	0.0697	0.0735	−0.0038	0.00880	0.00475	+0.00405
37.5	0.0862	0.0685	+0.0177	0.0130	0.00497	+0.0080
40.0	0.107	0.0645	+0.043	0.0190	0.00530	+0.0137
42.5	0.128	0.0605	+0.068	0.0260	0.00575	+0.0203
45.0	0.152	0.0570	+0.095	0.0340	0.00635	+0.0227
47.5	0.179	0.0535	+0.126	0.0460	0.00700	+0.0390
50.0	0.206	0.052	+0.154	0.057	0.00780	+0.049
52.5	0.234	0.049	+0.185	0.070	0.00870	+0.061
55.0	0.263	0.047	+0.216	0.087	0.00967	+0.077
57.5	0.292	0.045	+0.247	0.102	0.0108	+0.091
60.0	0.323	0.043	+0.280	0.120	0.0119	+0.108
62.5	0.355	0.0415	+0.314	0.140	–	–
65.0	0.383	0.040	+0.343	0.170	–	–
70.0	0.450	–	–	–	–	–
72.5	0.480	–	–	–	–	–
75.0	0.518	–	–	–	–	–

of E/p so long as the velocity gained by the ion from the field is considerably less than the average thermal velocity of the gas through which the ion moves.

To derive an expression for mobility of ions in a gas under an influence of electric field in the region of low values of E/p we assume that the ions are in thermal equilibrium with the gas molecules. Their drift velocity is small compared to the thermal velocity. If τ, the time interval between two successive collisions, is independent of E, then

$$\tau = \frac{\overline{\lambda}_i}{\overline{c}}$$

where $\overline{\lambda}_i$ is the ionic mean free path and $\overline{c}$ is the mean thermal velocity of the ion. During time τ the ion is accelerated by the field E with an acceleration

Table 5.6 *Experimental values of the ionization and attachment coefficients in SF_6 (temp. = 20°C)*

E/p_{20} V/cm. torr	p torr	$\bar{\alpha}/p \times 10^3$ cm^{-1} $torr^{-1}$	α/p cm^{-1} $torr^{-1}$	η/p cm^{-1} $torr^{-1}$	$\gamma \times 10^7$
115.0	5.2	−90	1.05	1.14	
125.0	5.2	200	1.32	1.12	
135.0	5.2	480	1.52	1.04	
145.0	5.2	760	1.73	0.97	
154.0	5.2	1000			
155.0	5.2	1050			
165.0	5.2	1300			
175.0	5.2	1550			
185.0	5.2	1850			
200.0	5.2	2250			
115.0	19.5	90	1.04	1.13	
120.0	19.5	50	1.18	1.13	
125.0	19.5	200	1.30	1.10	
135.0	19.5	505			1
116.0	50.2	−25			
118.0	50.2	8	1.15	1.14	
120.0	50.2	60	1.18	1.12	
122.0	50.2	115			
125.0	50.2	225			
126.0	50.2	240			3
116.0	99.1	−38			
118.0	99.1	10			
119.0	99.1	33			
120.0	99.1	56			
122.0	99.1	120			
115.0	202.0	75			
117.0	202.0	−16			
118.0	202.0	4			
119.0	202.0	29			
119.25	202.0	36			60
122.0	202.0	110			6
117.0	402.4	−30			
118.0	402.4	5			
118.5	402.4	16			75

$a = eE/m$, where m is the ionic mass and e is its charge. Therefore, in time τ it moves a distance

$$s = \frac{eE}{2m}\tau^2$$

and the drift velocity becomes

$$u = \frac{eE}{2m}\tau = \left(\frac{e\tau}{2m}\right)E = \left(\frac{e\overline{\lambda}_i}{2m\overline{c}}\right)E$$

and

$$K = \frac{u}{E} = \frac{e\overline{\lambda}_i}{2m\overline{c}}. \tag{5.56}$$

In deriving eqn (5.56) we assumed that $\overline{\lambda}_i$ is unaffected by the drift motion, that is all ions are moving with the same random velocity and all ions have the same mean free path $\overline{\lambda}_i$. To take the statistical distribution of mean free paths $\overline{\lambda}_i$ into account, let us assume that the ions are moving with an average velocity $\overline{c}$ in zig-zag projections of lengths which are distributed about the mean free path $\overline{\lambda}_i$. Then if a is the acceleration caused by the field E, the distance between two collisions is

$$s = \frac{1}{2}at^2 = \frac{1}{2}\frac{eE}{m}\left(\frac{x^2}{\overline{c}^2}\right).$$

x denotes the total distance travelled between these collisions. The average value of s is obtained by averaging x^2 over the distribution of free paths

$$\begin{aligned}\overline{s} &= \frac{eE}{2m}\frac{1}{\overline{c}^2}\int_0^\infty x^2 e^{-x/\overline{\lambda}_i}\frac{dx}{\overline{\lambda}_i} \Big/ \int_0^\infty e^{-x/\overline{\lambda}_i}\frac{dx}{\overline{\lambda}_i} \\ &= \frac{eE}{m\overline{c}^2}\overline{\lambda}_i^2.\end{aligned}$$

If the mean free time $\overline{\tau} = \overline{\lambda}_i/\overline{c}$

$$\overline{s} = \frac{eE}{m}\overline{\tau}^2,$$

the drift velocity

$$u = \frac{\overline{s}}{\overline{\tau}} = \frac{eE}{m\overline{c}}\overline{\lambda}_i,$$

and

$$K = \frac{u}{E} = \frac{e\overline{\lambda}_i}{m\overline{c}}. \tag{5.57}$$

Thus when the distribution of free paths is taken into account the expression for mobility is increased by a factor of 2. The expression (5.57) ignores the fact that after collision the ions may have initial velocities in the direction of field. Langevin[14] deduced a more exact expression which takes into account this effect of 'persistence of motion' and for an ion of mass m moving through a gas consisting of molecules of mass M the expression becomes

$$K = \frac{0.815e\bar{\lambda}}{Mc}\sqrt{\frac{m+M}{m}} \tag{5.58}$$

where c is the r.m.s. velocity of agitation of the gas molecules and is an approximation to the ionic mean free path ($\bar{\lambda} \approx \bar{\lambda}_i$). For condition of thermal equilibrium

$$\frac{mc_1^2}{2} = \frac{Mc^2}{2} = \frac{3}{2}kT.$$

With c_1 the r.m.s. velocity of the ions, k Boltzmann's constant and T absolute temperature, expression (5.58) can be written in the form

$$K = 0.815\frac{e\bar{\lambda}_i}{mc_1}\sqrt{\frac{m+M}{m}}. \tag{5.59}$$

For an electron $m \ll M$ this expression reduces to

$$K = 0.815\frac{e}{m}\frac{\bar{\lambda}_e}{c_1}. \tag{5.60}$$

Table 5.7 gives some experimentally determined mobilities for negative and positive ions. The presence of impurities is found to have a profound effect on the measured mobility. The effect is particularly large in the case of negative ions when measured in non-attaching gases such as helium or hydrogen for which the electrons are free if the gases are extremely pure. The ion and electron mobilities can be used for the determination of conductivity or resistivity of an ionized gas. In the simplest case when the concentrations of positive ions and electrons are equal

$$n_+ = n_e = n,$$

then the total current density

$$j = j_i + j_e = n(u_i + u_e)e$$

where u_i and u_e are the drift velocities of the ions and electrons respectively. In terms of mobilities, the current density (j) and the conductivity (σ) become

$$j = neE(K_e + K_i)$$

Table 5.7 *Mobility of singly charged gaseous ions at O°C and 760 Hg (in cm/sec/volts/cm) (taken from Cobine*[4]*)*

Gas	K^-	K^+	*Gas*	K^-	K^+
Air (dry)	2.1	1.36	H_2 (very pure)	7900.0	
Air (very pure)	2.5	1.8	HCl	0.95	1.1
A	1.7	1.37	H_2S	0.56	0.62
A (very pure)	206.0	1.31	He	6.3	5.09
Cl_2	0.74	0.74	He (very pure)	500.0	5.09
CCl_4	0.31	0.30	N_2	1.84	1.27
C_2H_2	0.83	0.78	N_2 (very pure)	145.0	1.28
C_2H_5Cl	0.38	0.36	NH_3	0.66	0.56
C_2H_5OH	0.37	0.36	N_2O	0.90	0.82
CO	1.14	1.10	Ne		9.9
CO_2	0.98	0.84	O_2	1.80	1.31
H_2	8.15	5.9	SO_2	0.41	0.41

and

$$\sigma = \frac{j}{E} = ne(K_e + K_i). \tag{5.61}$$

Since $K_e \gg K_i$, the conductivity is given approximately by

$$\sigma = neK_e. \tag{5.62}$$

In the presence of appreciable space charge $n_e \neq n_i$ the conductivity components must be considered separately.

Diffusion

In electrical discharges whenever there is a non-uniform concentration of ions there will be movement of ions from regions of higher concentration to regions of lower concentration. The process by which equilibrium is achieved is called diffusion. This process will cause a deionizing effect in the regions of higher concentrations and an ionizing effect in regions of lower concentrations. The presence of walls confining a given volume augments the deionizing effect as the ions reaching the walls will lose their charge. The flow of particles along the ion concentration gradient constitutes a drift velocity similar to that of charged particles in an electric field. Both diffusion and mobility result in mass motion described by drift velocity caused in one case by the net effect of unbalanced collision forces (ion concentration gradient) and in the other case by the electric field.

If we consider a container with gas in which the concentration varies in the x-direction, then taking a layer of unit area and thickness $\mathrm{d}x$ placed perpendicularly to the direction x, the number of particles crossing this area is proportional to the ion concentration gradient $\mathrm{d}n/\mathrm{d}x$. The flow of particles or flux in the x-direction is

$$\Gamma = -D\frac{\mathrm{d}n}{\mathrm{d}x} \tag{5.63}$$

The negative sign indicates that n increases and the rate of flow (Γ) must decrease in the direction of flow. The constant D is known as the diffusion coefficient. From kinetic theory it can be shown that $D = \bar{u} \cdot \bar{\lambda}/3$. With $\bar{u}$ being the mean thermal velocity, the rate of change of concentration in the layer $\mathrm{d}x$ is

$$\begin{aligned} \frac{\mathrm{d}}{\mathrm{d}t}(n\mathrm{d}x) &= \Gamma - \left(\Gamma + \frac{\mathrm{d}\Gamma}{\mathrm{d}x}\mathrm{d}x\right) \\ \frac{\mathrm{d}n}{\mathrm{d}t} &= D\frac{\mathrm{d}^2 n}{\mathrm{d}x^2}. \end{aligned} \tag{5.64}$$

For the three-dimensional case eqn (5.64) becomes

$$\frac{\partial n}{\partial t} = D\nabla^2 n \tag{5.65}$$

which is the general equation for diffusion.

5.2.8 Relation between diffusion and mobility

In most transport phenomena, both diffusion and mobility will be acting together. It is therefore important to establish a relation between the diffusion coefficient and mobility. Consider a cloud of singly charged particles diffusing through the gas. For simplicity let us take again the unidirectional case with particles diffusing in the x-direction at a rate of flow given by eqn (5.63). Then the ion velocity is equal to

$$u_i = \frac{\Gamma}{n_i} = -\frac{D}{n_i}\frac{\mathrm{d}n_i}{\mathrm{d}x}$$

where n_i is the ion concentration. Because n_i is directly proportional to p_i

$$u_i = -\frac{D}{p_i}\frac{\mathrm{d}p_i}{\mathrm{d}x} = -\frac{D}{p_i}f_i$$

the force acting on the ions in this volume.

Since there are N ions per unit volume, the force exerted on one ion is

$$f_i = \frac{1}{N}\frac{\mathrm{d}p_i}{\mathrm{d}x} = -\frac{p_i}{DN}u_i.$$

An ion subjected to E, the force acting on it opposite to drift motion is

$$f_e = eE = \frac{eu}{K}, \quad \text{with } u\text{-drift velocity of ion.}$$

In order that there is no net flow in the x-direction the force f_i must be balanced by f_e (oppositely directed, $u_i = u$) and $f_i = -f_e$:

$$\frac{D}{K} = \frac{P_i}{eN} = \frac{kTn_i}{en_i} = \frac{kT}{e}\left(n_i = \frac{p_i}{kT}\right) \tag{5.66}$$

In general the mobilities of negatively charged ions are higher than those of positive ones (Table 5.7) and consequently the negative ions will diffuse more rapidly. If the concentration of the diffusing particles is significant, the differential rate of diffusion will cause charge separation which will give rise to an electric field. The action of the field is such that it will tend to augment the drift velocity of the positive ions and retard that of negative ions, and the charge separation reaches a state of equilibrium in which the position and negative ions diffuse with the same velocity. This process is known as ambipolar diffusion. The average velocity of the diffusing ions may be obtained by considering the ion motion to be governed by the combined action of diffusion and mobility in the induced field E.

Then the velocity of the positive ions is given by

$$u^+ = -\frac{D^+\mathrm{d}n^+}{n^+\mathrm{d}x} + K^+E. \tag{5.67}$$

Similarly the velocity of negative ions is

$$u^- = -\frac{D^-\mathrm{d}n^-}{n^-\mathrm{d}x} - K^-E. \tag{5.68}$$

Eliminating E between eqns (5.67) and (5.68), and assuming $n^+ = n^- = n$,

$$\frac{\mathrm{d}n^+}{\mathrm{d}x} = \frac{\mathrm{d}n^-}{\mathrm{d}x} = \frac{\mathrm{d}n}{\mathrm{d}x} \quad \text{and} \quad u^+ = u^- = u.$$

The average velocity of the ions then becomes

$$\bar{u} = -\frac{D^+K^- + D^-K^+}{n(K^+ + K^-)}\frac{\mathrm{d}n}{\mathrm{d}x} \tag{5.69}$$

and the ambipolar diffusion coefficient for mixed ions may be written as

$$D_a = -\frac{D^+K^- + D^-K^+}{K^+ + K^-} \tag{5.70}$$

and since from eqn (5.66)

$$\frac{K^+}{D^+} = \frac{e}{kT^+} \quad \text{and} \quad \frac{K^-}{D^-} = \frac{e}{kT^-}$$

therefore, for the cases when $T_e = T^- \gg T^+$ and when $K_e = K^- \gg K^+$ we have

$$D_a \cong D^+\frac{T_e}{T^+} \cong D^-\frac{K^+}{K_e} \cong D_e\frac{K^+}{K_e} \cong \frac{kT_e}{e}. \tag{5.71}$$

If the electrons and ions are in equilibrium with the gas, that is all particles are at the same temperature, then we may put $D_eK_i = D_iK_e$ and the ambipolar diffusion coefficient becomes

$$D_a \approx \frac{2D_iK_e}{K_e} \approx 2D_i, \tag{5.72}$$

since $K_e \gg K_i$.

Finally, the field E between the space charges can be obtained by eliminating u from eqns (5.66) and (5.67), giving

$$E = -\frac{D^- - D^+}{K^- + K^+}\frac{1}{n}\frac{\mathrm{d}n}{\mathrm{d}x}. \tag{5.73}$$

Equations (5.71) and (5.72) are commonly used, although both are only approximated, but they demonstrate that D_a increases with T_e, that is with the random electron energy and that if electrons are at the same temperature as the gas, D_a is of the same order as D_i so that electrons are slowed much more than positive ions are accelerated. Diffusion processes are of particular importance in studying streamer discharge and spark channels.

5.3 Cathode processes – secondary effects

Electrodes, in particular the cathode, play a very important role in gas discharges by supplying electrons for the initiation, for sustaining and for the completion of a discharge. Under normal conditions electrons are prevented from leaving the solid electrode by the electrostatic forces between the electrons and the ions in the lattice. The energy required to remove an electron from a Fermi level is known as the work function (W_a) and is a characteristic of a given material. There are several ways in which the required energy may be supplied to release the electrons.

5.3.1 Photoelectric emission

Photons incident upon the cathode surface whose energy exceeds the work function ($hv > W_a$) may eject electrons from the surface. For most metals the critical frequency v_0 lies in the u.v. range. When the photon energy exceeds the work function, the excess energy may be transferred to electron kinetic energy according to the Einstein relation:

$$\tfrac{1}{2}mu_e^2 = hv = hv_0 \tag{5.74}$$

where m is the electron mass, u_e its velocity and hv_0 is the critical energy required to remove the electron and $hv_0 = W_a$ the work function.

Table 5.8 gives the work functions for several elements. The work function is sensitive to contamination which is indicated by the spread in the measured values shown in Table 5.8.

The spread is particularly large in the case of aluminium and metals which readily oxidize. In the presence of a thin oxide film, it has been shown by Malter[16] that positive ions may gather at the oxide layer without being neutralized, giving rise to a high field strength leading to augmented secondary emission. The effect is known as the Malter effect.

Table 5.8 *Work function for typical elements*[15]

Element	*Ag*	*Al*	*Cu*	*Fe*	W
W_a (eV)	4.74	2.98–4.43	4.07–4.7	3.91–4.6	4.35–4.6

5.3.2 Electron emission by positive ion and excited atom impact

Electrons may be emitted from metal surfaces by bombardment of positive ions or metastable atoms. To cause a secondary emission of an electron the impinging ion must release two electrons, one of which is utilized to neutralize the ion charge. The minimum energy required for a positive ion electron emission is twice the work function $W_K + W_p \geq 2W_a$, since the ion is neutralized by one electron and the other electron is ejected. W_K and W_p are the respective kinetic and potential energies of the incident ion. The electron emission by positive ions is the principal secondary process in the Townsend spark discharge mechanism.

Neutral excited (metastable) atoms or molecules incident upon the electrode surface are also capable of ejecting electrons from the surface.

5.3.3 Thermionic emission

In metals at room temperature the conduction electrons will not have sufficient thermal energy to leave the surface. If we consider the electrons as a gas at room temperature, then their average thermal energy is

$$\frac{mu_e^2}{2} = \frac{3kT}{2} = 3.8 \times 10^{-2}\,\text{eV},$$

which is much lower than the work function (Table 5.8). If, however, the metal temperature is increased to some 1500–2500 K, the electrons will receive energy from the violent thermal lattice vibrations sufficient to cross the surface barrier and leave the metal. The emission current is related to the temperature of the emitter by the Richardson[17] relation for thermionically emitted saturation current density:

$$J_s = \frac{4\pi mek^2}{h^3} T^2 \exp\left[-\frac{W_a}{kT}\right] \text{A/m}^2 \tag{5.75}$$

where e and m are the electronic charge and mass respectively, h is Planck's constant, k Boltzmann's constant, T the absolute temperature and W_a the surface work function.

Putting

$$A = \frac{4\pi mek^2}{h^3},$$

the above expression becomes

$$J_s = AT^2 \exp\left[-\frac{W_a}{kT}\right] \tag{5.76}$$

which shows that the saturation current density increases with decreasing work function and increasing temperature. On substitution of the constants m, e, k and h, $A = 120 \times 10^4\,\text{A}\,\text{m}^{-2}\,\text{degr}^{-2}$. The experimentally obtained values are lower than predicted by eqn (5.76). This discrepancy is attributed to the wave nature of the electrons. Although electrons may possess the required escape energy, some of them may be reflected back into the solid from the surface atoms or surface contaminants such as adsorbed gases. The effect may be taken into account by inserting the effective value $A_{\text{eff}} = A(1 - R)$ in the current density expression (5.76), where R is the reflection coefficient. In the presence of a strong electric field there will be a reduction in the work function as the Schottky[18] effect, discussed in the next section, and the thermionic emission will be enhanced.

5.3.4 Field emission

Electrons may be drawn out of a metal surface by very high electrostatic fields. It will be shown that a strong electric field at the surface of a metal may modify the potential barrier at the metal surface to such an extent that electrons in the upper level close to the Fermi level will have a definite probability of passing through the barrier. The effect is known as 'tunnel effect'. The fields required to produce emission currents of a few microamperes are of the order of 10^7–10^8 V/cm. Such fields are observed at fine wires, sharp points and submicroscopic irregularities with an average applied voltage quite low (2–5 kV). These fields are much higher than the breakdown stress even in compressed gases.

To derive an expression for the emission current let us consider an electron as it leaves the surface in the direction x as shown in Fig. 5.13. Its electric field can be approximated as that between a point charge and the equipotential planar surface. The field lines here are identical to those existing when an image charge of $+e$ is thought to exist at a normal distance of $-x$ on the other side of the equipotential metal surface. Applying Coulomb's law, the force on the electron in the x-direction is given by

$$F(x) = \frac{-e^2}{4\pi\varepsilon_0(2x)^2} = \frac{-e^2}{16\pi\varepsilon_0 x^2}.$$

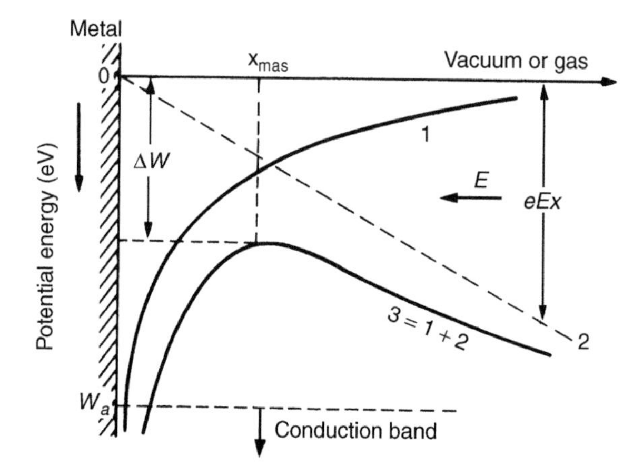

Figure 5.13 *Lowering of the potential barrier by an external field. 1. Energy with no field. 2. Energy due to field. 3. Resultant energy*

The potential energy at any distance x is obtained by integrating the above equation from ∞ to x.

$$W_{e1} = \frac{-e^2}{16\pi\varepsilon_0 x} \tag{5.77}$$

which gives a parabola shown by curve 1 of Fig. 5.13. The effect of the accelerating external field when applied at right angles to the cathode surface gives the electron a potential energy

$$W_E = -eEx \tag{5.78}$$

which is a straight line shown by Fig. 5.13 (curve 2). The total energy is then

$$W = W_a + W_E = -\left(\frac{-e^2}{16\pi\varepsilon_0 x}\right) - eEx \tag{5.79}$$

which is shown by the resultant curve 3 (Fig. 5.13). Thus a marked reduction ΔW in the potential barrier is obtained. The maximum reduction at x_m is obtained by differentiating eqn (5.79) or

$$\frac{dW}{dx} = \frac{e^2}{16\pi\varepsilon_0 x_m^2} - eE = 0$$

$$x_m = \sqrt{\frac{e}{16\pi\varepsilon_0 E}}.$$

Inserting this value into eqn (5.79) the lowering in the work function becomes

$$\Delta W = -e\sqrt{\frac{eE}{4\pi\varepsilon_0}}. \tag{5.80}$$

Hence, the effective value of the work function is

$$W_{\text{eff}} = W_a - \sqrt{\frac{eE}{4\pi\varepsilon_0}} \tag{5.81}$$

and the saturation current due to electron emission using eqn (5.76) in the presence of field E becomes

$$J_s = AT^2 \exp\left[-\frac{e}{kT}\left(W_a - \sqrt{\frac{eE}{4\pi\varepsilon_0}}\right)\right] \tag{5.81}$$

which is known as the Schottky's equation. If the current density in the absence of external field is J_0 (eqn (5.76)) then rearranging (5.81) we obtain

$$J_s = J_0 \exp\left[\frac{e}{kT}\left(W_a - \sqrt{\frac{eE}{4\pi\varepsilon_0}}\right)\right] = J_0 \exp\left[\frac{B\sqrt{E}}{T}\right]. \tag{5.82}$$

To obtain emission current J significantly higher than J_0, E must be of the magnitude of 10 MV/cm or higher. In practice a significant field emission current may be observed at somewhat lower fields. The effect has been explained by Fowler and Nordheim(19) who derived an expression for field emission on the basis of wave mechanics. These authors have shown that a few electrons in a metal will have an energy slightly above the Fermi level and thus will have a greater probability to penetrate the potential barrier 'tunnel effect'. The Fowler–Nordheim equation has the form

$$j = CE^2 \exp\left[-\frac{D}{E}\right] \tag{5.83}$$

where C and D are constants involving atomic constants. Equation (5.83) shows that field emission is independent of temperature, but this is valid only at low temperatures. At higher temperatures both thermionic and field emission will occur simultaneously.

5.3.5 Townsend second ionization coefficient γ

According to eqn (5.43) a graph of log I against gap length should yield a straight line of slope α if for a given pressure of p, E is kept constant. In his early measurements of current in parallel plate gaps Townsend(4) observed that at higher voltages the current increased at a more rapid rate than given by eqn (5.43) or (5.55). Figure 5.14 shows the kind of curves obtained by plotting log I against electrode separation at a constant pressure. To explain this departure from linearity Townsend postulated that a second mechanism must be affecting the current. He first considered liberation of electrons in the gas by collision of positive ions, and later the liberation of electrons from the cathode by positive ion bombardment according to the mechanism discussed earlier. On these assumptions he deduced the equation for the current in the self-sustained discharge. Other processes responsible for the upcurving of the (log $I - d$) graph, Fig. 5.14, include the secondary electron emission at the cathode by photon impact and photoionization in the gas itself

Following Townsend's procedure we consider the case for a self-sustained discharge where the electrons are produced at the cathode by positive ion bombardment.

Let n = number of electrons reaching the anode per second, n_0 number of electrons emitted from the cathode by (say) u.v. illumination, n_+ number of

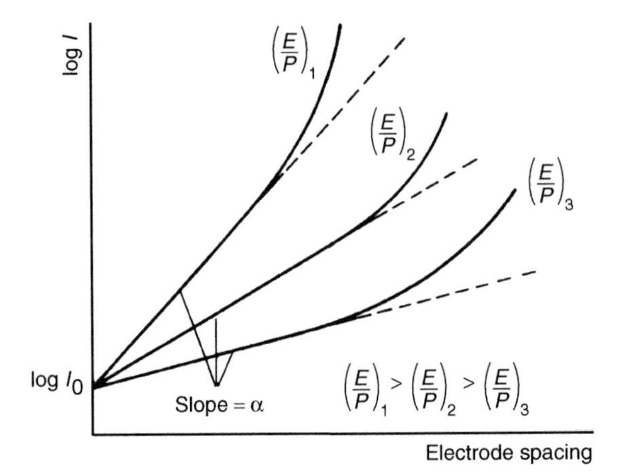

Figure 5.14 *Variation of gap current with electrode spacing in uniform field gaps*

electrons released from the cathode by positive ion bombardment, γ number of electrons released from the cathode per incident positive ion.

Then

$$n = (n_0 + n_+)e^{\alpha d}$$

and

$$n_+ = \gamma[n - (n_0 + n_+)].$$

Eliminating n_+

$$n = \frac{n_0 e^{\alpha d}}{1 - \gamma(e^{\alpha d} - 1)}.$$

or for steady state current

$$I = I_0 \frac{n_0 e^{\alpha d}}{1 - \gamma(e^{\alpha d} - 1)}. \tag{5.84}$$

Townsend's original expression was of the form

$$I = I_0 \frac{(\alpha - \beta)\, e^{(\alpha - \beta)d}}{\alpha - \beta e^{(\alpha - \beta)d}} \tag{5.85}$$

where β represents the number of ion pairs produced by a positive ion travelling a 1-cm path in the field direction and α, d, I and I_0 have the same significance as in eqn (5.84). Townsend's original suggestion for secondary ionization in the gas by positive ion impact does not work, because ions rapidly lose energy in elastic collisions according to eqn (5.35) and ordinarily are unable to gain sufficient energy from the field to cause ionization on collision with neutral atoms or molecules.

5.3.6 Secondary electron emission by photon impact

The case where the secondary emission arises from photon impact at the cathode may be expressed by the equation:[12]

$$I = I_0 \frac{\alpha e^{\alpha d}}{\alpha - \theta \eta g e^{(\alpha - \mu)d}} \tag{5.86}$$

where θ is the number of photons produced by an electron in advancing 1 cm in the direction of the field, μ is the average absorption coefficient for photons in the gas, g is a geometrical factor representing the fraction of photons that reach the cathode, and η is the fraction of the photons producing electrons at the cathode capable of leaving the surface.

In practice both positive ions and photons may be active at the same time in producing electrons at the cathode. Furthermore, metastable atoms may contribute to the secondary emission at the cathode. Which of the particular secondary mechanisms is predominant depends largely upon the experimental conditions in question. There may be more than one mechanism operating in producing the secondary ionization in the discharge gap and it is customary to express the secondary ionization by a single coefficient γ and represent the current by eqn (5.84), bearing in mind that γ may represent one or more of the several possible mechanisms ($\gamma = \gamma_I + \gamma_{ph} + \ldots$).

Experimental values of γ can be determined from eqn (5.84) by measurement of the current in the gap for various pressures, field strength and gap length and using the corresponding values of α. As would be expected from the considerations of the electron emission processes, the value of γ is greatly affected by the nature of the cathode surface. Low work function materials under the same experimental conditions will produce higher emission. The value of γ is relatively small at low values of E/p and will increase with E/p. This is because at higher values of E/p there will be a larger number of positive ions and photons of sufficiently high energy to eject electrons upon impact on the cathode surface.

Llewellyn Jones and Davies[20] have studied the influence of cathode surface layers on the breakdown characteristic of air and on the corresponding values of γ. Their data are included in Table 5.9 which shows a wide variation in the minimum breakdown voltage V_m and the accompanying variation in the values of γ. Influence of γ to breakdown strength is restricted to the 'Townsend breakdown mechanism', i.e. to low-pressure breakdown only as can be shown by the various breakdown criteria to be discussed in the next section.

Table 5.9

Gas	*Cathode*	V_m (*volts*)	E/p $\left(\frac{V}{cm.torr}\right)$	γ
Air contaminated with Hg vapour	Copper amalgam	460	720	0.004
	Mercury film on aluminium	390	885	0.014
	Mercury film on nickel	390	885	0.014
	Mercury film on staybrite steel	390	585	0.006
Air	Oxidized aluminium	416	905	0.01
	Oxidized nickel	421	957	0.01
Hydrogen (electrode treated by glow discharge)	Aluminium	243	200	0.1
	Aluminium deposited on nickel	212	200	0.15
	Nickel	289	180	0.075
	Nickel deposited on aluminiun	390	245	0.015
	Commercial aluminium	225	200	0.125
	Aluminium on staybrite steel	205	210	0.15
	Staybrite steel	274	190	0.075
	Steel deposited on aluminium	282	190	0.075

5.4 Transition from non-self-sustained discharges to breakdown

5.4.1 The Townsend mechanism

As the voltage between electrodes in a gas with small or negligible electron attachment increases, the electrode current at the anode increases in accordance with eqn (5.84)

$$I = I_0 \frac{e^{\alpha d}}{1 - \gamma(e^{\alpha d} - 1)}$$

or, introducing eqn (5.44) and $E = V/d$

$$\frac{I}{I_0} = \frac{e^{(pd)} \cdot f\left(\frac{V}{pd}\right)}{1 - \gamma\left[e^{(pd)} \cdot f\left(\frac{V}{pd}\right) - 1\right]}$$

until at some point there is a sudden transition from the dark current I_0 to a self-sustaining discharge. At this point the current (I) becomes indeterminate and the denominator in the above equation vanishes, i.e.

$$\gamma(e^{\alpha d} - 1) = 1.$$

If the electron attachment is taken into account (section 5.10), this equation becomes

$$\frac{\alpha\gamma}{\alpha - \eta}[e^{(\alpha-\eta)d} - 1] = 1$$

or approximately

$$\gamma e^{(\alpha-\eta)d} = \gamma e^{(\overline{\alpha}d)} = 1 \tag{5.87}$$

since

$$e^{\overline{\alpha}d} \gg 1 \quad \text{and} \quad \alpha \gg \eta$$

where $\overline{\alpha} = \alpha - \eta$ represents the effective ionization coefficient defined earlier in this chapter. The electron current at the anode equals the current in the external circuit. Theoretically the value of the current becomes infinitely large, but in practice it is limited by the external circuitry and, to a small extent, by the voltage drop within the arc. Equation (5.87) defines the conditions for onset of spark[(20)] and is called the Townsend criterion for spark formation or Townsend breakdown criterion. When $\gamma(e^{\overline{\alpha}d} - 1) = 1$, the number of ion pairs produced in the gap by the passage of one electron avalanche is sufficiently large that the resulting positive ions, on bombarding the cathode, are able to release one secondary electron and so cause a repetition of the avalanche process. The secondary electron may also come from a photoemission process (see eqn (5.86)). In either case electron avalanche will have a successor. The discharge is then self-sustaining and can continue in the absence of the source producing I_0, so that the criterion $\gamma(e^{\overline{\alpha}d} - 1) = 1$ can be said to define the sparking threshold. For $\gamma(e^{\overline{\alpha}d} - 1) > 1$ the ionization produced by successive avalanches is cumulative. The spark discharge grows more rapidly the more $\gamma(e^{\overline{\alpha}d} - 1)$ exceeds unity.

For $\gamma(e^{\overline{\alpha}d} - 1) < 1$ the current I is not self-sustained, i.e. on removal of the source producing the primary current I_0 it ceases to flow (see Fig. 5.14).

An alternative expression for the Townsend breakdown criterion is obtained by rewriting expression (5.87) in the form

$$\overline{\alpha}d = \ln\left(\frac{1}{\gamma} + 1\right) = K. \tag{5.88}$$

The right-hand side of this equation, K, can often be treated as being constant, due to the following phenomena. As mentioned earlier, the electron emission processes characterized by γ are greatly affected by cathode surface, as well as by gas pressure. However, γ is of very small value ($<10^{-}2 - 10^{-}3$) and therefore $1/\gamma$ is quite a high number. Therefore, $K = \ln(1/\gamma + 1)$ does not change too much and is for a Townsend discharge of the order of 8–10. As α is often very strongly dependent upon gas pressure p or field strength E, the exact value of K is of minor importance and may be treated as a constant for many conditions of p and E.

5.5 The streamer or 'Kanal' mechanism of spark

The growth of charge carriers in an avalanche in a uniform field $E_0 = V_0/d$ is described by the exponent $e^{\alpha d}$. This is valid only as long as the electrical field of the space charges of electrons and ions can be neglected compared to the external field E_0. In his studies of the effect of space charge of an avalanche on its own growth, Raether observed that when the charge concentration was higher than 10^6 but lower than 10^8 the growth of an avalanche was weakened.

When the ion concentration exceeded 10^8 the avalanche current was followed by a steep rise in current and breakdown of the gap followed. Both the underexponential growth at the lower concentration and rapid growth in the presence of the high concentration have been attributed to the modification of the originally uniform field (E_0) by the space charge field. Figure 5.15 shows diagramatically the electric field around an avalanche as it progresses along

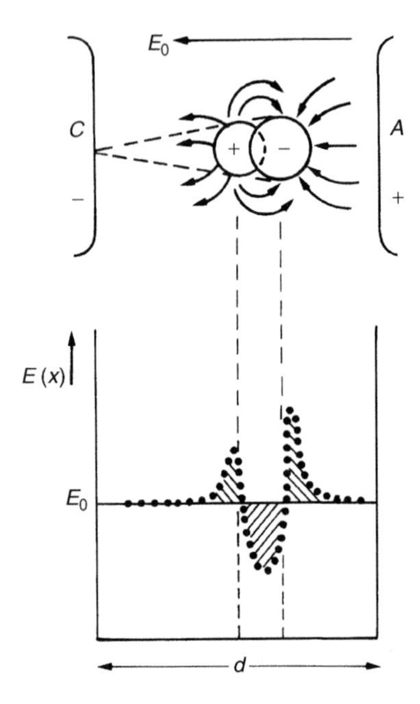

Figure 5.15 *Diagrammatic representation of field distortion in a gap caused by space charge of an electron avalanche*

the gap and the resulting modification to the original field (E_0). For simplicity the space charge at the head of the avalanche is assumed concentrated within a spherical volume, with the negative charge ahead because of the higher electron mobility. The field is enhanced in front of the head of the avalanche with field lines from the anode terminating at the head. Further back in the avalanche, the field between the electrons and the ions left behind reduced the applied field (E_0). Still further back the field between the cathode and the positive ions is enhanced again. The field distortion becomes noticeable with a carrier number $n > 10^6$. For instance, in nitrogen with $d = 2\,\text{cm}$, $p = 760\,\text{torr}$, $\alpha \approx 7$ and $E_0/p \approx 40\,\text{V/torr cm}$, the field distortion is about 1 per cent, leading to 5 per cent change in α. If the distortion of $\cong$1 per cent prevailed in the entire gap it would lead to a doubling of the avalanche size, but as the distortion is only significant in the immediate vicinity of the avalanche head it has still an insignificant effect. However, if the carrier number in the avalanche reaches $n \approx 10^8$ the space charge field becomes of the same magnitude as the applied field and may lead to the initiation of a streamer. The space charge fields play an important role in the mechanism of corona and spark discharges in non-uniform field gaps. For analytical treatment of space charge field distortion the reader is referred to reference 12.

In the Townsend spark mechanism discussed in the previous section the gap current grows as a result of ionization by electron impact in the gas and electron emission at the cathode by positive ion impact. According to this theory, formative time lag of the spark should be at best equal to the electron transit time t_i. In air at pressures around atmospheric and above ($pd > 10^3$ torr cm) the experimentally determined time lags have been found to be much shorter than t_i. Furthermore, cloud chamber photographs of avalanche development have shown[22] that under certain conditions the space charge developed in an avalanche is capable of transforming the avalanche into channels of ionization known as streamers that lead to rapid development of breakdown. From measurements of the prebreakdown current growth[23] and the minimum breakdown strength it has been found that the transformation from avalanche to streamer generally occurs when the charge within the avalanche head (Fig. 5.15) reaches a critical value of $n_0 \exp[\alpha x_c] \approx 10^8$ or $\alpha x_c \approx 18$–20, where x_c is the length of the avalanche path in field direction when it reaches the critical size. If x_c is larger than the gap length ($x_c > d$) then the initiation of streamers is unlikely. Typical cloud chamber photographs of electron avalanche and streamer development are shown in Figs 5.16(a) to (d). In (a) the discharge has been arrested before reaching the critical size ($\sim 10^8$), giving the avalanche the classical 'carrot' shape. In (b) the avalanche has grown beyond the critical size, its head has opened up indicating ionization around the original avalanche head and a cathode directed streamer starts. This continues (c, d) till a plasma channel connects cathode and anode. The early cloud chamber results have led Raether[22] to postulate the development of two

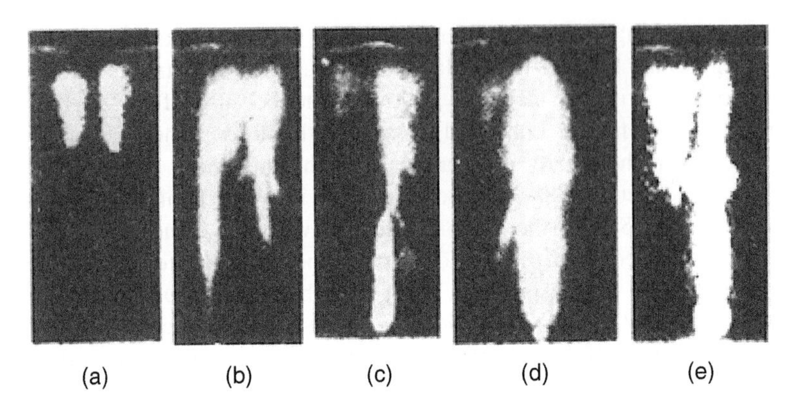

Figure 5.16 *Cloud chamber photographs showing development of the cathode directed streamers (with increasing pulse length): (a) avalanche near anode; (b) and (c) cathode directed streamer starts; (d) and (e) time period for plasma channel to connect cathode and anode*

types of streamers: (1) the 'anode directed streamer' describing the apparent growth of ionization and of the avalanche head, and (2) the 'cathode directed streamer' describing the additional discharge growth from the avalanche tail.

In later investigations, Wagner(24) has obtained streak photographs of 'avalanche streamer' development using an image intensifier. In these experiments the time and space resolved in radiation density which corresponds to the electron density is monitored. The observed radiation pattern together with the photocurrent growth is sketched in Figs 5.17(a) and (b). Region (a)–(b) corresponds to the development of avalanche with an approximate velocity of 10^8 cm/sec. The current growth is exponential. Beyond (b), after the avalanche has reached the critical size, there is an increase in the velocity of the avalanche head by about a factor of 10. In many cases almost simultaneously a second luminous front is observed proceeding towards the cathode with the same velocity as the anode directed growth. The current growth in this region is faster than exponential.

The observed short time lags together with the observations of discharge development have led Raether and independently Meek(25) and Meek and Loeb(26) to the advancement of the 'streamer' or 'Kanal' mechanism for spark formation, in which the secondary mechanism results from photoionization of gas molecules and is independent of the electrodes.

In the models developed by Raether and Meek it has been proposed that when the avalanche in the gap reaches a certain critical size the combined space charge field and externally applied field lead to intense ionization and excitation of the gas particles in front of the avalanche head. Instantaneous recombination between positive ions and electrons releases photons which

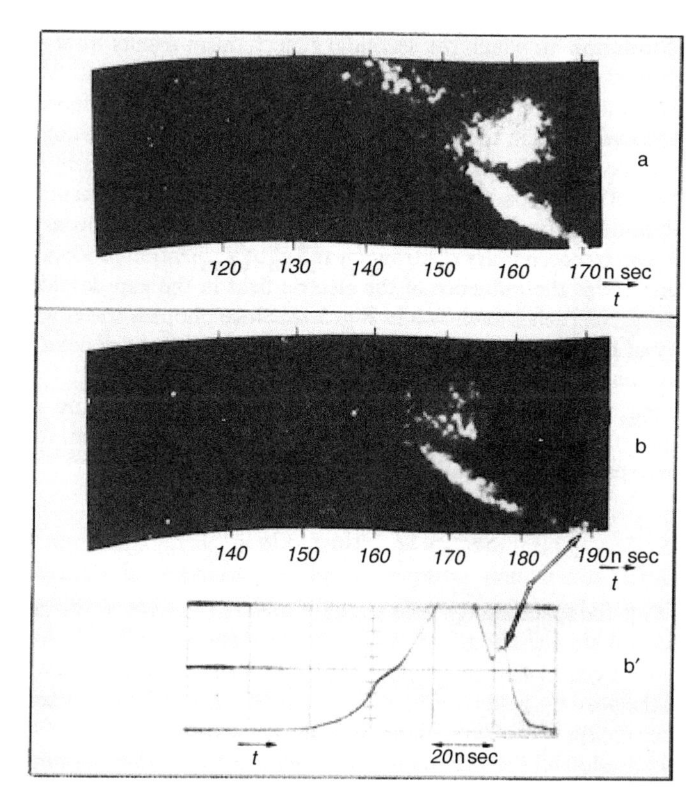

Figure 5.17 *Image intensifier photographs, and a photocurrent oscillogram showing the development of cathode directed streamers in N_2 (20 per cent CH_4) $p = 88.5$ torr. (a) and (b) progress of streamers after switching off external voltage; (b′) photocurrent oscillogram corresponding to (b)*[24]

in turn generate secondary electrons by the photoionization process. These electrons under the influence of the electric field in the gap develop into secondary avalanches as shown in Fig. 5.18. Since photons travel with the velocity of light, the process leads to a rapid development of conduction channel across the gap.

On the basis of his experimental observations and some simple assumptions Raether[33] developed an empirical expression for the streamer spark criterion of the form

$$\alpha x_c = 17.7 + \ln x_c + \ln \frac{E_r}{E} \tag{5.89}$$

where E_r is the space charge field strength directed radially at the head of avalanche as shown in Fig. 5.19, E is the externally applied field strength.

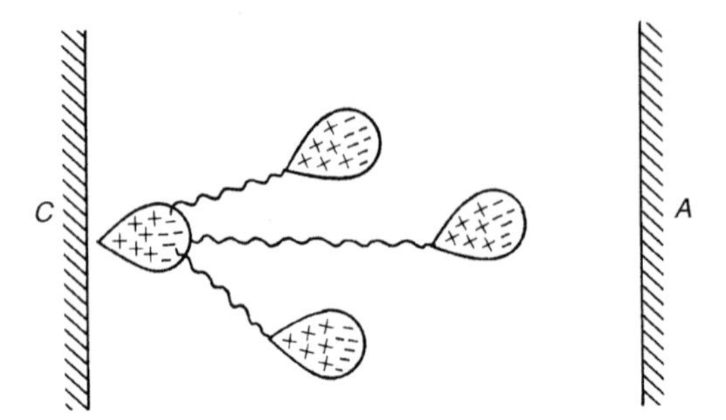

Figure 5.18 *Secondary avalanche formation by photoelectrons*

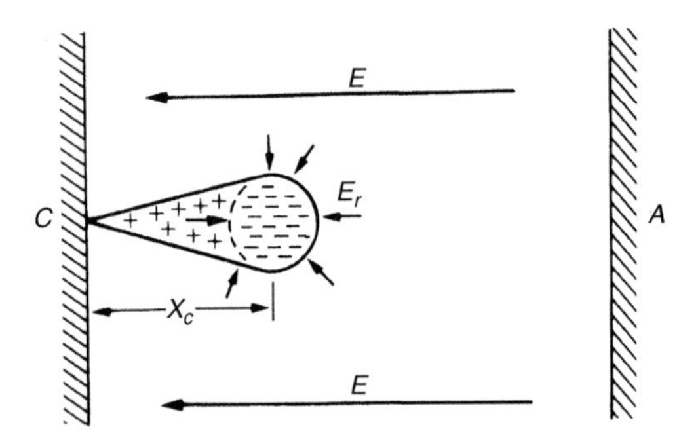

Figure 5.19 *Space charge field (E_r) around avalanche head*

The resultant field strength in front of the avalanche is thus $(E + E_r)$ while in the positive ion region just behind the head the field is reduced to a value $(E - E_r)$. It is also evident that the space charge increases with the avalanche length $(e^{\alpha x})$.

The condition for the transition from avalanche to streamer assumes that space charge field E_r approaches the externally applied field $(E_r \approx E)$, hence the breakdown criterion (eqn (5.89)) becomes

$$\alpha x_c = 17.7 + \ln x_c. \tag{5.90}$$

The minimum breakdown value for a uniform field gap by streamer mechanism is obtained on the assumption that the transition from avalanche to streamer occurs when the avalanche has just crossed the gap (d). Then Raether's empirical expression for this condition takes the form

$$\alpha d = 17.7 + \ln d. \tag{5.91}$$

Therefore the breakdown by streamer mechanism is brought about only when the critical length $x_c \geq d$. The condition $x_c = d$ gives the smallest value of α to produce streamer breakdown.

A similar criterion equation for the transition from avalanche to streamer has been developed by Meek.[25] As in Raether's case the transition is assumed to take place when the radial field about the positive space charge in an electron avalanche attains a value of the order of the externally applied field. Meek[25] has shown that the radial field produced by positive ions immediately behind the head of the avalanche can be calculated from the expression

$$E_r = 5.3 \times 10^{-7} \frac{\alpha e^{\alpha x}}{\left(\frac{x}{p}\right)^{1/2}} \text{ volts/cm} \tag{5.92}$$

where x is the distance (in cm) which the avalanche has progressed, p is the gas pressure in torr and α is the Townsend coefficient of ionization by electrons corresponding to the applied field E. As in Raether's model the minimum breakdown voltage is assumed to correspond to the condition when the avalanche has crossed the gap of length d and the space charge field E_r approaches the externally applied field. Substituting into eqn (5.92) $E_r = E$ and $x = d$ and rearranging gives

$$\alpha d + \ln \frac{\alpha}{p} = 14.5 + \ln \frac{E}{p} + \frac{1}{2} \ln \frac{d}{p}. \tag{5.93}$$

This equation is solved by trial and error using the experimentally determined relation between α/p and E/p. Values of α/p corresponding to E/p at a given pressure are chosen until the equation is satisfied.

Table 5.10 compares Meek's calculated and the measured values V_b for air according to eqn (5.93). At small d, the calculated values V_b are higher than the measured ones. The reverse is true at large d. In general, however, the deviation between theory and experiment should be regarded as not very large, in view of the various simplifying assumptions made by Meek,[25] especially those in order to determine the charge density and the tip radius of the avalanche. The avalanche radius was calculated on the basis of thermal diffusion using the relationship $r = \sqrt{3Dt}$ where D is thermal diffusion coefficient and t the time. The charge was assumed to be concentrated in a spherical volume which is only approximately correct. At the charge densities in question, ambipolar diffusion is likely to be important, but so far has been neglected.

In section 5.4 we have seen that the Townsend criterion for spark formation is satisfied when the product $\overline{\alpha}d$ reaches a value of 8–10 ($\overline{\alpha}d = \ln(1/\gamma + 1) =$ 8–10). The streamer criterion for spark formation, however, requires a value of 18–20, $\overline{\alpha}d = (\overline{\alpha}x_c) = \ln 10^8 \approx 20$ with $x_c \leq d$. Therefore under certain experimental conditions there will be a transition from the Townsend to streamer mechanism. This transition is brought about by increased pressure and gap length and in practice it occurs in the region of $pd \geq 1$–2 bar cm. The transition is indicated by a discontinuity in the formative time lag discussed in section 5.10. The streamer mechanism which relies on photoionization in the

Table 5.10 *Comparison of calculated and measured V_b values for air according to Meek's model*

Gap length cm	E/p V/cm torr	αd	V_b calculated kV	V_b measured kV
0.1	68.4	15.7	5.19	4.6
0.5	48.1	17.7	18.25	17.1
1.0	42.4	18.6	32.20	31.6
2.5	37	19.7	70.50	73
10	32.8	21.5	249	265
20	31.2	22.4	474	510

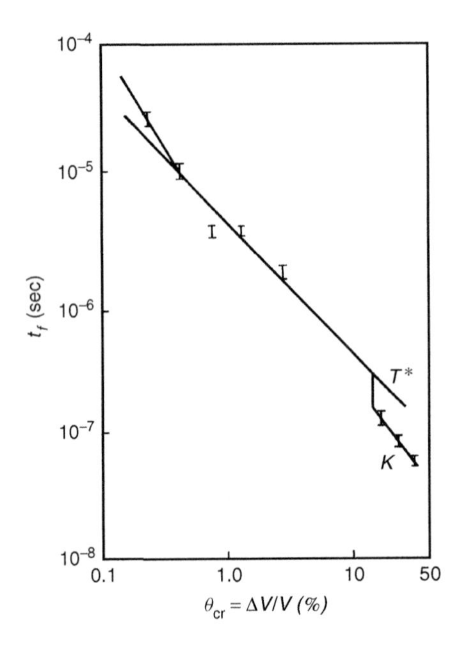

Figure 5.20 *Formative time lag in N_2 as function of overvoltage. $p = 500$ torr, $d = 2$ cm. T^* Townsend mechanism, K streamer mechanism*

gas requires a much shorter formative time than the Townsend mechanism in which the secondary mechanism is cathode dependent and is affected by the transit time of positive ions. Figure 5.20 compares the formative time lag(20) in nitrogen at $p = 500$ torr and $d = 2$ cm with the measured values, plotted as a function of percentage overvoltage ($\theta = \Delta V/V\%$). At lower overvoltages the formative time lag follows the Townsend mechanism up to a critical value $\theta_{crit} = \Delta V_{crit}/V$ at which the electron amplification within the avalanche reaches a value $e^{\alpha d} \geq 10^8$. Curve K has been calculated from the time (t_A) required to reach the critical size at various overvoltages,

$$t_A = \frac{x_{crit}}{v_-} = \frac{18}{\alpha v_-} \tag{5.94}$$

where v_- is the electron drift velocity. Curve T^* is obtained from the Townsend mechanism. No discontinuity is observed and the curve gives a too long formative time lag for the higher overvoltages.

Table 5.11 gives the critical overvoltages for several of the commonly used gases together with the corresponding pd values. The sudden change in the formative time lag usually takes place for values of some 10^{-7} sec.

Table 5.11 *Critical overvoltages for various gases*[27]

Gas	p *(torr)*	d *(cm)*	pd *(torr cm)*	θ_{crit} *(%)*
H_2	500	2	1000	16.6
N_2	500	2	1000	18.2
N_2	400	3	1200	17
Air	760	1	760	4.5

5.6 The sparking voltage–Paschen's law

The Townsend criterion, eqn (5.87), enables the breakdown voltage of the gap to be determined by the use of appropriate values $\overline{\alpha}/p$ and γ corresponding to the values E/p without ever taking the gap currents to high values, that is keeping them below 10^{-7} A, so that space charge distortions are kept to a minimum, and more importantly so that no damage to electrodes occurs. Good agreement has been found[28] between calculated and experimentally determined breakdown voltages for short or long gaps and relatively low pressures for which this criterion is applicable.

An analytical expression for breakdown voltage for uniform field gaps as a function of gap length and gas pressure can be derived from the threshold eqn (5.87) by expressing the ionization coefficient $\overline{\alpha}/p$ as a function of field strength and gas pressure. If we put $\overline{\alpha}/p = f(E/p)$ in the criterion equation we obtain

$$e^{f(E/p)pd} = \frac{1}{\gamma} + 1$$

or

$$f(E/p)pd = \ln\left(\frac{1}{\gamma} + 1\right) = K. \qquad (5.95)$$

For uniform field $V_b = Ed$, where V_b is the breakdown voltage,

$$e^{f(V_b/pd)pd} = K' = e^K \qquad (5.96)$$

or

$$V_b = F(pd)$$

which means that the breakdown voltage of a uniform field gap is a unique function of the product of pressure and the electrode separation for a particular gas and electrode material. Equation (5.96) is known as Paschen's law, and was established experimentally in 1889. Equation (5.96) does not imply that the sparking voltage increases linearly with the product *pd*, although it is found in practice to be nearly linear over certain regions. The relation between the sparking voltage and the product *pd* takes the form shown in Fig. 5.21 (solid curve). The breakdown voltage goes through a minimum value ($V_{b\min}$) at a particular value of the product ($pd_{\min}$).

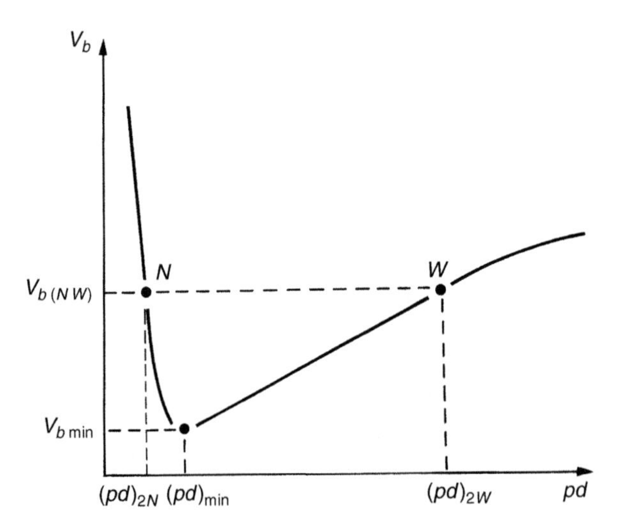

Figure 5.21 *The sparking voltage–pd relationship (Paschen's curve)*

Let us now examine graphically the relation of the Paschen's curve (Fig. 5.21) with the spark criterion eqn (5.88). If the experimental relationship between the ionization coefficient and the field strength ($\overline{\alpha}d/p = f(E/p)$) for a given gas is plotted we obtain a curve as shown in Fig. 5.22 (curve 1) with a limiting value (E/p), corresponding to the onset of ionization. Rearranging the Townsend criterion, eqn (5.88), and remembering that in uniform field $V = Ed$, where V is the applied voltage, gives

$$\frac{\overline{\alpha}}{p} = \frac{K}{V}\frac{E}{p}$$

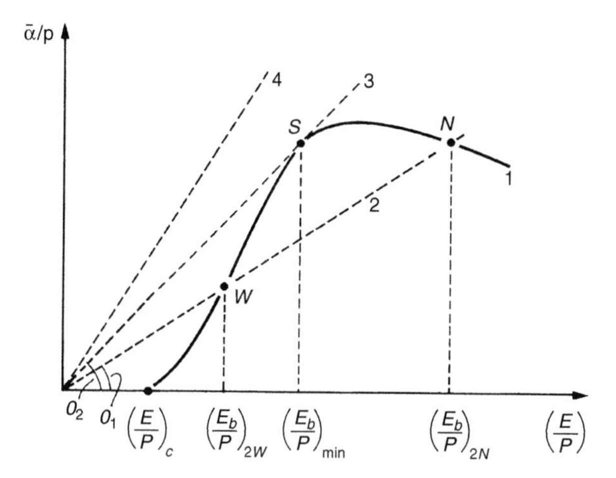

Figure 5.22 *Relation between the Townsend criterion for spark ($\alpha d = k$) and the function of $\alpha/p = (E/p)$*

and

$$\tan\theta = \frac{\overline{\alpha}/p}{E/p} = \frac{K}{V}. \tag{5.97}$$

Equation (5.97) gives for constant values of K straight lines of a slope ($\tan\theta$) depending upon the value of the applied voltage (V), curves (2, 3, 4) Fig. 5.22. At low values of V there is no intersection between the line (curve 4) and the curve $(\overline{\alpha}/p) = f(E/p)$. No breakdown therefore results with small voltages below Paschen's minimum irrespective of the value pd in eqn (5.96). At the higher applied voltage (V), there must exist two breakdown values at a constant pressure, one corresponding to the small value of gap length (d_1) and intersection at N and the other to the longer gap (d_2) intersection at W. The point S (tangent) gives the lowest breakdown value or the minimum sparking voltage. The breakdown voltages corresponding to the points W, N and S are indicated in the Paschen's curve in Fig. 5.21.

The existence of the minimum value in the breakdown voltage–gap length relation may be explained qualitatively by considering the efficiency of the ionization of electrons traversing the gap with different electron energies. Neglecting the secondary coefficient γ for values $pd > (pd)_{\min}$, electrons crossing the gap make more frequent collisions with the gas molecules than at $(pd)_{\min}$, but the energy gained between collisions is lower than at $(pd)_{\min}$. Hence, the probability of ionization is lower unless the voltage is increased. For $pd < (pd)_{\min}$ electrons cross the gap without making many collisions. The point $(pd)_{\min}$ corresponds to the highest ionization efficiency.

An analytical expression for the minimum values of $V_{b\min}$ and $(pd)_{\min}$ may be obtained by inserting in the criterion eqn (5.87) for $\bar{\alpha}/p$ the expression (5.47)

$$\bar{\alpha} = Ap\exp\left(-\frac{Bp}{E}\right)\left[= Ap\exp\left(-\frac{Bpd}{V}\right)\right]$$

and determining the minimum value of V.

Assuming that the coefficient γ remains constant, then

$$d = \frac{e^{Bpd/V_b}}{Ap}\ln\left(1 + \frac{1}{\gamma}\right).$$

Rearranging we obtain

$$V_b = \frac{Bpd}{\ln\dfrac{Apd}{\ln(1 + 1/\gamma)}} \tag{5.98}$$

Differentiating with respect to (pd) and equating the derivative to zero

$$\frac{dV_b}{d(pd)} = \frac{B}{\ln\dfrac{Apd}{\ln(1 + 1/\gamma)}} - \frac{B}{\left[\ln\dfrac{Apd}{\ln(1 + 1/\gamma)}\right]^2} = 0.$$

Therefore

$$\ln\frac{Apd}{\ln(1 + 1/\gamma)} = 1$$

and

$$(pd)_{\min} = \frac{e^1}{A}\ln\left(1 + \frac{1}{\gamma}\right). \tag{5.99}$$

Substitution into eqn (5.98) gives

$$V_{b(\min)} = 2.718\frac{B}{A}\ln\left(\frac{1}{\gamma} + 1\right). \tag{5.100}$$

This equation could be used for the calculation of the minimum sparking constants ($V_{b\min}$, $(pd)_{\min}$) if the correct values of A and B are used for the simulation of the real dependency of $(\overline{\alpha}/p = f(E/p)$ in the vicinity of

$$\frac{d^2(\alpha/p)}{d(E/p)^2} = 0.$$

In practice, the sparking constants ($V_{b\min}$ and $(pd)_{\min}$) are measured values, and some of these are shown in Table 5.12. For example, by inserting in eqn (5.100) the values for the constants $A = 12$, $B = 365$ and $\gamma = 0.02$ that are commonly quoted in the literature, we obtain for the minimum breakdown voltage for air $V_{b\min} = 325$ V which agrees well with the experimental value quoted in Table 5.12. It should be noted, however, that these values are sometimes strongly dependent upon the cathode material and cathode conditions, according to eqns (5.94) and (5.95), in which the real value of γ is significant.

Table 5.12 *Minimum sparking constants for various gases*[29]

Gas	$(pd)_{\min}$ *torr cm*	$V_{b\min}$ *volts*
Air	0.55	352
Nitrogen	0.65	240
Hydrogen	1.05	230
Oxygen	0.7	450
Sulphur hexafluoride	0.26	507
Carbon dioxide	0.57	420
Neon	4.0	245
Helium	4.0	155

The measured minimum sparking voltage in any gas is dependent upon the work function of the cathode material. A minimum sparking voltage as low as 64 V has been observed by Cueilleron[30] in neon between cesium-coated electrodes at a gas pressure of 26 torr.

The breakdown voltage for uniform field gaps in air over a wide range of pressures and gap lengths may be calculated by combining the Schumann's relation with the criterion eqn (5.88). Schumann[31] has shown that over a wide but restricted range of E/p, α/p may be expressed as

$$\frac{\overline{\alpha}}{p} = C\left[\left(\frac{E}{p}\right) - \left(\frac{E}{p}\right)_c\right]^2 \tag{5.101}$$

where E and E_c are field strengths, E_c being the limiting value of E at which effective ionization starts. p is pressure and C is a constant.

Dividing eqn (5.88) by pd and combining with eqn (5.101) we obtain

$$\frac{K}{pd} = C\left[\left(\frac{E}{p}\right) - \left(\frac{E}{p}\right)_c\right]^2$$

or

$$\frac{E}{p} = \left(\frac{E}{p}\right)_c + \sqrt{\frac{K/C}{pd}}$$

and the expression for the breakdown voltage V_b becomes

$$V_b = \left(\frac{E}{p}\right)_c pd + \sqrt{\frac{K}{C}}\sqrt{pd}. \qquad (5.102)$$

Inserting the values for the constants E_c and K/c which were determined by Sohst[32] and Schröder[33] for homogeneous field gaps at $p = 1$ bar; 20°C; $E_c = 24.36$ (kV/cm); $(K/C) = 45.16\,(\text{kV})^2/\text{cm}$; eqn (5.102) becomes

$$V_b = 6.72\sqrt{pd} + 24.36(pd)\,\text{kV}. \qquad (5.103)$$

The calculated breakdown voltages, using eqn (5.103) for uniform field gaps in air for a range of the product pd from 10^{-2} to 5×10^2 (bar cm) are compared with the available experimental data in Fig. 5.23. The calculated

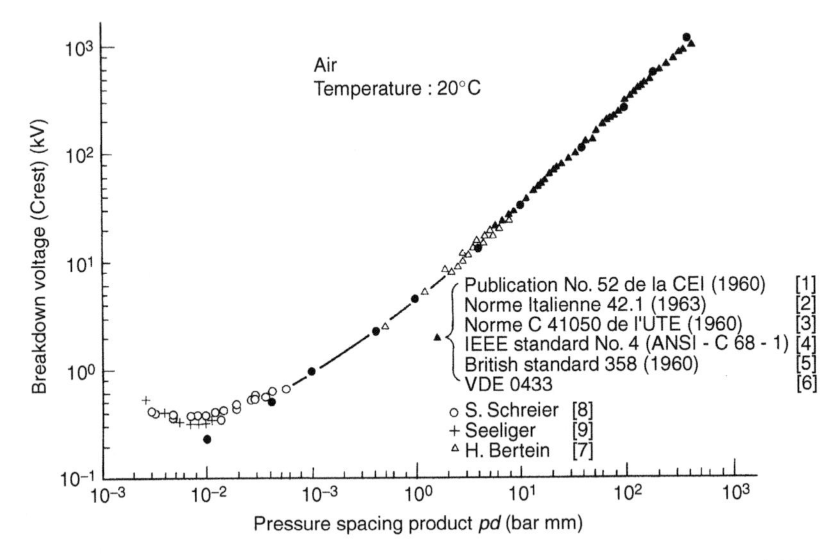

Figure 5.23 *Paschen curve for air in log–log scale. Temperature 20°C. (• calculated $V_B = 6.72\sqrt{pd} + 24.4(pd)$)*

and the measured data agree well except at the very low values of the product pd. In this region in which the E/p values are high the Schumann's quadratic relationship (eqn (5.102)) no longer holds, but this region is of little practical interest.

It is often more convenient to use the gas density δ instead of the gas pressure p in the Paschen's eqn (5.96), since in the former case account is taken for the effect of temperature at constant pressure on the mean free path in the gas. The number of collisions by an electron in crossing the gap is proportional to the product δd and γ.

Atmospheric air provides the basic insulation for many practical h.v. installations (transmission lines, switchyards, etc.). Since the atmospheric conditions (temperature and pressure) vary considerably in time and locations, the breakdown characteristics of various apparatus will be affected accordingly. For practical purposes, therefore, the breakdown characteristics can be converted to standard atmospheric conditions ($p = 760$ torr $= 1.01$ bar and $t = 20°\text{C} = 293$ K). Correction for the variation in the ambient conditions is made by introducing the relative density defined as

$$\delta = \frac{p}{760}\frac{293}{273+t} = 0.386\frac{p}{273+t}. \tag{5.104}$$

The breakdown voltage at standard conditions multiplied by this factor gives the breakdown voltage corresponding to the given ambient conditions approximately

$$V_b(\delta) = \delta V_b(\delta = 1) \tag{5.105}$$

Paschen's law is found to apply over a wide range of the parameter value up to 1000–2000 torr cm. At higher products, however, the breakdown voltage (in non-attaching gases) is found to be somewhat higher than at smaller spacing for the same values of pd. This departure is probably associated with the transition from the Townsend breakdown mechanism to the streamer mechanism, as the product pd is increased above a certain value. We have seen that the streamer breakdown criterion is satisfied at higher values of $\overline{\alpha}d$ than the Townsend criterion, i.e. the value of the constant K in eqn (5.88) will increase from about 8–10 to 18–20. At very low pressure deviations from Paschen's law are observed when the breakdown mechanism ceases to be influenced by the gas particles and becomes electrode dominated (vacuum breakdown).

5.7 Penning effect

Paschen's law is not applicable in many gaseous mixtures. The outstanding example is the neon–argon mixture. A small admixture of argon in neon reduces the breakdown strength below that of pure argon or neon as shown in

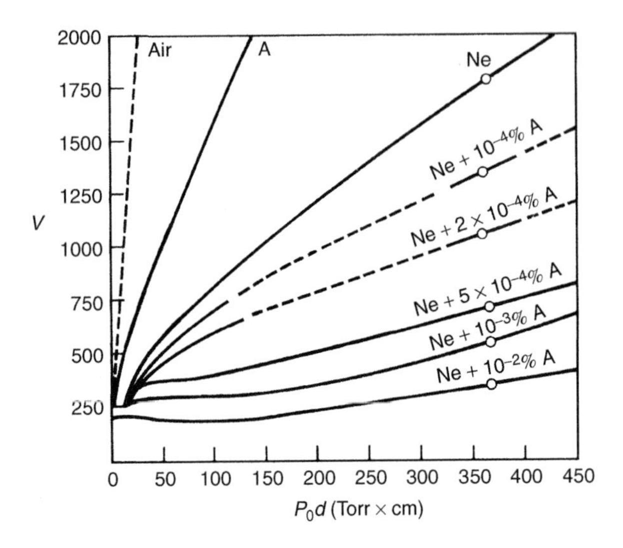

Figure 5.24 *Breakdown voltage curves in neon–argon mixtures between parallel plates at 2-cm spacing at 0°C*

Fig. 5.24. The reason[34] for this lowering in the breakdown voltage is that the lowest excited state of neon is metastable and its excitation potential (16 eV) is about 0.9 eV greater than the ionization potential of argon. The metastable atoms have a long life in neon gas, and on hitting argon atoms there is a very high probability of ionizing them. The phenomenon is known as the Penning effect.

5.8 The breakdown field strength (E_b)

For uniform field gaps the breakdown field strength in a gas may be obtained from eqn (5.98) by dividing both sides of this equation by (pd), then

$$\frac{V_b}{(pd)} = \frac{E_b}{p} = \frac{B}{\ln \dfrac{Apd}{\ln(1+1/\gamma)}} \tag{5.106}$$

We note that for a constant gas pressure p the breakdown field strength (E_b) decreases steadily with the gap length (d). Furthermore, the field strength to pressure ratio (E_b/p) is only dependent upon the product of (pd). Equation (5.106) also shows that the breakdown field strength (E_b) for a constant gap length increases with the gap pressures but at a rate slightly lower than directly proportional, as the pressure also affects the denominator in the expression.

Qualitatively the decrease in the pressure related breakdown field strength (E_b/p) with increasing pd may easily be understood by considering the relationship between the ionization coefficient $\overline{\alpha}$ and the field strength

$$\frac{\overline{\alpha}}{p} = f\left(\frac{E}{p}\right)$$

plotted in Fig. 5.25 and applying the Townsend criterion equation to different values of (pd) as shown. The breakdown criterion of eqn (5.88) can be written as

$$\frac{\overline{\alpha}}{p}(pd) = k.$$

Assuming first that this equation is satisfied for a small product $(pd)_2$ at $(E_b/p)_2$ when $\overline{\alpha}/p$ reaches the point A (Fig. 5.25) and then increasing in value of the product to $(pd)_1$, the criterion equation will now be satisfied at a lower value of $\overline{\alpha}/p(B)$ giving a reduced breakdown strength $(E_b/p)_1$. By repeating this procedure for other values of pd we obtain a functional relationship between the breakdown strength E_b/p and the product pd as shown in Fig. 5.26. Thus, the curve obtained in Fig. 5.26 is in qualitative agreement with experimental data.

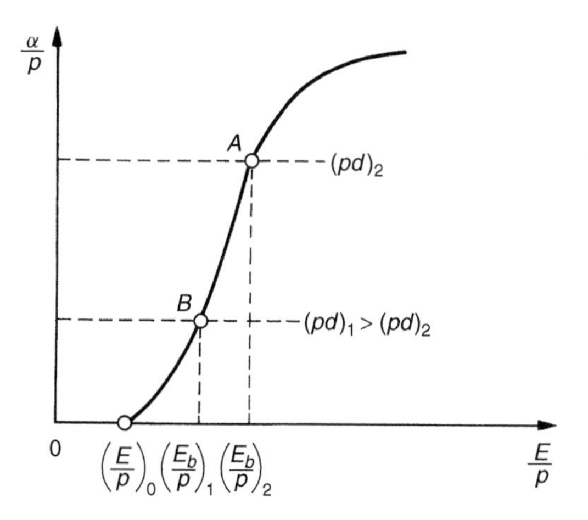

Figure 5.25 *Functional relationship between ionization coefficient α and breakdown field strength at different values of pd*

Calculations of the breakdown strength (E_b) and the pressure related breakdown field strength (E/p) using eqn (5.106) yield data that are in agreement with the experimental values over a limited range of pressures and gap length. For air a much closer agreement with the experimental data may be obtained

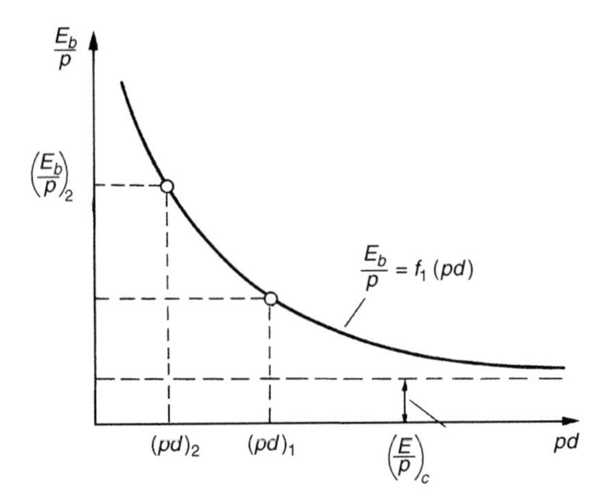

Figure 5.26 *Functional relationship between the breakdown field strength and the product pd*

using the Schumann's eqn (5.103). Dividing both sides of eqn (5.103) by the product pd gives

$$\frac{V_b}{pd} = \frac{E_b}{p} = \frac{6.72}{\sqrt{pd}} + 24.36 \frac{kV}{\text{cm bar}}. \tag{5.107}$$

The breakdown field strength (E_b), when calculated for air at standard temperature and pressure for gap lengths extending from 1 mm to 100 mm using eqn (5.107), agrees well with experimental values.

5.9 Breakdown in non-uniform fields

In non-uniform fields, e.g. in point-plane, sphere-plane gaps or coaxial cylinders, the field strength and hence the effective ionization coefficient $\overline{\alpha}$ vary across the gap. The electron multiplication is governed by the integral of $\overline{\alpha}$ over the path ($\int \overline{\alpha} dx$). At low pressures the Townsend criterion for spark takes the form

$$\gamma \left[\exp \left(\int_0^d \overline{\alpha} dx \right) - 1 \right] = 1 \tag{5.108}$$

where d is the gap length. The integration must be taken along the line of the highest field strength. The expression is valid also for higher pressures if the field is only slightly non-uniform. In strongly divergent fields there will be at first a region of high values of E/p over which $\alpha/p > 0$. When the field falls below a given strength E_c the integral $\int \overline{\alpha} dx$ ceases to exist. The Townsend mechanism then loses its validity when the criterion relies solely on the γ effect, especially when the field strength at the cathode is low.

In reality breakdown (or inception of discharge) is still possible if one takes into account photoionization processes.

The criterion condition for breakdown (or inception of discharge) for the general case may be represented by modifying the expression (5.90) to take into account the non-uniform distribution of $\overline{\alpha}$ or

$$\exp \int_0^{x_c<d} \overline{\alpha} \mathrm{d}x = N_{cr} \tag{5.109}$$

where N_{cr} is the critical electron concentration in an avalanche giving rise to initiation of a streamer (it was shown to be approx. 10^8), x_c is the path of avalanche to reach this size and d the gap length. Hence eqn (5.109) can be written as

$$\int_0^{x_c<d} \overline{\alpha} \mathrm{d}x = \ln N_{cr} \approx 18 - 20. \tag{5.109a}$$

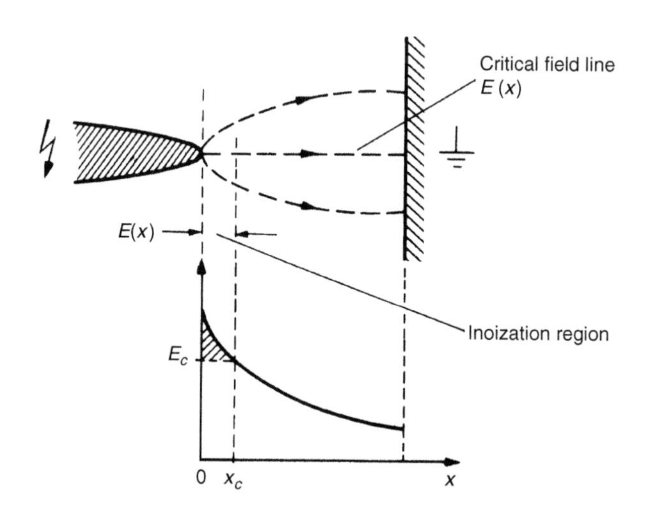

Figure 5.27 *Electric field distribution in a non-uniform field gap*

Figure 5.27 illustrates the case of a strongly divergent field in a positive point-plane gap. Equation (5.109a) is applicable to the calculation of breakdown or discharge inception voltage, depending on whether direct breakdown occurs or only corona. The difference between direct breakdown and corona inception will be discussed in detail in the next section.

For the special case of a coaxial cylindrical geometry in air, an empirical relation based on many measurements of the critical field strength E_c (corona inception) for different diameters of the inner conductor ($2r$) and relative air density δ was developed by Peek[35] of the form:

$$\frac{E_c}{\delta} = 31.53 + \frac{9.63}{\sqrt{\delta r}} \tag{5.110}$$

where E_c is in kV/cm, r in cm and δ is the relative air density defined by eqn (5.104). For values of $(\delta r) > 1$ cm this expression gives higher values than experimentally observed. More recently Zaengl *et al.*[36] have developed an analytical expression based upon eqns (5.109) and (5.101) replacing the Peek's empirical eqn (5.110) for calculating the corona inception voltage given as

$$\left(\frac{E_c}{\delta}\right)^2 - 2\left(\frac{E_c}{\delta}\right) E_0 \ln\left[\frac{1}{E_0}\left(\frac{E_c}{\delta}\right)^2\right] - E_0^2 = \frac{K/C}{(\delta r)}. \tag{5.111}$$

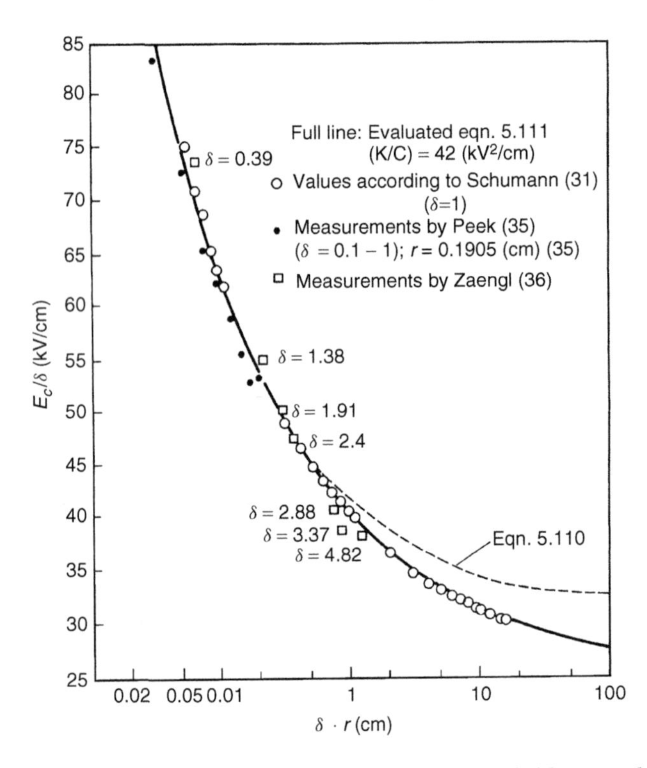

Figure 5.28 *Variation of corona inception field strength (E_c/δ) with δr for coaxial cylindrical geometry in air*

In this expression the constants E_c and K/C have the same significance as in the Schumann's eqn (5.103), but the best agreement of the calculated values with many of the known measured values (a.c. and d.c.) of positive polarity is obtained with the constant $K/C = 42\,(\text{kV/cm})^2$ as compared to $K/C = 45.16\,(\text{kV/cm})^2$ used in eqn (5.103). Figure 5.28 compares the calculated values (E_c/δ) plotted as function of the product (δr) using eqn (5.111) (solid curve), with the measured values by Peek and those measured by Schumann. The dotted curve indicates the calculated values obtained using the original empirical expression of Peek (eqn (5.110)). It is seen that for the product less than $\delta r < 1$ the values obtained from the Peek's empirical expression are in good agreement with experimental observations, but a deviation is observed for conductors of larger radius, due to the fact that the original expression was based on measurements on conductors of small size.

Equation (5.111) also shows that the critical field strength E_c for a coaxial arrangement is independent of the radius of the outer cylinder (R). This is true as long as the field strength $E(R)$ does not exceed (δE_0).

5.10 Effect of electron attachment on the breakdown criteria

In section 5.2 it was shown that there are a number of gases in which the molecules readily attach free electrons forming negative ions, having a similar mass as the neutral gas molecules. They are, therefore, unable to ionize neutral particles under field conditions in which electrons will readily ionize. The ionization by electron collision is then represented by the effective ionization coefficient $(\overline{\alpha} = \alpha - \eta)$.

In the presence of attachment, the growth of current in a gap when the secondary coefficient γ is included in eqn (5.84) is given by the relation[37]

$$I = I_0 \frac{\dfrac{\alpha}{\alpha - \eta}[\exp(\alpha - \eta)d - (\eta/\alpha)]}{\left\{1 - \dfrac{\alpha\gamma}{\alpha - \eta}[\exp(\alpha - \eta)d - 1]\right\}} \tag{5.112}$$

where α and γ are the primary and secondary ionization coefficients and η is the attachment coefficient as defined earlier in this chapter. It was also shown that in a given gas both coefficients α and η are dependent only on the field strength E and the gas pressure

$$\left[\frac{\alpha}{p} = f\left(\frac{E}{p}\right)\right], \quad \left[\frac{\eta}{p} = f\left(\frac{E}{p}\right)\right]$$

For a self-sustained discharge in an attaching gas the denominator in eqn (5.112) will tend to zero, and as a result we obtain the Townsend criterion for attaching gases

$$(\alpha - \eta)d = \ln\left[\frac{(\alpha - \eta)}{\alpha}\frac{1}{\gamma} + 1\right]. \tag{5.113}$$

As the difference of $(\alpha - \eta)/p$ for low values of E/p becomes negative, eqn (5.113) can only be valid for values $\alpha > \eta$. This means that a critical value of a pressure-dependent field strength exists for which $\alpha = \eta$ and $(E/p) \rightarrow (E/p)_0$. Therefore, in the presence of attachment no breakdown can take place in accordance with processes so far considered for this or lower values of E/p.

Of special engineering applications interest is sulphur hexafluoride (SF_6), which has a large attachment coefficient (and thus exhibits high dielectric strength) and has been widely applied in gas-insulated power equipment. In the early 1950s Geballe and Harrison and Geballe and Reeve(10,37) studied the values of α and η for SF_6 (see Table 5.6) over a wide range of E/p and found these fit well into the following linear equation

$$\frac{\overline{\alpha}}{p} = \frac{\alpha - \eta}{p} = k\left[\frac{E}{p} - \left(\frac{E}{p}\right)_0\right] \tag{5.114}$$

where α and η are the ionization and attachment coefficients, and k is obtained from Figure 5.29, which has a numerical value of

$$k = 27.7\,\text{kV}^{-1} \quad \text{or} \quad 2.77 \times 10^{-2}\,\text{V}^{-1}$$

and

$$(E/p)_0 = 88.5\,\text{kV/(cm bar)} \quad \text{or} \quad 118\,\text{V/(cm Torr)} \quad \text{at which } \overline{\alpha} = 0 \quad \text{or} \quad \alpha = \eta.$$

The linear relationship of eqn 5.114 (Fig 5.29) remains valid within

$$75 < \frac{E}{P} < 200\frac{\text{kV}}{\text{cm bar}}$$

If eqn (5.114) is combined with the simplest breakdown criterion eqn (5.88), which is

$$\overline{\alpha d} = \ln\left(\frac{1}{\gamma} + 1\right) = k$$

We obtain for the breakdown field strength for SF_6 in uniform fields

$$\frac{E_b}{p} = \left(\frac{E}{p}\right)_0 + \frac{K}{k(pd)}\left[\frac{\text{kV}}{\text{cm}}\right] \tag{5.115}$$

where $((E/p))_0 = 88.5\,\text{kV/cm bar}$, $k = 27.7\,\text{kV}^{-1}$, and $K = 8 \ldots 10$ for the Townsend mechanism and $18 \ldots 20$ for the streamer mechanism. Substituting

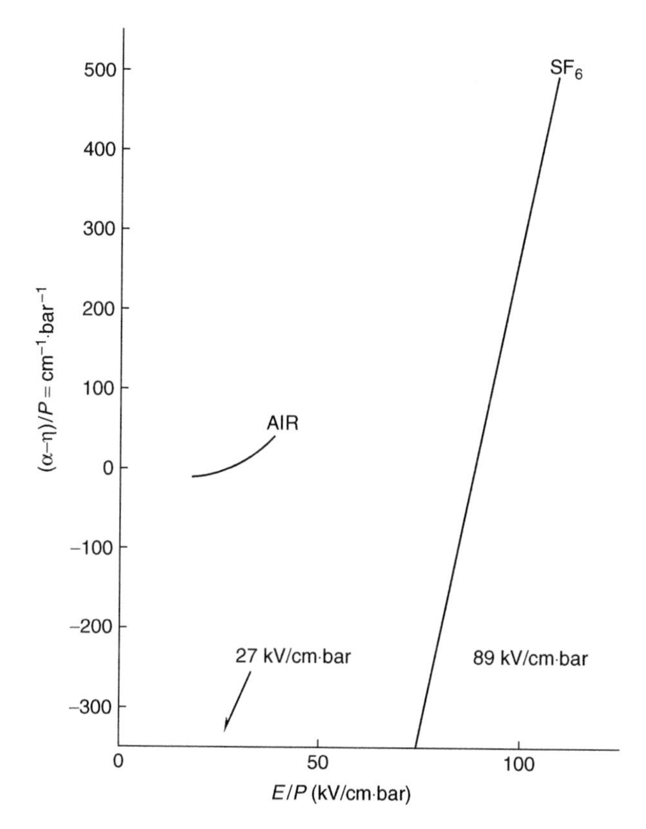

Figure 5.29 $(\alpha - \eta)/P \sim E/P$ *relationship in* SF_6

for the constants ($K = 18$) gives for a gas pressure p (in bar) and a gap distance d (in cm):

$$E_b = 88.5 + \frac{0.65}{d}[\text{kV/cm}] \tag{5.116}$$

The steep increase of $(\alpha - \eta)/p$ in Fig. 5.29 with pressure dependent field strength E/p (the gradient k is much larger than in air) accounts for the strong influence of local field distortions upon the breakdown strength.

For a uniform field gap eqn (5.116) converts to the Paschen dependency

$$V_b = 0.65 + 88.5\,(pd)[\text{kV}] \tag{5.117}$$

This gives good agreement with measured values for the approximate voltage range

$$1\,\text{kV} \le V_b \le 250\,\text{kV} \quad \text{in } pd \text{ range} \quad 0.04 \le pd \le 3\ \text{bar}\cdot\text{cm}$$

At higher values of pd, V_b is slightly lower than predicted by eqn (5.117) and follows the equation

$$V_b = 40 + 68pd \text{ [kV] } (pd \text{ in bar} \cdot \text{cm}) \tag{5.118}$$

Zaengl has shown[38] that since all the secondary feedback processes represented by the coefficient can be quite sensitive to gas pressure, to electrode surface conditions or even impurities, this reduction is largely due to 'microscopic field effects' generated by protrusions at the electrodes (called the electrode effect). In uniform field gaps the dielectric strength given by eqns (5.116) and (5.117) is approximately three times that of air, as shown in Fig. 5.30

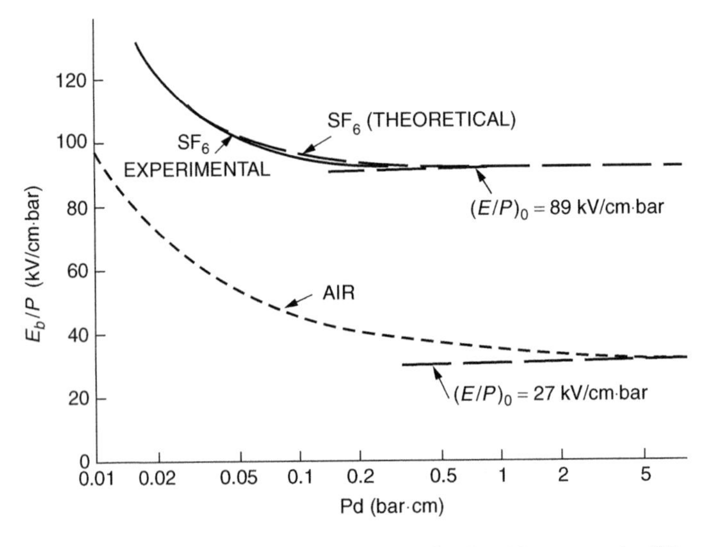

Figure 5.30 *Breakdown field strength of uniform gap in SF_6 and air*

A number of refined equations for predicting breakdown of SF_6 have been proposed by various research workers, notably by the group working at CRIEPI, Japan (Nitta and Takuma and Associates).[39,40,41] All are based on the assumption that the streamer mechanism prevails, that is, the criterion for spark is satisfied when the space charge in the avalanche changes the avalanche to a streamer. Further information can be found in reference 42.

5.11 Partial breakdown, corona discharges

In uniform field and quasi-uniform field gaps the onset of measurable ionization usually leads to complete breakdown of the gap. In non-uniform fields various manifestations of luminous and audible discharges are observed long before the complete breakdown occurs. These discharges may be transient or steady state and are known as 'coronas'. An excellent review of the subject may be found in a book by Loeb.[(43)] The phenomenon is of particular importance in h.v. engineering where non-uniform fields are unavoidable. It is responsible for considerable power losses from h.v. transmission lines and often leads to deterioration of insulation by the combined action of the discharge ions bombarding the surface and the action of chemical compounds that are formed by the discharge. It may give rise to interference in communication systems. On the other hand, it has various industrial applications such as high-speed printing devices, electrostatic precipitators, paint sprayers, Geiger counters, etc.

The voltage gradient at the surface of the conductor in air required to produce a visual a.c. corona in air is given approximately by the Peek's expression (5.110).[(35)]

There is a distinct difference in the visual appearance of a corona at wires under different polarity of the applied voltage. Under positive voltage, a corona appears in the form of a uniform bluish-white sheath over the entire surface of the wire. On negative wires the corona appears as reddish glowing spots distributed along the wire. The number of spots increases with the current. Stroboscopic studies show that with alternating voltages a corona has about the same appearance as with direct voltages. Because of the distinctly different properties of coronas under the different voltage polarities it is convenient to discuss separately positive and negative coronas.

In this section a brief review of the main features of corona discharges and their effect on breakdown characteristics will be included. For detailed treatment of the basic fundamentals of this subject the reader is referred to other literature sources.[(43)]

5.11.1 Positive or anode coronas

The most convenient electrode configurations for the study of the physical mechanism of coronas are hemispherically capped rod-plane or point-plane gaps. In the former arragement, by varying the radius of the electrode tip, different degrees of field non-uniformity can be readily achieved. The point-plane arrangement is particularly suitable for obtaining a high localized stress and for localization of dense space charge.

In discussing the corona characteristics and their relation to the breakdown characteristics it is convenient to distinguish between the phenomena that occur under pulsed voltage of short duration (impulse corona), where no space charge is permitted to drift and accumulate, and under long lasting (d.c.) voltages (static field corona).

Under impulse voltages at a level just above ionization threshold, because of the transient development of ionization, the growth of discharge is difficult to monitor precisely. However, with the use of 'Lichtenberg figures' techniques,[44] and more recently with high-speed photographic techniques, it has been possible to achieve some understanding of the various discharge stages preceding breakdown under impulse voltages.

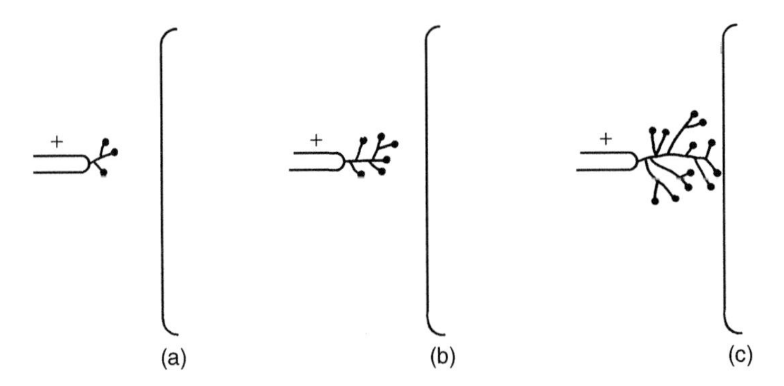

Figure 5.31 *Schematic illustration of the formation of streamers under impulse voltage-progressive growth with increasing pulse duration-positive rod-plane gap*

The observations have shown that when a positive voltage pulse is applied to a point electrode, the first detectable ionization is of a filamentary branch nature, as shown diagrammatically in Fig. 5.31(a). This discharge is called a streamer and is analogous to the case of uniform field gaps at higher *pd* values. As the impulse voltage level is increased, the streamers grow both in length and their number of branches as indicated in Figs 5.31(b) and (c). One of the interesting characteristics is their large number of branches which never cross each other. The velocity of the streamers decreases rapidly as they penetrate the low field region. Figure 5.32 shows velocities of impulse streamers recorded in air in a 2.5-cm gap under two different values of voltage. The actual mechanism of the transition from streamer to final breakdown is complex, and several models have been developed[4] to explain this transition, but because of space limitation the reasons will not be discussed here.

When the voltage is applied for an infinitely long time (e.g. under d.c. or 60 Hz) the ionization products will have sufficient time to wander in the gap and accumulate in space, causing a distortion in the original field.

To study this phenomenon, let us choose the rod-plane gap with the rod tip of radius of (say) 1 cm as shown in Fig. 5.33 and study the various discharge modes together with the breakdown characteristics for this arrangement in atmospheric air. Then if the gap length is small (less than about 2 cm) and

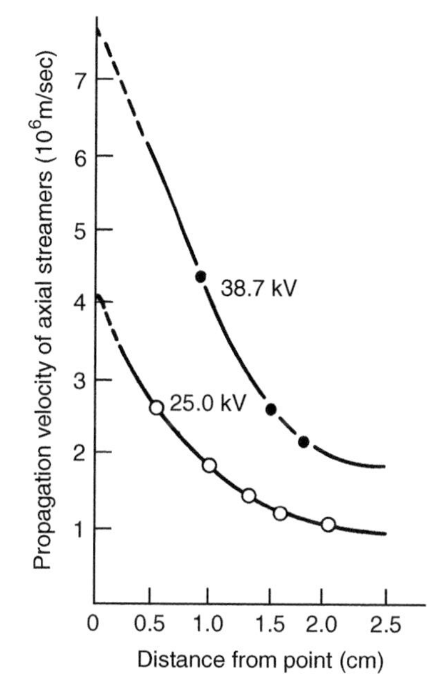

Figure 5.32 *Streamer velocity in a gap of 2.5 cm under two different voltages of fast rise in air*[43]

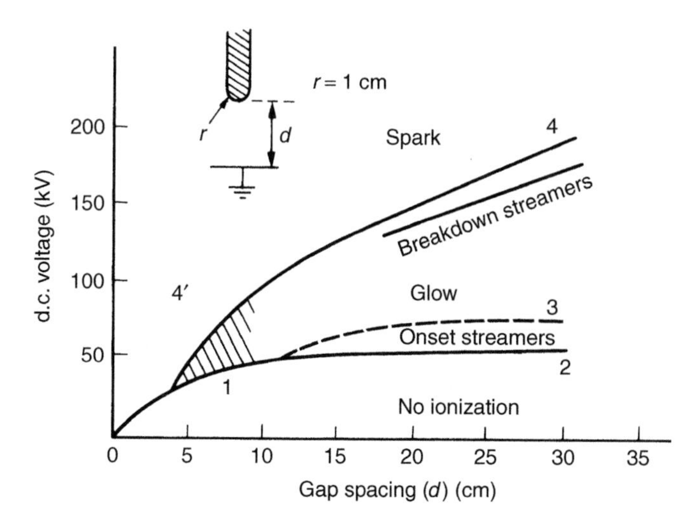

Figure 5.33 *Threshold curves for various modes of anode corona and for spark breakdown for a hemispherically capped anode and plate cathode*

the voltage is gradually raised no appreciable ionization is detected up to breakdown. As the gap is increased, the field distribution becomes more inhomogeneous, and on increasing the voltage at first a transient slightly branched filamentary discharge appears. These discharges have been shown to be identical with those observed under impulse voltages and are also called streamers. Under steady state the streamer develops with varying frequencies, giving rise to currents that are proportional to their physical length. These streamers are sometimes called onset streamers or burst pulses.

When the voltage is increased further, the streamers become more frequent, until the transient activity stops, the discharge becomes self-sustained and a steady glow appears close to the anode. This glow gives rise to continuous but fluctuating current. A further increase in voltage increases the luminosity of the glow both in area and in the intensity. It should be noted that glow corona develops only in the presence of negative ions. On increasing the voltage still further, new and more vigorous streamers appear which ultimately lead to complete breakdown of the gap.

The onsets of the various discharge modes observed, as the gap length is increased, are illustrated schematically in Fig. 5.33 together with the corresponding discharge characteristics. At the smaller spacing when the voltage is still reasonably uniform the streamer is capable of penetrating the weaker field, reaching the cathode and initiating breakdown in the same manner as in uniform field gaps. This condition is shown by curve 1 of Fig. 5.33. With the larger spacing above 10 cm, streamers appear that do not cross the gap (shown by curve 2). Curve 3 represents transition from streamers to steady glow corona without sparking. At the larger spacings there is a considerable spread in the voltage at which breakdown streamers develop preceding the complete breakdown of the gap. The dashed area represents the region of uncertain transitions; portion 1 indicates the onset of streamers followed immediately by transition to spark. If, however, the gap is increased to a point where glow is established and then reduced keeping the voltage constant, the glow discharge will have stabilized the gap against breakdown at a voltage that otherwise would have broken down. If the voltage is then raised, a spark is induced by glow corona (curve 4), but if it is lowered, a streamer breakdown is induced. By decreasing the gap further to lower values and increasing the voltage at the various points the glow-corona sparking voltage characteristic can be projected backwards as shown by curve 4. Thus if a steady corona glow is established, the sparking voltage is raised and the lower breakdown by streamer is suppressed.

5.11.2 Negative or cathode corona

With a negative polarity point-plane gap under static conditions above the onset voltage the current flows in very regular pulses as shown in Fig. 5.34(b), which indicates the nature of a single pulse and the regularity with which the pulses are repeated. The pulses were studied in detail by Trichel[(45)] and are named after their discoverer as 'Trichel pulses'. The onset voltage is practically independent of the gap length and in value is close to the onset of streamers under positive voltage for the same arrangement. The pulse frequency increases with the voltage and depends upon the radius of the cathode, the gap length and the pressure. The relationship between the pulse frequency and the gap voltage for different gap lengths and a cathode point of 0.75 mm radius in atmospheric air is shown in Fig. 5.34(a). A decrease in pressure decreases the frequency of the Trichel pulses.

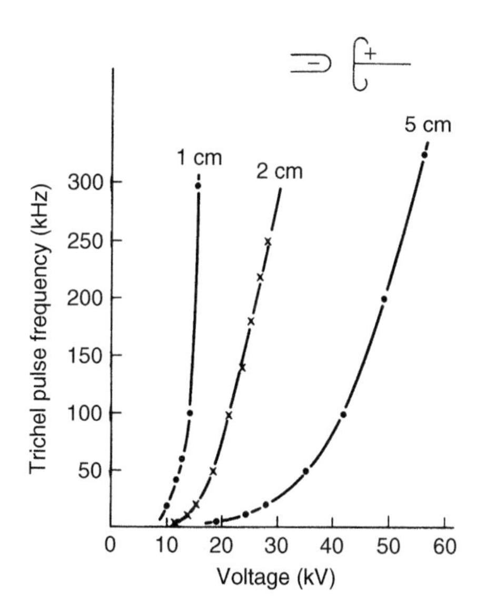

Figure 5.34 *Trichel pulse frequency–voltage relationship for different gap lengths in air ($r = 0.75$ mm)*

Figure 5.35 illustrates the onset voltage of different negative coronas plotted as a function of electrode separation for a typical example of a cathode of 0.75 mm radius. The lowest curve gives the onset voltage for Trichel pulses not greatly affected by the gap length. Raising the voltage does not change the mode of the pulses over a wide voltage range. Eventually at a much higher voltage a steady glow discharge is observed, but the transition from Trichel pulses to glow discharge is not sharply defined and is therefore shown as a broad transition region in Fig. 5.35. On increasing the voltage further, the glow discharge persists until breakdown occurs. It should be noted that breakdown under negative polarity occurs at considerably higher voltage than under positive voltage, except at low pressures; therefore, under alternating

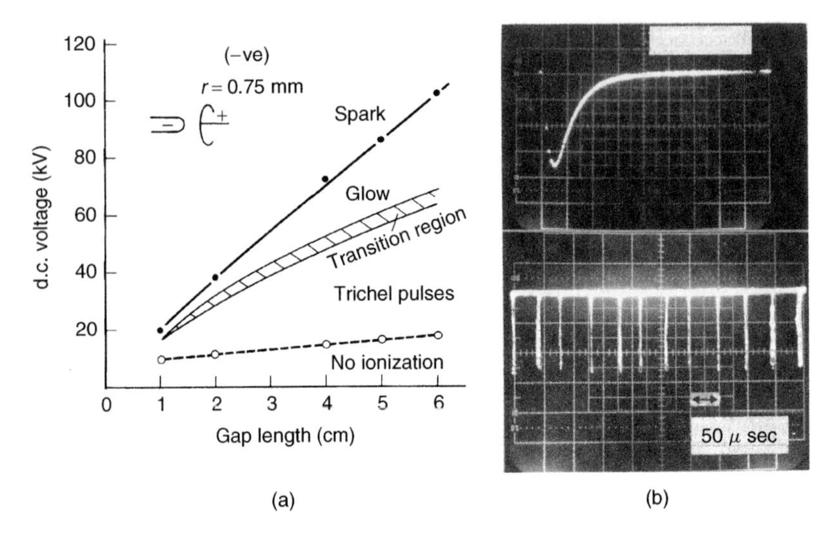

Figure 5.35 *Negative rod-plane breakdown and corona characteristics in atmospheric air (a) discharge modes, (b) pulse pattern*

power frequency voltage the breakdown of non-uniform field gap invariably takes place during the positive half-cycle of the voltage wave.

5.12 Polarity effect – influence of space charge

It was shown in Fig. 5.33 that in non-uniform field gaps in air the appearance of the first streamer may lead to breakdown or it may lead to the establishment of a steady state corona discharge which stabilizes the gap against breakdown. Accordingly we may have a corona stabilized or direct breakdown. This subject is been discussed in section 5.11. Whether direct or corona stabilized breakdown occurs depends on factors such as the degree of field non-uniformity, gas pressure, voltage polarity and the nature of the gas. For example, in air the corona stabilized breakdown will extend to higher pressures than in SF_6 due to the relatively immobile SF_6 ions (Figs 5.36 and 5.37).

Figure 5.36 compares the positive and negative point-plane gap breakdown characteristics measured in air as a function of gas pressure. At very small spacing the breakdown characteristics for the two polarities nearly coincide and no corona stabilized region is observed. As the spacing is increased, the positive characteristics display the distinct high corona breakdown up to a pressure of approximately 7 bar, followed by a sudden drop in breakdown strengths. Under the negative polarity the corona stabilized region extends to much higher pressures.

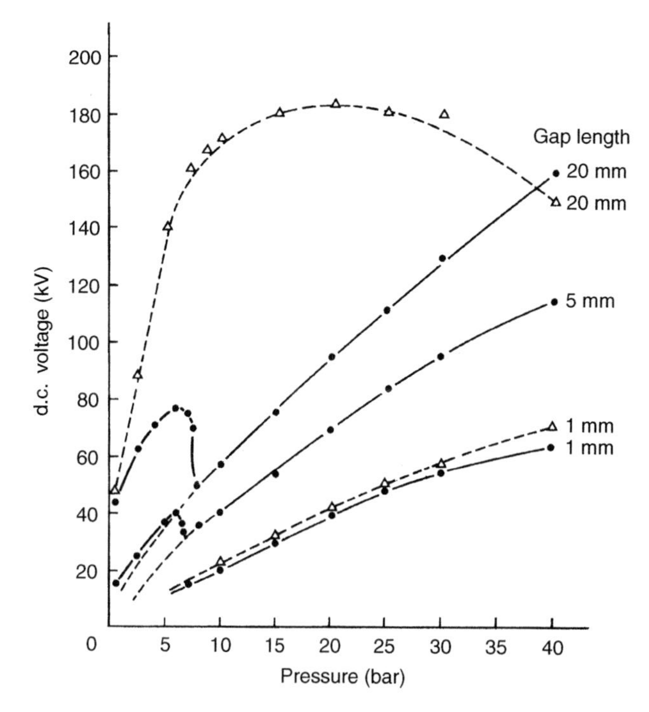

Figure 5.36 *Point-plane breakdown and corona inception characteristics in air: — positive point, - - negative point (radius of curvature of point $r = 1$ mm)*

A practical non-uniform field geometry that is frequently used in the construction of h.v. apparatus is the coaxial cylindrical arrangement. By properly choosing the radial dimensions for the cylinders it is possible to optimize such a system for the maximum corona-free breakdown.

Let us consider a system of two coaxial cylinders with inner and outer radii r_i and r_o respectively. Then it can be readily shown that in the interelectrode space at radial distance r the field strength is given by

$$E_r = \frac{V}{r \ln \dfrac{r_o}{r_i}}$$

where V is the applied voltage. Since breakdown or corona onset will follow when the voltage stress at the smaller wire reaches the breakdown stress (E_b) we can write the above equation as

$$V_b = E_b r_i \ln \frac{r_o}{r_i}. \tag{5.119}$$

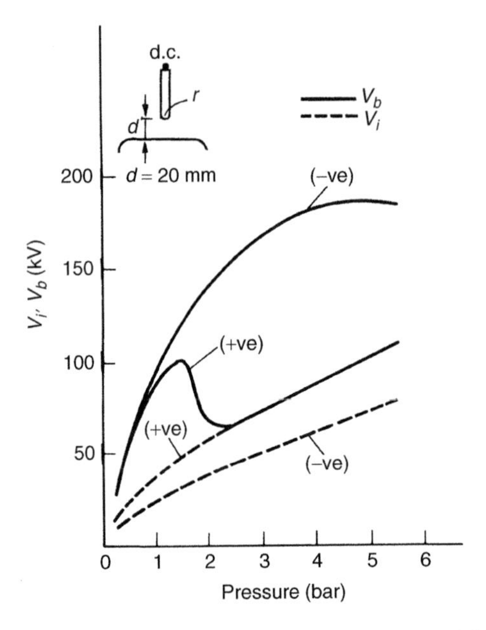

Figure 5.37 *D.C. corona inception and breakdown voltage in SF_6–rod-plane gap $d = 20$ mm; $r = 1$ mm*[46]

The maximum breakdown voltage for the system is obtained by differentiating eqn (5.119) with respect to r_i. In eqn (5.119) E_b is the breakdown (or corona inception) field strength of the system. It was shown earlier that this field strength depends upon the gas density as well as the radius $r = r_i$ of the inner conductor. Neglecting this dependency, which would hold approximately for not too small radii r_i and/or strongly attaching gases (with a steep increase of $\overline{\alpha}/p = f(E/p)$), we may assume that E_b is a constant value. Then, keeping r_o, constant this condition gives the optimal design for the system.

$$\frac{dV_b}{dr_i} = E_b\left(\ln\frac{r_o}{r_i} - 1\right) = 0$$

or

$$\frac{r_o}{r_i} = e$$

and

$$(V_b)_{\max} = E_b r_i. \tag{5.120}$$

Figure 5.38 shows the functional relationship between the breakdown voltage and the radius of the inner cylinder for a fixed radius r_o of the outer cylinder.

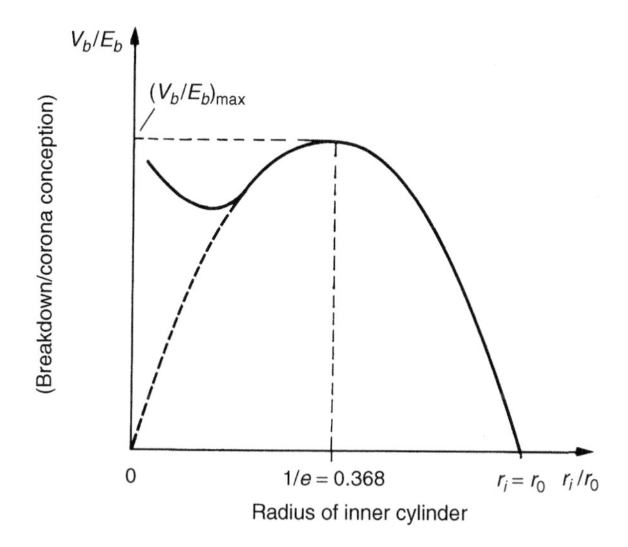

Figure 5.38 *Relationship between breakdown voltage and inner radius in a coaxial cylinder system*

The maximum breakdown voltage is also indicated. The dotted curve indicates quantitatively the corona onset voltage and the solid curve the breakdown voltage.

At low pressures the breakdown voltage is usually lower when the smaller electrode is negative. The effect is due to the higher field at the cathode so that γ is greater and therefore a lower value is needed for $\exp(\int_o^d \alpha dx)$ to satisfy the sparking criterion equation. Figure 5.39 shows the direct breakdown voltage characteristics for nitrogen at low pressures between a wire and coaxial cylinder. At higher pressures the order of the characteristics is reversed. The large polarity effect at the higher pressure can be qualitatively explained by considering the role of the space charge of the prebreakdown current.

If we consider the case of a positive point-plane gap shown in Fig. 5.40(a) then an ionization by electron collision takes place in the high field region close to the point. Electrons because of their higher mobility will be readily drawn into the anode, leaving the positive space charge behind. The space charge will cause a reduction in the field strength close to the anode and at the same time will increase the field further away from it. The field distortion caused by the positive space charge is illustrated in Fig. 5.40(b). The dotted curve represents the original undistorted field distribution across the gap while the solid curve shows the distorted field. The high field region is in time moving further into the gap extending the region for ionization. The field strength at the tip of the space charge may be high enough for the initiation of a cathode-directed streamer which subsequently may lead to complete breakdown. With

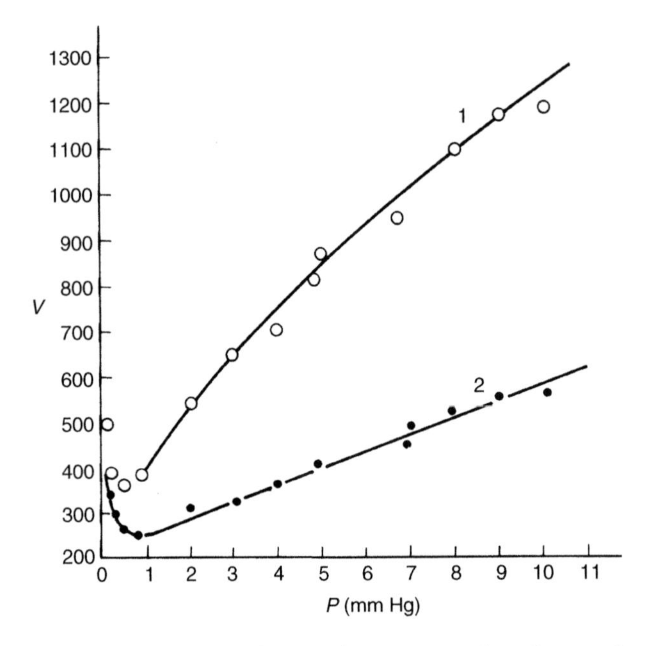

Figure 5.39 *Breakdown voltage curves for nitrogen between a wire and a coaxial cylinder (radii 0.083 and 2.3 cm respectively); curve 1 refers to a positive wire, curve 2 to a negative wire*

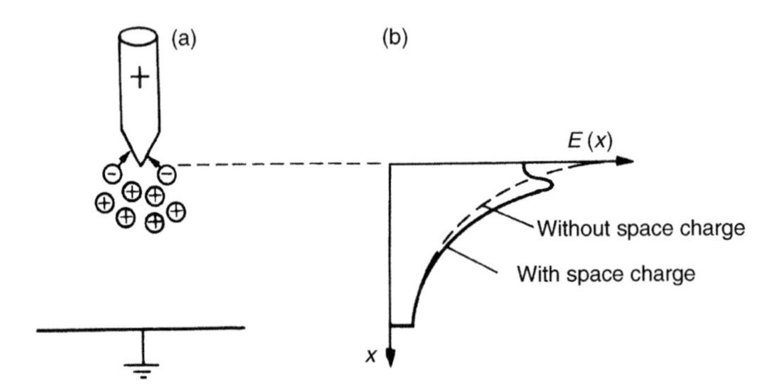

Figure 5.40 *(a) Space charge build-up in positive point-plane gap. (b) Field distortion by space charge*

the negative point (Fig. 5.40) the electrons are repelled into the low field region and in the case of attaching gases become attached to the gas molecules and tend to hold back the positive space charge which remains in the space between the negative charge and the point. In the vicinity of the point the field is grossly enhanced, but the ionization region is drastically reduced.

The effect is to terminate ionization. Once ionization ceases, the applied field sweeps away the negative and positive ion space charge from the vicinity of the point and the cycle starts again after the clearing time for the space charge. To overcome this retarding action of the ions a higher voltage is required, and hence negative breakdown voltage is higher than the positive breakdown voltage in gaps with marked asymmetrical fields.

Mathematically at any given time the voltage across the gap is given by the field integral $\int E(x)\,dx = V$. Integration of the space charge distorted field in Figs 5.40 and 5.41 respectively shows immediately that

$$V_b(+\text{point}) < V_b(-\text{point}).$$

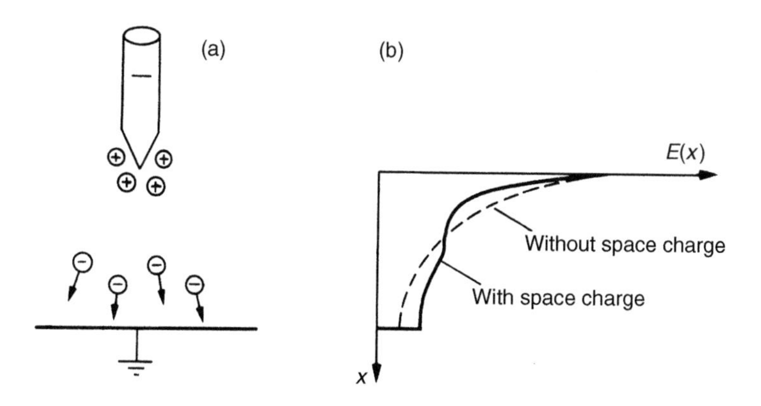

Figure 5.41 *(a) Space charge build-up in negative point-plane gap. (b) Field distortion by space charge*

5.13 Surge breakdown voltage–time lag

For the initiation of breakdown an electron must be available to start the avalanche. With slowly rising voltages (d.c. and a.c.) there are usually sufficient initiatory electrons created by cosmic rays and naturally occurring radioactive sources. Under surge voltages and pulses of short duration, however, the gap may not break down as the peak voltage reaches the lowest breakdown value (V_s) unless the presence of initiatory electrons is ensured by using artificial irradiation. V_s is a voltage which leads to breakdown of the gap after a long time of application. With weak irradiation the peak value may have to be greatly increased so that the voltage remains above the d.c. value (V_s) for long intervals of time. Figure 5.42 illustrates the breakdown on a step-function voltage pulse; V_p represents the peak value of a step voltage applied at time $t = 0$ to a gap that breaks down under V_s after a long time.

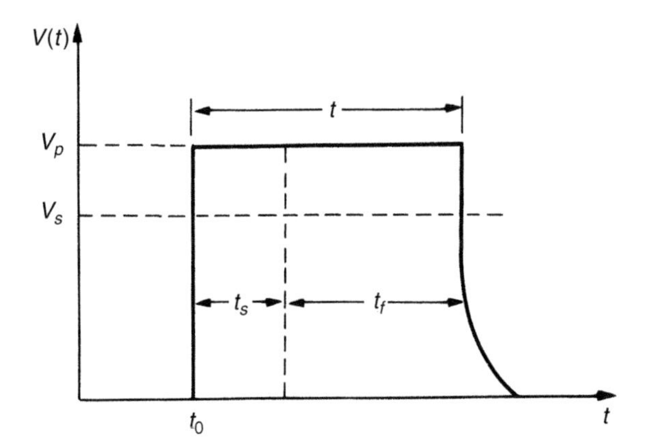

Figure 5.42 *Time lag components under a step voltage. V_s minimum static breakdown voltage; V_p peak voltage; t_s statistical time lag; t_f formative time lag*

The time which elapses between the application of voltage to a gap sufficient to cause breakdown and the breakdown is called the time lag (t). It consists of two components: one is the time which elapses during the voltage application until a primary electron appears to initiate the discharge and is known as the statistical time lag (t_s); and the other is the time required for the breakdown to develop once initiated and is known as the formative time lag (t_f).

The statistical time lag depends upon the amount of preionization in the gap. This in turn depends upon the size of the gap and the radiation producing the primary electrons. The appearance of such electrons is usually statistically distributed. The techniques generally used for irradiating gaps artificially, and thereby reducing the statistical time lag, include the use of u.v. light, radioactive materials and illumination by auxiliary sparks. The statistical time will also be greatly reduced by the application of an overvoltage ($V_p - V_s$) to the gap.

The formative time lag (t_f) depends essentially upon the mechanism of spark growth in question. In cases when the secondary electrons arise entirely from electron emission at the cathode by positive ions, the transit time from anode to cathode will be the dominant factor determining the formative time. The formative time lag increases with the gap length and the field non-uniformity, but it decreases with the applied overvoltage.

5.13.1 Breakdown under impulse voltages

An impulse voltage is a unidirectional voltage which rises rapidly to a maximum value and then decays slowly to zero. The exact definition of a standard impulse voltage was presented in Chapter 2.

When an impulse voltage of a peak value higher than V_s is applied to a gap, as shown in Fig. 5.43, there is a certain probability but not a certainty that breakdown will follow. For breakdown it is essential that the spark develops during the interval of overvoltage $[V(t) - V_s]$ duration, i.e. the overvoltage duration must exceed the time lag $[t < (t_2 - t_1)]$. For a given impulse voltage waveshape the overvoltage duration will increase with the voltage amplitude (V_p).

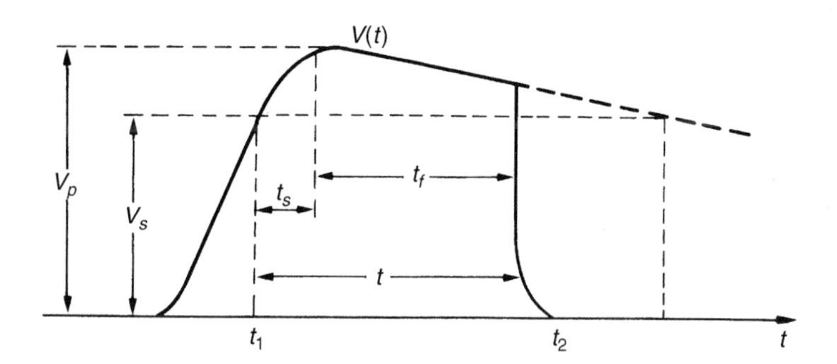

Figure 5.43 *Breakdown under impulse voltage*

Because of the statistical nature of the time lags, when a given number of impulses of an amplitude V_p, exceeding the static value V_s, are applied to a gap only a certain percentage will lead to breakdown. We therefore obtain a breakdown probability P for each given applied maximum impulse voltage V_p as a function of V_p. This subject will be discussed in Chapter 8.

5.13.2 Volt–time characteristics

When an impulse voltage of sufficiently high value is applied to a gap, breakdown will result on each voltage application. The time required for the spark development (time lag) will depend upon the rate of rise of voltage and the field geometry. Therefore, for each gap geometry it is possible to construct a volt–time characteristic by applying a number of impulses of increasing amplitude and noting oscillographically the time lag. A schematic plot of such a characteristic is shown in Fig. 5.44. In uniform and quasi-uniform field gaps the characteristic is usually sharply defined and it rises steeply with increasing the rate of rise of the applied voltage. In non-uniform field gaps, however, due to larger scatter in the results, the data fall into a dispersion band as shown in Fig. 5.45. The time to breakdown is less sensitive to the rate of voltage rise. Hence, quasi-uniform field gaps (sphere–sphere) have often been used as protective devices against overvoltages in electric power

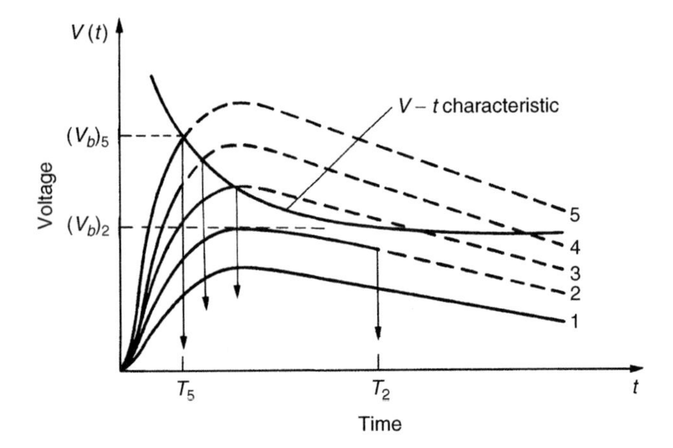

Figure 5.44 *Impulse 'volt–time' characteristics*

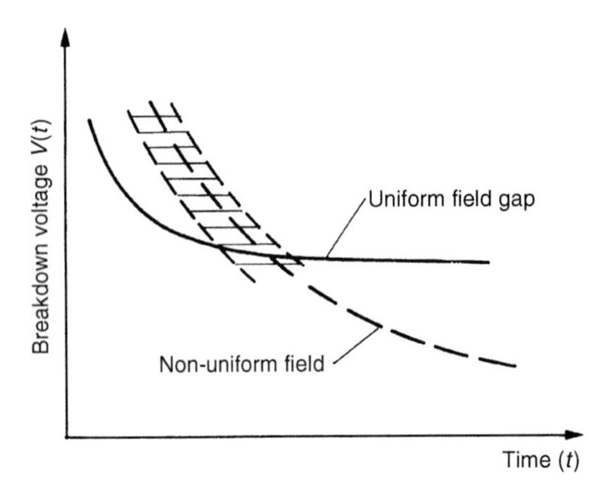

Figure 5.45 *Schematic diagram of volt–time characteristics for uniform and non-uniform field gaps*

systems. The volt–time characteristic is an important practical property of any insulating device or structure. It provides the basis for establishing the impulse strength of the insulation as well as for the design of the protection level against overvoltages and will be discussed in Chapter 8.

5.13.3 Experimental studies of time lags

Numerous investigators have studied time lags in the past. In the techniques generally used either a constant voltage is applied to an irradiated gap and a spark is initiated by a sudden illumination of the gap from a nearby spark, or an overvoltage is suddenly applied to a gap already illuminated.

In the former case the time lag is measured from the flash until breakdown occurs, while in the latter the time lag is measured between the voltage application and the gap breakdown. The overvolted conditions may be obtained either by superimposing a step voltage pulse upon a direct voltage already applied to the gap or by using an impulse voltage of a suitably short front duration. The measured time lags for given experimental conditions are usually presented graphically by plotting the average time lags against the overvoltage. The latter is defined as the percentage ratio of the voltage in question to the minimum direct voltage which will cause breakdown. In the case when an impulse voltage is used on its own, the time lags are plotted against the impulse ratio defined as the ratio of the applied impulse voltage to the minimum direct breakdown voltage.

The measured values are affected by factors such as the intensity of the background irradiation, the nature and the condition of the electrode surface, the gap length, the electron affinity of the gas, etc.

With a gap illuminated from an intense u.v. source, time lags down to 10^{-8} sec and shorter have been recorded in highly overvolted gaps.(47) Figure 5.46 shows time lags of spark breakdown for short gaps with the

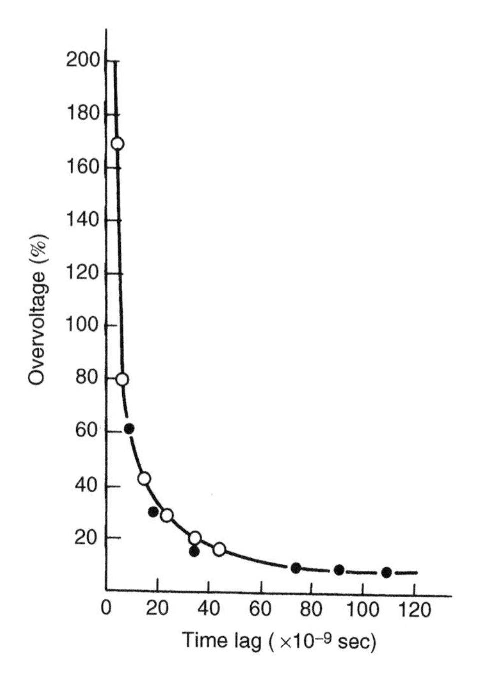

Figure 5.46 *Time lag of spark gap as a function of overvoltage for short gap between spheres with intense u.v. illumination of the cathode in air*

cathode irradiated by a quartz mercury lamp, obtained by Brayant and Newman,[47] between spheres in air. Fisher and Benderson[48] studied time lags in air between uniform field electrodes in slightly overvolted conditions and the results obtained for four gap lengths are shown in Fig. 5.47. These authors used different gas pressures and found that in the range of pressure from 760 Torr down to about 200 Torr the results were independent of the gas pressure.

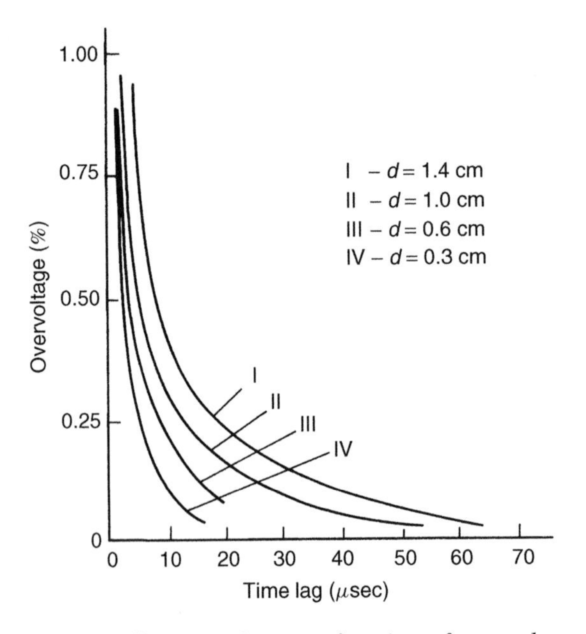

Figure 5.47 *Time lag as a function of overvoltage for four gap lengths in air. The curves represent the average data for all pressures between atmospheric and 200 mm Hg*

Long and highly scattered time lags have been observed in strongly electronegative gases under irradiated conditions. Figure 5.48 compares time lags observed in SF_6 with those obtained in air under similar experimental conditions. It was impossible[49] to attribute these long time lags to the shortage of initiatory electrons. It was suggested that the long time lags are associated with the complex nature of the growth of spark in the highly electron attaching gases.

An alternative method for presenting time lags has been developed by Laue[50] and Zuber.[51] These authors showed that the time lag in spark gaps may be represented in the form

$$\frac{n}{N} = \mathrm{e}^{-\int_0^t \rho_1 \rho_2 \beta \, \mathrm{d}t} \tag{5.121}$$

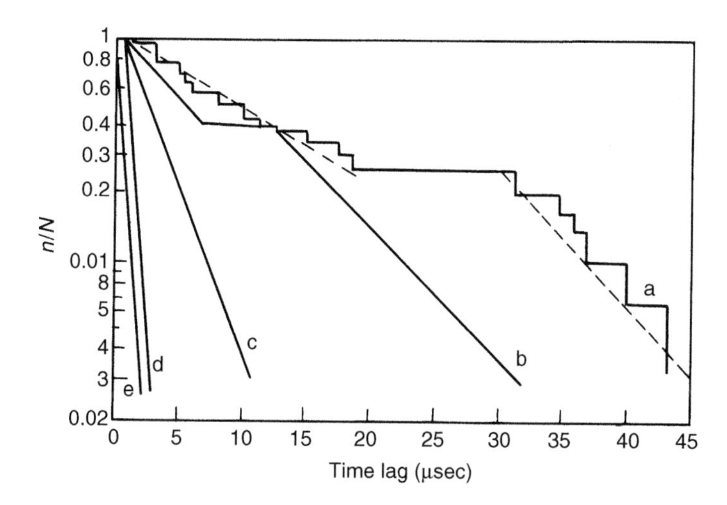

Figure 5.48 *Time lag distribution in SF_6 and air. Overvoltages: (a) 10%; (b) 15%; (c) 20%; (d) 25% for SF_6; (e) 5.3% for air*

where N represents the total number of time lags observed, n is the number of time lags of length greater than t, β the rate at which electrons are produced in the gap by irradiation, ρ_1 the probability of an electron appearing in a region of the gap where it can initiate a spark, and ρ_2 the probability that an electron at a given field strength will lead to the development of the spark. The factor ρ_1 is a function of the gap length and the gas density, while ρ_2 is a function of the applied field. The factor β is dependent on the source of irradiation. Providing that the primary current in the gap is constant and the applied field remains constant with respect to time, eqn (5.121) can be written as:

$$\frac{n}{N} = \mathrm{e}^{-kt}. \tag{5.122}$$

Equation (5.122) gives a linear relation between $\ln \frac{n}{N}$ and time t. The method gives a truer representation of the results in the case of highly scattered results.

References

1. B. Loeb. *The Kinetic Theory of Gases*. Wiley, New York, 1963, Chapter 2.
2. E.W. McDaniel. *Collision Phenomena in Ionised Gases*. Wiley, New York, 1964.
3. D. Ramp and P. Englander-Golden. *J. Chem. Phys.* **43** (1965), p. 1964.
4. J.S. Townsend. *Electricity in Gases*. Oxford Press, 1914.
5. H. Raether. *Z. Phys.* **117** (1941), pp. 375, 524.
6. A. von Angel. *Ionised Gases* (2nd edn), p. 181. Clarendon Press, 1965.
7. F.H. Sanders. *Phys. Rev.* **44** (1932), p. 667.
8. K. Marsch. *Arch. Elektrotechnik* **26** (1932), p. 598.

9. R. Geballe and M.A. Harrison. *Phys. Rev.* **85** (1952), p. 372.
10. G.K. Kachickas and L.H. Fischer. *Phys. Rev.* **91** (1953), p. 775.
11. M.H. Saba. *Phil. Mag.* **40** (1920), p. 472.
12. L.B. Loeb. *Basic Processes of Gaseous Electronics*. University of California Press, 1955.
13. L.B. Loeb. Formation of negative ions. *Encyclopedia of Physics*, Vol. 16, p. 445. Springer, Berlin, 1956.
14. P. Langevin. *Ann. Chim. Phys.* **8** (1905), p. 238.
15. G.L. Weissler. Photoelectric emission from solids. *Encyclopedia of Physics* Vol. 16, p. 342.
16. L. Matter. *Phys. Rev.* **49** (1936), p. 879.
17. O.W. Richardson. *The Emission of Electricity from Hot Bodies*. Longmans Green, London, 1921.
18. W. Schottky. *Ann. Phys.* **44** (1914), p. 1011.
19. R.H. Fowler and L.W. Nordheim. *Proc. Roy. Soc. London* **119** (1928), p. 173; **124** (1929), p. 699.
20. F. Llewellyn Jones and D.E. Davies. *Proc. Phys. Soc.* **B64** (1951), p. 397.
21. J.M. Meek and J.D. Craggs. *Electrical Breakdown of Gases*. Clarendon Press, Oxford, 1953.
22. H. Raether. *Electron Avalanches and Breakdown in Gases*. Butterworths, London, 1964.
23. H. Raether. *Z. Phys.* **112** (1939), p. 464.
24. K.H. Wigner. *Z. Phys.* **189** (1966), p. 466.
25. J.M. Meek. *Phys. Rev.* **57** (1940), p. 722.
26. L.B. Loeb and J.M. Meek. *The Mechanism of Electric Spark*. Stanford University Press, 1940.
27. K. Dehne, W. Khörman and H. Lenne. Measurement of formative time lags for sparks in air, H_2, and N_2. *Dielectrics* **1** (1963), p. 129.
28. D.H. Hale. *Phys. Rev.* **56** (1948), p. 1199.
29. J.J. Thomson and G.P. Thomson. *Conduction of Electricity through Gases*. 2 vols. New York: Dover Publications Inc. 1969 (paper edition).
30. J. Cueilleron. *C.R. Acad. Sci. Paris* **226** (1948), p. 400.
31. W.O. Schumann. *Arch. fur Elektrotechnik* **12** (1923), p. 593.
32. H. Sohst. *Zeitsch. fur Angew. Physik* **14** (1962), p. 620.
33. G.A. Schöder, *Zeitsch. fur Angew. Physik* **13** (1961), p. 296.
34. F.M. Penning. *Physica* **1** (1934), p. 1028.
35. F.W. Peek. *Dielectric Phenomena in High Voltage Engineering* (2nd edn.), McGraw-Hill, New York, 1920.
36. W.S. Zaengl and N.U. Nyffenegger. *Proc. 3rd Int. Conf. on Gas Discharges*, 1974, p. 303.
37. R. Geballe and M.L. Reeves. *Phys. Rev.* **92** (1953), p. 867.
38. W.S. Zaengl. *Proc. 10th Symposium on Electrical Insulating Materials*, Tokyo, 1977, p. 13.
39. T. Nitta and Y. Sakata. *IEEE Trans. EI* **89** (1971), pp. 1065–1071.
40. H. Fujinami *et al.*, *IEEE Trans. EI* **18**(4) (1983), pp. 429–435.
41. T. Takuma. *IEEE Trans. EI* **21**(6) (1986), pp. 855–867.
42. W.S. Zaengl. Electronegative Gases, Present State of Knowledge and Application, Future Prospects. Nordic Symposium on Electric Insulation, June 13–15, 1988, Trondheim, pp. I1–I39.
43. L.B. Loeb. *Electrical Coronas*. University of California Press, 1965.
44. E. Nasser. *IEEE Spectrum* **5** (1968), p. 127.
45. G.W. Trichel. *Phys. Rev.* **55** (1939), p. 382.
46. R.L. Hazel and E. Kuffel. *Trans. IEEE Pas.* **95** (1976), p. 178.
47. J.M. Bryant and M. Newman. *Trans. AIEE* **59** (1940), p, 813.
48. L.H. Fischer and B. Benderson. *Phys. Rev.* **81** (1951), p. 109.
49. E. Kuffel and R. M. Radwan. *Proc. IEE* **113** (1966), p. 1863.
50. M. von Laue. *Ann. Phys. Lpz.* **76** (1925), p. 261.
51. K. Zuber. *Ann. Phys. Lpz.* **76** (1925), p. 231.

Chapter 6

Breakdown in solid and liquid dielectrics

6.1 Breakdown in solids

Solid insulation forms an integral part of high voltage structures. The solid materials provide the mechanical support for conducting parts and at the same time insulate the conductors from one another. Frequently practical insulation structures consist of combinations of solids with liquid and/or gaseous media. Therefore, the knowledge of failure mechanisms of solid dielectrics under electric stress is of great importance.

In gases the transport of electricity is limited to positive and negative charge carriers, and the destruction of insulating properties involves a rapid growth of current through the formation of electron avalanches. The mechanism of electrical failure in gases is now understood reasonably clearly. This is not the case for solid insulation. Although numerous investigators have studied the breakdown of solids for nearly a century now, and a number of detailed theories have been put forward which aimed to explain quantitatively the breakdown processes in solids, the state of knowledge in this area is still very crude and inconclusive.

Electrical conduction studies in solids are obscured by the fact that the transport phenomena besides electronic and ionic carriers include also currents due to the slower polarization processes such as slow moving dipoles (orientation polarization) and interfacial polarization (see Chapter 7, Section 7.1). Electrical methods are unable to distinguish between the conduction currents and the currents due to polarization having a longer time constant than the duration of a particular experiment. At low stresses and normal temperatures conduction by free electrons and ions in solids is exceptional. Examples in which the conduction is believed to be of the simple electrolytic type at room temperature and above are glasses. In this case the conduction–temperature relation is found to be of the form

$$\sigma = A \exp\left[-\frac{u}{kT}\right]$$

where A and u are empirical constants. Ceramics also develop a significant conductivity at higher temperatures that may be electronic or ionic.

As the stress in solids is increased and approaches the breakdown stress, the current is found to increase exponentially, but does not vary so markedly with time for steady voltage.[(1)]* This increased current at high stresses is generally believed to result from the injection of carriers from an electrode or from electron multiplication in the bulk of the material or both. In addition, if impurities or structural defects are present they may cause local allowed energy levels (traps) in the forbidden band, and electrons may pass through the insulator by jumping from one trap to another (hopping effect).

From the electrodes the electrons are believed to be ejected by either the 'Schottky's emission effect' or the 'field emission effect' (tunnelling) discussed already in Chapter 5. Once injected into the material the electron multiplication is thought to be analogous to that in a gas discharge. Under certain strictly controlled experimental conditions the breakdown of solids may therefore be accomplished by a process similar to gas breakdown. Under normal industrial conditions, however, the same solid materials are found to exhibit a wide range of dielectric strength, depending upon the conditions of the environment and the method of testing. The measured breakdown voltage is influenced by a large number of external factors such as temperature, humidity, duration of test, whether a.c., d.c., or impulse voltage is applied, pressure applied to the electrodes, discharges in the ambient medium, discharges in cavities and many other factors. The fundamental mechanisms of breakdown in solids are understood much less clearly than those in gases; nevertheless, several distinct mechanisms have been identified and treated theoretically.[(2–4)]

In this section the presently accepted breakdown mechanisms will be discussed briefly in a qualitative manner. No conduction mechanism will be discussed here and the reader is referred to reference 6. Broadly speaking the mechanism of failure and the breakdown strength changes with the time of voltage application and for discussion purposes it is convenient to divide the time scale of voltage application into regions in which different mechanisms operate, as shown in Fig 6.1.

6.1.1 Intrinsic breakdown

If the material under test is pure and homogeneous, the temperature and environmental conditions are carefully controlled, and the sample is so stressed that there are no external discharges. With undervoltages applied for a short time the electric strength increases up to an upper limit which is called the intrinsic electric strength. The intrinsic strength is a property of the material and temperature only. Experimentally the intrinsic strength is rarely reached, but numerous attempts have been made to measure it for various materials. To achieve the highest strength the sample is so designed that there is a high

* Superscript numbers are to references at the end of the chapter.

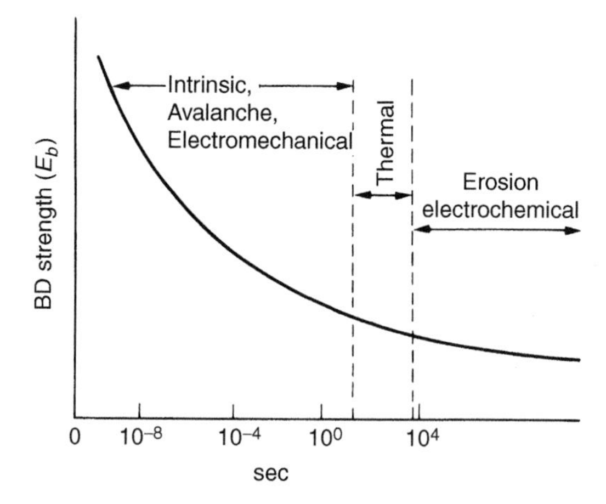

Figure 6.1 *Mechanisms of failure and variation of breakdown strength in solids with time of stressing*

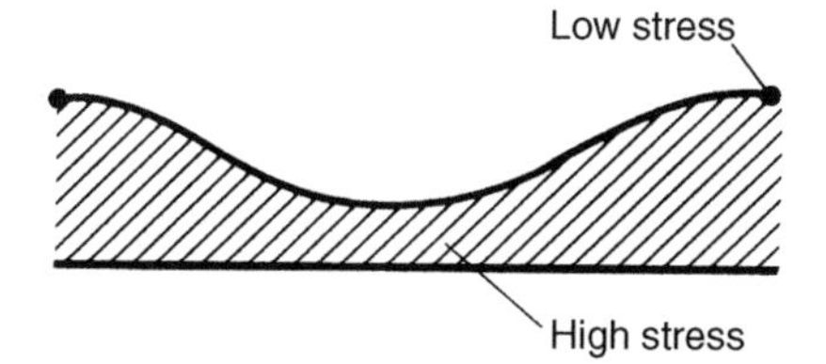

Figure 6.2 *Electrode arrangement used for measuring intrinsic breakdown in solids*

stress in the centre of the solid under test and too low stress at the edges which cause discharge in the medium as shown in Fig. 6.2.

The intrinsic breakdown is accomplished in times of the order of 10^{-8} sec and has therefore been postulated to be electronic in nature. The stresses required for an intrinsic breakdown are well in excess of 10^6 V/cm. The intrinsic strength is generally assumed to be reached when electrons in the insulator gain sufficient energy from the applied field to cross the forbidden energy gap from the valence to the conduction band. The criterion condition is formulated by solving an equation for the energy balance between the gain of energy by conduction electrons from the applied field and its loss to the lattice. Several models have been proposed in an attempt to predict the critical value of the field which causes intrinsic breakdown, but no completely satisfactory solution has yet been obtained. The models used by various workers differ from each other in the proposed mechanisms of energy transfer from conduction electrons to the lattice, and also by the assumptions made concerning the distribution of conduction electrons. In pure homogeneous dielectric materials the conduction and the valence bands are separated by a large energy gap, and at room temperature the electrons cannot acquire sufficient thermal energy to make transitions from valence to conduction band. The conductivity in perfect dielectrics should therefore be zero. In practice, however, all crystals contain some imperfections in their structures due to missing atoms, and more frequently due to the presence of foreign atoms (impurities). The impurity atoms may act as traps for free electrons in energy levels that lie just below the conduction band, as illustrated schematically in Fig. 6.3.

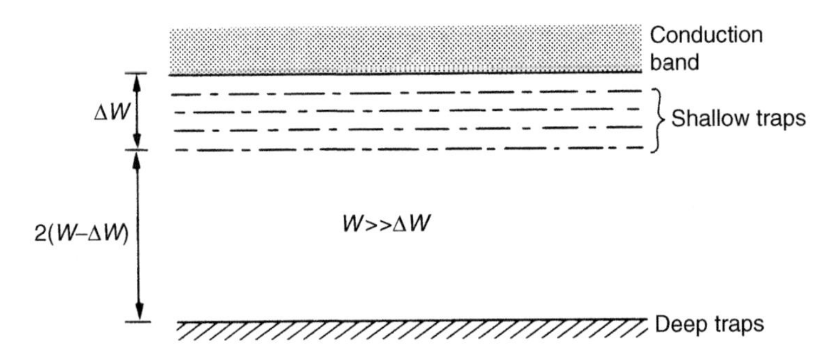

Figure 6.3 *Schematic energy level diagram for an amorphous dielectric*

At low temperatures the trap levels will be mostly filled with electrons caught there as the crystal was cooled down during its manufacture. At room temperature some of the trapped electrons will be excited thermally into the conduction band, because of the small energy gap between the trapping levels and the conduction level. An amorphous crystal will therefore have some free conduction electrons.

When a field is applied to a crystal the conduction electrons gain energy from it, and due to collisions between them the energy is shared by all electrons. For a stable condition this energy must be somehow dissipated. If there are relatively few electrons such as in pure crystals, most of the energy will be transferred to the lattice by electron–lattice interaction. In steady state conditions the electron temperature (T_e) will be nearly equal to the lattice temperature (T).

In amorphous dielectrics the electron interactions predominate, the field raises the energy of the electrons more rapidly than they can transfer it to the lattice, and the electron temperature T_e will exceed the lattice temperature T. The effect of the increased electron temperature will be a rise in the number of trapped electrons reaching the conduction band. This increases the material's conduction and as the electron temperature continues to increase a complete breakdown is eventually reached known as 'high-temperature breakdown'.

Neglecting for the moment the details of the mechanism of energy transfer and assuming electronic conduction in solids, for an applied field E the rate of energy gained by electrons from the field will be a function of the field strength E and the lattice temperature T. The rate at which this energy is transferred to the lattice will depend only on T. In addition, both rates will depend on parameters describing the conduction electrons. If we denote these parameters collectively by α, then for the steady state conditions the energy equation for conduction electrons may be written as

$$A(E, T, \alpha) = B(T, \alpha) \tag{6.1}$$

where the l.h.s. represents the rate of energy gain by electrons from the field, and the r.h.s., the rate of energy transfer from electrons to lattice. Equation (6.1) can be physically satisfied for values of E below a certain critical value E_c, and this value has been considered by several workers as the intrinsic critical field. The value of E_c can be found by identifying correctly the parameters α describing the conduction electrons and then solving eqn (6.1) for the critical field strength E_c.

For a pure homogeneous dielectric Fröhlich developed the so-called 'high energy' breakdown criterion, based on the assumption that the dielectric is destroyed by an infinitely large multiplication of electrons in the conduction band. In this model the critical field strength (E_c) in the energy balance eqn (6.1) is obtained by first identifying the parameter α with the electron energy (W_e) such that the balance equation is satisfied and then calculating the critical field strength.

The functional relationship between the parameters in eqn (6.1) is shown schematically in Fig. 6.4, which shows the average rate of energy gain from the field for various field strengths and the rate of energy loss to the lattice.

For the critical field criterion, eqn (6.1) becomes

$$A(E_c, T, I) = B(T, I) \tag{6.2}$$

where I is the ionization energy corresponding to the transition of an electron from a valence band to a conduction band. From Fig. 6.4 it is seen that for an electron to remain accelerated and thus lead to instability at any given field it should find itself with an energy which brings it above the curve B so that it gains energy more rapidly than it loses. Equation (6.2) enables us to determine the critical field strength E_c that is required to cause collision ionization from valence to conduction band. For field strength exceeding E_c the electrons gain energy more rapidly from the field than they lose to the lattice and breakdown will result. The above mechanism applies to pure solids in which the equilibrium is controlled by collisions between electrons and the lattice vibrations.

Fröhlich and Paranjape[(5)] have extended this model to amorphous materials in which the concentration of conduction (or trapped) electrons is high enough

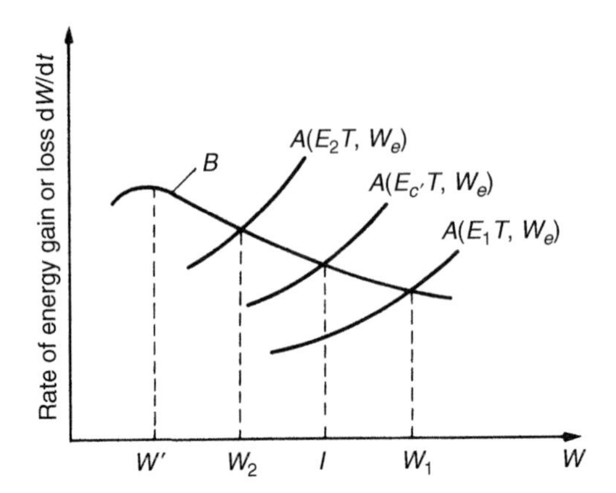

Figure 6.4 *The average rate of energy gain $A(E, T, W_e)$ from an applied field for various field strengths and the average rate of energy loss to lattice $B(W_L, T)$*

to make electron–electron collisions the dominant factor. In this case it is necessary to calculate the electron temperatureT_e which will be higher than the lattice temperature T.

The energy balance eqn (6.1) will then take the form

$$A(E, T_e, T) = B(T_e, T). \tag{6.3}$$

This relationship is plotted schematically in Fig. 6.5 in which the family of curves plotted for various values of E represents the l.h.s. of the equation and

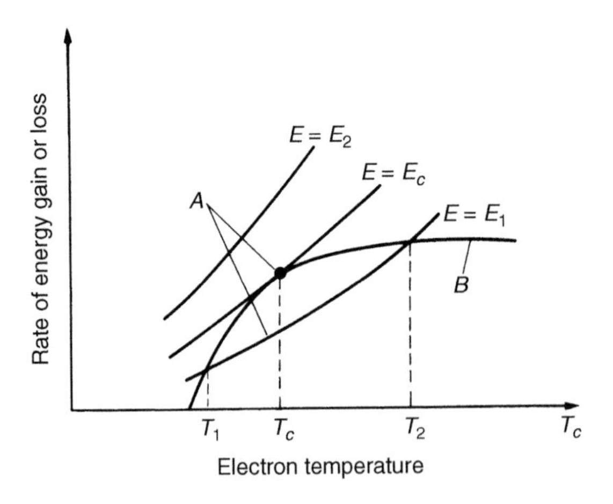

Figure 6.5 *Rate of energy gain and loss for h.t. intrinsic breakdown model*

the single curve represents the r.h.s. The intersections give possible solutions for the various electron temperatures.

For the analytical expressions for the critical field strength (E_c) for both of the above two models the reader should refer to reference 6.

To date there has been no direct experimental proof to show whether an observed breakdown is intrinsic or not, except for plastic materials such as polyethylene and so conceptually it remains an ideal mechanism identified as the highest value obtainable after all secondary effects have been eliminated.

6.1.2 Streamer breakdown

Under certain controlled conditions in strictly uniform fields with the electrodes embedded in the specimen, breakdown may be accomplished after the passage of a single avalanche. An electron entering the conduction band of the dielectric at the cathode will drift towards the anode under the influence of the field gaining energy between collisions and losing it on collisions. On occasions the free path may be long enough for the energy gain to exceed the lattice ionization energy and an additional electron is produced on collision. The process is repeated and may lead to the formation of an electron avalanche similar to gases. Seitz(7) suggested that breakdown will ensue if the avalanche exceeds a certain critical size and derived an expression for a single avalanche breakdown strength. The concept is similar to the streamer theory developed by Raether, and Meek and Loeb for gases discussed earlier.

6.1.3 Electromechanical breakdown

Substances which can deform appreciably without fracture may collapse when the electrostatic compression forces on the test specimen exceed its mechanical compressive strength. The compression forces arise from the electrostatic attraction between surface charges which appear when the voltage is applied. The pressure exerted when the field reaches about 10^6 V/cm may be several kN/m^2. Following Stark and Garton,(8) if d_0 is the initial thickness of a specimen of material of Young's modulus Y, which decreases to a thickness of d (m) under an applied voltage V, then the electrically developed compressive stress is in equilibrium with the mechanical compressive strength if

$$\varepsilon_0\varepsilon_r\frac{V^2}{2d^2} = Y\ln\left(\frac{d_0}{d}\right) \tag{6.4}$$

or

$$V^2 = d^2\frac{2Y}{\varepsilon_0\varepsilon_r}\ln\left(\frac{d_0}{d}\right)$$

where ε_0 and ε_r are the permittivity of free space and the relative permittivity of the dielectric.

Differentiating with respect to d we find that expression (6.4) has a maximum when $d/d_0 = \exp[-1/2] = 0.6$. Therefore, no real value of V can produce a stable value of d/d_0 less than 0.6. If the intrinsic strength is not reached at this value, a further increase in V makes the thickness unstable and the specimen collapses. The highest apparent strength is then given by

$$E_a = \frac{V}{d_0} = 0.6\left[\frac{Y}{\varepsilon_0\varepsilon_r}\right]^{1/2} \tag{6.5}$$

This treatment ignores the possibility of instability occurring in the lower average field because of stress concentration at irregularities, the dependence of Y on time and stress, and also on plastic flow.

6.1.4 Edge breakdown and treeing

In practical insulation systems, the solid material is stressed in conjunction with one or more other materials. If one of the materials is, for example, a gas or a liquid, then the measured breakdown voltage will be influenced more by the weak medium than by the solid.

A cross-section of a simplified example is shown in Fig. 6.6 which represents testing of a dielectric slab between sphere-plane electrodes. Ignoring the field distribution, i.e. assuming a homogeneous field, if we consider an

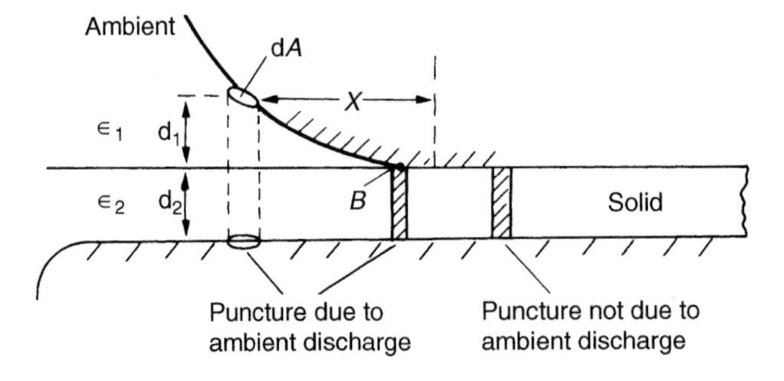

Figure 6.6 *Breakdown of solid specimen due to ambient discharge-edge effect*

elementary cylindrical volume of end area dA spanning the electrodes at distance x as shown in Fig. 6.5, then on applying the voltage V between the electrodes, according to Section 4.3.1 a fraction V_1 of the voltage appears across the ambient given by

$$V_1 = \frac{Vd_1}{d_1 + \left(\frac{\varepsilon_1}{\varepsilon_2}\right) d_2} \tag{6.6}$$

here d_1 and d_2 represent the thickness of the media 1 and 2 in Fig. 6.6 and ε_1 and ε_2 are their respective permittivities. For the simple case when a gaseous dielectric is in series with a solid dielectric stressed between two parallel plate electrodes, the stress in the gaseous part will exceed that of the solid by the ratio of permittivities or $E_1 = \varepsilon_r E_2$. For the case shown in Fig. 6.6, the stress in the gaseous part increases further as x is decreased, and reaches very high values as d_1 becomes very small (point B). Consequently the ambient breaks down at a relatively low applied voltage. The charge at the tip of the discharge will further disturb the applied local field and transform the arrangement to a highly non-uniform system. The charge concentration at the tip of a discharge channel has been estimated to be sufficient to give a local field of the order of 10 MV/cm, which is higher than the intrinsic breakdown field. A local breakdown at the tips of the discharge is likely, therefore, and complete breakdown is the result of many such breakdown channels formed in the solid and extending step by step through the whole thickness.

The breakdown event in solids in general is not accomplished through the formation of a single discharge channel, but assumes a tree-like structure as shown in Fig. 6.7 which can be readily demonstrated in a laboratory by applying an impulse voltage between point-plane electrodes with the point embedded in a transparent solid, e.g. plexiglass. The tree pattern shown in Fig. 6.7 was recorded by Cooper[9] with a $1/30 - \mu$sec impulse voltage of the same amplitude. After application of each impulse the channels were observed with a microscope and new channels were recorded. Not every impulse will produce a channel. The time required for this type of breakdown under alternating voltage will vary from a few seconds to a few minutes.

The tree-like pattern discharge is not limited specifically to the edge effect but may be observed in other dielectric failure mechanisms in which non-uniform field stresses predominate.

6.1.5 Thermal breakdown

When an insulation is stressed, because of conduction currents and dielectric losses due to polarization, heat is continuously generated within the dielectric. In general, the conductivity (σ) increases with temperature, conditions of instability are reached when the rate of heating exceeds the rate of cooling and the specimen may undergo thermal breakdown. The situation is illustrated graphically in Fig. 6.8 in which the cooling of a specimen is represented by the straight line and the heating at various field strengths by curves of increasing

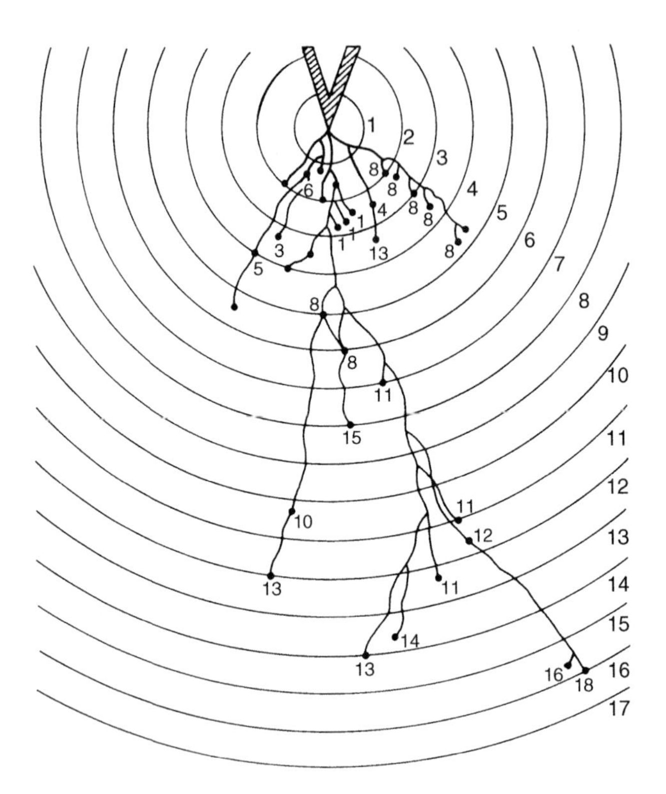

Figure 6.7 *Breakdown channels in plexiglass between point-plane electrodes. Radius of point = 0.01 in; thickness 0.19 in. Total number of impulses = 190. Number of channels produced = 16; (n) point indicates end of nth channel. Radii of circles increase in units of 10^{-2} in*

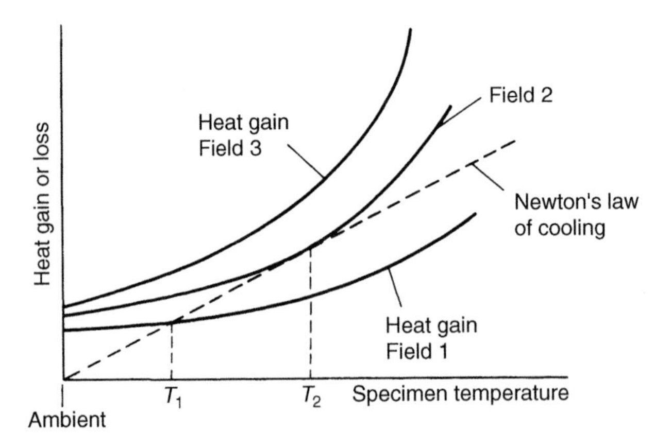

Figure 6.8 *Thermal stability or instability under different applied fields*

slope. Field (1) is in equilibrium at temperature T_1, field (2) is in a state of unstable equilibrium at T_2 and field (3) does not reach a state of equilibrium at all. To obtain the basic equation for thermal breakdown let us consider a cube of face area A (m^2) within dielectric. Assume that the heat flow in the x-direction is as shown in Fig. 6.9, then the

$$\text{heat flow across face (1)} = KA\frac{dT}{dx} \quad (K\text{-thermal conductivity}).$$

$$\text{heat flow across face (2)} = KA\frac{dT}{dx} + KA\frac{d}{dx}\left(\frac{dT}{dx}\right)\Delta x.$$

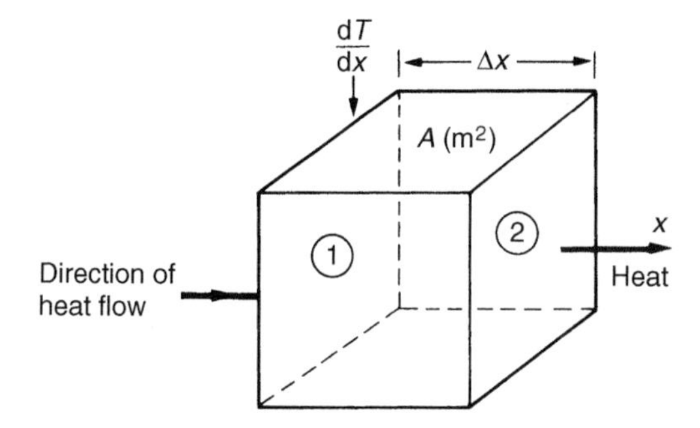

Figure 6.9 *Heat input and output, cubical specimen*

The second term represents the heat input into the block.

Hence

$$\text{heat flow/volume } K\frac{d}{dx}\left(\frac{dT}{dx}\right) = \text{div } (K \text{ grad } T).$$

The conservation of energy requires that heat input into the element must be equal to the heat conducted away, plus the heat used to raise the temperature T of the solid or

$$\text{heat generated} = \text{heat absorbed} + \text{heat lost to surroundings},$$

i.e.

$$C_v\frac{dT}{dt} + \text{div } (K \text{ grad } T) = \sigma E^2 \tag{6.7}$$

where C_v is the thermal capacity of the dielectric, σ is the electrical conductivity and in the case of alternating voltage the heat is generated primarily as a result of dipole relaxation and the conductivity is replaced by $\omega\varepsilon_0\varepsilon_r''$ where ε_0 represents permittivity of free space and ε_r'' the imaginary component of the complex relative permittivity of the material.

Calculation of the critical thermal situation involves the solution of eqn (6.7). In solving it, one assumes that a critical condition arises and the insulation properties are lost, when at some point in the dielectric the temperature exceeds a critical temperature T_c. The solution gives the time required to reach T_c for a given field and boundary condition. The equation cannot be solved analytically for the general case since C_v, K and σ may be all functions of temperature (T) and σ may also depend upon the applied field. We consider two extreme cases for the solution of eqn (6.7).

Case 1. This assumes a rapid build-up of heat so that heat lost to surroundings can be neglected and all heat generated is used in raising the temperature of the solid. We obtain an expression for 'impulse thermal breakdown' and eqn (6.7) reduces to

$$C_v \frac{dT}{dt} = \sigma E^2.$$

To obtain the critical field E_c, assume that we apply a ramp function field. Then

$$E = \left(\frac{E_c}{t_c}\right) t$$

and

$$\sigma E^2 = C_v \frac{dT}{dE}\frac{dE}{dt}.$$

For the conductivity, we can assume

$$\sigma = \sigma_0 \exp\left[-\frac{u}{kT}\right].$$

σ_0 is here the conductivity at ambient temperature T_0. Substituting for σ and rearranging, we get

$$\int_0^{E_c} \frac{t_c}{E_c}\frac{\sigma_0}{C_v} E^2 dE = \int_{T_0}^{T_c} \exp\left(\frac{u}{kT}\right) dT.$$

For the case when

$$u \gg kT$$

and

$$T_c > T_0 \quad (T_c - \text{ critical temperature})$$

the solution of the r.h.s. is

$$\int_{T_0}^{T_c} \exp\left(\frac{u}{kT}\right) dT \rightarrow T_0^2 \frac{k}{u} \exp\left(\frac{u}{kT_0}\right)$$

and that of the l.h.s. is

$$\int_0^{E_c} \frac{t_c}{E_c}\frac{\sigma_0}{C_v}E^2\mathrm{d}E \rightarrow \frac{1}{3}t_c\frac{\sigma_0}{C_v}E_c^2.$$

Therefore

$$E_c = \left[\frac{3C_v kT_0^2}{\sigma_0 u t_c}\right]^{0.5} \exp\left(\frac{u}{2kT_0}\right). \tag{6.8}$$

It is seen that reaching the critical condition requires a combination of critical time and critical field and that the critical field is independent of the critical temperature T_c due to the fast rise in temperature.

Case 2 concerns minimum thermal voltage, i.e. the lowest voltage for thermal breakdown. For this case we assume a thick dielectric slab that is constrained to ambient temperature at its surfaces by using sufficiently large electrodes as shown in Fig. 6.10.

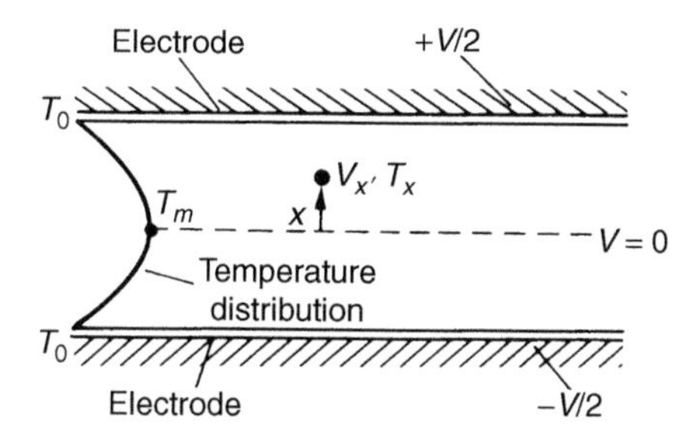

Figure 6.10 *Arrangement for testing a dielectric for minimum thermal breakdown voltage*

On application of voltage, after some time, a temperature distribution within the dielectric will be established with the highest temperature at the centre (T_1), that at the surface remaining at ambient temperature. On increasing the voltage to a new higher value, an equilibrium will be established at a higher central temperature (T_2). If the process is continued, a thermal runaway will eventually result as shown in Fig. 6.11.

To calculate the minimum thermal voltage, let us consider a point inside the dielectric distance x from the centre, and let the voltage and temperature at that point be V_x and T_x respectively. For this case we assume that all the heat generated in the dielectric will be carried away to its surroundings through the electrodes. Neglecting the term $C_v(\mathrm{d}T/\mathrm{d}t)$, eqn (6.6) becomes

$$\sigma E^2 = \frac{\mathrm{d}}{\mathrm{d}x}\left(K\frac{\mathrm{d}T}{\mathrm{d}x}\right).$$

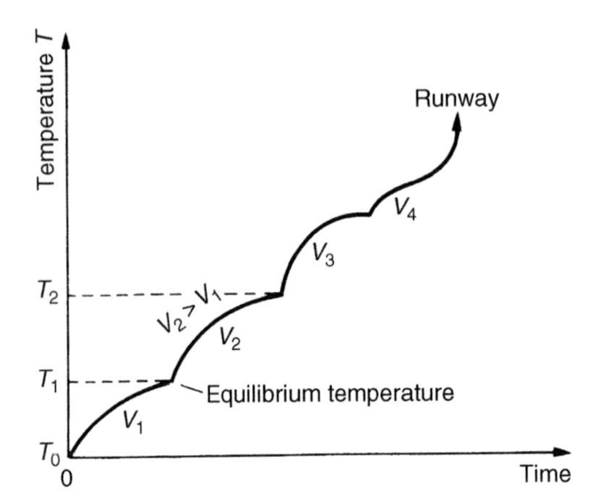

Figure 6.11 *Temperature–time relationship for slow thermal stressing under various applied voltages*

Using the relations of $\sigma E = j$ and $E = \partial V/\partial x$ (j-current density), and inserting in the above equation, we obtain

$$-j\frac{\partial V}{\partial x} = \frac{\mathrm{d}}{\mathrm{d}x}\left(K\frac{\mathrm{d}T}{\mathrm{d}x}\right).$$

Integrating to an arbitrary point x in the dielectric

$$-j\int_0^{V_x} \mathrm{d}V = \int_0^x \frac{\mathrm{d}}{\mathrm{d}x}\left(K\frac{\mathrm{d}T}{\mathrm{d}x}\right)\mathrm{d}x$$

$$-jV_x = K\frac{\mathrm{d}T}{\mathrm{d}x}$$

or

$$V_x\sigma\frac{\mathrm{d}V}{\mathrm{d}x} = K\frac{\mathrm{d}T}{\mathrm{d}x}.$$

Substituting for $\sigma = \sigma_0 \exp[-u/kT]$, and integrating from the centre of the dielectric to the electrode,

$$\int_0^{V_c/2} V_x \mathrm{d}V = \frac{K}{\sigma_0}\int_{T_0}^{T_c} \exp\left[\frac{u}{kT}\right]\mathrm{d}T$$

$$V_c^2 = 8\frac{K}{\sigma_0}\int_{T_0}^{T_c} \exp\left[\frac{u}{kT}\right]\mathrm{d}T \tag{6.9}$$

Equation (6.9) gives the critical thermal breakdown voltage, where T_c is the critical temperature at which the material decomposes and the calculation assumes that T_c corresponds to the centre of the slab. The voltage is independent of the thickness of the specimen, but for thin specimens the thermal breakdown voltage becomes thickness dependent and is proportional to the square root of the thickness tending asymptotically to a constant value for thick specimens. Under alternating fields the losses are much greater than under direct fields. Consequently the thermal breakdown strength is generally lower for alternating fields, and it decreases with increasing the frequency of the supply voltage. Table 6.1 shows thermal breakdown values for some typical dielectrics under alternating and direct voltages at 20°C. These results correspond to a thick slab of material.

The thermal breakdown is a well-established mechanism, therefore the magnitude of the product $\varepsilon \tan \delta$ which represents the loss is a very essential parameter for the application of insulation material.

Table 6.1 *Thermal breakdown voltages for some typical dielectrics* (20°*C*)

	Material	*Thermal voltage in MV/cm*	
		d.c.	*a.c.*
Crystals:	Mica muscovite	24	7–18
Rock salts		38	1.4
Quartz:	Perpendicular to axis	12 000	–
	Parallel to axis	66	–
	Impure	–	2.2
Ceramics:	H.V. steatite	–	9.8
	L.F. steatite	–	1.5
	High-grade porcelain	–	2.8
Organic materials:	Capacitor paper	–	3.4–4
	Ebonite	–	1.45–2.75
	Polythene	–	3.5
	Polystyrene	–	5
	Polystyrene at 1 MHz	–	0.05
	Acrylic resins		0.3–1.0

6.1.6 Erosion breakdown

Practical insulation systems often contain cavities or voids within the dielectric material or on boundaries between the solid and the electrodes. These cavities are usually filled with a medium (gas or liquid) of lower breakdown strength than the solid. Moreover, the permittivity of the filling medium is frequently lower than that of the solid insulation, which causes the field intensity in the cavity to be higher than in the dielectric. Accordingly, under normal working stress of the insulation system the voltage across the cavity may exceed the breakdown value and may initiate breakdown in the void.

Figure 6.12 shows a cross-section of a dielectric of thickness d containing a cavity in the form of a disc of thickness t, together with an analogue circuit. In the analogue circuit the capacitance C_c corresponds to the cavity, C_b corresponds to the capacitance of the dielectric which is in series with C_c, and C_a is the capacitance of the rest of the dielectric. For $t \ll d$, which is usually the case, and assuming that the cavity is filled with gas, the field strength across C_c is given by the expression

$$E_c = \varepsilon_r E_a$$

where ε_r is the relative permittivity of the dielectric.

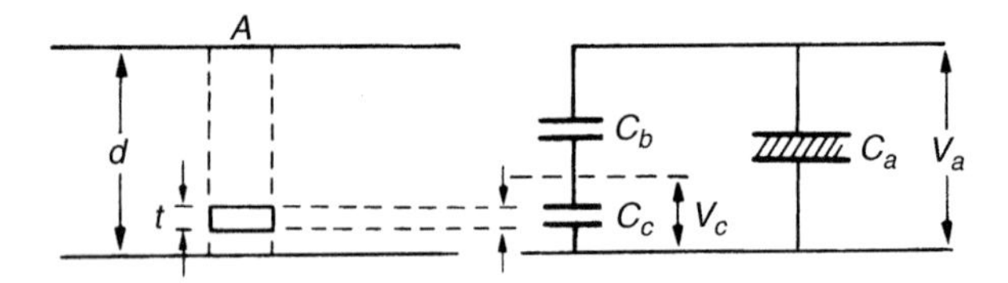

Figure 6.12 *Electrical discharge in cavity and its equivalent circuit*

For the simple case of a disc-shaped dielectric in solid shown in Fig. 6.12, the discharge inception voltage applied across the dielectric can be expressed in terms of the cavity breakdown stress. Assuming that the gas-filled cavity breakdown stress is E_{cb}, then treating the cavity as series capacitance with the healthy part of the dielectric we may write

$$C_b = \frac{\varepsilon_0 \varepsilon_r A}{d - t}$$

$$C_c = \frac{\varepsilon_0 A}{t}.$$

The voltage across the cavity is

$$V_c = \frac{C_b}{C_c + C_b} V_a = \frac{V_a}{1 + \dfrac{1}{\varepsilon_r}\left(\dfrac{d}{t} - 1\right)}.$$

Therefore the voltage across the dielectric which will initiate discharge in the cavity will be given by

$$V_{ai} = E_{cb}t\left\{1 + \frac{1}{\varepsilon_r}\left(\frac{d}{t} - 1\right)\right\}. \tag{6.10}$$

In practice a cavity in a material is often nearly spherical, and for such a case the internal field strength is

$$E_c = \frac{3\varepsilon_r E}{\varepsilon_{rc} + 2\varepsilon_r} = \frac{3E}{2} \tag{6.11}$$

for $\varepsilon_r \gg \varepsilon_{rc}$, where E is in the average stress in the dielectric, under an applied voltage V_a when V_c reaches breakdown value V^+ of the gap t, the cavity may break down. The sequence of breakdowns under sinusoidal alternating voltage is illustrated in Fig. 6.13. The dotted curve shows qualitatively the voltage that would appear across the cavity if it did not break down. As V_c reaches the value V^+, a discharge takes place, the voltage V_c collapses and the gap extinguishes. The voltage across the cavity then starts increasing again until it reaches V^+, when a new discharge occurs. Thus several discharges may take place during the rising part of the applied voltage. Similarly, on decreasing the applied voltage the cavity discharges as the voltage across it reaches V^-. In this way groups of discharges originate from a single cavity and give rise to positive and negative current pulses on raising and decreasing the voltage respectively. For measurements of discharges refer to Chapter 7.

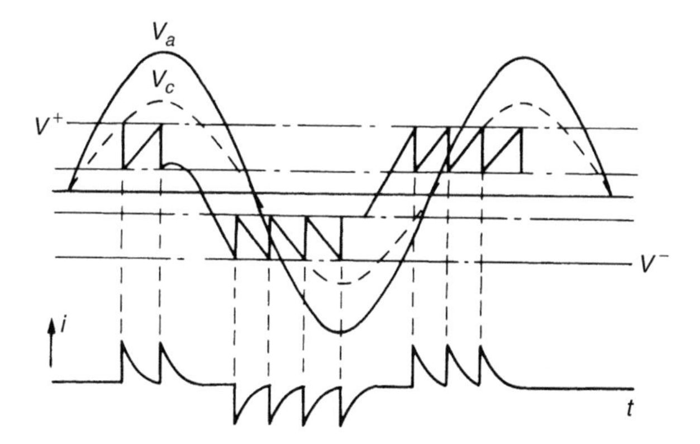

Figure 6.13 *Sequence of cavity breakdown under alternating voltages*

When the gas in the cavity breaks down, the surfaces of the insulation provide instantaneous cathode and anode. Some of the electrons impinging upon the anode are sufficiently energetic to break the chemical bonds of the insulation surface. Similarly, bombardment of the cathode by positive ions may cause damage by increasing the surface temperature and produce local thermal instability. Also channels and pits are formed which elongate through the insulation by the 'edge mechanism'. Additional chemical degradation may result from active discharge products, e.g. O_3 or NO_2, formed in air which may cause deterioration. Whatever is the deterioration mechanism operating, the net effect is a slow erosion of the material and a consequent reduction of the breakdown strength of the solid insulation.

When the discharges occur on the insulation surface, the erosion takes place initially over a comparatively large area. The erosion roughens the surface and slowly penetrates the insulation and at some stage will again give rise to channel propagation and 'tree-like' growth through the insulation.

For practical application it is important that the dielectric strength of a system does not deteriorate significantly over a long period of time (years). In practice, however, because of imperfect manufacture and sometimes poor design, the dielectric strength (e.g. in cables) decreases with the time of voltage application (or the life) and in many cases the decrease in dielectric strength (E_b) with time (t) follows the empirical relationship

$$tE_b^n = \text{const} \tag{6.12}$$

where the exponent 'n' depends upon the dielectric material, the ambient conditions, and the quality of manufacture. Figure 6.14 illustrates the case for

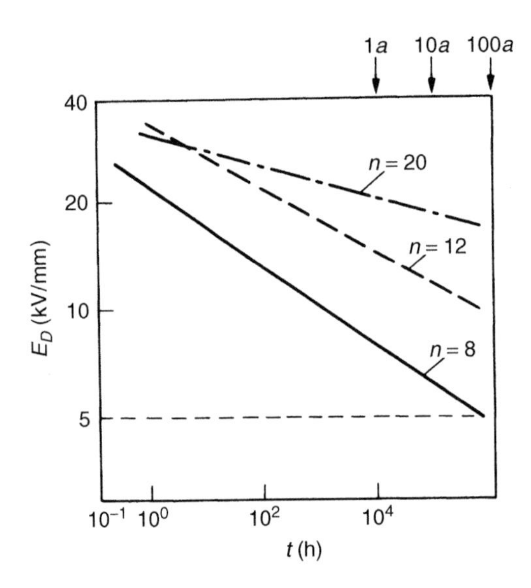

Figure 6.14 *Lifetime (t) stress relationship of polyethylene m.v. cables determined by different manufacturers*[10]

several m.v. polyethylene cables produced by different manufacturers. The breakdown strength has been plotted against time on a log–log scale.

In earlier years when electric power distribution systems used mainly paper-insulated lead-covered cables (PILC) on-site testing specifications called for tests under d.c. voltages. Typically the tests were carried out at 4 to 4.5 V_0. The tests helped to isolate defective cables without further damaging good cable insulation. With the widespread use of extruded insulation cables of higher dielectric strength, the test voltage levels were increased to 5–8 V_0. In the 1970s premature failures of extruded dielectric cables that were factory tested under d.c. voltage at specified levels were noted.[26] Hence on-site testing of cables under very low frequency (VLF), ~0.1 Hz, has been adopted. The subject has recently been reviewed by Gnerlich[10] and will be further discussed in Chapter 8.

6.1.7 Tracking

Tracking is the formation of a permanent conducting path, usually carbon, across a surface of insulation and in most cases the conduction path results from degradation of the insulation. For tracking to occur the insulation must contain some organic substance.

In an outdoor environment insulation will in time become covered with contaminant which may be of industrial or coastal origin. In the presence of moisture the contamination layer gives rise to leakage current which heats the surface and causes interruption in the moisture film; small sparks are drawn between the separating moisture films. This process acts effectively as an extension to the electrodes. The heat resulting from the small sparks causes carbonization and volatilization of the insulation and leads to formation of permanent 'carbon track' on the surface. The phenomenon of tracking severely limits the use of organic insulation in the outdoor environment. The rate of tracking depends upon the structure of the polymers and it can be drastically slowed down by adding appropriate fillers to the polymer which inhibit carbonization.

Moisture is not essential to tracking. The conducting path may arise from metallic dust; for example, in oil-immersed equipment with moving parts which gradually wear and deposit on the surface.

6.2 Breakdown in liquids

The general state of knowledge on the electrical breakdown in liquids is less advanced than is in case of gases or even solids. Many aspects of liquid breakdown have been investigated over the last decades, but the findings and conclusions of the many workers cannot be reconciled and so produce a general theory applicable to liquids, as the independent data are at variance and sometimes contradictory. The principal reason for this situation is the lack of comprehensive theory concerning the physical basis of the liquid state which would form the skeleton structure in which observations could be compared and related.

Comprehensive reviews of the published data on the subject have been made periodically and the more recent ones include the reviews of Lewis,[11] Sharbaugh and Watson,[12] Swann,[13] Kok,[14] Krasucki,[15] Zaky and Hawley,[16] and Gallagher.[17] The work falls broadly into two schools of thought. On the one hand there are those who attempt to explain the breakdown of liquids on a model which is an extension of gaseous breakdown, based on the avalanche ionization of the atoms caused by electron collision in the applied field. The electrons are assumed to be ejected from the cathode into the liquid by either a field emission, in which case they are assumed to tunnel out through the surface aided by the field, or by the field enhanced thermionic (Schottky's) effect. This type of breakdown mechanism has been considered to apply to homogeneous liquids of extreme purity, and does not apply to commercially exploited liquid insulation. Conduction studies in highly pure liquids showed that at low fields the conduction is largely ionic due to dissociation of impurities and increases linearly with the field strength. This conduction saturates at intermediate fields. At high field, as we approach breakdown, the conduction increases more rapidly and tends to be unstable. It is believed that this increased current results from electron emission at the cathode by one or both of the above mechanisms, and possibly by field aided dissociation of molecules in the liquid.

It has long been recognized that the presence of foreign particles in liquid insulation has a profound effect on the breakdown strength of liquids. In one approach it has been postulated[14] that the suspended particles are polarizable and are of higher permittivity than the liquid. As a result they experience an electrical force directed towards the place of maximum stress. With uniform field electrodes the movement of particles is presumed to be initiated by surface irregularities on the electrodes, which give rise to local field gradients. The accumulation of particles continues and tends to form a bridge across the gap which leads to initiation of breakdown.

The impurities can also be gaseous bubbles of lower breakdown strength than the liquid, in which case on breakdown of the bubble the total breakdown of the liquid may be triggered. A mathematical model for bubble breakdown has been proposed by Kao.[18]

6.2.1 Electronic breakdown

Both the field emission and the field-enhanced thermionic emission mechanisms discussed earlier have been considered responsible for the current at the cathode. Conduction studies in insulating liquids at high fields show that most experimental data for current fit well the Schottky-type equation (eqn (5.81)[19–Chapter 5]) in which the current is temperature dependent. Breakdown measurements carried out over a wide range of temperatures, however, show little temperature dependence. This suggests that the cathode process is field emission rather than thermionic emission. It is possible that the return of positive ions and particularly positively charged foreign particles to the cathode will cause local field enhancement and give rise to local electron emission.

Once the electron is injected into the liquid it gains energy from the applied field. In the electronic theory of breakdown it is assumed that some electrons gain more energy from the field than they lose in collisions with molecules. These electrons are accelerated until they gain sufficient energy to ionize molecules on collisions and initiate avalanche.

The condition for the onset of electron avalanche is obtained by equating the gain in energy of an electron over its mean free path to that required for ionization of the molecule.

$$eE\lambda = ch\upsilon \tag{6.13}$$

where E is the applied field, λ the electron mean free path, $h\upsilon$ the quantum of energy lost in ionizing the molecule and c an arbitrary constant.

Typical strengths for several highly pure liquids are included in Table 6.2.

Table 6.2 *Electric strength of highly purified liquids*

Liquid	*Strength (MV/cm)*
Hexane	1.1–1.3
Benzene	1.1
Good oil	~1.0–4.0
Silicone	1.0–1.2
Oxygen	2.4
Nitrogen	1.6–1.88

The electronic theory satisfactorily predicts the relative magnitude of breakdown strength of liquids, but the observed formative time lags are much longer than predicted by electronic theory.[18]

6.2.2 Suspended solid particle mechanism

Solid impurities may be present in the liquid either as fibres or as dispersed solid particles. Let us consider a spherical particle of radius r and permittivity

ε to be suspended in dielectric liquid of permittivity ε_{liq}. Then in a field the particle will become polarized and it will experience a force given by

$$F_e = \varepsilon_{liq.} r^3 \frac{\varepsilon - \varepsilon_{liq.}}{\varepsilon + 2\varepsilon_{liq.}} E \text{ grad } E. \tag{6.14}$$

This force is directed towards a place of maximum stress if $\varepsilon > \varepsilon_{liq.}$ but for bubbles $\varepsilon < \varepsilon_{liq.}$, it has the opposite direction. The force given by eqn (6.14) increases as the permittivity of the suspended particle (ε) increases, and for a conducting particle for which $\varepsilon \to \infty$ the force becomes

$$F_e = F_\infty = r^3 E \text{ grad } E. \tag{6.15}$$

Thus the force will urge the particle to move to the strongest region of the field.

In a uniform field gap or sphere gap of small spacing the strongest field is in the uniform region.

In this region grad E is equal to zero so that the particle will remain in equilibrium there. Accordingly, particles will be dragged into the uniform field region. If the permittivity of the particle is higher than that of the medium, then its presence in the uniform field region will cause flux concentration at its surface. Other particles will be attracted into the region of higher flux concentration and in time will become aligned head to tail to form a bridge across the gap. The field in the liquid between the particles will be enhanced, and if it reaches critical value breakdown will follow.

The movement of particles by electrical force is opposed by viscous drag, and since the particles are moving into the region of high stress, diffusion must also be taken into account. For a particle of radius r slowly moving with a velocity v in a medium of viscosity η, the drag force is given by Stokes relation

$$F_{drag} = 6\pi r \eta v(x) \tag{6.16}$$

Equating the electrical force with the drag force ($F_e = F_{drag}$) we obtain

$$v_E = \frac{r^2 E}{6\pi\eta} \frac{dE}{dx} \tag{6.17}$$

where v_E is the velocity of the particle towards the region of maximum stress. If the diffusion process is included, the drift velocity due to diffusion will be given by the equation

$$v_d = -\frac{D}{N}\frac{dN}{dx} = -\left(\frac{kT}{6\pi r\eta}\right)\frac{dN}{Ndx}. \tag{6.18}$$

The relation on the r.h.s. of the equation follows from the Stokes–Einstein relation $D = kT/67\pi r\eta$, where k is Boltzmann's constant and T is the absolute temperature. Equating v_E with v_d gives

$$\frac{r^2}{6\pi r\eta}E\frac{\mathrm{d}E}{\mathrm{d}x} = -\left(\frac{kT}{6\pi r\eta r N}\right)\frac{\mathrm{d}N}{\mathrm{d}x}. \tag{6.19}$$

This introduces breakdown strength dependence in time on concentration of particles N, their radii and the liquid viscosity. The critical value of transverse field $E(x)$, the equilibrium value above which breakdown will occur sooner or later, can be obtained from integration of eqn (6.19).

$$\left[\frac{r^2E^2}{2}\right]_{E=E_{(\infty)}}^{E=E_{(x)}} = \left[-\frac{kT}{r}\ln N\right]_{N=N_{(\infty)}}^{N=N_{(x)}}$$

$$\frac{N_{(x)}}{N_{(\infty)}} = \exp\left[r^3\frac{\{E_{(x)}^2 - E_{(\infty)}^2\}}{2kT}\right]. \tag{6.20}$$

If the increase in the electrostatic energy when the particles drift towards a place of maximum stress is much smaller than their kinetic energy, i.e. $r^3\left\{E_{(x)}^2 - E_{(\infty)}^2\right\} \ll 2kT$, the life of the insulation is infinite. The criterion for breakdown resulting from movement of particles towards the high stress region corresponds to the condition

$$r^3\left[E_{(x)}^2 - E_{(\infty)}^2\right] = 2kT. \tag{6.21}$$

If we consider the case where the initial non-uniformity of field is caused by a hemispherical hump on the electrode, discussed earlier in Chapter 4, and assume that an applied field E_0 will lead to breakdown after a long time of application, then the maximum stress at the tip of the sphere is $3E_0$, or in general the maximum stress is gE_0, where g is a geometrical factor. Then eqn (6.21) can be written as

$$r^3[g^2 - 1]E_0^2 = \tfrac{1}{4}kT. \tag{6.22}$$

For $g = 3$ we obtain

$$r^3E_0^2 = \tfrac{1}{4}kT. \tag{6.23}$$

A more complete theory gives a relation which takes into account the permittivities and is of the form

$$\frac{\varepsilon - \varepsilon_{\text{liq.}}}{\varepsilon + 2\varepsilon_{\text{liq.}}}r^2E_0^2 = \frac{1}{4}kT.$$

Equation (6.23) gives a breakdown strength E_0 after a long time as a function of the size of the suspended impurities. This relationship has been checked experimentally and reasonable agreement has been obtained with calculations.

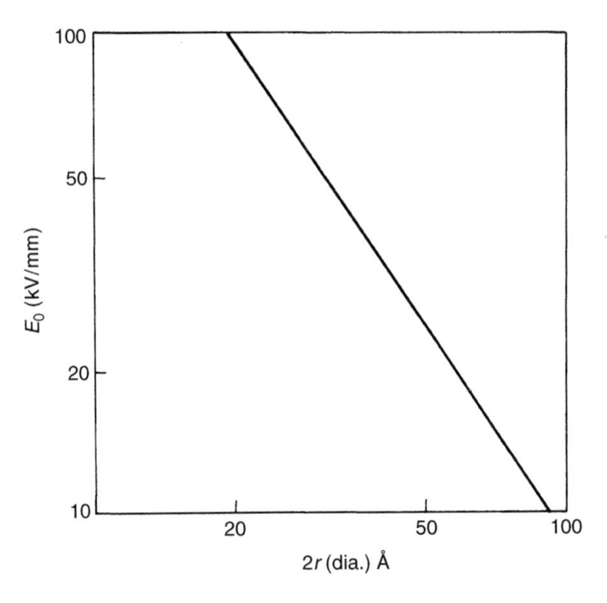

Figure 6.15 *Breakdown strength E_0 after a long duration of time as a function of the diameter $2r$ of foreign particles of high permittivity, with $T = 300$ K*[18]

Figure 6.15 shows a plot of eqn (6.23) for a range of sizes up to 50 A in radius at temperature $T = 300\,\text{K}$, for the case where $\varepsilon_{\text{liq.}} \ll \varepsilon$.

6.2.3 Cavity breakdown

Insulating liquids may contain gaseous inclusions in the form of bubbles. The processes by which bubbles are formed include:

(i) gas pockets on the electrode surface,
(ii) changes in temperature and pressure,
(iii) dissociation of products by electron collisions giving rise to gaseous products,
(iv) liquid vaporization by corona-type discharges from points and irregularities on the electrodes.

The electric field in a spherical gas bubble[18] which is immersed in a liquid of permittivity $\varepsilon_{\text{liq.}}$ is given by:

$$E_b = \frac{3E_0}{\varepsilon_{\text{liq.}} + 2} \tag{6.24}$$

where E_0 is the field in the liquid in the absence of the bubble. When the field E_b becomes equal to the gaseous ionization field, discharge takes place which will lead to decomposition of the liquid and breakdown may follow. Kao[18] has developed a more accurate expression for the 'bubble' breakdown field strength which is of the form

$$E_0 = \frac{1}{(\varepsilon_1 - \varepsilon_2)} \left\{ \frac{2\pi\sigma(2\varepsilon_1 + \varepsilon_2)}{r} \left[\frac{\pi}{4} \sqrt{\left(\frac{V_b}{2rE_0} \right) - 1} \right] \right\}^{1/2} \tag{6.25}$$

where σ is the surface tension of the liquid, ε_1 and ε_2 are the permittivities of the liquid and the bubble respectively, r is the initial radius of the bubble (initially spherical, which is assumed to elongate under the influence of the field), and V_b is the voltage drop in the bubble. This expression indicates that the critical electric field strength required for breakdown of liquid depends upon the initial size of the bubble which is affected by the external pressure and temperature. A strong dependence of liquid breakdown strength upon the applied hydrostatic pressure has been observed experimentally.[22]

Commercial insulating liquids cannot readily be subjected to highly elaborated purification treatment, and the breakdown strength will usually depend upon the nature of impurities present.

6.2.4 Electroconvection and electrohydrodynamic model of dielectric breakdown

The importance of electroconvection in insulating liquids subjected to high voltages was not appreciated until recently. Most of the work comes from Felici and his coworkers.[19,20,21,22] In highly purified dielectric liquids subjected to high voltage, electrical conduction results mainly from charge carriers injected into the liquid from the electrode surface. The resulting space charge gives rise to Coulomb's force, which under certain conditions causes hydrodynamic instability yielding convecting current. It follows that whenever conduction in a fluid is accompanied by a significant space charge formation, convection motion is very likely to occur. Lacroix *et al.*[19] have studied the conditions under which turbulent motion sets in. Using parallel plate electrodes and controlled injection current, they showed that the onset of instability is associated with a critical voltage. They observed that as the applied voltage is increased near the critical voltage the motion at first exhibits a structure of hexagonal cells. With a further increase in voltage the motion becomes turbulent. Thus the interaction between electric field and space charge gives rise to forces creating an eddy motion of the liquid. It has been shown that at voltages close to breakdown the speed of this motion approaches a value given by $\sqrt{\varepsilon/\rho}/E$ where ε is the permittivity of the liquid, ρ the specific mass and E the electric field strength. In liquids the ratio of this speed to ionic drift velocity (KE), K being the mobility, $M = \sqrt{\varepsilon/\rho}/K$, is always larger than unity and the ratio sometimes is very much larger than unity (see Table 6.3). M is considered to play a dominant role in the theory of electroconvection.

Thus, the charge transport will be largely by liquid motion and not by ionic drift. The key condition for the instability onset is that local flow velocity $u(=\sqrt{\varepsilon/\rho}/E)$ exceeds the ionic drift velocity ($u > KE$).

Table 6.3 $M = \sqrt{\varepsilon/\rho}/K$

Medium	*Ion*	*Relative permittivity*	*M number*
Methanol	H^+	33.5	4.1
Ethanol	Cl^-	25	26.5
Nitrobenzene	Cl^-	35.5	22
Propylene carbonate	Cl^-	69	51
Transformer oil	H^+	2.3	~200
Air N.T.P.	$O_2{}^-$	1.0	2.3×10^{-2}

The experimental values for M for various fluid media and common ions obtained by Lacroix *et al.*[19] are included in Table 6.3. The table also contains the value for air at NTP. It is seen that in this case $M \ll 1$ and the rate of electroconvection is negligible. Experiments show that electroconvection is prevalent in all experimental settings in dielectric liquids subjected to electric fields irrespective of the gap geometries, provided the applied voltage is high enough. This is true even in thoroughly deionized liquids because of the adequate supply of ions by the high field processes at the electrodes.

Cross *et al.*,[23] have studied electric stress-induced motion in transformer oil under d.c. and 60 Hz stresses. Using high-speed schlieren photography, they found that the turbulent motion was due to injection of positive charges from one electrode. This was confirmed for both d.c. and 60 Hz stresses. They also observed that the delay time in the onset of instability is related to the condition for the injection or creation of charges at the electrode surface. The time delay was found to decrease rapidly with increasing the field strength ranging from a few seconds at 10^6 V/m to a few milliseconds at 6×10^6 V/m. Also as the temperature of the liquid increased, the time delay for the given field decreased. Under 60 Hz voltage the time delay was found to reach a minimum value approximately 4 msec, which is to be expected. A 60 Hz wave requires 4.17 msec to reach the peak. From these observations and calculations Cross *et al.* concluded that under these conditions instability occurs when the injection strength, which is the ratio of the space charge field to the applied field, reaches a large enough value for a critical voltage to develop across the space charge layer within one half-cycle period. The lowest value of the critical voltage occurs where space charge limited conditions prevail at the injecting electrode.

6.4 Static electrification in power transformers

Static electrification (SE) in transformers is an interfacial phenomenon, which involves oil, paper and transformer board. Its physical mechanism involves a source of charge and region of excessive charge accumulation. Extensive investigations about this phenomena have been made during recent years.[25] When oil is forced through the tank and coolers, it acquires an electrostatic charge, i.e. it contains an equal number of positive and negative ions. When the oil passes the paper and solid insulation in the windings, the insulation becomes negatively charged and the oil positively charged with the charge separation occurring at the oil-insulated interface (Fig. 6.16). The earliest reports on this phenomenon were in the 1970s from Japan, where number

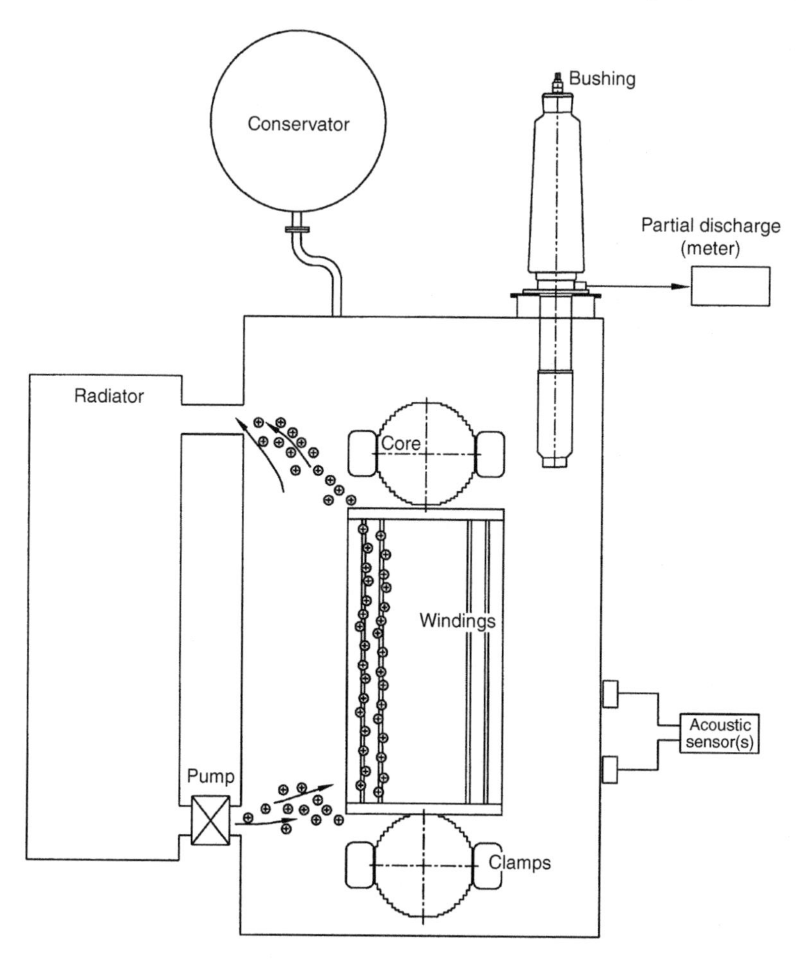

Figure 6.16 *Schematic of flow electrification density in transformers*

of h.v. large transformer failures occurred. And later quite a few SE-related incidents were also reported in the USA and other countries. It is believed that transformers of large rating (e.g. >100 MVA) are most likely affected by SE because they possess greater amounts of insulation and require larger oil flow volumes than transformers of smaller ratings. As different oils have different electrostatic charging tendencies (ECT), oil additives might be a way to reduce oil ECT. As an alternative to the additive, used oil can be regenerated because new oil exhibits a lower ECT than aged oil. On the other hand, operation practices are also of great importance. SE incidents can be caused by poor operating practices such as increasing forced oil cooling capacity beyond manufacturer's recommendations, or having more forced oil cooling in operation than the load on the transformer justifies.

References

1. D.M. Taylor and T.J. Lewis. *J. Phys.* **D4** (1971), p. 1346.
2. A. von Hippel. *Ergebn. Exakt. Naturw.* **14** (1935), p. 79.
3. H. Fröhlich. *Proc. Roy. Soc.* **160** (1937), p. 230; **A188** (1947), pp. 521, 532.
4. R. Stratton. *Progress in Dielectrics*, Vol. 3, p. 235. Haywood, London, 1961.
5. H. Fröhlich and B.V. Paranjape. *Proc. Phys. Soc. London* **B69** (1956), p. 21.
6. J.J. O'Dwyer. *The Theory of Electrical Conduction and Breakdown in Solid Dielectrics.* Clarendon Press, Oxford, 1973.
7. F. Seitz. *Phys. Rev.* **73** (1979), p. 833.
8. K.H. Stark and C.G. Garton. *Nature*, London **176** (1955), p. 1225.
9. R. Cooper. *Int. J. of Elec. Eng. Education* **1** (1963), p. 241.
10. H.R. Gnerlich. Field Testing of HV Power Cables Under VLF Voltages. *IEEE Electrical Insulation Magazine*, Vol. 11, No. 5, 1995, p. 13.
11. T.J. Lewis. *Progress in Dielectrics*, Vol. 1. Haywood, London; Wiley, New York, 1959.
12. A.H. Sharbaugh and P.K. Watson, *Progress in Dielectrics*, Vol. 4, 1962.
13. D.W. Swann. *Brit. J. Appl. Phys.* **135** (1962), p. 208.
14. J.A. Kok. *Electrical Breakdown in Insulating Liquids*. Philips Tech. Library, 1961.
15. Z. Krasucki. Breakdown of commercial liquid and liquid solid dielectrics, in *High Voltage Technology* (Alston), p. 129. Oxford University Press, 1968.
16. A.A. Zaky and R. Hawley. *Conduction and Breakdown in Mineral Oils*. Pergamon Press, Oxford, 1973.
17. T.J. Gallagher. *Simple Dielectric Liquids*. Clarendon Press, Oxford, 1975.
18. K.C. Kao. *Trans. AIEEE Elec. Ins.* Vol. E1-11 (1976), pp. 121–128.
19. J.C. Lacroix, P.A. Hen and E.J. Hopfinger. *J. Fluid Mech.* **69** (1975), p. 539.
20. N.J. Felici. *Direct Current* **2** (1971), p. 147.
21. N.J. Felici and J.C. Lacroix. *J. Electrost.* **5** (1978), p. 135.
22. J.K. Nelson. Dielectric Fluids in Motion. *IEEE Electrical Insulation Magazine*, Vol. 10, No. 11 994, pp. 16–28.
23. J.D. Cross, M. Nakans and S. Savanis. *J. Electrost.* **7** (1979), p. 361.
24. CIGRE Report, JWG 12/15.13, Task Force 02, August 1997.
25. Peyraque *et al.* Static Electrification and Partial Discharges induced by Oil Flow in Power Transformers. *IEEE Transactions Dielectric and Elec.* Insul. Vol. 2, No. 1, 1995, pp. 40–45.
26. G.S. Eager, Jr, *et al.* Effect of DC Testing Water Tree Deteriorated Cables and a Primary Evaluation of VLF as alternative. *IEEE Transaction on Power Delivery*, Vol. 7, N, July 1992.

Chapter 7

Non-destructive insulation test techniques

This chapter is dedicated to test techniques, which provide information about the quality of insulation systems which form part of an equipment or apparatus. The tests as described here take advantage of well-known or desirable electrical properties of either a specific dielectric material or an insulation system as formed by a combination of different (fluid and/or solid) materials. Although also mechanical or chemical tests are often applied to assess the insulation quality, such tests are not taken into account.

Tests related to electrical properties are usually based on measurements of insulation resistance or (d.c.) resistivity as well as capacitance and loss factors, which are dependent on the frequency of the a.c. voltages applied. As the techniques for the measurement of d.c. resistances are well known to electrical engineers, they are neglected. Another group of non-destructive tests on insulation systems is based on the detection and quantification of 'partial discharges' or PDs, a measurement technique already applied over the past five decades, but still a topic of research and increasing application. This latter topic is strongly tied with Chapter 5, gaseous discharges and gas breakdown.

The tests related to electrical properties are often assumed to be quite simple and 'standard'. This is true if only tests with d.c. and power frequency are considered. During recent years, however, it was recognized that the dynamic electrical properties are very essential to quantify or at least to indicate the ageing phenomena of insulation and thus to use the results as an essential diagnostic tool for equipment already in service for a long time. The individual, partly very specific methods used to quantify the changes of the dielectric properties can only be mentioned and not be described in detail in this chapter. It is essential to introduce a short description of the basic of 'dielectric relaxation' processes. We therefore, start, this chapter with an introduction to 'dynamic electrical properties'.

7.1 Dynamic properties of dielectrics

In contrast to Chapters 5 and 6, which primarily dealt with the maximum resistance to destructive breakdown of gaseous, liquid and solid dielectrics in high electrical fields, we will now examine the situation occurring when such materials are exposed to much lower field stresses, thereby avoiding any destructive or non-linear effects. Gases are generally not referred to as 'materials', because the distance between the adjacent molecules is so large and the number of atoms or molecules per unit volume is so low, that they are not able to withstand mechanical forces. However, gases are 'dielectrics' in the sense of (electrical) insulators, and are used to prevent the flow of current.

Inherent in any dielectric material within an electric field are the well-known effects of 'dielectric polarization', which are well documented in the literature (see, e.g., A.K. Jonscher, 1983,[(1)]* W. von Münch, 1987[(2)]). In understanding these effects, it is useful to review some of the fundamental aspects of dielectric polarization.[(1)]

At an atomic level, all matter consists of negative and positive charges balancing each other in microscopic as well as in more macroscopic scales, in the absence of any unipolar charge having been deposited within the matter before. Macroscopically, some local space charge may be present, but even in that condition an overall charge neutrality exists. While such local space charges may have been produced by, e.g., thermal excitation or through the absorption of light, an equal number of positively charged ions and detached electrons will be present, and these processes of ionization and recombination are usually in equilibrium.

As soon as the matter is stressed by even a very weak 'macroscopic' or external electric field as, e.g., generated by a voltage across some electrodes between which the dielectric is deposited, very different kinds of dipoles become excited even within atomic scales. Local charge imbalance is thus 'induced' within the neutral species (atoms or molecules) as the 'centres of gravity' for the equal amounts of positive and negative charges, $\pm q$, become separated by a small distance d, thus creating a dipole with a 'dipole moment' $p = qd$, which can also be related to the 'local' electric field E acting in close vicinity of the species. The relation between the dipole moment, p, and the electric field, E, is given by $p = \alpha E$, where α is the 'polarizability' of the material under consideration. Note that p, d and E are vectors, which is not marked here. As the distance d will be different for different materials, so is their polarizability. Due to chemical interactions between dissimilar atoms forming molecules, many molecules will have a constant distance d between the charge centres thus forming 'permanent dipole moments', which, however, are generally distributed irregularly within the matter as long as no external field is applied. (Note that any kind of 'permanent polarization' such as that occurring in electrets or ferroelectrics is not considered here.) The macroscopic effect of the 'polarizability' of individual materials is ultimately manifested in a general relation between the *macroscopic polarization* **P** and the number

* Superscript numbers are to references at the end of the chapter.

of the polarized species N per unit volume of the matter. These relationships are quite well known, but not treated here.

The following highlights of the polarization processes should be sufficient to gain an understanding of the main effects producing polarization. *Electronic polarization* is effective in every atom or molecule as the centre of gravity of the electrons surrounding the positive atomic cores will be displaced by the action of the electric field. This process is extremely fast and thus effective up to optical frequencies. *Ionic polarization* refers to matter containing molecules which will form ions, which are not separated by low electric fields or working temperatures. *Dipolar polarization* belongs to matter containing molecules with permanent dipole moments, the local distribution of which is governed statistically due to action of thermal energies. Under the influence of E, the dipoles will only partly be oriented so that a linear dependency of $\mathbf{P}$ with E can still be assumed. Ionic and dipolar polarization are also quite fast effects and can follow a.c. frequencies up to GHz or MHz. *Interfacial polarization* is effective in insulating materials composed from different dielectric materials such as oil-impregnated paper. The mismatch of the products of permittivity and conductivity for the different dielectrics forces moveable charges to become attached on interfaces. This phenomenon is quite often very slow and in general active up to power frequencies. Finally, *polarization by hopping charge carriers,*[(1)] a mechanism more recently postulated,[(3),(4)] may occur.[(3),(4)] This type of polarization process is based on the well-known hopping processes of electronic charges in amorphous and disordered non-metallic solids, in which direct current conduction generally takes place by the hopping of ions. A prerequisite of d.c. conduction is the presence of a continuous connected network of hopping sites, so that the charges are able to traverse the physical dimension of the dielectric. If, however, the matter is very strongly disordered, the normal concept of band conduction by free charge carriers must be replaced by very localized sites, which are surrounded by very high potential wells which cannot be surpassed by electrons.

In summary, dielectric polarization is the result of a relative shift of positive and negative charges in the matter under consideration. This shift is produced by an electric field, provoking either 'induced polarization' of individual atoms and/or ions, an orientation of any permanent dipoles, the build-up of charges at interfaces between quite different dielectrics, or the creation of dipoles at localized hopping sites. During all of these processes, the electric field is therefore not able to force the charges to escape from the matter, which would lead to electric conduction.

For any matter, which is isotropic and homogeneous at least in macroscopic scales, we may therefore write the following general relation between the (macroscopic) polarization $\mathbf{P}$ and the field $\mathbf{E}$ as:

$$\mathbf{P} = \varepsilon_0 \chi \mathbf{E} + \text{higher order terms in } \mathbf{E}. \tag{7.1}$$

Here, ε_0 is the permittivity of free space (=8.85419×10^{-12} (As/Vm)) and χ is the *susceptibility* of the matter, which is a dimensionless number and with a value of zero for vacuum and/or free space. From equation (7.1) we see that χ accounts for all kinds of polarization processes acting within the dielectric. Note, that the vectors **E** and **P** have the same direction in isotropic materials. The additional higher order terms in equation (7.1) can be neglected under the assumption that the dielectric response of the material remains linear, i.e. as long as the magnitude of the exciting electric field is not too large.

Dynamic properties of dielectrics can be defined and also measured in the frequency or time domain. We will start to define the properties in the time domain and proceed to the frequency domain definitions.

7.1.1 Dynamic properties in the time domain

In any vacuum-insulated electrode arrangement, the 'dielectric displacement' or 'dielectric flux density' (or 'electrical induction') **D** is proportional to the applied electric field **E**. The relation between the two quantities is

$$\mathbf{D} = \varepsilon_0 \mathbf{E}$$

or, if the electric field is generated by a time-varying voltage,

$$\mathbf{D}(t) = \varepsilon_0 \mathbf{E}(t) \tag{7.2}$$

where $\varepsilon_0 = 8.85419 \times 10^{-12}$ As/Vm is the permittivity of free space or vacuum, a number with dimensions converting the unit for electric field (V/m) to that of area charges (As/m^2). As **E** is a vector, **D** is also a vector usually assumed to exist within the space in which the electrostatic field is present. One should note, however, that the electric displacement D represents the (positive and negative) electric charges per unit area as deposited at the surface of the electrodes which are the origin – sources and sinks – of all electric field lines. The origin of D and E is usually provided by a voltage source connected to the electrodes of the electrode arrangement under consideration. If the voltage is time-dependent as already assumed in eqn (7.2), both D and E are of identical time dependency with no time delay between their magnitudes. The so-called 'displacement current' released from the voltage source as necessary to maintain the area charge density at the electrodes is then only governed by $\mathrm{d}Q/\mathrm{d}t$, if Q is the sum or integral of all charges deposited on each of the electrodes.

If the vacuum is replaced by any kind of isotropic dielectric material, the displacement is obviously increased by the (macroscopic) *polarization* **P** of this dielectric, which was already defined in eqn (7.1), resulting in:

$$\mathbf{D}(t) = \varepsilon_0 \mathbf{E}(t) + \mathbf{P}(t). \tag{7.3}$$

From equation (7.1) and the explanations previously given, we know that P will be a vector in the direction of E, as isotropic materials are assumed. (Further on, we can therefore avoid to indicate this by bold letters.) The time dependency of P, however, will not be the same as that of E, as the different polarization processes will have different delays with respect to the appearance of E. This delay is obviously caused by the time-dependent behaviour of the susceptibility $\chi \ldots$

The time delay between $E(t)$ and $P(t)$ may best be understood with the following considerations. Let us assume, that the (macroscopic) electric field E within the matter is excited by an ideal voltage step at time $t = 0$ and that its time evolution is marked by E_0. The dielectric material is then characterized by its *susceptibility* $\chi(t)$ as a response in the time domain. This parameter covers the formation and evolution of the different kinds of polarization processes including those that develop within extremely short times (e.g. electronic polarization) as well as those which are much slower or even very slow (e.g. interfacial polarization). For $t \leq 0$, the magnitude of susceptibility is still zero.

Figure 7.1 illustrates the situation. For this special case of excitation, P becomes

$$P_0(t)/E_0 = \varepsilon_0 \chi(t) 1(t) \tag{7.4}$$

where $\chi(t)$ and $P_0(t)$ represent 'step response (SR) functions'. The factor $1(t)$ is used to indicate the unit step.

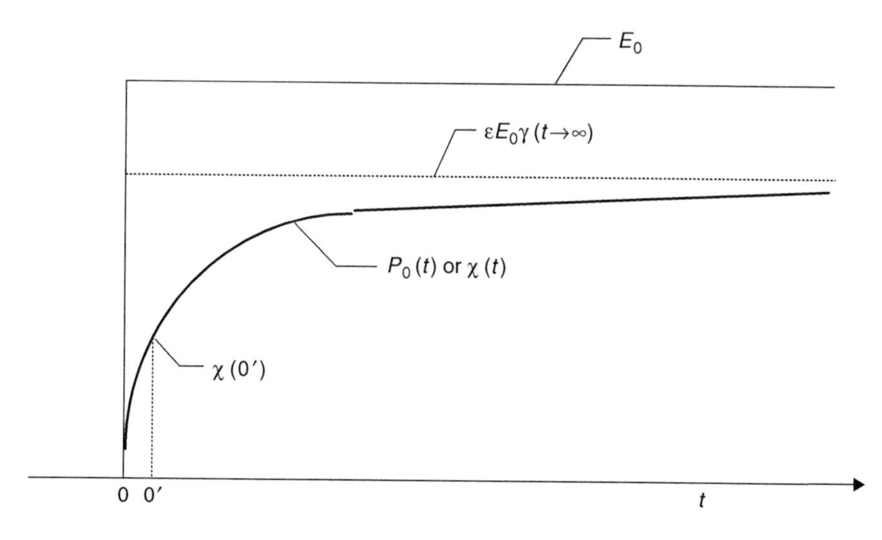

Figure 7.1 *Polarization of a dielectric material subjected to an electrical step field of magnitude* E_0

As known from general circuit theory, the time-dependent polarization $P(t)$ can be calculated for any other time-dependent excitation $E(t)$ of the system since the special solution for the SR is already known. This can be done using 'Duhamel's integral' or by convolution in the time domain. As not everybody may be familiar with Duhamel's integral, we first display it in general terms.

If $a(t)$ is a general force acting on a system or network and $a(t) = 0$ for $t < 0$, and if $b(t)$ is the effect of this force at any place of the system, and if $b_1(t)$ is the known effect at that place for the unit step of $a(t) = 1(t)$, then the following result applies for $t \geq 0$:

$$b(t) = \frac{\mathrm{d}}{\mathrm{d}t}\int_0^t a(z)b_1(t-z)\,\mathrm{d}z = b_1(0)a(t) + \int_0^t a(z)\frac{\mathrm{d}b_1(t-z)}{\mathrm{d}t}\,\mathrm{d}z. \tag{7.5}$$

Identifying now 'a' with 'E', 'b' with 'P' and 'b_1' with the step response function of eqn (7.4), for an arbitrary time-dependent electrical field $E(t)$ we now obtain the polarization $P(t)$ as

$$P(t) = \varepsilon_0\chi(0)E(t) + \varepsilon_0\int_0^t E(\tau)\frac{\mathrm{d}\chi(t-\tau)}{\mathrm{d}t}\,\mathrm{d}\tau. \tag{7.6}$$

In this equation, the derivative of the susceptibility $\chi(t)$ appears. This function is given by

$$f(t) = \mathrm{d}\chi(t)/\mathrm{d}t \tag{7.7}$$

and is the *dielectric response (or relaxation) function* of the dielectric material. This function is, as shown by Fig. 7.1, a monotonically decaying function for time scales accessible to usual measurements. Thus eqn (7.6) may be rewritten as

$$P(t) = \varepsilon_0\chi(0)E(t) + \varepsilon_0\int_0^t E(\tau)f(t-\tau)\,\mathrm{d}\tau. \tag{7.6a}$$

The first term of this equation is zero, since when a true ideal step excitation is assumed, $\chi(0) = 0$. For actual measurements, however, the fast or very fast polarization phenomena cannot be defined and then this term will be finite. This fact can be taken into account by finding the actual value of this term for a delayed time instant $t = 0'$ (Fig. 7.1) and then quantifying this value by $\chi(0')$. This represents a value effective for high or very high frequencies. More detailed explanation of this is given in section 7.1.3.

The polarization $P(t)$ is not an observable magnitude by itself, but it produces a main part of the displacement current in a test object. Up to now, we have not yet considered any conductivity of the dielectric, which is not involved in polarization. As already postulated by Maxwell in 1891, an electrical field $E(t)$ applied to a dielectric generates a current density $j(t)$, which can be written as a sum of conduction and total displacement current:

$$j(t) = \sigma_0 E(t) + \frac{\mathrm{d}D(t)}{\mathrm{d}t}. \tag{7.8}$$

Here σ_0 *represents the 'pure' or effective d.c. conductivity* of the material. Using the relations (7.3), (7.6a) and (7.8), the current density can now be expressed as:

$$\begin{aligned} j(t) &= \sigma_0 E(t) + \varepsilon_0(1 + \chi(0'))\frac{\mathrm{d}E(t)}{\mathrm{d}t} + \varepsilon_0 \frac{\mathrm{d}}{\mathrm{d}t}\int_{0'}^{t} f(t-\tau)E(\tau)\,\mathrm{d}\tau \\ &= \sigma_0 E(t) + \varepsilon_0(1 + \chi(0'))\frac{\mathrm{d}E(t)}{\mathrm{d}t} + \varepsilon_0 f(0')E(t) + \varepsilon_0 \int_{0'}^{t} \frac{\mathrm{d}f(t-\tau)}{\mathrm{d}t}E(\tau)\,\mathrm{d}\tau. \end{aligned} \tag{7.9}$$

The factor $(1 + \chi(0'))$ in these equations is the real part of the relative permittivity (ε_r) for 'high' frequencies, which is already effective at a time instant $0'$ as defined before. These equations are thus the basis for the experimental measurement of the dielectric response function $f(t)$.

Determination of the dielectric response function from polarization and depolarization (relaxation) currents

Together with a known geometry, the current densities can easily be converted to currents in test equipment or a test cell, as shown later. Therefore we may still use eqn (7.9) for further considerations related to currents.

If a step voltage at time $t = 0$ is applied and maintained for a long time (minutes, hours), a **polarization** (or charging, absorption) *current* can be monitored. The amplitude of this current will change by orders of magnitude with time. In accordance with eqn (7.9), in which the time-variable field must be replaced by a step-like 'charging' field of magnitude E_c, this current is due to a current density of

$$j_{\mathrm{pol.}}(t) = \sigma_0 E_c + \varepsilon_0(1 + \chi(0'))E_c\delta(t) + \varepsilon_0 E_c f(t). \tag{7.10}$$

Here, $\delta(t)$ is the delta function which will produce an extremely large current pulse coincident with the sudden increase of the voltage. The magnitude of this current pulse will, in general, not be measured. This second term is thus related to the displacement currents due to the sum of vacuum capacitance of the test object and the capacitance related to the 'high-frequency' susceptibility of the dielectric used. The first term is due to the d.c. conductivity of the material and determines the current after a (more or less) long time, for which the last term, which quantifies the response function completely, becomes negligible.

Polarization current measurements can finally be stopped if the current becomes stable. Immediately afterwards, the *depolarization* (or discharging, desorption) *current* can be measured by a subsequent short-circuiting of the

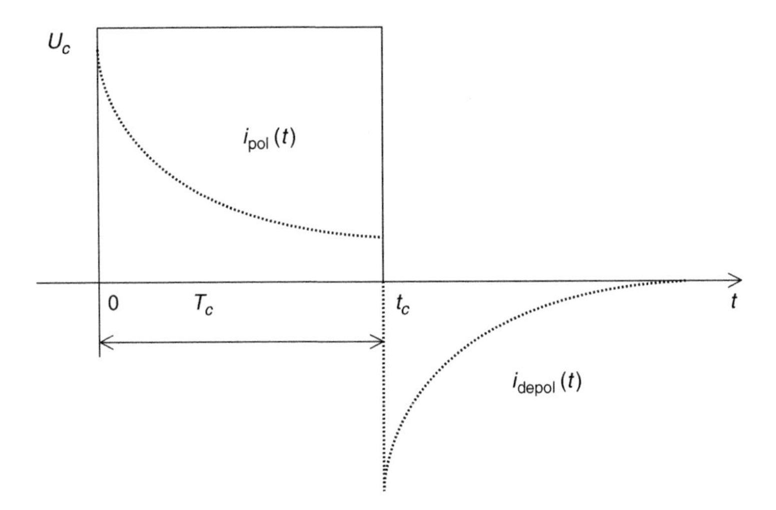

Figure 7.2 *Principle of relaxation current measurement*

sample, see Fig. 7.2. According to the superposition principle and neglecting immediately the second term in eqn (7.10), we get

$$j_{\text{dep}}(t) = -\varepsilon_0 E_c[f(t) - f(t + T_c)] \tag{7.11}$$

where T_c is the time during which the step voltage was applied to the test object. The second term in this equation can be neglected, if measurements with large values of T_c have been made for which the final value of the polarization current was already reached. Then the depolarization current becomes directly proportional to the dielectric response function.

An example of a recently performed relaxation current measurement is shown in Fig. 7.3. The data were obtained during investigations concerning the dielectric response of oil-paper (transformer or pressboard) insulation with different moisture content (m.c.). Preparation and test conditions of the samples can be found within the original publication.[5] In this case, all measurements started 1 s after voltage application (i_{pol}) and short-circuit (i_{depol}) respectively. From the selected results it can be seen that the final value of the polarization current will be reached only for the higher moisture content samples in spite of the fact that the measurements lasted up to 200 000 s, i.e. about 56 hours. Representation of such results in log–log scale is paramount due to the large change of the quantities.

The effect of depolarization currents is illustrated by the recharging of h.v. capacitors. When such capacitors are stressed for a long time with d.c. voltage and then briefly discharged through short-circuiting they will recharge to quite a high voltage when the short-circuit is removed. This 'return' or 'recovery'

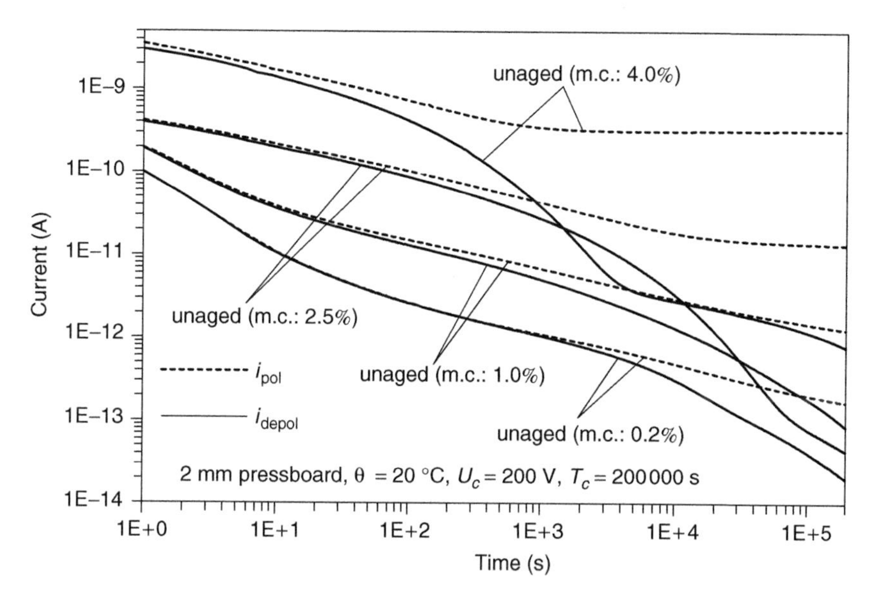

Figure 7.3 *Relaxation currents of unaged samples with different moisture contents*

voltage effect is due to depolarization currents. In the case of h.v. capacitors, it is dangerous and may cause severe accidents.

Specific response functions

The relaxation currents from Figs 7.2 and 7.3 decay monotonically, but do not follow simple relationships which can be expressed by adequate analytic functions. Nevertheless, many researchers have attempted to represent experimental data through such functions which can sometimes be related to at least idealized physical models of the polarization processes. We will here only mention some analytic functions to which reference is often made. For further studies see reference 1.

The simplest response is related to Debye and is represented by

$$f(t) = B\exp(-t/\tau)$$

where B is an amplitude for $t = 0$ and τ is a single relaxation time. Only some pure and simple liquids will follow this response at least within certain time regions. This response is also typical for 'interfacial polarization', i.e. a series combination of two different dielectrics between which a barrier is present adjacent to a bulk conducting material. The barrier attracts the charges and appears as a capacitance, whereas the bulk behaves like a series resistance.

Another extreme is provided by the 'general response' expression[1] which also involves the somewhat simpler 'power law' expression. Such a response is expressed by

$$f(t) = \frac{B}{\left(\frac{t}{\tau}\right)^n + \left(\frac{t}{\tau}\right)^m}$$

and will appear in the log–log scale as the superposition of two straight decaying lines. The simpler power law expression neglects one term in the denominator to form only a single straight line dependency. The polarization processes in polyethylene for instance can in general be simulated by such a behaviour. The processes can then be related to the physical process of diffusion or injected charges. A special form of an electric network, the uniform distributed R-C line, can simulate such a response. In section 7.1.3 some further hints to simulating networks are provided.

7.1.2 Dynamic properties in the frequency domain

The dielectric properties of dielectrics can also be measured and quantified in the frequency domain, i.e. with a.c. voltages as a function of frequency. The transition to the frequency domain from the time domain can be executed by means of Laplace or Fourier transformation. This is shown by means of eqn (7.9) in the following revised form, in which the response for the total current within a test specimen for an ideal voltage step starting at $t = 0$ is considered:

$$j(t) = \sigma_0 E(t) + \varepsilon_0 \frac{\mathrm{d}E(t)}{\mathrm{d}t} + \varepsilon_0 \frac{\mathrm{d}}{\mathrm{d}t} \int_0^t f(t-\tau) E(\tau)\,\mathrm{d}\tau. \tag{7.12}$$

With $j(t) \Rightarrow j(p)$; $E(t) \Rightarrow E(p)$; $E'(t) \Rightarrow pE(p)$; $f(t) \Rightarrow F(p)$; and by considering the convolution of the last term in this equation we get, for the present, formally with p being the Laplace operator:

$$j(p) = \sigma_0 E(p) + \varepsilon_0 pE(p) + \varepsilon_0 pF(p)E(p).$$

As p for the given conditions is the complex frequency, $j\omega$, we can reduce the equation to

$$\underline{j}(\omega) = \underline{E}(\omega)[\sigma_0 + j\omega\varepsilon_0(1 + \underline{F}(\omega)]. \tag{7.13}$$

Thus it becomes obvious that the dielectric response function $f(t)$ is the time domain of the frequency dependent susceptibility $\underline{\chi}(\omega)$, which is defined as the Fourier transform of the dielectric response function $f(t)$:

$$\underline{\chi}(\omega) = \underline{F}(\omega) = \chi'(\omega) - j\chi''(\omega) = \int_0^\infty f(t)\exp(-j\omega t)\,\mathrm{d}t. \tag{7.14}$$

From this equation in the frequency domain the following relationships for the susceptibility in time and frequency domain apply:

$$\chi(t \to \infty) \Leftrightarrow \chi'(\omega \to 0) \quad \text{and} \quad \chi(t \to 0) \Leftrightarrow \chi'(\omega \to \infty).$$

Now, in the frequency domain the polarization can be written as:

$$\underline{P}(\omega) = \varepsilon_0 \underline{\chi}(\omega) \underline{E}(\omega). \tag{7.15}$$

Equation (7.8) expressed in the frequency domain becomes:

$$\underline{j}(\omega) = \sigma_0 \underline{E}(\omega) + j\omega \underline{D}(\omega). \tag{7.16}$$

Using eqns (7.2) and (7.14) the current density is rewritten as:

$$\begin{aligned} \underline{j}(\omega) &= \{\sigma_0 + i\omega\varepsilon_0[1 + \chi'(\omega) - j\chi''(\omega)]\}\underline{E}(\omega) \\ &= \{\sigma_0 + \varepsilon_0\omega\chi''(\omega) + i\omega\varepsilon_0[1 + \chi'(\omega)]\}\underline{E}(\omega). \end{aligned} \tag{7.17}$$

The displacement can now be expressed by the relative but complex dielectric permittivity of the material $\varepsilon_r(\omega)$ with the relation:

$$\underline{D}(\omega) = \varepsilon_0 \underline{\varepsilon}_r(\omega) \underline{E}(\omega) = \varepsilon_0[1 + \chi'(\omega) - j\chi''(\omega)]\underline{E}(\omega) \tag{7.18}$$

where:

$$\underline{\varepsilon}_r(\omega) = \varepsilon_r'(\omega) - j\varepsilon_r''(\omega) = (1 + \chi'(\omega)) - j\chi''(\omega). \tag{7.19}$$

For a practical determination of the dielectric response from eqn (7.17), a bridge or any other instrument *cannot* distinguish between the current contribution of d.c. conductivity and that of dielectric loss. This means that the effective measured relative dielectric permittivity $\tilde{\underline{\varepsilon}}_r(\omega)$ is different from the relative permittivity $\varepsilon_r(\omega)$ defined in eqns (7.17) and (7.18). If the effective relative dielectric permittivity $\tilde{\underline{\varepsilon}}_r(\omega)$ is defined from the following relation

$$\underline{j}(\omega) = j\omega\varepsilon_0 \tilde{\underline{\varepsilon}}_r(\omega) \underline{E}(\omega). \tag{7.20}$$

Therefore:

$$\begin{aligned} \tilde{\underline{\varepsilon}}_r(\omega) &= \varepsilon_i'(\omega) - i[\varepsilon_r''(\omega) + \sigma_0/\varepsilon_0\omega] \\ &= 1 + \chi'(\omega) - j[\chi''(\omega) + \sigma_0/\varepsilon_0\omega]. \end{aligned} \tag{7.21}$$

Then the dissipation factor tan δ (see section 7.2) will be:

$$\tan\delta(\omega) = \frac{\varepsilon_r''(\omega) + \sigma_0/\varepsilon_0\omega}{\varepsilon_r'(\omega)}. \tag{7.22}$$

The real part of eqn (7.21) defines the capacitance of a test object, while the imaginary part represents the losses. Both quantities are dependent on frequency, which sometimes is not realized. We display, therefore, the results of frequency-dependent measurements as made on the test samples which were used in Fig. 7.3 for the measurements of relaxation currents. The results, again reproduced from literature,[(5)] are shown in Figs 7.4 and 7.5. Figure 7.4 shows the capacitance of the specimens and Fig. 7.5 their dissipation factors over a frequency range of nearly 8 decades. The measurements were taken

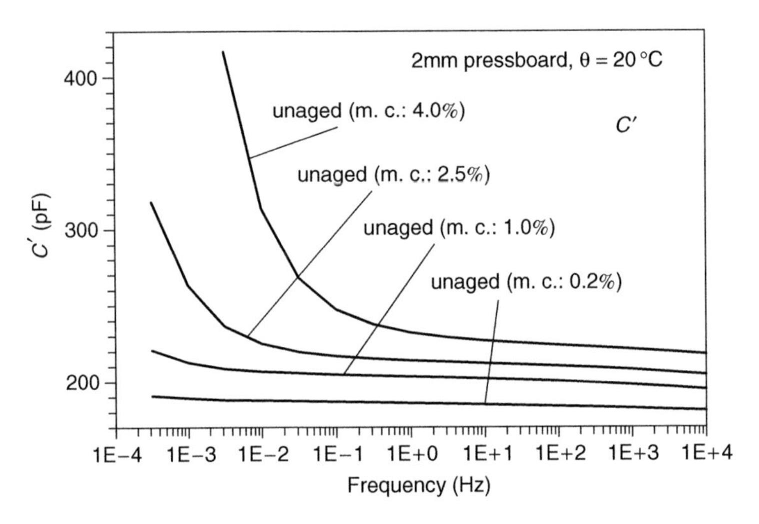

Figure 7.4 *Real part of the complex capacitance of pressboard samples in dependence on frequency*

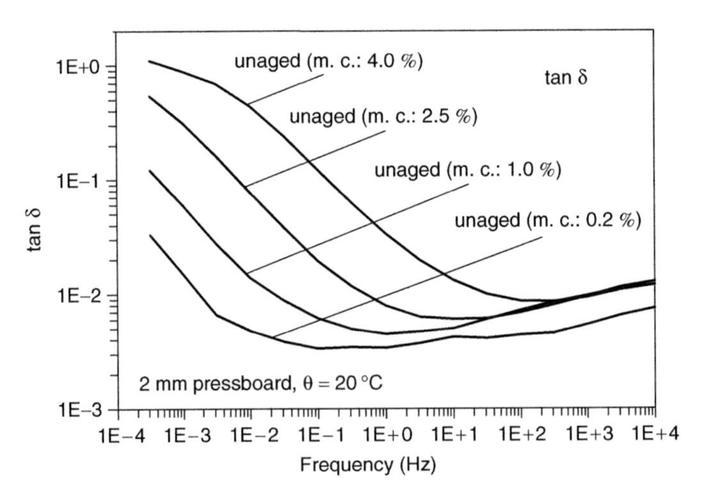

Figure 7.5 *Dissipation factor tan δ of pressboard samples in dependence on frequency*

using a 'dielectric spectrometer' at an a.c. voltage of about 3 volts. Note that the influence of moisture is much more significant at the lower frequencies. Its influence at power frequencies (50/60 Hz), where the measurements are usually made, is not as pronounced.

Measurements in the frequency domain become very lengthy if many individual values for very low frequencies are considered. At least three cycles of an a.c. voltage are in general necessary to quantify the amplitudes and phase shift between voltage and currents. Therefore, about 3000 seconds are necessary to get a single value of C and $\tan\delta$ for a frequency of 1 MHz. Since the results of relaxation current measurements can be converted in the frequency domain and vice versa, both methods complement each other.

7.1.3 Modelling of dielectric properties

Modelling or simulating dielectric properties through the use of equivalent electrical circuits has been practised for decades.(6) As an introduction, we will derive one of the models which can immediately be detected from the relaxation currents as treated in sections 7.1.1 and 7.1.2 respectively. For convenience, first a formal transition from current densities and electric fields to currents and voltages as applied to the terminals of a system shall be made.

Let us rewrite eqn (7.17) first:

$$\underline{j}(\omega) = \{\sigma_0 + \varepsilon_0\omega\chi''(\omega) + j\omega\varepsilon_0[1 + \chi'(\omega)]\}\underline{E}(\omega) \tag{7.17a}$$

and assume an ideal plate capacitor, in which the area of the plates is A and the gap distance is d. Without any dielectric within the gap, the 'vacuum capacitance' of this object becomes $C_{\text{vac}} = \varepsilon_0 A/d$. As the dielectric shall have the d.c. conductivity σ_0, the d.c. resistance becomes $R_0 = d/(\sigma_0 A)$. As the voltage applied to the terminals is $V = d.E$ and the current is $I = jA$, we can introduce these equations into eqn (7.17a) and obtain:

$$I(\omega) = U(\omega)\left\{\frac{1}{R_0} + \omega C_{\text{vac}}\chi''(\omega) + j\omega C_{\text{vac}}[1 + \chi'(\omega)]\right\}. \tag{7.23}$$

This equation represents an equivalent circuit comprising a single resistor R_0 in parallel with a single capacitor C_{vac}. But in addition, two other elements are in parallel: another resistor of magnitude $1/[\omega C_{\text{vac}}\chi''(\omega)]$ producing additional losses and a second capacitor, whose vacuum capacitance is multiplied by $\chi'(\omega)$. Both additional circuit elements are strongly dependent on frequency, as $\chi''(\omega)$ as well as $\chi'(\omega)$ depend on ω, and the lossy term $1/[\omega C_{\text{vac}}\chi''(\omega)]$ decreases additionally by $1/\omega$. This f dependency can be simulated either by a multiple series connection of parallel RC circuits or a formal conversion of the two elements into one series RC circuit, in which again both elements are frequency dependent. This single RC circuit can then again be split up in a multiple arrangement of parallel RC elements, which represent as a whole the frequency dependence of the terms of the dielectric. We display the second type of equivalent circuit in Fig. 7.6.

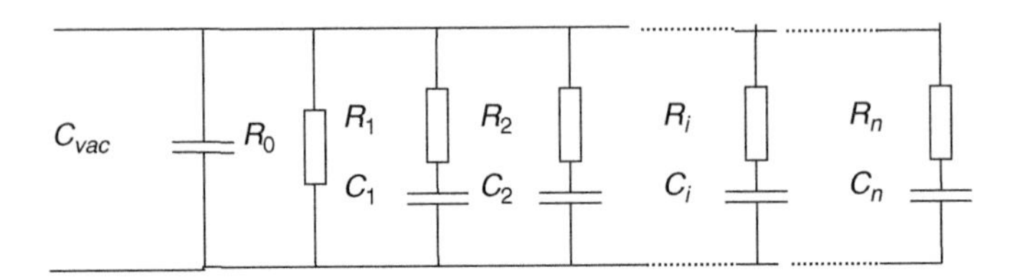

Figure 7.6 *Equivalent circuit to model a linear dielectric*

That this type of circuit will completely simulate both types of relaxation currents can well be recognized from the behaviour of this circuit in time domain, an example of which was shown in Fig. 7.3: the depolarization currents and thus also the response function $f(t)$ is a monotonic decaying function as well as the polarization currents. As every R_iC_i element produces – for charging as well as for discharging the circuit – an exponential (increasing or decreasing) current, the depolarization current is treated as a superposition of such exponentials. For this model, all circuit parameters can therefore be determined by measured quantities! The d.c. resistance R_0 can be approximated from the applied step voltage U_c and the difference between polarization and depolarization currents for the highest available time values. The individual elements R_i, C_i with the corresponding time constants $\tau_i = R_iC_i$ can then be determined by fitting the depolarization current with the equation

$$i_{\text{depol.}}(t) = \sum_{i=1}^{n} A_i \exp(-t/\tau_i), \tag{7.24}$$

where

$$A_i = U_c[1 - \exp(-T_c/\tau_i)]/R_i \quad \text{for } i = 1 \ldots n \tag{7.25}$$

and T_c is the duration of the time, during which the sample was charged. If the measured currents are not recorded within a very short time of the application of the step voltage, or the short-circuit after polarization, then C_{vac} must be replaced by a capacitance effective at the time instant at which the currents are available. For more information see the relevant literature.[5,7]

The basic idea in applying this simple model is to facilitate all further calculations with respect to the frequency domain or even the calculation of a 'polarization spectrum' belonging to a very special procedure of recovery voltage measurements, which will be briefly explained in section 7.1.4.

To complete this section, some additional hints are provided. For the equivalent circuit of Fig. 7.6, the complex capacitance $\underline{C}(\omega)$ can be calculated according to eqn (7.23) from its complex admittance, $\underline{Y}(\omega)$ as:

$$\underline{C}(\omega) = \frac{\underline{Y}(\omega)}{i\omega} = C_{?\,\mathrm{Hz}} + \frac{1}{j\omega R_0} + \sum_{j=1}^{n} \frac{C_i}{1 + j\omega R_i C_i} \tag{7.26}$$

Instead of C_{vac} some larger capacitance $C_{?\,\mathrm{Hz}}$ is taken into account as explained before. The real and imaginary parts of $\underline{C}(\omega)$ are then given as:

$$C'(\omega) = C_{?\,\mathrm{Hz}} + \sum_{i=1}^{n} \frac{C_i}{1 + (\omega R_i C_i)^2} \tag{7.27}$$

and

$$C''(\omega) = \frac{1}{\omega R_0} + \sum_{i=1}^{n} \frac{\omega R_i C_i^2}{1 + (\omega R_i C_i)^2}. \tag{7.28}$$

Finally, $\tan\delta(\omega)$ can be written as:

$$\tan\delta(\omega) = \frac{\dfrac{1}{\omega R_0} + \displaystyle\sum_{i=1}^{n} \frac{\omega R_i C_i^2}{1 + (\omega R_i C_i)^2}}{C_{?\,\mathrm{Hz}} + \displaystyle\sum_{i=1}^{n} \frac{C_i}{1 + (\omega R_i C_i)^2}}. \tag{7.29}$$

7.1.4 Applications to insulation ageing

As already mentioned in the introduction to this chapter, the application of measurements related to the dynamic properties of dielectrics has recently undergone an increase in use. The background and motivation of such applications is due to the need for utilities to employ insulation monitoring and diagnostic systems with the goal of reducing costs, assessing the performance of existing apparatus, and maintaining and enhancing safety and reliability. It is now generally believed that 'condition-based monitoring' or 'condition maintenance' will replace the 'periodic maintenance' practices applied to date. One of the primary drives in this direction is the large amount of aged and expensive equipment forming the backbone of most modern power systems. The increasingly popular move towards a competitive electricity supply market coupled with the ageing infrastructure calls for new techniques to retain aged plant in service as long as possible.

It is not possible to discuss this new philosophy in more detail in this text. Many publications including those available through CIGRE or IEEE outline the continued development of condition monitoring over the past ten years. Here, only some hints on the application of the more recently developed methods can be provided with special relevance to the dynamic properties of dielectrics and insulation systems. One should note, however, that the application of 'dielectric measurements' can be traced back to the last century and that an overview about the history and the state of the art of such measurements up to about 1990 may be found in the literature.[(11)]

Well-established, conventional methods

Ageing effects in electric insulation are always caused by changes of the chemical structure of the dielectric matter. These changes produce mechanical degradation (which usually cannot be detected within sealed equipment), release chemical 'ageing' products, and alter electrical properties leading to in-service electrical breakdown. Examples of conventional methods used for in-service monitoring of power transformers with standard oil-paper insulation are: Dissolved Gas Analysis (DGA);[8] Oil Parameter Analysis;[9] measurement of $C \tan \delta$ at power frequency (see section 7.2); measurement of insulation resistance; and measurement of the 'polarization index'.

New methods

New methods for transformer in-service monitoring include: Furan Analysis and HPLC (High Performance Liquid Chromatography) to quantify chemical ageing products and the Dielectric Response Analysis (DRA), the fundamentals of which have been treated above. The methods related to DRA are briefly explained below.

Dielectric response analysis (DRA)

This method is based on the measurements of polarization as well as depolarization currents and is thus sometimes described as the 'PDC method'. The background and the measurement procedure are described at the end of section 7.1.1. It should, however, again be noted that the dielectric response function $f(t)$, if quantified for times larger than several milliseconds, also provides access to the frequency-dependent dielectric parameters starting from ultra-low frequencies up to power frequencies.

Recovery voltage

The measurement of recovery or return voltages is another method to quantify the dielectric response of materials. The principle of the measurement can be traced back to the last century. With reference to Fig. 7.7, it can be explained as follows: a constant voltage U_c charges the test object for $0 \leq t < t_1$; after a relatively short period between $t_1 \leq t < t_2$ during which the sample is short-circuited, the test object is left in open-circuit condition. Then for times $t \geq t_2$

a recovery voltage $u_r(t)$ caused by residual polarization is built up across the test object, resulting in repolarization. If the voltmeter recording $u_r(t)$ has an extremely high input impedance, the test object remains charged until it discharges through its internal resistance. If the dielectric response function of the test object is known, the time dependence of recovery voltage for $t \geq t_2$ can be derived from eqn (7.9).[1,5,11] The advantage of this method is that it includes a self-calibration with respect to the capacitance of the test object, but for numerical evaluations the response function has to be measured as shown previously.

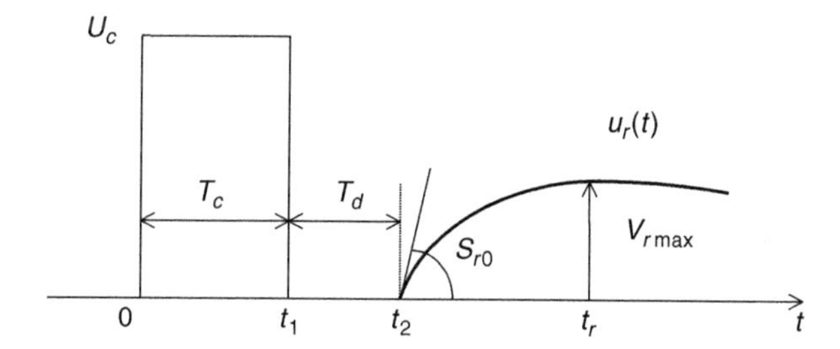

Figure 7.7 *Principle of recovery voltage measurement*

The 'polarization spectrum'

The so-called 'polarization spectrum'[11] is a quantity derived from a special measurement procedure of recovery voltages. With reference to Fig. 7.7, it is determined from the peak values (at time instants t_r) of many recovery voltages, which vary depending on charging duration T_c. These recovery voltage maxima are then presented as a function of increasing charging duration. Normally the charging time T_c is chosen to be twice as long as the discharge duration, T_d. It can be shown that the peak of the 'polarization spectrum' will be reached for a very dominant time constant τ of a Debye process already mentioned in section 7.1.1. This measurement method, for which a commercial equipment exists, has often been applied in practice during recent years with the goal of identifying the moisture content of the pressboard within power transformers. The interpretation of the results, however, is in general very difficult. Misleading results can be obtained if the moisture content is quantified in accordance with instructions provided by the manufacturer.

7.2 Dielectric loss and capacitance measurements

In section 7.1 it was explained that all kinds of dielectrics or insulation materials and systems can be characterized by its inherent polarization phenomena, which in the frequency domain can be expressed by a capacitance C and a magnitude of power dissipation (dielectric loss) as quantified by the dissipation or loss factor tan δ. Whereas these quantities within a wide frequency range are of utmost interest for new materials or even for the quality control of well-known insulation materials on receipt after delivery, nearly every high-voltage equipment prior to delivery to the customer will undergo a test related to 'C tan δ' for a final quality control. Such tests are in general only made with a frequency for which the equipment is designed. The reasons for this are as follows. First, too high losses at least during a.c. test voltages may cause thermal breakdown, see Chapter 6, section 6.1. Secondly, the manufacturer of a specific equipment knows the typical magnitudes which can be tolerated and which provide information about the quality of the newly manufactured equipment. Finally, such tests are in general made in dependence of the test voltage applied; both magnitudes, capacitance and dissipation factor shall be essentially constant with increasing voltage, as insulation systems are linear systems and any 'tip-up' of the tan δ with voltage level, called 'ionization knee', is a preliminary indication of 'partial discharges' (discussed in section 7.3) within the system.

The measurement of these dielectric properties with power frequency belongs, therefore, to standard testing procedures, for which ancillary principles are used. New measurement equipment is in general based on these principles, but nowadays is supplemented with microprocessor control and evaluation supports or software.

The various laboratory techniques for electrical insulation measurement have earlier been reviewed by Baker(13) and for detailed descriptions the reader is advised to refer to that publication. Additional information about the earlier techniques is also provided by Schwab.(14)

7.2.1 The Schering bridge

Still one of the most commonly used methods for measuring 'loss tangent', tan δ, and capacitance with high precision is the high-voltage Schering bridge, originally patented by P. Thomas in 1915 and introduced to h.v. measurements by H. Schering in 1920.(15,p.212) The basic circuit arrangement is shown in Fig. 7.8.

The bridge measures the capacitance C_X and loss angle δ (i.e. tan δ) of a capacitor or any capacitance of a specimen by comparing it with a gas-filled standard capacitor C_N which has very low and nearly negligible loss over a wide frequency range and can be built for test voltages up to the megavolt range (see Chapter 3, section 3.5.4). The 'X' h.v. arm of the bridge consists of a sample, the dielectric loss and capacitance of which are to be measured. On account of the dielectric loss the current through the capacitor leads the voltage by an angle $(90 - \delta)$ which is only slightly less than 90°. This current

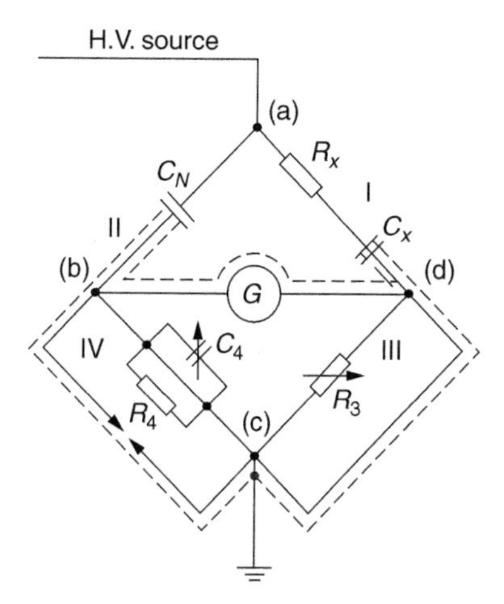

Figure 7.8 *The high-voltage Schering bridge*

produces a voltage drop of usually less than 100 V across the variable resistor R_3 of the low-voltage part of the bridge. The elements R_4 and C_4 of this part are necessary to balance the bridge. As seen from the circuit, the balance conditions are represented by considering that for the frequency applied a pure capacitance C_x is connected in series with a hypothetical resistance R_X, the power dissipated in the resistance simulating the power loss in the actual capacitor.

For the Schering bridge, the balance conditions are always derived for this series $R_X - C_X$ equivalent circuit. The derivation is shown below. For any series equivalent circuit the dissipation factor tan δ is defined by the following equation:

$$\tan \delta_s = \omega R_s C_s. \tag{7.30}$$

The balance conditions obtained when the indicator (null detector) 'G' shows zero deflection in Fig. 7.8 are:

$$\frac{Z_{ab}}{Z_{bc}} = \frac{Z_{ad}}{Z_{dc}},$$

where

$$Z_{ad} = R_x - j\frac{1}{\omega C_x}, \quad Z_{ab} = -i\frac{1}{\omega C_N},$$

$$Z_{bc} = \frac{R_4[-j(1/\omega C_4)]}{R_4 - j(1/\omega C_4)}, \quad Z_{dc} = R_3$$

By separation of the real and imaginary terms we get:

$$C_x = C_N \frac{R_4}{R_3}, \tag{7.31}$$

$$R_X = R_3 \frac{C_4}{C_N} \tag{7.32}$$

Substituting C_N in eqn (7.31) from eqn (7.32) and multiplying by ω we obtain the dissipation factor according eqn (7.30):

$$\omega C_x R_x = \omega C_4 R_4 = (\tan\delta)_X. \tag{7.33}$$

If the results shall be expressed in terms of a parallel equivalent circuit for the test object, for which a parallel arrangement of a capacitance C_P and a resistor R_P is assumed, the following equations shall be taken into account. For a parallel equivalent circuit, the dissipation is

$$\tan\delta_p = \frac{1}{\omega R_P C_P}. \tag{7.34}$$

On the condition that the losses in the two circuits must be equal, the quantities of series and parallel equivalent circuits may be converted to each other by:

$$C_p = \frac{C_s}{1 + (\tan\delta_s)^2}; \tag{7.35}$$

$$R_p = R_s\left(1 + \frac{1}{1 + (\tan\delta_s)^2}\right) \tag{7.36}$$

In practice R_3 is a variable resistance and is usually in the form of a four-decade box. Its maximum value is limited to about 10 000 Ω in order to keep the effects of any stray capacitance relatively small. R_4 is made constant and in general realized as a multiple of $(1000/\pi)$ Ω thus making possible a direct reading of tan δ, provided a constant value of the frequency is indicated. C_4 is variable. To exclude from Z_{bc} and Z_{dc} and the galvanometer branches any currents due to inter-capacity between the h.v. and l.v. arms, except those flowing through Z_{ab} and Z_{ad}, the bridge is fully double screened as indicated in Fig. 7.8, in which only one screen is sketched. The l.v. branches are usually protected with spark gaps against the appearance of high voltages in the event of failure of Z_{I} or Z_{II}.

In Fig. 7.8 the network is earthed at the l.v. end of the transformer supplying the high voltage, and by this also the bridge is earthed at (c). Under balance conditions, both sides of the null detector (G) are at the same potential, but the shield is earthed. Therefore partial stray capacitances appear across the branches III and IV, and depending upon the length of the leads to C_N and C_x, these partial capacitances can assume values over a wide limit. These capacitances can be measured and thus their influence on the dissipation factor can be calculated. If C_I is the partial capacitance of branch I, and C_{II} that of branch II, the calculation shows

$$\tan\delta = R\omega(C_4 + C_{II}) - R_3\omega C_I. \quad (7.37)$$

This procedure is time consuming and inconvenient, and there are methods available to overcome this effect. The principle of the method may easiest be described by the 'Wagner earth' as shown in Fig. 7.9 applied to the Schering bridge, but introduced by K.W. Wagner for another bridge already in 1911.[17] In this arrangement an additional arm Z is connected between the l.v. terminal of the four-arm bridge and earth. Together with the stray capacitance of the h.v. busbar to earth the arrangement becomes equivalent to a six-arm bridge and a double balancing procedure is required which can be achieved either by using two detectors or a switch arrangement which enables the detector to be switched on into either sets of arms. At balance the terminals of the detector are at earth potential and capacitances between the terminals and screens having no potential difference between them do not affect the balance conditions. Both the detector and the l.v. leads must be screened. The capacitances between the leads and screens are in parallel with the impedance Z and as such do not contribute to the balance conditions.

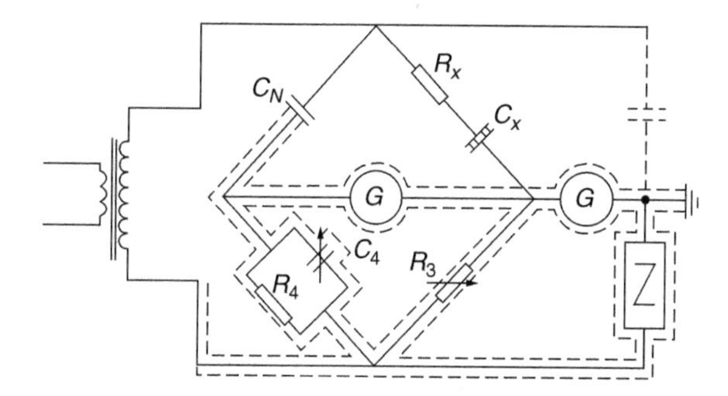

Figure 7.9 *Bridge incorporating 'Wagner earth'*

This method, however, is rarely used today, as operational amplifiers for automatic balancing of 'Wagner earth' may be used. The basic circuit is shown in Fig. 7.10. Although the bridge may well be earthed at (c), the potentials of the screens are shifted to the potential of the detector branch by a high-quality amplifier with unity voltage gain. The shields of the leads to C_x and C_N are not grounded, but connected to the output of the amplifier, for which operational amplifiers can conveniently be used. The high input impedance and very low output impedance of the amplifier do not load the detector branch and keep the screen potential at any instant at an artificial 'ground'.

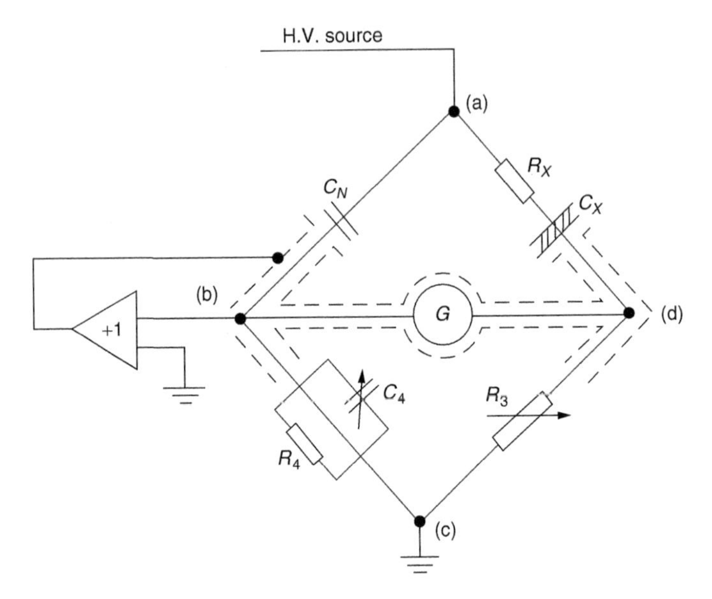

Figure 7.10 *Automatic 'Wagner earth' (dividing screen technique)*

A second screen, which is earthed, may be added to prevent disturbances by neighbouring voltage sources.

Measurement of large capacitance

When the capacitance to be measured is large, a variable resistance R_3 in eqn (7.31) capable of passing large currents would be required. To maintain a high value of R_3 it may be shunted by another resistor (A) as shown with the simplified circuit in Fig. 7.11. An additional resistor (B) is put in series with R_3 to protect it from excessive currents should it accidentally be set to a very low value. With this arrangement it can be shown[(1)] that the specimen's capacitance and loss tangent become respectively:

$$C_x = C_N\left(\frac{R_4}{R_3}\right)\left[1+\left(\frac{B}{A}\right)+\left(\frac{R_3}{A}\right)\right] \tag{7.38}$$

and

$$\tan\delta = \omega C_N R_4\left(\frac{B}{R_3}\right). \tag{7.39}$$

For more sophisticated circuits see reference 14.

The Schering bridge principle is suitable for measurements at frequencies up to some 100 kHz, if the circuit elements are properly designed. Common Schering bridges for power frequencies may be used at frequencies up to about

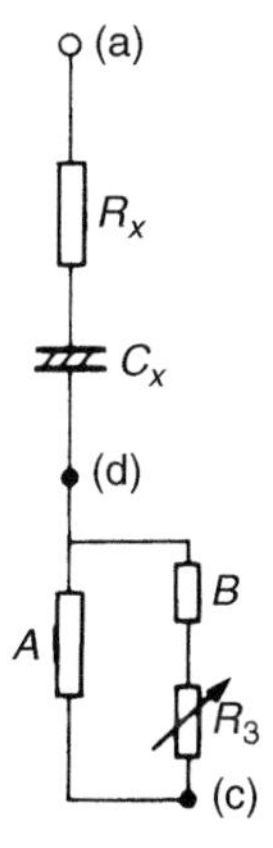

Figure 7.11 *Shunt arrangement for measurement of large capacitance (compare with Fig. 7.8)*

10 kHz only. At higher frequencies it becomes necessary to use a special high-frequency Schering bridge or substitution and resonance methods.(13) Measuring uncertainty is in general as low as 10^{-5}, if the bridge ratio R_3/R_4 is 1. With other ratios the uncertainty is typically 10^{-4}. The uncertainty of the standard capacitor used as reference will contribute to these values. The range of measurements is in the interval of 1 pF–100 μF.

7.2.2 Current comparator bridges

The shortcomings of the Schering bridge, among which also the possible changes of the circuit elements with temperature and ageing can be mentioned, stimulated quite early the search for improved forms of bridge circuits, which have been based on 'inductive coupling' or 'ampere-ratio arms'. But only in the late 1950s has the technology of such new circuits been successful in demonstrating the advantages of such new circuits, in which the current in the test piece and that in the standard capacitor is compared by means of a magnetic toroidal core on which two uniformly distributed coils with different numbers of turns are wound such as to force zero flux conditions. Although the first ideas to apply this principle may be traced back to Blumlein in 1928, the fundamental circuit arrangement for the measurement of capacitance and losses in h.v. capacitors was introduced by Glynne.(18) A prerequisite of applying the 'ampere-ratio arm' principle was the availability of new magnetic material with a very high initial permeability with which such 'current comparator bridges' could be made.(19)

The basic circuit of the 'Glynne bridge' is shown in Fig. 7.12. The main part of the bridge circuit consists of a three-winding current comparator which is carefully and heavily shielded against magnetic stray fields and protected against mechanical vibrations. Thus, the particular merit in this arrangement is that there is no net m.m.f. across windings 1 and 2 at balance conditions. Furthermore, the stray capacitance across the windings and that of the screened l.v. leads does not enter in the balance expression since there is no voltage drop on the comparator windings apart from their d.c. resistance. This enables long leads to be used without 'Wagner earth'. The sensitivity of the bridge is higher than that of the Schering bridge.

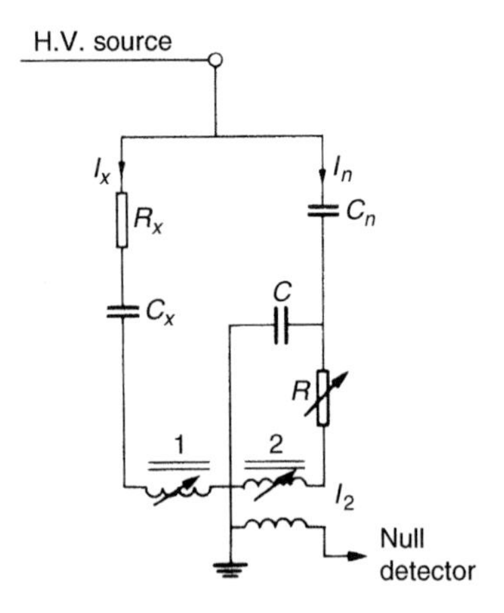

Figure 7.12 *Current comparator (Glynne) bridge*

The balance is indicated by zero voltage induced in the detector coil and corresponds to the conditions when $I_X N_1 = I_N N_2$ where N_1 and N_2 are the number of turns in series with the sample (C_X) and the standard capacitor (C_N) respectively, and I_X and I_2 are the corresponding currents flowing in C_X and N_N. Again a series equivalent circuit is assumed for the specimen under test.

For a current I_N in the standard capacitor the voltage developed across the R-C arm is given by:

$$V = \frac{I_N R}{1 + j\omega CR}$$

The portion of current I_2 in coil 2 is

$$I_2 = \frac{I_N}{1 + j\omega CR},$$

and for a unity total applied voltage

$$I_2 = \frac{1}{[\{R/(1 + j\omega CR) + (1/j\omega C_N)\}(1 + j\omega CR)]} = \frac{j\omega C_N}{1 + j\omega(C_N + C)R},$$

therefore:

$$C_x = C_N\left(\frac{N_2}{N_1}\right) \quad (7.40)$$

and

$$\tan\delta = \omega R(C_N + C). \quad (7.41)$$

The capacitance and phase angle balance are obtained by making N_1, N_2 and R variable. The uncertainty and sensitivity of this type of bridge is better than that indicated for the Schering bridge; the working frequency range is about 50 Hz to 1 kHz.

Nowadays, bridges with fully automatic self-balancing are preferred especially if only unskilled personnel are used or series measurements have to be performed within a production process. There are many solutions available. Older methods used servo-motor-driven potentiometers controlled from a feedback loop. The capability of electronic circuits provides many different solutions,[20] the most recent of which takes advantage of microcomputer control. Figure 7.13 displays such a typical circuit as published by Osvàth and Widmer in 1986.[21] Raw balancing of the bridge is realized by relays, the fine balancing and loss factor compensation are made by electronic circuits. The whole system is controlled by a microprocessor, which simulates the manual procedure of balancing. Newest developments of this circuit provide continuous automatic balancing within less than 1 second as well as manual balancing.[50] The sensitivity and uncertainty of this type of bridge is equal to the best h.v. $C \tan\delta$ bridges.

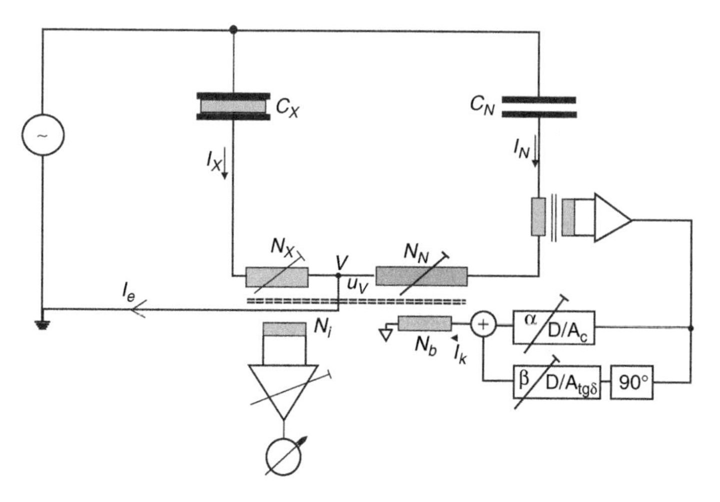

Figure 7.13 *Automatic high-voltage current comparator bridge*

7.2.3 Loss measurement on complete equipment

It is often required to measure the dielectric loss on specimens one side of which is permanently earthed. There are two established methods used for such measurement. One is the inversion of a Schering bridge, shown in Fig. 7.14, with the operator, ratio arms and null detector inside a Faraday cage at high potential. The system requires the cage to be insulated for the full test voltage and with suitable design may be used up to the maximum voltage available. There are, however, difficulties in inverting physically the h.v. standard capacitor and it becomes necessary to mount it on a platform insulated for full voltage.

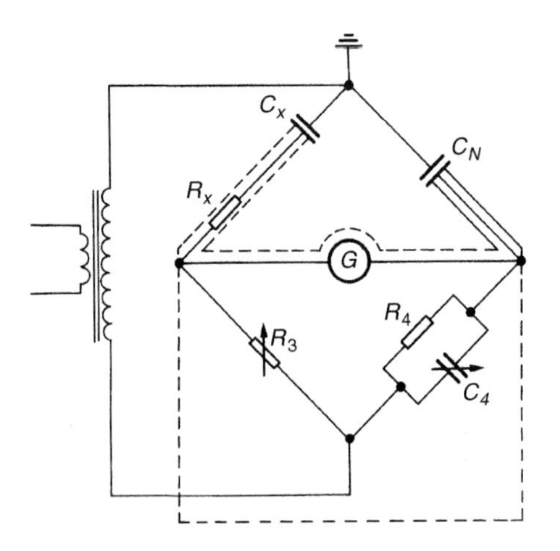

Figure 7.14 *High-voltage bridge with Faraday cage*

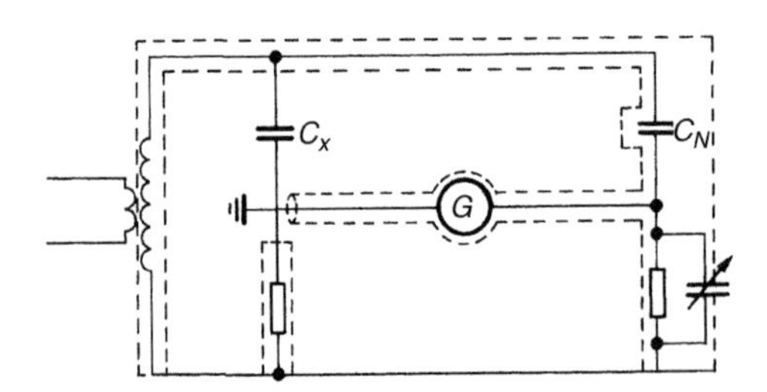

Figure 7.15 *Fully screened bridge*

An alternative method, though limited to lower voltages, employs an artificial earth which differs in potential from a true earth by the voltage developed across each of the l.v. arms as shown in Fig. 7.15. The artificial earth screen intercepts all the field from high potential to earth except in the specimen. It thus requires screening of the h.v. lead and presents difficulties at voltages in excess of about 5 kV.

7.2.4 Null detectors

The null detector (G) for ancient bridges was simply a vibration galvanometer of high mechanical Q factor. Although their application is well justified, the sensitivity to mechanical noise (if present) and the limited electrical sensitivity present some disadvantages. Since a few decades more sensitive electronic null detectors are commonly used. The possible high sensitivity, however, cannot be utilized in general, as noise voltages from the circuit, or electromagnetically induced voltages from the stray fields of the h.v. circuit, disturb the balance. This electronic null detector reads the voltage across the detector branch. As the balance equations of the bridge are only valid for a particular fixed frequency, the unavoidable harmonic content of the high input voltage of the bridge results in higher harmonic voltages across the null detector, for which the bridge is not balanced. A very pronounced pass-band characteristic is therefore necessary to attenuate these harmonics.

A very much improved balance is possible using electronic null detectors, which are also sensitive to the phase. Bridges may only slowly converge, i.e. the magnitude of the detector branch voltage may only slightly change within the individual settings of R_3 and C_4 in the Schering bridge or R in the transformer ratio-arm bridge. In the use of phase-sensitive null detectors, the balance condition is indicated in terms of magnitude and phase. With a reference voltage in phase with the (high) source voltage, these values describe Lissajou figures at the screen of a CRO used for the display. In this way the balancing procedure is always known and the balance is obtained much faster.

7.3 Partial-discharge measurements

What is a 'partial discharge'? Let us use the definition given in the International Standard of the IEC (International Electrotechnical Commission) related to partial discharge measurements, see reference 31:

> Partial discharge (PD) is a localized electrical discharge that only partially bridges the insulation between conductors and which may or may not occur adjacent to a conductor.

This definition is supplemented by three notes, from which only notes 1 and 2 shall be cited:

> NOTE 1 – Partial discharges are in general a consequence of local electrical stress concentrations in the insulation or on the surface of the insulation.

Generally such discharges appear as pulses of duration of much less than 1 μs. More continuous forms may, however, occur, as for example the so-called pulse-less discharges in gaseous dielectrics. This kind of discharge will normally not be detected by the measurement methods described in this standard.

NOTE 2 – 'Corona' is a form of partial discharge that occurs in gaseous media around conductors which are remote from solid or liquid insulation. 'Corona' should not be used as a general term for all forms of PD.

No further explanations are necessary to define this kind of phenomena: PDs are thus localized electrical discharges within any insulation system as applied in electrical apparatus, components or systems. In general PDs are restricted to a part of the dielectric materials used, and thus only partially bridging the electrodes between which the voltage is applied. The insulation may consist of solid, liquid or gaseous materials, or any combination of these. The term 'partial discharge' includes a wide group of discharge phenomena: (i) internal discharges occurring in voids or cavities within solid or liquid dielectrics; (ii) surface discharges appearing at the boundary of different insulation materials; (iii) corona discharges occurring in gaseous dielectrics in the presence of inhomogeneous fields; (iv) continuous impact of discharges in solid dielectrics forming discharge channels (treeing).

The significance of partial discharges on the life of insulation has long been recognized. Every discharge event causes a deterioration of the material by the energy impact of high energy electrons or accelerated ions, causing chemical transformations of many types. As will be shown later, the number of discharge events during a chosen time interval is strongly dependent on the kind of voltage applied and will be largest for a.c. voltages. It is also obvious that the actual deterioration is dependent upon the material used. Corona discharges in air will have no influence on the life expectancy of an overhead line; but PDs within a thermoplastic dielectric, e.g. PE, may cause breakdown within a few days. It is still the aim of many investigations to relate partial discharge to the lifetime of specified materials. Such a quantitatively defined relationship is, however, difficult to ensure. PD measurements have nevertheless gained great importance during the last four decades and a large number publications are concerned either with the measuring techniques involved or with the deterioration effects of the insulation.

The detection and measurement of discharges is based on the exchange of energy taking place during the discharge. These exchanges are manifested as: (i) electrical pulse currents (with some exceptions, i.e. some types of glow discharges); (ii) dielectric losses; (iii) e.m. radiation (light); (iv) sound (noise); (v) increased gas pressure; (vi) chemical reactions. Therefore, discharge detection and measuring techniques may be based on the observation of any of the above phenomena. The oldest and simplest method relies on listening to the acoustic noise from the discharge, the 'hissing test'. The sensitivity is, however, often low and difficulties arise in distinguishing between discharges and extraneous noise sources, particularly when tests are carried out on factory premises. It is also well known that the energy released by PD will increase the dissipation factor; a measurement of tan δ in dependency of voltage applied displays an 'ionization knee', a bending of the otherwise straight dependency (see section 7.2). This knee, however, is blurred and not pronounced, even with an appreciable amount of PD, as the additional losses generated in very localized sections can be very small in comparison to the volume losses resulting from polarization processes. The use of optical techniques is limited to discharges within transparent media and thus not applicable in most cases. Only modern acoustical detection methods utilizing ultrasonic transducers can successfully be used to localize the discharges.[(22–25)] These very specialized methods are not treated here. Summaries of older methods can be found in the book of Kreuger.[(26)] More recent developments may be found in reference 45.

The most frequently used and successful detection methods are the electrical ones, to which the new IEC Standard is also related. These methods aim to separate the impulse currents linked with partial discharges from any other phenomena. The adequate application of different PD detectors which became now quite well defined and standardized within reference 31, presupposes a fundamental knowledge about the electrical phenomena within the test samples and the test circuits. Thus an attempt is made to introduce the reader to the basics of these techniques without full treatment, which would be too extensive. Not treated here, however, are non-electrical methods for PD detection.

7.3.1 The basic PD test circuit

Electrical PD detection methods are based on the appearance of a 'PD (current or voltage) pulse' at the terminals of a test object, which may be either a simple dielectric test specimen for fundamental investigations or even a large h.v. apparatus which has to undergo a PD test. For the evaluation of the fundamental quantities related to a PD pulse we simulate the test object, as usual, by the simple capacitor arrangement as shown in Fig. 7.16(a), comprising solid or fluid dielectric materials between the two electrodes or terminals A and B, and a gas-filled cavity. (A similar arrangement was used in Chapter 6, see Fig. 6.12.) The electric field distribution within this test object is here simulated by some partial capacitances, which is possible as long as no space charges disturb this distribution. Electric field lines within the cavity are represented by C_c and those starting or ending at the cavity walls form the two capacitances C_b' and C_b'' within the solid or fluid dielectric. All field lines outside the cavity are represented by $C_a = C_a' + C_a''$. Due to realistic geometric dimensions involved, and as $C_b = C'_b C''_b/(C'_b + C''_b)$, the magnitude of the capacitances will then be controlled by the inequality

$$C_a \gg C_c \gg C_b. \tag{7.42}$$

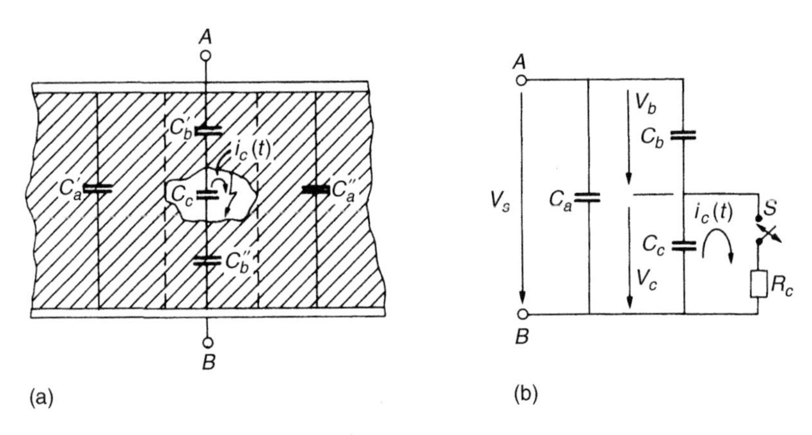

Figure 7.16 *Simulation of a PD test object. (a) Scheme of an insulation system comprising a cavity. (b) Equivalent circuit*

This void will become the origin of a PD if the applied voltage is increased, as the field gradients in the void are strongly enhanced by the difference in permittivities as well as by the shape of the cavity. For an increasing value of an a.c. voltage the first discharge will appear at the crest or rising part of a half-cycle. This discharge is a gas discharge (see Chapter 5) creating electrons as well as negative and positive ions, which are driven to the surfaces of the void thus forming dipoles or additional polarization of the test object. This physical effect reduces the voltage across the void significantly. Within our model, this effect is causing the cavity capacitance C_c to discharge to a large extent. If the voltage is still increasing or decreasing by the negative slope of an a.c. voltage, new field lines are built up and hence the discharge phenomena are repeated during each cycle (see Fig. 6.13 in Chapter 6). If increasing d.c. voltages are applied, one or only a few partial discharges will occur during the rising part of the voltage. But if the voltage remains constant, the discharges will stop as long as the surface charges as deposited on the walls of the void do not recombine or diffuse into the surrounding dielectric.

These phenomena can now be simulated by the equivalent circuit of this scheme as shown in Fig. 7.16(b). Here, the switch S is controlled by the voltage V_c across the void capacitance C_c, and S is closed only for a short time, during which the flow of a current $i_c(t)$ takes place. The resistor R_c simulates the time period during which the discharge develops and is completed. This discharge current $i_c(t)$, which cannot be measured, would have a shape as governed by the gas discharge process and would in general be similar to a Dirac function, i.e. this discharge current is generally a very short pulse in the nanosecond range.

Let us now assume that the sample was charged to the voltage V_a but the terminals A, B are no longer connected to a voltage source. If the switch S is closed and C_c becomes completely discharged, the current $i_c(t)$ releases a charge $\delta q_c = C_c \delta V_c$ from C_c, a charge which is lost in the whole system as assumed for simulation. By comparing the charges within the system before and after this discharge, we receive the voltage drop across the terminal δV_a as

$$\delta V_a = \frac{C_b}{C_a + C_b} \delta V_c \tag{7.43}$$

This voltage drop contains no information about the charge δq_c, but it is proportional to $(C_b \delta V_c)$, a magnitude vaguely related to this charge, as C_b will increase with the geometric dimensions of the cavity.

δV_a is clearly a quantity which could be measured. It is a negative voltage step with a rise time depending upon the duration of $i_c(t)$. The magnitude of the voltage step, however, is quite small, although δV_c is in a range of some 10^2 to 10^3 V; but the ratio C_b/C_a will always be very small and unknown according to eqn (7.42). Thus a direct detection of this voltage step by a measurement of the whole input voltage would be a tedious task. The detection circuits are therefore based upon another quantity, which can immediately be derived from a nearly complete circuit shown in Fig. 7.17. The test object, Fig. 7.16(a), is now connected to a voltage source V, in general an a.c. power supply. An impedance Z, comprising either only the natural impedance of the lead between voltage source and the parallel arrangement of C_K and C_t or enlarged by a PD-free inductance or filter, may disconnect the 'coupling capacitor' C_K and the test specimen C_t from the voltage source during the short duration PD phenomena only. Then C_K is a storage capacitor or quite a stable voltage source during the short period of the partial discharge. It

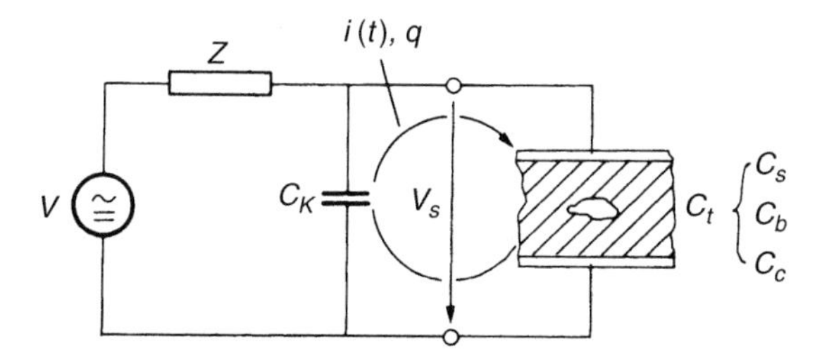

Figure 7.17 *The PD test object C_t within a PD test circuit*

releases a charging current or the actual 'PD current pulse' $i(t)$ between C_K and C_t and tries to cancel the voltage drop δV_a across $C_t \approx (C_a + C_b)$. If $C_K \gg C_t$, δV_a is completely compensated and the charge transfer provided by the current pulse $i(t)$ is given by

$$q = \int i(t) = (C_a + C_b)\delta V_a \tag{7.44}$$

With eqn (7.43), this charge becomes

$$q = C_b \delta V_c \tag{7.45}$$

and is the so-called *apparent charge of a PD pulse*, which is the most fundamental quantity of all PD measurements. The word 'apparent' was introduced because this charge again is not equal to the amount of charge locally involved at the site of the discharge or cavity C_c. This PD quantity is much more realistic than δV_a in eqn (7.43), as the capacitance C_a of the test object, which is its main part of C_t, has no influence on it. And even the amount of charge as locally involved during a discharge process is of minor interest, as only the number and magnitude of their dipole moments and their interaction with the electrodes or terminals determine the magnitude of the PD current pulse.

The condition $C_K \gg C_a(\cong C_t)$ is, however, not always applicable in practice, as either C_t is quite large, or the loading of an a.c. power supply becomes high and the cost of building such a large capacitor, which must be free of any PD, is not economical. For a finite value of C_K the charge q or the current $i(t)$ is reduced, as the voltage across C_K will also drop during the charge transfer. Designating this voltage drop by δV_a^*, we may compute this value by assuming that the same charge $C_b \delta V_c$ has to be transferred in the circuits of Figs 7.16(b) and 7.17. Therefore

$$\delta V_a(C_a + C_b) = \delta V^*(C_a + C_b + C_K). \tag{7.46}$$

Introducing eqn (7.43) as well as eqn (7.45), we obtain

$$\delta V^* = \frac{C_b}{C_a + C_b + C_K}\delta V_c = \frac{q}{C_a + C_b + C_K}. \tag{7.47}$$

Again, δV^* is a difficult quantity to be measured. The charge transferred from C_K to C_t by the reduced current $i(t)$ is, however, equal to $C_K \delta V^*$; it is related to the real value of the apparent charge q which then can be measured by an integration procedure, see section 7.3.3. If we designate this measured quantity as q_m, then

$$q_m = C_K \delta V^* = \frac{C_K}{C_a + C_b + C_K} q \approx \frac{C_K}{C_a + C_K} q$$

or

$$\frac{q_m}{q} \cong \frac{C_K}{C_a + C_K} \approx \frac{C_K}{C_t + C_K}. \tag{7.48}$$

The relationship q_m/q indicates the difficulties arising in PD measurements for test objects of large capacitance values C_t. Although C_K and C_t may be known, the ability to detect small values of q will decrease as all instruments capable of integrating the currents $i(t)$ will have a lower limit for quantifying q_m. Equation (7.48) therefore sets limits for the recording of 'picocoulombs' in large test objects. During actual measurements, however, a calibration procedure is needed during which artificial apparent charge q of well-known magnitude is injected to the test object, see section 7.3.7.

A final, critical note is made with reference to the definition of the apparent charge q as given in the new IEC Standard 60270.(31) The original text of this definition is:

> *apparent charge q* of a *PD pulse* is that unipolar charge which, if injected within a very short time between the terminals of the test object in a specified test circuit, would give the same reading on the measuring instrument as the PD current pulse itself. The *apparent charge* is usually expressed in picocoulombs.
>
> This definition ends with:
>
> NOTE – The *apparent charge* is not equal to the amount of charge locally involved at the site of the discharge and which cannot be measured directly.

This definition is an indication of the difficulties in understanding the physical phenomena related to a PD event. As one of the authors of this book has been chairman of the International Working Group responsible for setting up this new standard, he is familiar with these difficulties and can confirm that the definition is clearly a compromise which could be accepted by the international members of the relevant Technical Committee of IEC. The definition is correct. It relates to a calibration procedure of a PD test and measuring circuit, as already mentioned above. The 'NOTE', however, is still supporting the basically wrong assumption that a certain amount or number of charges at the site of the discharge should be measured. As already mentioned: it is not the number of charges producing the PD currents, but the number of induced dipole moments which produce a sudden increase in the capacitance of the test object. With section 7.1, this phenomenon is much more plausible.

7.3.2 PD currents

Before discussing the fundamentals of the measurement of the apparent charge some remarks concerning the PD currents $i(t)$ will be helpful, as much of the research work has been and is still devoted to these currents, which are difficult to measure with high accuracy. The difficulties arise for several reasons.

If V is an a.c. voltage, the main contribution of the currents flowing within the branches C_K and C_t of Fig. 7.17 are displacement currents $C(\mathrm{d}V/\mathrm{d}t)$, and both are nearly in phase. The PD pulse currents $i(t)$ with crest values in the range of sometimes smaller than 10^{-4} A, are not only small in amplitude, but also of very short duration. If no stray capacitance in parallel to C_K were present, $i(t)$ would be the same in both branches, but of opposite polarity. For accurate measurements, a shunt resistor with matched coaxial cable may be introduced in the circuit as shown in Fig. 7.18. The voltage across the CRO (or transient recorder) input is then given by $V_m(t) = (i_t + i)Z_0R/(R + Z_0)$. Only if the capacitance of the test object is small, which is a special case, will the voltages referring to the PD currents $i(t)$ be clearly distinguished from the displacement currents $i_t(t)$.

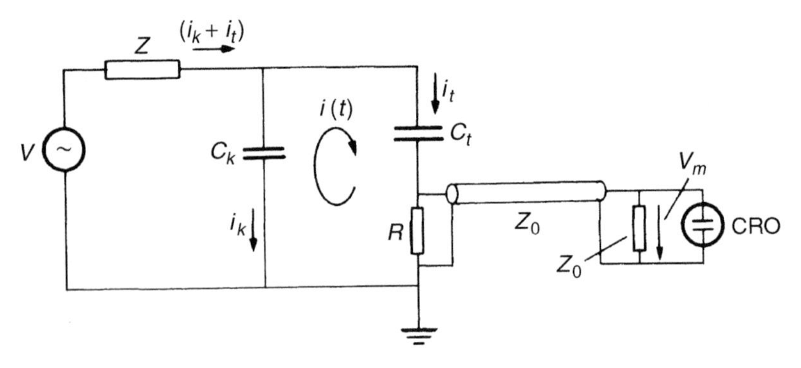

Figure 7.18 *Measurement of PD current $i(t)$ – low sensitivity circuit*

Improvements are possible by inserting an amplifier (e.g. active voltage probe) of very high bandwidth at the input end of the signal cable. In this way the signal cable is electrically disconnected from R. High values of R, however, will introduce measuring errors, which are explained with Fig. 7.19. A capacitance C of some 10 pF, which accounts for the lead between C_t and earth as well as for the input capacitance of the amplifier or other stray capacitances, will shunt the resistance R and thus bypass or delay the very high-frequency components of the current $i(t)$. Thus, if $i(t)$ is a very short current pulse, its shape and crest value are heavily distorted, as C will act as an integrator. Furthermore, with R within the discharge circuit, the current pulse will be lengthened, as the charge transfer even with $C = 0$ will be delayed by a time constant $RC_tC_K/(C_t + C_K)$. Both effects are influencing the shape of the original current pulse, and thus the measurement of $i(t)$ is a tedious task and is only made for research purposes.

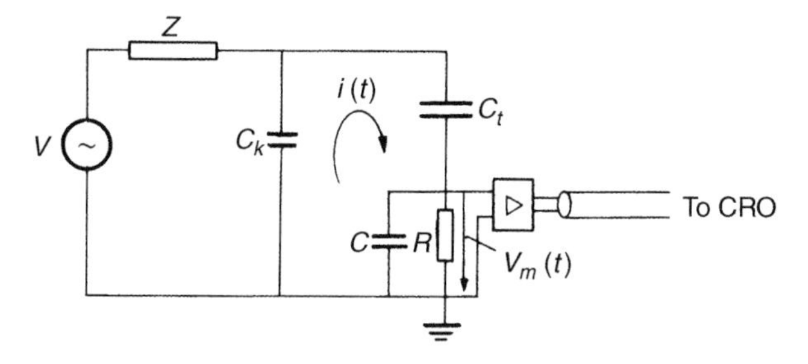

Figure 7.19 *Measurement of PD currents – high sensitivity circuit*

All measured data on current shapes published in many papers are suffering from this effect. One may, however, summarize the results by the following statements. Partial discharge currents originated in voids within solids or liquids are very short current pulses of less than a few nanoseconds duration. This can be understood, as the gas discharge process within a very limited space is developed in a very short time and is terminated by the limited space for movement of the charge carriers. Discharges within a homogeneous dielectric material, i.e. a gas, produce PD currents with a very short rise time ($\lesssim$5 nsec) and a longer tail. Whereas the fast current rise is produced by the fast avalanche processes (see Chapter 5), the decay of the current can be attributed to the drift velocity of attached electrons and positive ions within the dielectric. Discharge pulses in atmospheric air provide in general current pulses of less than about 100 nsec duration. Longer current pulses have only been measured for partial discharges in fluids or solid materials without pronounced voids, if a number of consecutive discharges take place within a short time. In most of these cases the total duration of $i(t)$ is less than about 1 μsec, with only some exceptions e.g. the usual bursts of discharges in insulating fluids.

All these statements refer to test circuits with very low inductance and proper damping effects within the loop $C_K - C_t$. The current $i(t)$, however, may oscillate, as oscillations are readily excited by the sudden voltage drop across C_t. Test objects with inherent inductivity or internal resonant circuits, e.g. transformer or reactor/generator windings, will always cause oscillatory PD current pulses. Such distortions of the PD currents, however, do not change the transferred charge magnitudes, as no discharge resistor is in parallel to C_K or C_t. If the displacement currents $i_t(t)$ or $i_K(t)$ are suppressed, the distorted PD currents can also be filtered, integrated and displayed.

7.3.3 PD measuring systems within the PD test circuit

In sections 7.3.1 and 7.3.2 the evolution of the PD current pulses and measurement procedures of these pulses have been broadly discussed. To quantify the 'individual apparent charge magnitudes' q_i for the repeatedly occurring PD pulses which may have quite specific statistical distributions, a measuring system must be integrated into the test circuit which fulfils specific requirements. Already at this point it shall be mentioned that under practical environment conditions quite different kinds of disturbances (background noise) are present, which will be summarized in a later section.

Most PD measuring systems applied are integrated into the test circuit in accordance with schemes shown in Figs 7.20(a) and (b), which are taken from the new IEC Standard[31] which replaces the former one as issued in 1981.[32] Within these 'straight detection circuits', the coupling device 'CD' with its input impedance Z_{mi} forms the input end of the measuring system. As indicated in Fig. 7.20(a), this device may also be placed at the high-voltage terminal side, which may be necessary if the test object has one terminal earthed. Optical links are then used to connect the CD with an instrument instead of a connecting cable 'CC'. Some essential requirements and explanations with reference to these figures as indicated by the standard are cited here:

the coupling capacitor C_k shall be of low inductance design and should exhibit a sufficiently low level of partial discharges at the specified test voltage to allow the measurement of the specified partial discharge magnitude. A higher level of partial discharges can be tolerated if the measuring system is capable of separating the discharges from the test object and the coupling capacitor and measuring them separately;

the high-voltage supply shall have sufficiently low level of *background noise* to allow the specified partial discharge magnitude to be measured at the specified test voltage;

high-voltage connections shall have sufficiently low level of *background noise* to allow the specified partial discharge magnitude to be measured at the specified test voltage;

an impedance or a filter may be introduced at high voltage to reduce *background noise* from the power supply.

The main difference between these two types of PD detection circuits is related to the way the measuring system is inserted into the circuit. In Fig. 7.20(a), the CD is at ground potential and in series to the coupling capacitor C_k as it is usually done in praxis. In Fig. 7.20(b), CD is in series with the test object C_a. Here the stray capacitances of all elements of the high-voltage side to ground potential will increase the value of C_k providing a somewhat higher sensitivity for this circuit according to eqn (7.48). The disadvantage is the possibility of damage to the PD measuring system, if the test object fails.

The new IEC Standard defines and quantifies the *measuring system characteristics*. The most essential ones will again be cited and further explained below:

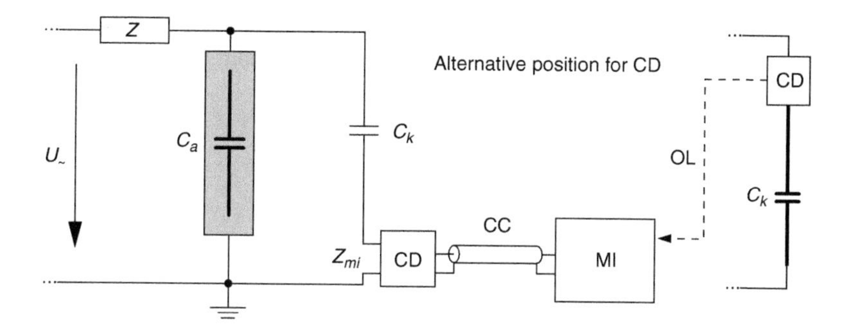

(a) Coupling device CD in series with the coupling capacitor

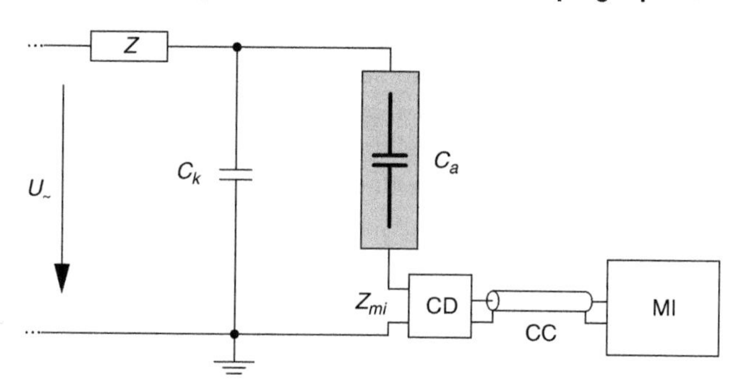

(b) Coupling device CD in series with the test object

$U_{\sim}$ high-voltage supply
Z_{mi} input impedance of measuring system
CC connecting cable
OL optical link
C_a test object
C_k coupling capacitor
CD coupling device
MI measuring instrument
Z filter

Figure 7.20 *Basic partial discharge test circuits – 'straight detection'*

The *transfer impedance* $Z(f)$ is the ratio of the output voltage amplitude to a constant input current amplitude, as a function of frequency f, when the input is sinusoidal.

This definition is due to the fact that any kind of output signal of a measuring instrument (MI) as used for monitoring PD signals is controlled by a voltage, whereas the input at the CD is a current.

The *lower and upper limit frequencies* f_1 *and* f_2 are the frequencies at which the transfer *impedance* $Z(f)$ has fallen by 6 dB from the peak pass-band value.

Midband frequency f_m *and bandwidth* Δf: for all kinds of measuring systems, the midband frequency is defined by:

$$f_m = \frac{f_1 + f_2}{2} \tag{7.49}$$

and the bandwidth by:

$$\Delta f = f_2 - f_1; \tag{7.50}$$

The *superposition error* is caused by the overlapping of transient output pulse responses when the time interval between input current pulses is less than the duration of a single output response pulse. Superposition errors may be additive or subtractive depending on the *pulse repetition rate* n of the input pulses. In practical circuits both types will occur due to the random nature of the pulse repetition rate.

This rate 'n' is defined as the ratio between the total number of PD pulses recorded in a selected time interval and the duration of the time interval.

The pulse resolution time T_r is the shortest time interval between two consecutive input pulses of very short duration, of same shape, polarity and charge magnitude for which the peak value of the resulting response will change by not more than 10 per cent of that for a single pulse. The pulse resolution time is in general inversely proportional to the bandwidth Δf of the measuring system. It is an indication of the measuring system's ability to resolve successive PD events.

The integration error is the error in apparent charge measurement which occurs when the upper frequency limit of the PD current pulse amplitude-spectrum is lower than (i) the upper cut-off frequency of a wideband measuring system or (ii) the mid-band frequency of a narrow-band measuring system.

The last definition of an 'integration error' will need some additional explanation. PD measuring systems quantifying apparent charge magnitudes are band-pass systems, which predominantly are able to suppress the high power frequency displacement currents including higher harmonics. The lower frequency limit of the band-pass f_1 and the kind of 'roll-off' of the band-pass control this ability. Adequate integration can thus only be made if the 'pass-band' or the flat part of the filter is still within the constant part of the amplitude frequency spectrum of the PD pulse to be measured. Figure 7.21,

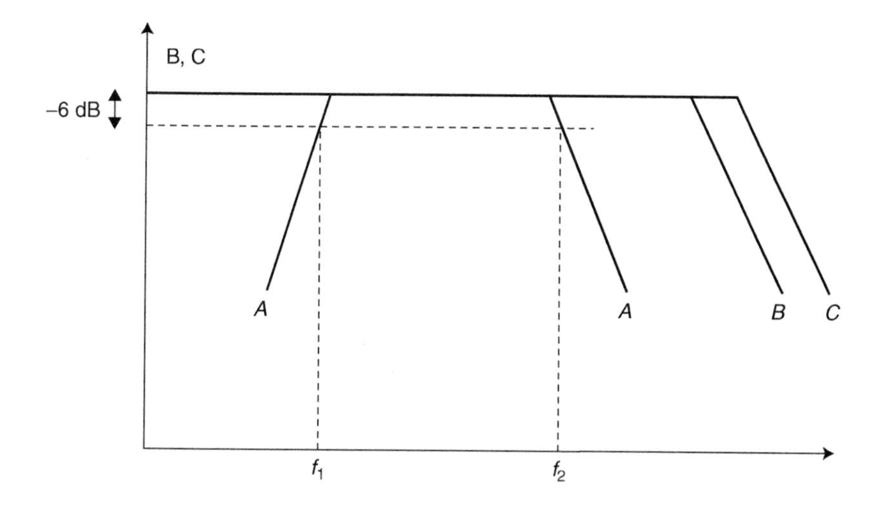

A band-pass of the measuring system

B amplitude frequency spectrum of the PD pulse

C amplitude frequency spectrum of calibration pulse

f_1 lower limit frequency

f_2 upper limit frequency

Figure 7.21 *Correct relationship between amplitude and frequency to minimize integration errors for a wide-band system*(31)

again taken from the new standard, provides at least formal information about correct relationships. More fundamental information may be found within some specific literature.(28,29)

Now we can proceed to explain the basic types of PD instruments to see how the requirements can be fulfilled.

7.3.4 Measuring systems for apparent charge

The following types of measuring systems all comprise the already mentioned subsystems: coupling device (CD), transmission system or connecting cable (CC), and a measuring instrument (MI), see Fig. 7.20. In general the transmission system, necessary to transmit the output signal of the CD to the input of the MI, does not contribute to the measuring system characteristics as both ends are matched to the characteristics of both elements. The CC will thus not be considered further.

The input impedance Z_{mi} of the CD or measuring system respectively will have some influence on the waveshape of the PD current pulse $i(t)$ as already mentioned in the explanation of Fig. 7.19. A too high input impedance will delay the charge transfer between C_a and C_k to such an extent that the upper limit frequency of the amplitude frequency spectrum would drop to unacceptable low values. Adequate values of Z_{mi} are in the range of 100 Ω.

In common with the first two measuring systems for apparent charge is a *newly* defined 'pulse train response' of the instruments to quantify the 'largest repeatedly occurring PD magnitude', which is taken as a measure of the 'specified partial discharge magnitude' as permitted in test objects during acceptance tests under specified test conditions. Sequences of partial discharges follow in general unknown statistical distributions and it would be useless to quantify only one or very few discharges of large magnitude within a large array of much smaller events as a specified PD magnitude. For further information on quantitative requirements about this pulse train response, which was not specified up to now and thus may not be found within in earlier instruments, reference is made to the standard.[31]

Wide-band PD instruments

Up to 1999, no specifications or recommendations concerning permitted response parameters have been available. Now, the following parameters are recommended. In combination with the CD, wide-band PD measuring systems, which are characterized by a *transfer impedance* $Z(f)$ having fixed values of the lower and upper limit frequencies f_1 and f_2, and adequate attenuation below f_1 and above f_2, shall be designed to have the following values for f_1, f_2 and Δf:

$$\begin{aligned} 30\,\text{kHz} \le f_1 &\le 100\,\text{kHz}; \\ f_2 &\le 500\,\text{kHz}; \\ 100\,\text{kHz} \le \Delta f &\le 400\,\text{kHz}. \end{aligned} \tag{7.51}$$

The response of these instruments to a (non-oscillating) PD current pulse is in general a well-damped oscillation as shown below. Both the apparent charge q and – with some reservation – the polarity of the PD current pulse can be determined from this response. The pulse resolution time T_r is small and is typically 5 to 20 μs.

Figure 7.22 shows the typical principle of such a system. The coupling devices CD (Fig. 7.20) are passive high-pass systems but behave more often as a parallel R-L-C resonance circuit (Fig. 7.22(a)) whose quality factor is relatively low. Such a coupling impedance provides two important qualities. At first, a simple calculation of the ratio output voltage V_0 to input current I_i in dependency of frequency (=transfer impedance $Z(f)$) would readily demonstrate an adequate suppression of low- and high-frequency currents in the neighbourhood of its resonance frequency. For a quality factor of $Q = 1$, this attenuation is already −20 dB/decade and could be greatly increased close to resonance frequency by increasing the values of Q. Secondly, this parallel circuit also performs an integration of the PD currents $i(t)$, as this circuit is already a simple band-pass filter and can be used as an integrating device. Let us assume that the PD current pulse $i(t)$ would not be influenced by the test circuit and would be an extremely short duration pulse as simulated by a Dirac function, comprising the apparent charge q. Then the calculation of the output voltage $V_0(t)$ according to Fig. 7.22(a) results in:

$$V_0(t) = \frac{q}{C}e^{-\alpha t}\left[\cos\beta t - \frac{\alpha}{\beta}\sin\beta t\right] \tag{7.52}$$

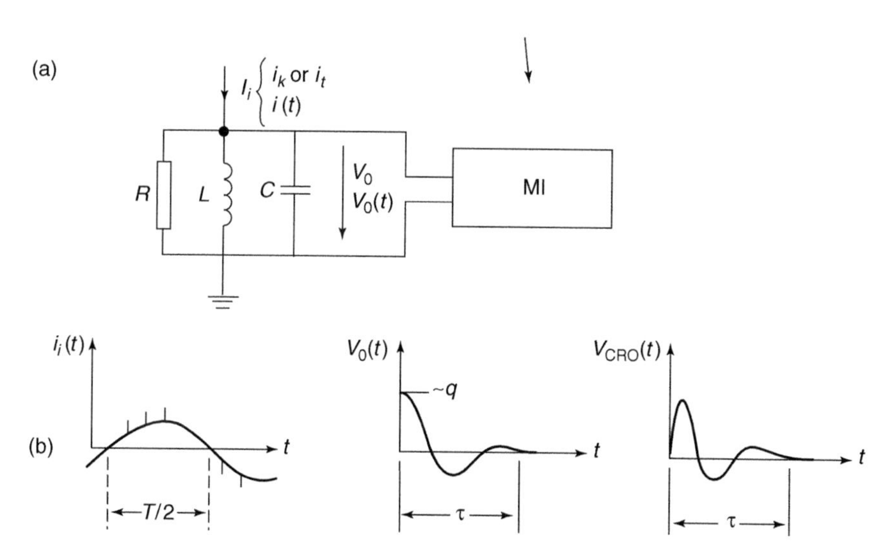

Figure 7.22 *Principle of 'wide-band' PD measuring system. (a) Simplified equivalent circuit for the CD and amplifier. (b) Typical time-dependent quantities within (a) (T = period of power frequency; $\tau \approx$ pulse resolution time T_r)*

where

$$\alpha = \frac{1}{2RC}; \quad \beta = \sqrt{\frac{1}{LC} - \alpha^2} = \omega_0\sqrt{1 - \alpha^2 LC}.$$

This equation displays a damped oscillatory output voltage, whose amplitudes are proportional to q. The integration of $i(t)$ is thus performed instantaneously ($t = 0$) by the capacitance C, but the oscillations, if not damped, would heavily increase the 'pulse resolution time T_r' of the measuring circuit and cause 'superposition errors' for too short time intervals between consecutive PD events (see definitions above). With a quality factor of $Q = 1$, i.e. $R = \sqrt{L/C}$, a very efficient damping can be achieved, as then $\alpha = \omega_0/2 = \pi f_0$. For a resonance frequency f_0 of typically 100 kHz, and an approximate resolution time of $T_r \cong t = 3/\alpha$, this time becomes about 10 μsec. For higher Q values, T_r will be longer, but also the filter efficiency will increase and therefore a compromise is necessary. The resonance frequency f_0 is also influenced by the main test circuit elements C_k and C_a, as their series connection contributes to C. The 'RLC input units' must therefore be changed according to specimen capacitance to achieve a bandwidth or resonance frequency f_0 within certain limits. These limits are postulated by the bandwidth Δf of the additional band-pass amplifier connected to this resonant circuit to increase the sensitivity and thus to provide again an integration. These amplifiers are typically designed for lower and upper limit frequencies of some 10 kHz and some 100 kHz respectively, and sometimes the lower limit frequency range may also be switched from some 10 kHz up to about 150 kHz to further suppress power frequencies. In general the fixed limit frequencies are thus within a frequency band in general not used by radio stations, and higher than the harmonics of the power supply voltages. The band-pass amplifier has in general variable amplification to feed the 'CRO' (reading device!) following the amplifier with adequate magnitudes during calibration and measurement. For a clearer understanding the time-dependent quantities (input a.c. current with superimposed PD signals, voltages before and after amplification) see Fig. 7.22(b).

Finally, the amplified discharge pulses are in general displayed by an (analogue or digital) oscilloscope superimposed on a power frequency elliptic timebase, as shown in Fig. 7.23. The magnitude of the individual PD pulses is then quantified by comparing the pulse crest values with those produced during a calibration procedure, see section 7.3.7. With this type of reading by individual persons it is not possible to quantify the standardized 'pulse train response' which quantifies the 'largest repeatedly occurring PD magnitude'. Correct readings are, however, possible by applying additional analogue peak detection circuits or digital peak detection software prepared to follow the specified pulse train response.

The pattern on the CRO display can often be used to recognize the origin of the PD sources. (Instead of a simple CRO display digital acquisition of PD quantities and up-to-date methods for evaluation are used now, see section 7.3.8.) A typical pattern of Trichel pulses can be seen in Fig. 7.23(a). Figure 7.23(c) is typical for the case for which the pulse resolution time of the measuring system including the test circuit is too large to distinguish between individual PD pulses.

It was clearly shown that even the response of such 'wide-band PD instruments' provided no more information about the original shape of the input PD current pulse as indicated in Fig. 7.22(b) and confirmed by the pattern

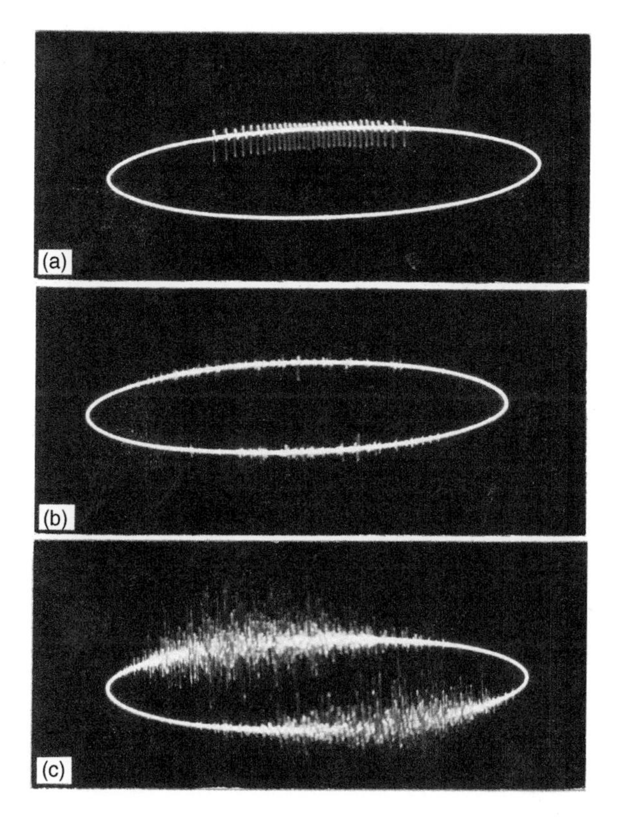

Figure 7.23 *Elliptical display. (a) Point plane ('Trichel pulses!'). (b) Void discharges at inception. (c) Void discharges at twice inception voltage*

of the Trichel pulses in Fig. 7.23(a). Figure 7.24 further confirms this statement. Here, two kinds of recorded responses – Figs 7.24(a) and (b) – of two consecutive calibration pulses ('double pulse') are shown within a time scale of microseconds. A comparison of both recorded responses shows their differences with respect to a (positive) short and lengthened input pulse, which has some significant influence on the peak value of the undershoot after the first excursion of the response which indicates the polarity of the input signal. Polarity detection by digital PD acquisition systems may thus be difficult.

Narrow-band PD instruments

It is well known that radio transmission or radiotelephony may be heavily disturbed by high-frequency interference voltages within the supply mains to which receivers are connected or by disturbing electromagnetic fields picked up by the aerials.[27] It was also early recognized that corona discharges at h.v. transmission lines are the source of such disturbances. The measurement

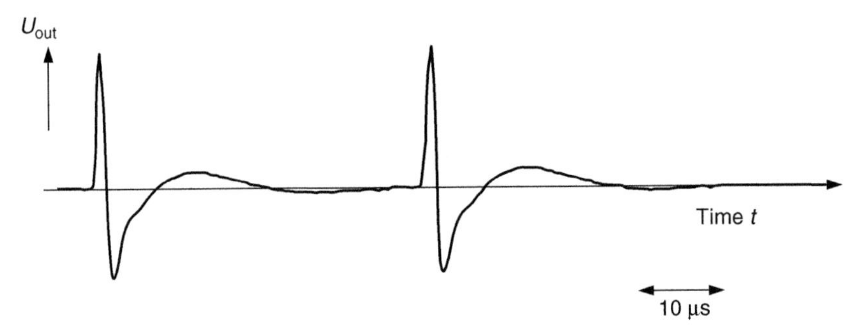

(a) Short-duration input pulse

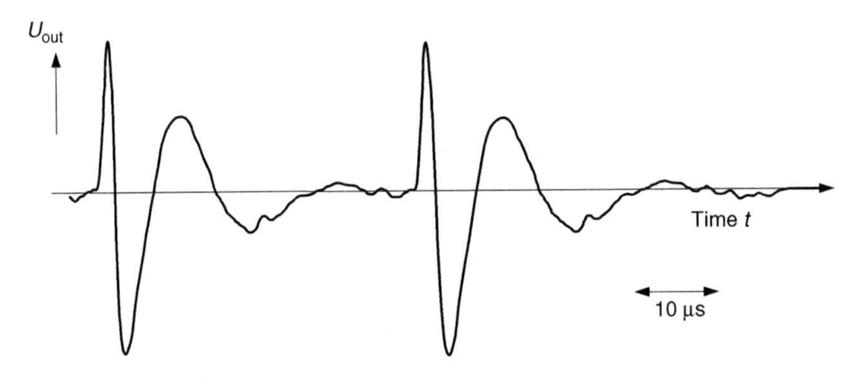

(b) Lengthened input pulse

Figure 7.24 *Output voltage signals U_{out} of a wide-band PD detector with $\Delta f = 45 \ldots 440$ kHz for two different input pulses*

of 'radio noise' in the vicinity of such transmission lines is thus an old and well-known technique which several decades ago triggered the application of this measurement technique to detect insulation failures, i.e. partial discharges, within h.v. apparatus of any kind.

The methods for the measurement of radio noise or radio disturbance have been subjected to many modifications during the past decades. Apart from many older national or international recommendations, the latest 'specifications for radio disturbance and immunity measuring apparatus and methods' within a frequency range of 10 kHz to 1000 MHz are now described in the CISPR Publication 16-1.[30] As defined in this specification, the expression 'radio disturbance voltage (RDV)', earlier termed as 'radio noise', 'radio influence' or 'radio interference' voltages, is now used to characterize the measured disturbance quantity.

Narrow-band PD instruments, which are now also specified within the new IEC Standard(31) for the measurement of the apparent charge, are very similar to those RDV meters which are applied for RDV measurements in the frequency range 100 kHz to 30 MHz. The PD instruments are characterized by a small bandwidth Δf and a mid-band frequency f_m, which can be varied over a wider frequency range, where the amplitude frequency spectrum of the PD current pulses is in general approximately constant. The recommended values for Δf and f_m for PD instruments are

$$9\,\text{kHz} \leq \Delta f \leq 30\,\text{kHz}; \text{ and}$$
$$50\,\text{kHz} \leq f_m \leq 1\,\text{MHz}. \tag{7.53}$$

It is further recommended that the transfer impedance $Z(f)$ at frequencies of $f_m \pm \Delta f$ should already be 20 dB below the peak pass-band value.

Commercial instruments of this type may be designed for a larger range of mid-band frequencies; therefore the standard provides the following note for the user. 'During actual apparent charge measurements, mid-band frequencies $f_m > 1\,\text{MHz}$ should only be applied if the readings for such higher values do not differ from those as monitored for the recommended values of f_m.' This statement denotes that only the constant part of the PD current amplitude frequency spectrum is an image of the apparent charge. As shown below in more detail, the response of these instruments to a PD current pulse is a transient oscillation with the positive and negative peak values of its envelope proportional to the apparent charge, independent of the polarity of this charge. Due to the small values of Δf, the pulse resolution time T_r will be large, typically above 80 μs.

The application of such instruments often causes some confusion for the user. A brief description of their basic working principle and their use in PD measurements will help make things clearer. Figure 7.25 displays the relevant situation and results.

In general, such instruments are used together with coupling devices providing high-pass characteristics within the frequency range of the instrument. Power frequency input currents including harmonics are therefore suppressed and we may assume that only the PD current pulses converted to PD voltage pulses are at the input of the amplifying instrument, which resembles closely a selective voltmeter of high sensitivity (or a superheterodyne-type receiver) which can be tuned within the frequency range of interest. Such a narrow-band instrument is again a quasi-integration device for input voltage pulses. To demonstrate this behaviour, we assume (Fig. 7.25(a)) an input voltage $v_1(t) = V_0 \exp(-t/T)$, i.e. an exponentially decaying input pulse which starts suddenly with amplitude V_0 (see Fig. 7.25(b)). The integral of this pulse, $\int_0^\infty v_1(t)\,dt$, is $V_0 T$ and is thus a quantity proportional to the apparent charge q of a PD current pulse. The complex frequency spectrum of this impulse is then given by applying the Fourier integral

$$V_1(j\omega) = \int_0^{\infty} v_1(t)\exp(-j\omega t)\,dt = \frac{V_0 T}{1 + j\omega T} = \frac{S_0}{1 + j\omega T} \tag{7.53}$$

and the amplitude frequency spectrum $|V_1(i\omega)|$ by

$$|V_1(j\omega)| = \frac{V_0 T}{\sqrt{1 + (\omega T)^2}} = \frac{S_0}{\sqrt{1 + (\omega T)^2}} \tag{7.54}$$

where S_0 is proportional to q. From the amplitude frequency spectrum, sketched in Fig. 7.25(c), it is obvious that the amplitudes decay already to

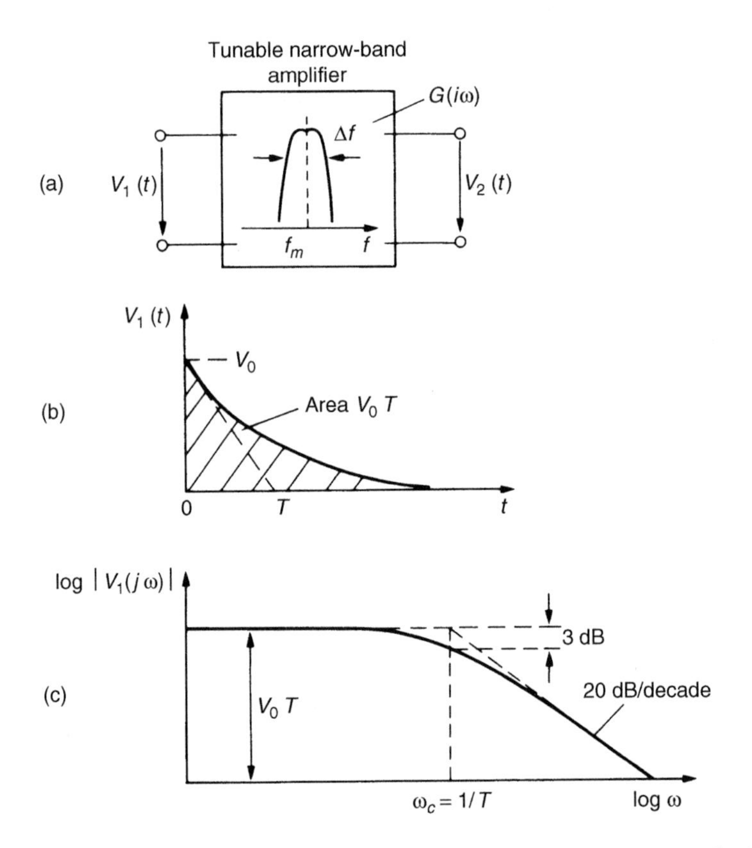

Figure 7.25 *Narrow-band amplifiers: some explanations to the impulse response. (a) Block diagram. (b) Input voltage $V_1(t)$, see text. (c) Amplitude frequency spectrum from $V_1(t)$. (d) Idealized transfer function of narrow-band amplifier. (e) Computed impulse response according to eqn (7.56) for $f = 150$ kHz and $\Delta f \cong 9$ kHz*

(d)

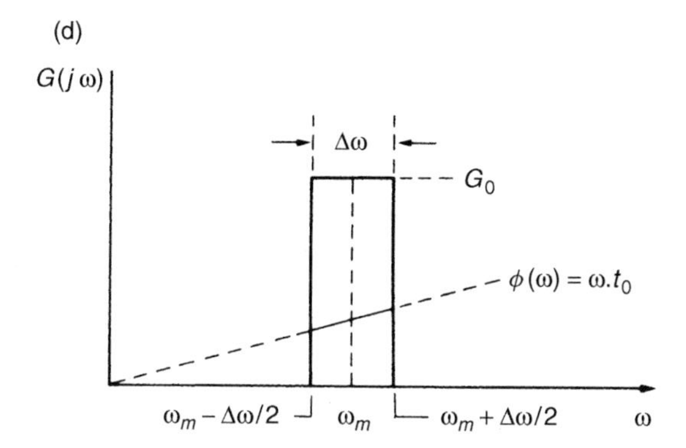

(e)

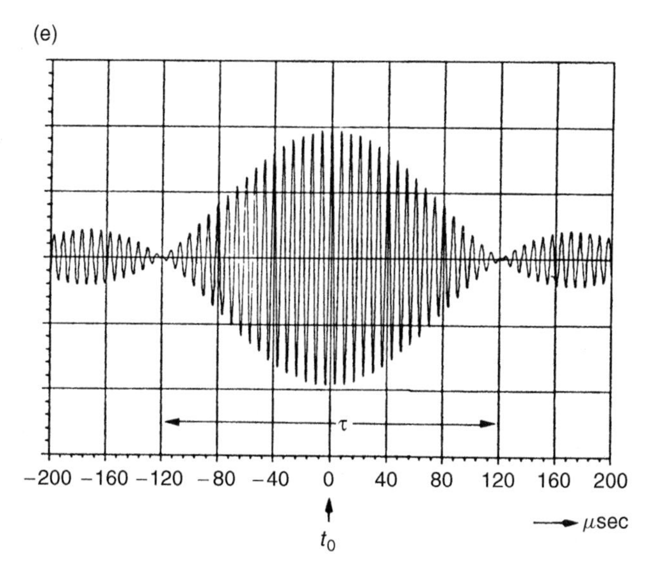

Figure 7.25 *(continued)*

$-3\,\text{dB}$ or more than about 30 per cent for the angular frequency of $\omega_c > 1/T$. This critical frequency f_c is for $T = 0.1\,\mu\text{sec}$ only 1.6 MHz, a value which can be assumed for many PD impulses. As the indication of a narrow-band instrument, if tuned to f_m, will be proportional to the relevant amplitude of this spectrum at f_m the recommendations of the new standard can well be understood. If the input PD current pulse is, however, distorted by oscillations, the amplitude frequency spectrum would also be distorted by maxima and minima which can then be recorded by tuning f_m.

If the narrow-band instrument is tuned to the constant part of the spectrum which is proportional to q, we may also assume a Dirac pulse or delta function of magnitude $V_0T = S_0$ to calculate its output voltage $V_2(t)$. As the spectrum of a Dirac pulse is constant for all frequencies, the response $v_2(t)$ is then proportional to S_0 at any frequency f_m. The impulse response of the instrument is then of course dependent upon the exact (output/input voltage) transfer function $G(j\omega)$ of the system; we may, however, approximate the actual band-pass characteristic by an idealized one as shown in Fig. 7.25(d), with a mid-band angular frequency ω_m, an angular bandwidth $\Delta\omega$ and the constant amplitude or 'scale factor' G_0 within $\omega_m \pm (\Delta\omega/2)$. For such ideal band-pass systems and especially narrow-band amplifiers the phase shift $\phi(\omega)$ may well be assumed to be linear with frequency as indicated, at least within the band-pass response. With this approximation no phase distortion is assumed, and t_0 (see Fig. 7.25(d)) is equal to the delay time of the system. The impulse response with S_0 as input pulse appearing at $t = 0$ can then be evaluated[47,48] from

$$v_2(t) = \frac{1}{\pi}\int_{\omega_m-\Delta\omega/2}^{\omega_m+\Delta\omega/2} S_0 G_0 \cos[\omega(t - t_0)]\,d\omega \tag{7.55}$$

This integral can easily be solved; the result is

$$v_2(t) = \frac{S_0 G_0 \Delta\omega}{\pi}\mathrm{si}\left[\frac{\Delta\omega}{2}(t - t_0)\right]\cos\omega_m(t - t_0) \tag{7.56}$$

where $\mathrm{si}(x) = \sin(x/x)$.

Equation (7.56) shows an oscillating response whose main frequency is given by $f_m = \omega_m/2\pi$, the amplitudes are essentially given by the $\mathrm{si}(x)$ function which is the envelope of the oscillations. A calculated example for such a response is shown in Fig. 7.25(e). The maximum value will be reached for $t = t_0$ and is clearly given by

$$V_{2\max} = \frac{S_0 G_0 \Delta\omega}{\pi} = 2S_0 G_0 \Delta f \tag{7.57}$$

where Δf is the idealized bandwidth of the system. Here, the two main disadvantages of narrow-band receivers can easily be seen: first, for $\Delta\omega \ll \omega_m$ the positive and negative peak values of the response are equal and therefore the polarity of the input pulse cannot be detected. The second disadvantage is related to the long duration of the response. Although more realistic narrow-band systems will effectively avoid the response amplitudes outside of the first zero values of the $(\sin x)/x$ function, the full length τ of the response, with τ as defined by Fig. 7.25(e), becomes

$$\tau = \frac{2}{\Delta f} = \frac{4\pi}{\Delta\omega}, \tag{7.58}$$

being quite large for small values of Δf, due to the actual definition of the 'pulse resolution time T_r' as defined before. This quantity is about 10 per cent smaller than τ, but still much larger than for wide-band PD detectors.

Simple narrow-band detectors use only RLC resonant circuits with high quality factors Q, the resonance frequency of which cannot be tuned. Although then their responses are still quite similar to the calculated one (eqn 7.56), we show such a response for a 'double pulse' in Fig. 7.26, taken from a

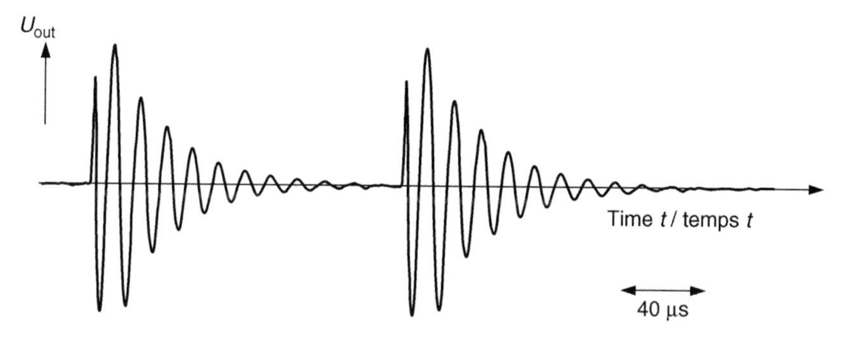

Figure 7.26 *Response of a simple narrow-band circuit with $\Delta f = 10$ kHz; $f_m = 75$ kHz*

commercial PD instrument. As the time scale is shown and data for the frequencies applied are provided, no further explanations are necessary. High-quality tunable detectors apply the heterodyne principle. Responses for such instruments can be taken from a RDV meter in front of the 'psophometric weighting circuit' (see Fig. 7.29) as shown in Fig. 7.27. Here again a situation (Fig. 7.27(b)) is displayed for which superposition errors occurs.

Radio disturbance (interference) meters for the detection of partial discharges

As instruments such as those specified by the International Special Committee on Radio Disturbance (Comité International Spécial des Perturbation Radioélectrique, CISPR) of IEC[30] or similar organizations are still in common use for PD detection, the possible application of an 'RDV' or 'RIV' meter is still mentioned within the new standard.[31] New types of instruments related to the CISPR Standard are often able to measure 'radio disturbance voltages, currents and fields' within a very large frequency range, based on different treatment of the input quantity. Within the PD standard, however, the expression 'Radio Disturbance Meter' is only applied for a specific radio disturbance (interference) measuring apparatus, which is specified for a frequency band of 150 kHz to 30 MHz (band B) and which fulfils the requirements for a so-called 'quasi-peak measuring receivers'.

In Fig. 7.28 a block diagram of such a simple RIV meter is sketched and compared with the principle of a narrow-band PD instrument as described and

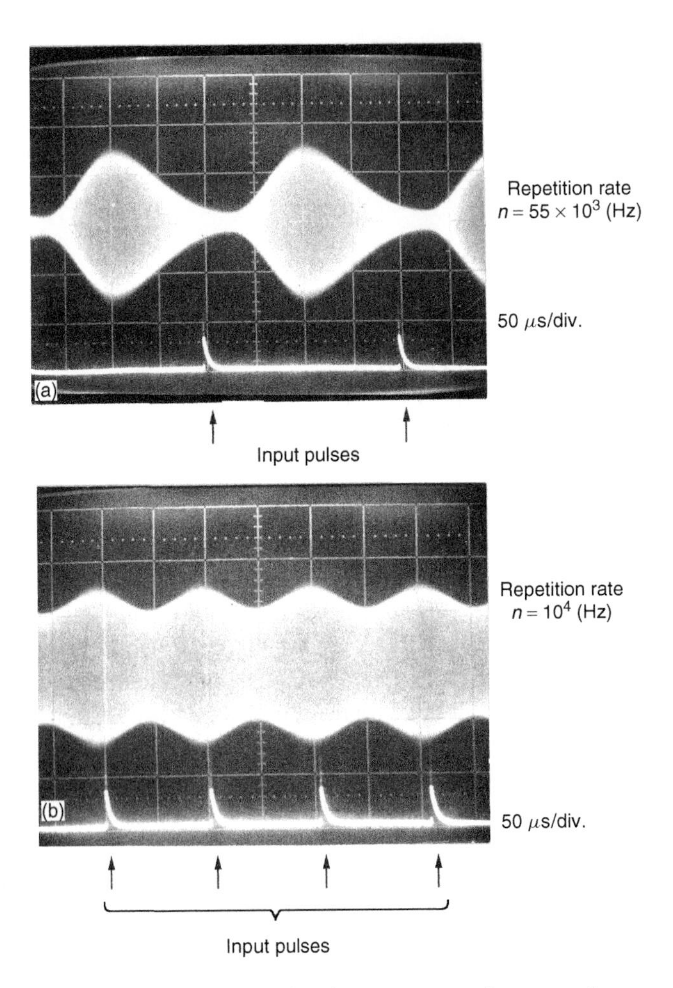

Figure 7.27 *Measured pulse response of an actual narrow-band detector (RIV meter). Signals taken from the intermediate frequency amplifiers for repetitive input signals (a) with adequate and (b) inadequate time distances. Bandwidth* $\Delta f \cong 9$ *kHz*

discussed before. The main difference is only the 'quasi-peak' or 'psophometric weighting circuit' which simulates the physiological noise response of the human ear. As already mentioned within the introduction of this section, forthcoming PD instruments will be equipped with a similar, but different circuit with a 'pulse train response' quantifying the 'largest repeatedly occurring PD magnitudes'. Within the block diagram of Fig. 7.28, the simplified coupling device as indicated by a resistance shunted by the inductance L forms a transfer impedance Z_m with a high-pass characteristic which for RDV meters

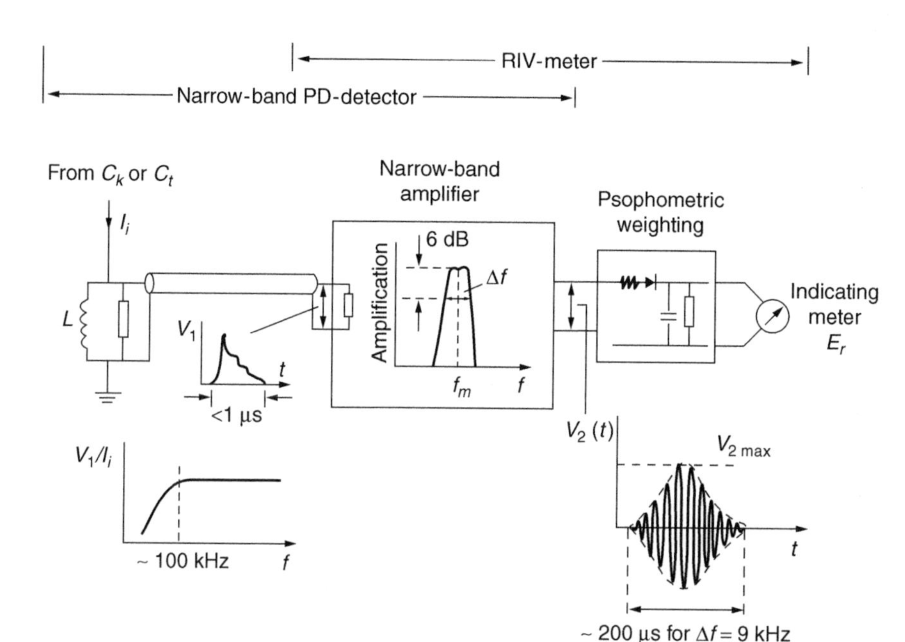

Figure 7.28 *Block diagram of a quasi-peak RIV meter including weighting circuit compared with PD narrow-band PD detector*

have standardized values. Based on the derivations as already made for the calculation of eqn (7.57) we can now easily quantify the differences of both types of meter.

The quasi-peak RDV meters are designed with a very accurately defined overall pass-band characteristic fixed at $\Delta f = 9$ kHz. They are calibrated in such a way that the response to Dirac type of equidistant input pulses providing each a volt–time area of $0.316\,\mu$Vs at a pulse repetition frequency (N) of 100 Hz is equal to an unmodulated sine-wave signal at the tuned frequency having an e.m.f. of 2 mV r.m.s. as taken from a signal generator driving the same output impedance as the pulse generator and the input impedance of the RIV meter. By this procedure the impulse voltages as well as the sine-wave signal are halved. As for this repetition frequency of 100 Hz the calibration point shall be only 50 per cent of $V_{2\max}$ in eqn (7.57), the relevant reading of the RDV meter will be

$$E_{\mathrm{RDV}} = \frac{1}{2\sqrt{2}} 2 S_0 G_0 \Delta f = \frac{S_0 G_0 \Delta f}{\sqrt{2}} \tag{7.59}$$

As $G_0 = 1$ for a proper calibration and $\Delta f = 9$ kHz, $S_0 = 158\,\mu$Vs, the indicated quantity is $S_0 \Delta f/\sqrt{2} = 1$ mV or 60 dB (μV), as the usual reference quantity is 1 μV. RDV meters are thus often called 'microvolt meters'!

This response is now weighted by the 'quasi-peak measuring circuit' with a specified electrical charging time constant τ_1 (=1 ms), an electrical discharging time constant τ_2 (=160 ms) and by an output voltmeter, which, for conventional instruments, is of moving coil type, critically damped and having a mechanical time constant τ_3 (=160 ms). This procedure makes the reading of the output voltmeter dependent on the pulse repetition frequency N. This non-linear function $f(N)$ as available from reference 30 or 31 is shown in Fig. 7.29 and is only accurate if the input pulses are equidistant *and* of equal amplitudes! It can be seen that for $N > 1000$ the function $f(N)$ would saturate to a value of 2, for which, however, superposition errors occur.

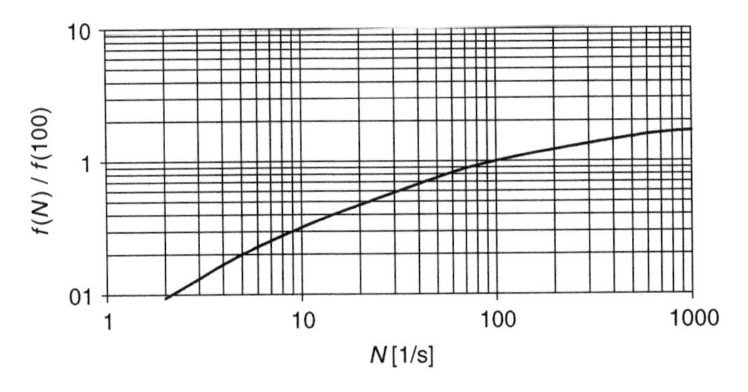

Figure 7.29 *Variation of CISPR radio interference meter reading with repetition frequency N, for constant input pulses*

With this function $f(N)$ we can now finalize the reading of an RIV meter by taking the transfer impedance Z_m of our CD in eqn (7.59) into account, which converts input PD currents into input voltages $v_1(t)$. For RDV meters, this transfer impedance, the real value of which $|Z_m|$ is constant for the frequency range under consideration, the quantity S_0 in eqn (7.59) may then be written as

$$S_0 = \int v_1(t)\,\mathrm{d}t = |Z_m| \int i_1(t)\,\mathrm{d}t = |Z_m|\,q, \tag{7.60}$$

where q is the measured charge quantity for an impulse current $i_1(t)$. Now eqn (7.59) becomes

$$E_{\mathrm{RDV}} = \frac{G_0}{\sqrt{2}} q \Delta f\,|Z_m|\,f(N) \tag{7.61}$$

With this equation conversion factors between the measured charge q and the indicated voltage by an RDV meter can be calculated. For $N = 100$ equidistant pulses of equal magnitude ($f(N) = 1$), $\Delta f = 9\,\mathrm{kHz}$, correct calibration ($G_0 = 1$) and a reading of 1 mV ($=E_{\mathrm{RDV}}$) or 60 dB, charge magnitudes of 1 (or 2.6) nC for $|Z_m| = 150$ (or 60) Ω can be calculated. These relationships have also been confirmed experimentally.[34,35] Instead of eqn (7.61) the new standard[31] displays in Annex D a reading in which the first term of eqn (7.61) is generalized, namely

$$U_{\mathrm{RDV}} = \frac{q \Delta f Z_m f(N)}{k_i}$$

where

N = pulse repetition frequency,
$f(N)$ = the non-linear function of N (see Fig. 7.29),
Δf = instrument bandwidth (at 6 dB),
Z_m = value of a purely resistive measuring input impedance of the instrument,
k_i = the scale factor for the instrument ($=q/U_{\mathrm{RDV}}$)

As, however, the weighting of the PD pulses is different for narrow-band PD instruments and quasi-peak RDV meters, there is no generally applicable conversion factor between readings of the two instruments. The application of RDV meters is thus not forbidden; but if applied the records of the tests should include the readings obtained in microvolts and the determined apparent charge in picocoulombs together with relevant information concerning their determination.

Ultra-wide-band instruments for PD detection

The measurement of PD current pulses as briefly treated in section 7.3.2 belongs to this kind of PD detection as well as any similar electrical method to quantify the intensity of PD activities within a test object. Such methods need coupling devices with high-pass characteristics which shall have a pass-band up to frequencies of some 100 MHz or even higher. Records of the PD events are then taken by oscilloscopes, transient digitizers or frequency selective voltmeters especially spectrum analysers. For the location of isolated voids with partial discharges in cables a bandwidth of about some 10 MHz only is useful, whereas tests on GIS (gas-insulated substations or apparatus) measuring systems with ‘very high’ or even ‘ultra-high’ frequencies (VHF or UHF methods for PD detection) can be applied. This is due to the fact that the development of any partial discharge in sulphur hexafluoride is of extremely short duration providing significant amplitude frequency spectra up to the GHz region. More information concerning this technique can be found in the literature.[54,55,56]

As none of these methods provides integration capabilities, they cannot quantify apparent charge magnitudes, but may well be used as a diagnostic tool.[26]

7.3.5 Sources and reduction of disturbances

Within the informative Annex G of the IEC Standard[31] sources and suggestions regarding the reduction of disturbances are described in detail. A citation of some of the original text together with some additional information is thus adequate.

> Quantitative measurements of PD magnitudes are often obscured by interference caused by disturbances which fall into two categories:
>
> Disturbances which occur even if the test circuit is not energized. They may be caused, for example, by switching operations in other circuits, commutating machines, high-voltage tests in the vicinity, radio transmissions, etc., including inherent noise of the measuring instrument itself. They may also occur when the high-voltage supply is connected but at zero voltage.
>
> Disturbances which only occur when the test circuit is energized but which do not occur in the test object. These disturbances usually increase with increasing voltage. They may include, for example, partial discharges in the testing transformer, on the high-voltage conductors, or in bushings (if not part of the test object). Disturbances may also be caused by sparking of imperfectly earthed objects in the vicinity or by imperfect connections in the area of the high voltage, e.g. by spark discharges between screens and other high-voltage conductors, connected with the screen only for testing purposes. Disturbances may also be caused by higher harmonics of the test voltage within or close to the bandwidth of the measuring system. Such higher harmonics are often present in the low-voltage supply due to the presence of solid state switching devices (thyristors, etc.) and are transferred, together with the noise of sparking contacts, through the test transformer or through other connections, to the test and measuring circuit.

Some of these sources of disturbances have already been mentioned in the preceding sections and it is obvious that up to now numerous methods to reduce disturbances have been and still are a topic for research and development, which can only be mentioned and summarized here.

The most efficient method to reduce disturbances is *screening and filtering*, in general only possible for tests within a shielded laboratory where all electrical connections running into the room are equipped with filters. This method is expensive, but inevitable if sensitive measurements are required, i.e. if the PD magnitudes as specified for the test objects are small, e.g. for h.v. cables.

Straight PD-detection circuits as already shown in Fig. 7.20 are very sensitive to disturbances: any discharge within the entire circuit, including h.v. source, which is not generated in the test specimen itself, will be detected by the coupling device CD. Therefore, such 'external' disturbances are not rejected. Independent of screening and filtering mentioned above, the testing transformer itself should be PD free as far as possible, as h.v. filters or inductors as indicated in Fig. 7.20 are expensive. It is also difficult to avoid any partial discharges at the h.v. leads of the test circuit, if the test voltages are very high. A basic improvement of the straight detection circuit may therefore become necessary by applying a 'balanced circuit', which is similar to a Schering bridge. In Fig. 7.30 the coupling capacitor C_K and test specimen C_t form the h.v. arm of the bridge, and the l.v. arms are basically analogous to a Schering bridge. As C_K is not a standard capacitor but should be PD free, the dissipation factor $\tan\delta_K$ may also be higher than that of C_t, and therefore the capacitive branch of the l.v. arm may be switched to any of the two arms. The bridge can then be adjusted for balance for all frequencies at which $\tan\delta_K = \tan\delta_t$. This condition is best fulfilled if the same insulation media are used within both capacitors. The use of a partial discharge-free sample for C_K of the same type as used in C_t is thus advantageous. If the frequency dependence of the dissipation factors is different in the two capacitors, a complete balance within a larger frequency range is not possible. Nevertheless, a fairly good balance can be reached and therefore most of the sinusoidal or transient voltages appearing at the input ends of C_K and C_t cancel out between the points 1 and 2. A discharge within the test specimen, however, will contribute to voltages of opposite polarity across the l.v. arms, as the PD current is flowing in opposite directions within C_K and C_t.

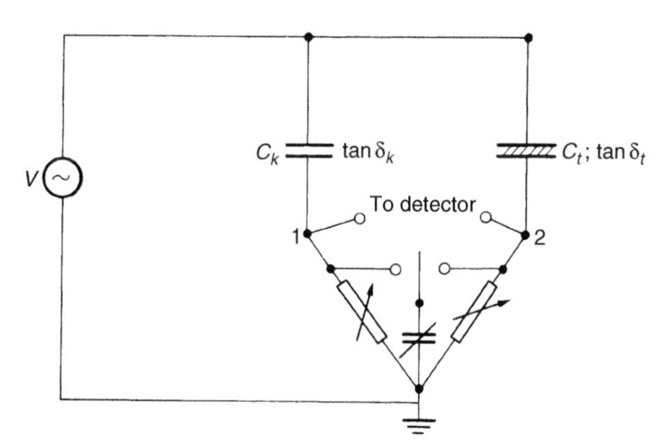

Figure 7.30 *Differential PD bridge (balanced circuit)*

Polarity discrimination methods take advantage of the effect of opposite polarities of PD pulses within both arms of a PD test circuit. Two adequate coupling devices CD and CD_1 as shown in Fig. 7.31 transmit the PD signals to the special measuring instrument MI, in which a logic system performs the comparison and operates a gate for pulses of correct polarity. Consequently only those PD pulses which originate from the test object are recorded and quantified. This method was proposed by I.A. Black.[37,38]

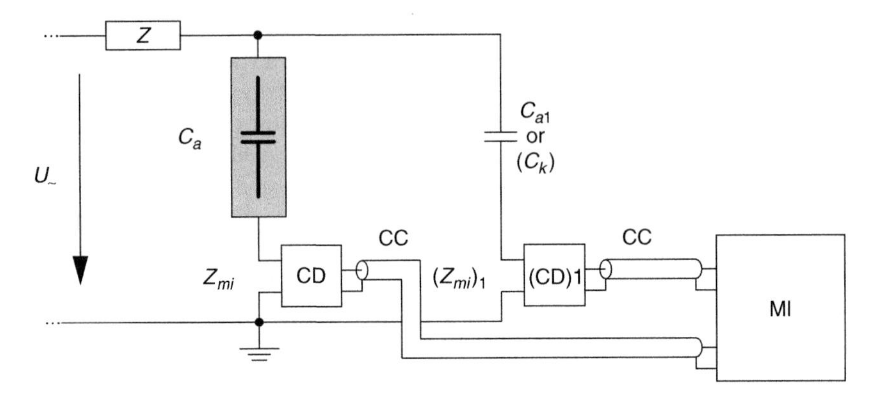

Figure 7.31 *Polarity discrimination circuit*

Another extensively used method is the *time window method* to suppress interference pulses. All kinds of instruments may be equipped with an electronic gate which can be opened and closed at preselected moments, thus either passing the input signal or blocking it. If the disturbances occur during regular intervals the gate can be closed during these intervals. In tests with alternating voltage, the real discharge signals often occur only at regularly repeated intervals during the cycles of test voltage. The time window can be phase locked to open the gate only at these intervals.

Some more sophisticated methods use digital acquisition of partial discharge quantities, to which the final section 7.3.8 is devoted.

7.3.6 Other PD quantities

The measurement of the 'apparent charge q' as the fundamental PD quantity is widely acknowledged and used today, and only the 'largest repeatedly occurring magnitudes' of this kind are usually specified. Individual charge magnitudes q_i are different, however, as well as the number of partial discharges recorded within a selected reference time interval. But the deterioration process within an insulation system is certainly a result of all discharges and is not limited to the maximum values only. Much research work has been related to the measurement of all single PD impulses and to the evaluation of the results on a statistical basis. Such measuring systems are known as PD pulse analysers and depending on the performance of the detection and analysing systems, the number of pulses, the pulse intervals or the amplitudes of the individual pulses may be recorded and stored (section 7.3.8).

Such additional quantities related to PD pulses, although already mentioned in earlier standards, will be much more used in future and thus their definitions are given below with brief comments only:

(a) The phase angle ϕ_i and time t_i of occurrence of a PD pulse is

$$\phi_i = 360(t_i/T) \tag{7.62}$$

where t_i is the time measured between the preceding positive going transition of the test voltage through zero and the PD pulse. Here T is the period of the test voltage.

(b) The *average discharge current I* is the sum of the absolute values of individual apparent charge magnitudes qi during a chosen reference time interval T_{ref} divided by this time interval, i.e.:

$$I = \frac{1}{T_{\text{ref}}}(|q_1| + |q_2| + \ldots + |q_i|) \tag{7.63}$$

This current is generally expressed in coulombs per second or in amperes. By this definition a quantity is available which includes all individual PD pulses as well as the pulse repetition rate n. The measurement of this quantity is possible based upon either linear amplification and rectification of the PD discharge currents, by processing the output quantities of the apparent charge detectors by integration and averaging or by digital post processing. This average discharge current has not been investigated extensively up to now, although early investigations show quite interesting additional information about the impact on the lifetime of insulation.(31)

(c) *The discharge power P* is the average pulse power fed into the terminals of the test object due to apparent charge magnitudes q_i during a chosen reference time interval T_{ref}, i.e.:

$$P = \frac{1}{T_{\text{ref}}}(q_1u_1 + q_2u_2 + \ldots + q_iu_i) \tag{7.64}$$

where $u_1, u_2, \ldots, u_i$ are instantaneous values of the test voltage at the instants of occurrence t_i of the individual apparent charge magnitudes q_i. This quantity is generally expressed in watts. In this equation the sign of the individual values must be strictly observed, which is often difficult to fulfil. Narrow-band PD instruments are not able to quantify the polarity of PD events and even the response of wide-band instruments may not be clear, see Fig. 7.26. In the vicinity of the test voltage zero PD pulses and instantaneous voltage are often different in polarity!(36,45) As discharge energy is directly related to discharge power, this quantity is always directly related to insulation decomposition.(46)

(d) The *quadratic rate D* is the sum of the squares of the individual apparent charge magnitudes q_i during a chosen reference time interval T_{ref} divided by this time interval, i.e.:

$$D = \frac{1}{T_{\text{ref}}}(q_1^2 + q_2^2 + \ldots + q_m^2) \tag{7.65}$$

and is generally expressed in (coulombs)^2 per second. Although this quantity appears to have no advantages compared to the measurement of the maximum values of q only[32], some commercially available, special instruments record this quantity.

7.3.7 Calibration of PD detectors in a complete test circuit

The reasons why any PD instrument providing continuously variable sensitivity must be calibrated in the complete test circuit have mainly be explained within sections 7.3.1 to 7.3.3. Even the definition of the 'apparent charge q' is based on a routine calibration procedure, which shall be made with each new test object. Calibration procedures are thus firmly defined within the standard.[31]

A calibration of measuring systems intended for the measurement of the fundamental quantity q is made by injecting short duration repetitive current pulses of well-known charge magnitudes q_0 across the test object, whatever test circuit is used. For an example, see Fig. 7.32. These current pulses are generally derived from a calibrator which comprises a generator producing step voltage pulses (see 'G') of amplitude V_0 in series with a precision capacitor C_0. If the voltages V_0 also remain stable and are exactly known, repetitive calibration pulses with charge magnitudes of $\boldsymbol{q_0=V_0C_0}$ are injected. A short rise time of 60 ns is now specified for the voltage generator to produce current pulses with amplitude frequency spectra which fit the requirements set by the bandwidth of the instruments and to avoid integration errors if possible.

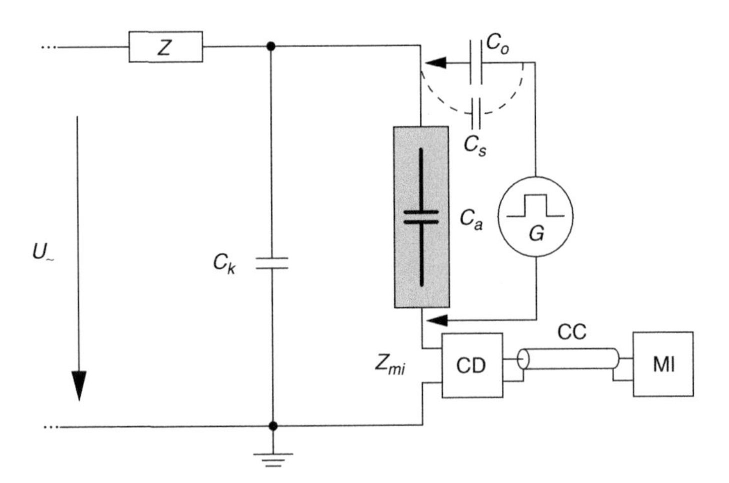

Figure 7.32 *The usual circuit for the calibration of a PD measuring instrument MI within the complete test circuit. For identification of circuit elements see text and Fig. 7.20*

Whereas further details for the calibration procedures shall not be discussed here, the new philosophy in reducing measuring errors during PD tests will be presented.[52]

It has been known for some time that measuring uncertainties in PD measurements are large. Even today, PD tests on identical test objects performed with different types of commercially available systems will provide different results even after routine calibration performed with the same calibrator. The main reasons for this uncertainty are the different transfer impedances (bandwidth) of the measuring systems, which up to 1999 have never been well defined and quantified. The new but not very stringent requirements[31] related to this property will improve the situation; together with other difficulties related to disturbance levels measuring uncertainties of more than about 10 per cent may, however, exist. The most essential part of the new philosophy concerns the calibrators, for which – up to now – no requirements for their performance exist. Tests on daily used commercial calibrators sometimes display deviations of more than 10 per cent of their nominal values. Therefore routine type, and performance tests on calibrators have been introduced with the new standard. At least the first of otherwise periodic performance tests should be traceable to national standards, this means they shall be performed by an accredited calibration laboratory. With the introduction of this requirement it can be assumed that the uncertainty of the calibrator charge magnitudes q_0 can be assessed to remain within ±5 per cent or 1 pC, whichever is greater, from its nominal values. Very recently executed intercomparison tests on calibrators performed by accredited calibration laboratories showed that impulse charges can be measured with an uncertainty of about 3 per cent.[42]

7.3.8 Digital PD instruments and measurements

Between 1970 and 1980 the state of the art in computer technology and related techniques rendered the first application of digital acquisition and processing of partial discharge magnitudes.[39,40,41] Since then this technology was applied in numerous investigations generally made with either instrumentation set up by available components or some commercial instruments equipped with digital techniques. One task for the working group evaluating the new IEC Standard was thus concerned with implementing some main requirements for this technology. It is again not the aim of this section to go into details of digital PD instruments, as too many variations in designing such instruments exist. Some hints may be sufficient to encourage further reading.

Digital PD instruments are in general based on analogue measuring systems or instruments for the measurement of the apparent charge q (see section 7.3.4) followed by a digital acquisition and processing system. These digital parts of the system are then used to process analogue signals for further evaluation, to store relevant quantities and to display test results. It is possible that in the near future a digital PD instrument may also be based on a high-pass coupling device and a digital acquisition system without the analogue signal processing front end. The availability of cheap but extremely fast flash A/D converters and digital signal processors (DSPs) performing signal integration is a prerequisite for such solutions.

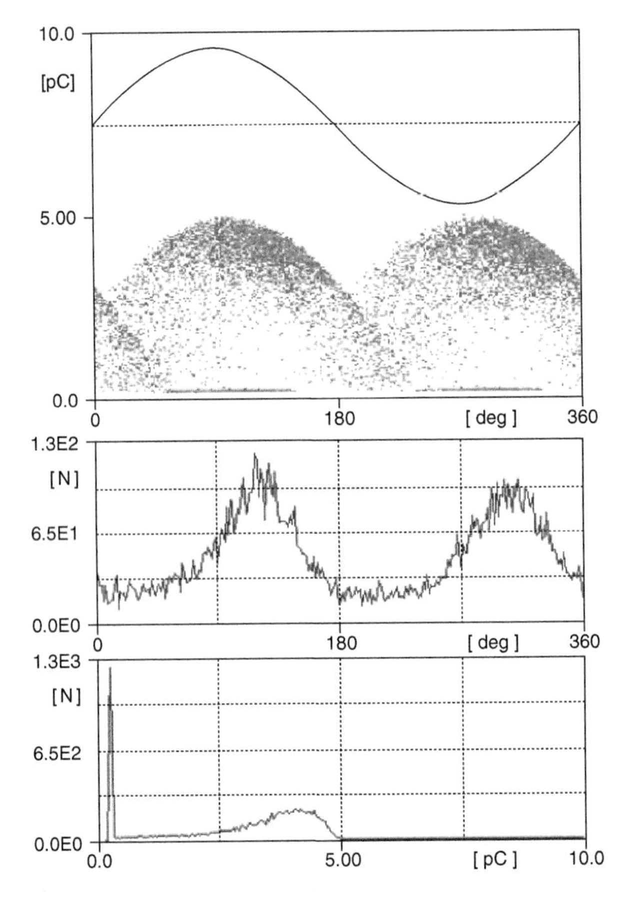

Figure 7.33 *The pattern of a phase-resolved PD measurement for a moving metal particle within a GIS. Further details see text (courtesy FKH, Zurich, Switzerland)*

The main objective of applying digital techniques to PD measurements is based on recording in real time at least most of consecutive PD pulses quantified by its apparent charge q_i occurring at time instant t_i and its instantaneous values of the test voltage u_i occurring at this time instant t_i or, for alternating voltages, at phase angle of occurrence ϕ_i within a voltage cycle of the test voltage. As, however, the quality of hard- and software used may limit the accuracy and resolution of the measurement of these parameters, the new standard[(31)] provides some recommendations and requirements which are relevant for capturing and registration of the discharge sequences.

One of the main problems in capturing the output signals of the analogue front end correctly may well be seen from Figs 7.24 and 7.26, in which three output signals as caused by two consecutive PD events are shown. Although none of the signals is distorted by superposition errors, several peaks of each signal with different polarities are present. For the wideband signals, only the first peak value shall be captured and recorded including polarity, which is not easy to do. For the narrow-band response for which polarity determination is not necessary, only the largest peak is proportional to the apparent charge. For both types of signals therefore only one peak value shall be quantified, recorded and stored within the pulse resolution time of the analogue measuring system. Additional errors can well be introduced by capturing wrong peak values which add to the errors of the analogue front end.

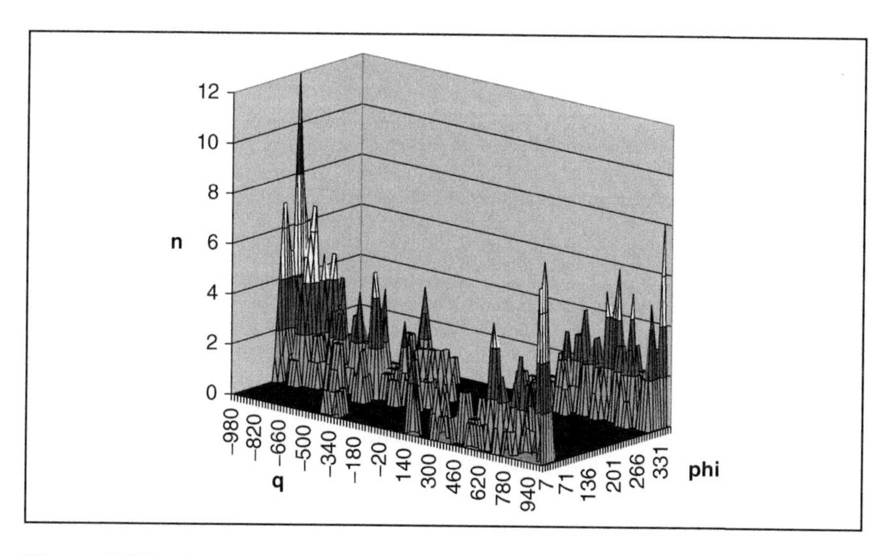

Figure 7.34 *An example of a $\phi - q - n$ diagram. On-site PD measurements performed on an h.v. cable, heavy partial discharges at a terminator (courtesy Presco AG, Weiningen, Switzerland)*

Further aims of PD instruments are related to post-processing of the recorded values. Firstly, the so-called '$\phi_i - q_i - n_i$' patterns as available from the recorded and stored data in which n_i is the number of identical or similar PD magnitudes recorded within short time (or phase) intervals and an adequate total recording duration can be used to identify and localize the origin of the

PDs based on earlier experience(see, e.g., 49,51,53,57,59,67) and/or even to establish physical models for specific PD processes.(60 and cited references) If recorded raw data are too much obscured by disturbances, quite different numerical methods may also be applied to reduce the disturbance levels.[61,62]

We end this chapter with two records of results from PD tests made with digital PD instrument. In Figs 7.33 and 7.34 typical test results of phase resolved PD measurement for a moving metal particle within a GIS and on-site PD measurements performed on HV cable (at a terminator) are shown.

For further reading about PD measurements and their applications see references 63 to 66.

References

1. A.K. Jonscher, *Dielectric Relaxation in Solids*. Chelsea Dielectrics Press, London, 1983.
2. Waldemar von Münch, *Elektrische und magnetische Eigenschaften der Materie*. B.G. Teubner, Stuttgart, 1987.
3. A.K. Jonscher and R.M. Hill *Physics of Thin Films*, Vol. 8, 1975, pp. 169–249.
4. N.F Mott and E.A. Davies *Electronic Processes in Non-cristalline Materials*. Oxford University Press 1979.
5. Der Houhanessian, Vahe. Measurement and Analysis of Dielectric Response in Oil-Paper Insulation Systems. Thesis Swiss Federal Institute of Technology, Diss. ETH No. 12, 832, 1998.
6. Bernhard Gross. Dielectric relaxation functions and models. *J. Appl. Phys.* **67**(10), May 1990, pp. 6399–6404.
7. Der Houhanessian, Vahe and Zaengl, Walter S. Application of relaxation current measurements to on-site diagnosis of power transformers. 1997 IEEE Annual Report, Conference on El. Insulation and Dielectric Phenomena, Minneapolis, October 1997, pp. 45–51.
8. J. Jalbert and R. Gilbert Decomposition of Transformer Oils: A New Approach for the Determination of Dissolved Gases. *IEEE Trans. PD* Vol. 12, pp. 754–760, 1997.
9. IEC Standard 61181 (1993). Impregnated insulating materials – Application of dissolved gas analysis (DGA) to factory tests on electrical equipment.
10. E. Ildstad, P. Thärning and U. Gäfvert Relation between voltage return and other methods for measurement of dielectric response. 1994 IEEE Symposium on El. Insulation, June 1994, Baltimore, USA.
11. A. Bognar, L. Kalocsai, G. Csepes, E. Nemeth and J. Schmidt Diagnostic tests of high voltage oil-paper insulating systems (in particular transformer insulation) using DC dielectrometrics. CIGRE 1990 Session, paper 15/33-08.
12. A. Van Roggen An Overview of Dielectric Measurements. *IEEE Trans. on El. Insulation* Vol. 25, 1990, pp. 95–106.
13. W.P. Baker. *Electrical Insulation Measurements*. Newnes International Monographs on Electrical Engineering and Electronics, 1965.
14. A.J. Schwab. *High Voltage Measurement Techniques*. M.I.T Press, 1972.
15. See Chapter 4 in L. Schnell (ed.). *Technology of Electrical Measurements*. John Wiley and Sons Ltd, 1993, or: B. Hague. *Alternating-Current Bridge Methods* (5th ed). Pitman & Sons, London, 1962.
16. W. Brueckel. The Commutable Schering-Bridge of Tettex AG. TETTEX Information No. 1 (1983).

17. K.W. Wagner. Zur Messung dielektrischer Verluste mit der Wechselstrombrücke (To the measurement of dielectric losses with the a.c. bridge). *Elektrotechnische Zeitschrift* Vol. 32 (1911), p. 1001.
18. W.B. Baker. Recent developments in 50c/s bridge networks with inductively coupled ratio arms for capacitance and loss-tangent measurements. *Proceedings of IEE* Part A, Vol. 109 (1962), pp. 243–247.
19. P.N. Miljanic, N.L. Kusters and W.J.M. Moore. The development of the current comparator, a high-accuracy a.c. ratio measuring device. *AIEE Transactions*, Pt I (Communication and Electronics), Vol. 81 (Nov. 1962), pp. 359–368.
20. O. Peterson. A self-balancing high-voltage capacitance bridge. *IEEE Trans. on Instr. and Meas.* Vol. IM13 (1964), pp. 216–224.
21. P. Osvath and S. Widmer. A High-Voltage High-Precision Self-Balancing Capacitance and Dissipation Factor Measuring Bridge. *IEEE Trans. on Instr. and Meas.* Vol. IM35, (1986), pp. 19–23.
22. L. Lundgaard, M. Runde and B. Skyberg. Acoustic diagnosis of gas insulated substations; a theoretical and experimental basis. *IEEE Trans. on Power Delivery* Vol. 5 (1990), pp. 1751–1760.
23. P. Moro and J. Poittevin. Localization des décharges partielles dans le transformateurs par détection des ondes ultrasonores emises. *Rev. Générale de l'Electricité* **87** (1978), pp. 25–35.
24. E. Howells and E.T. Norton. Detection of partial discharges in transformers using emission techniques. *Trans. IEEE* **PAS97** (1978), pp. 1538–1549.
25. R.T. Harrold. Acoustic waveguides for sensing and locating electrical discharges in H.V. power transformers and other apparatus. *Trans. IEEE* **PAS98** (1979), pp. 449–457.
26. F.H. Kreuger. *Discharge Detection in High Voltage Equipment.* Heywood, London, 1964; Elsevier, New York, 1964, Temple Press Books, 1964.
27. NEMA Publication No. 107 (1940). *Methods of Measuring Radio Noise.*
28. L. Satish and W.S. Zaengl. An effort to find near-optimal band-pass filter characteristics for use in partial discharge measurements. *European Transactions on Electrical Power Engineering*, ETEP 4, No. 6 (1994), pp. 557–563.
29. W.S. Zaengl, P. Osvath and H.J. Weber. Correlation between the bandwidth of PD detectors and its inherent integration errors. 1986 IEEE Intern. Symp. on Electr. Insul. (ISEI), Washington DC. June 1986, Conference Rec. pp. 115–121.
30. CISPR 16-1 (1993). Specifications for radio disturbance and immunity measuring apparatus and methods – Part 1: Radio disturbance and immunity measuring apparatus.
31. IEC Standard 60270 (Third edition, 2000). Partial Discharge Measurements. International Electrotechnical Commission (IEC), Geneva, Switzerland.
32. Partial Discharge Measurements. IEC Standard 270, 1981.
33. H. Polek. Kompensator-Messbrücke für Kapazitäts- und Verlustfaktormessung mit Registrierung. *Elektrotechnische Zeitschrift ETZ*, Vol. 76 (1955), pp. 822–826.
34. E.M. Dembinski and J.L. Douglas. Calibration of partial-discharge and radio interference measuring circuits. *Proc. IEE* **115** (1968), pp. 1332–1340.
35. R.T. Harrold and T.W. Dakin. The relationship between the picocoulomb and microvolt for corona measurements on hv transformers and other apparatus. *Trans. IEEE* **PAS92** (1973), pp. 187–198.
36. Th. Praehauser. Lokalisierung von Teilentladungen in Hochspannungs-apparaten. *Bull. SEV.* **63** (1972), pp. 893–905.
37. British Patent No. 6173/72. Improvements in or relating to High Voltage Component Testing Systems.
38. I.A. Black. A pulse discrimination system for discharge detection in electrically noisy environments. 2nd Int. High Voltage Symposium (ISH), Zurich, Switzerland, 1975, pp. 239–243.
39. R. Bartnikas. Use of multichannel analyzer for corona pulse-height distribution measurements on cables and other electrical apparatus. *Trans. IEEE* **IM-22** (1973), pp. 403–407.

40. S. Kärkkainen. Multi-channel pulse analyzer in partial discharge studies. 2nd Int. High Voltage Symposium (ISH), Zurich, 1975, pp. 244–249.
41. K. Umemoto, E. Koyanagy, T. Yamada and S. Kenjo. Partial discharge measurement system using pulse-height analyzers. 3rd Int. Symp. on High Voltage Engg. (ISH), Milan, 1979, Report 43.07.
42. K. Schon and W. Lucas. Intercomparison of impulse charge measurements. EU Synthesis Report on Project SMT4-CT95-7501, 1998.
43. B. Kübler. Investigation of partial discharge measuring techniques using epoxy resin samples with several voids. IEEE Intern. Symp. on El. Insulation, 1978, 78 CH 1287-2EI (see also Ph.D. Thesis, Techn. University Braunschweig, Germany, 1978).
44. J. Carlier *et al.* Ageing under voltage of the insulation of rotating machines: influence of frequency and temperature. CIRGRE-Rapport No. 15-06, 1976.
45. R. Bartnikas and E.J. McMahon. Corona measurement and interpretation. *Engineering Dielectrics* Vol. 1, ASTM STP 669, 1979.
46. F. Viale *et al.* Study of a correlation between energy of partial discharges and degradation of paper-oil-insulation. CIGRE Session 1982, report 15-12.
47. K. Kuepfmueller. *Die Systemtheorie der elektrischen Nachrichten-uebertragung.* S. Hirzel Verlag, Stuttgart, 1968.
48. A. Papoulis. *The Fourier Integral and its Applications.* McGraw-Hill, 1962.
49. A.G. Millar *et al.* Digital acquisition, storage and processing of partial discharge signals. 4th Int. Symp. on High Voltage Engg. (ISH), Athens 1983, Report 63.01.
50. Presco AG, Weiningen, Switzerland. Capacitance and Dissipation Factor Measuring Bridge TG-1Mod. Company brochure 1999.
51. L. Satish and W.S. Zaengl. Artificial neural networks for recognition of 3-d partial discharge pattern. *IEEE Trans. on Dielectrics and El. Insulation* Vol. 1 (1994), pp. 265–275.
52. Calibration procedures for analog and digital partial discharge measuring instruments. *Electra* No. 180 (Oct. 1998), pp. 123–143.
53. Partial discharge measurement as a diagnostic tool. *Electra* No. 181 (Dec. 1998), pp. 25–51.
54. Insulation co-ordination of GIS: Return of experience, on site tests and diagnostic techniques. *Electra* No. 176 (Feb. 1998), pp. 67–97.
55. J.S. Pearson, O. Farish, B.F. Hampton *et al.* PD diagnostics for gas insulated substations. *IEEE Trans. on Dielectrics and El. Insulation* Vol. 2 (1995), pp. 893–905.
56. B.F. Hampton *et al.* Experience and progress with UHF diagnostics in GIS. CIGRE Session 1992, Report 15/23-03.
57. E. Gulski *et al.* Experiences with digital analysis of discharges in high voltage components. *IEEE Electrical Insulation Magazine* Vol. 15, No. 3 (1999), pp. 15–24.
58. G.C. Stone. Partial discharge part XXV: Calibration of PD measurements for motor and generator windings – why it cant't be done. *IEEE Electrical Insulation Magazine* Vol. 14, No. 1 (1998), pp. 9–12.
59. International Conference on Partial Discharge. University of Kent in Canterbury, UK, Sept. 1993. Conference Proceedings No. 378, IEE 1993.
60. K. Wu, Y. Suzuoki, T. Mizutani and H. Xie. A novel physical model for partial discharge in narrow channels. *IEEE Trans. on Dielectrics and Electr. Insulation* Vol. 6 (1999), pp. 181–190.
61. Beierl *et al.* Intelligent monitoring and control systems for modern AIS and GIS substations. CIGRE Session 1998, paper 34–113.
62. U. Köpf and K. Feser. Noise suppression in partial discharge measurements. 8th Int. Symp. on HV Engineering, Yokohama, Japan (1995), paper No. 63.02.
63. IEEE Committee Report. Digital techniques for partial discharge measurements. *IEEE Trans. on Power Delivery* Vol. 7 (1992), pp. 469–479 (with 105 references).
64. IEEE Committee Report. Partial discharge testing of gas insulated substations. *IEEE Trans. on Power Delivery* Vol. 7 (1992), pp. 499–506 (with 25 references).

65. Special Issue: The Volta Colloquium on Partial Discharge Measurements. *IEEE Trans. on Electrical Insulation* Vol. 27, No. 1 (Feb. 1992), with 13 original papers related to PD.
66. Special Issue: Partial Discharge Measurement and Interpretation. *IEEE Trans. on Dielectrics and Electrical Insulation* Vol. 2, No. 4 (Aug. 1995), with 19 original papers related to PD.
67. M. Hoof, B. Freisleben and R. Patsch. PD source identification with novel discharge parameters using counterpropagation neural networks. *IEEE Trans. on Dielectrics and El. Insulation* Vol. 4 (1997), pp. 17–32.

Chapter 8

Overvoltages, testing procedures and insulation coordination

Power systems are always subjected to overvoltages that have their origin in atmospheric discharges in which case they are called external or lightning overvoltages, or they are generated internally by connecting or disconnecting the system, or due to the systems fault initiation or extinction. The latter type are called internal overvoltages. This class may be further subdivided into (i) temporary overvoltages, if they are oscillatory of power frequency or harmonics, and (ii) switching overvoltages, if they are heavily damped and of short duration. Temporary overvoltages occur almost without exception under no load or very light load conditions. Because of their common origin the temporary and switching overvoltages occur together and their combined effect has to be taken into account in the design of h.v. systems insulation.

The magnitude of the external or lightning overvoltages remains essentially independent of the system's design, whereas that of internal or switching overvoltages increases with increasing the operating voltage of the system. Hence, with increasing the system's operating voltage a point is reached when the switching overvoltages become the dominant factor in designing the system's insulation. Up to approximately 300 kV, the system's insulation has to be designed to withstand primarily lightning surges. Above that voltage, both lightning and switching surges have to be considered. For ultra-h.v. systems, 765 kV and above switching overvoltages in combination with insulator contamination become the predominating factor in the insulation design.[(1)]* For the study of overvoltages occurring in power systems, a thorough knowledge of surge propagation laws is needed which can be found in a number of textbooks[(2,3)] and will not be discussed here.

8.1 The lightning mechanism

Physical manifestations of lightning have been noted in ancient times, but the understanding of lightning is relatively recent. Franklin carried out experiments on lightning in 1744–1750, but most of the knowledge has been obtained over the last 50 to 70 years. The real incentive to study lightning came when electric transmission lines had to be protected against lightning. The methods

* Superscript numbers are to references at the end of the chapter.

include measurements of (i) lightning currents, (ii) magnetic and electromagnetic radiated fields, (iii) voltages, (iv) use of high-speed photography and radar.

Fundamentally, lightning is a manifestation of a very large electric discharge and spark. Several theories have been advanced to explain accumulation of electricity in clouds and are discussed in references 4, 5 and 6. The present section reviews briefly the lightning discharge processes.

In an active thunder cloud the larger particles usually possess negative charge and the smaller carriers are positive. Thus the base of a thunder cloud generally carries a negative charge and the upper part is positive, with the whole being electrically neutral. The physical mechanism of charge separation is still a topic of research and will not be treated here. As will be discussed later, there may be several charge centres within a single cloud. Typically the negative charge centre may be located anywhere between 500 m and 10 000 m above ground. Lightning discharge to earth is usually initiated at the fringe of a negative charge centre.

To the eye a lightning discharge appears as a single luminous discharge, although at times branches of variable intensity may be observed which terminate in mid-air, while the luminous main channel continues in a zig-zag path to earth. High-speed photographic technique studies reveal that most lightning strokes are followed by repeat or multiple strokes which travel along the path established by the first stroke. The latter ones are not usually branched and their path is brightly illuminated.

The various development stages of a lightning stroke from cloud to earth as observed by high-speed photography is shown diagrammatically in Fig. 8.1

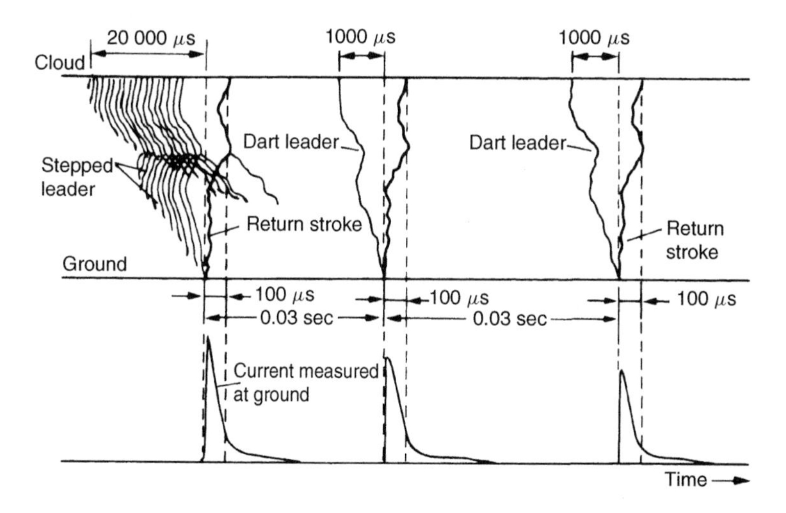

Figure 8.1 *Diagrammatic representation of lightning mechanism and ground current*[3]

together with the current to ground. The stroke is initiated in the region of the negative charge centre where the local field intensity approaches ionization field intensity ($\cong$30 kV/cm in atmospheric air, or $\sim$10 kV/cm in the presence of water droplets).

During the first stage the leader discharge, known as the 'stepped leader', moves rapidly downwards in steps of 50 m to 100 m, and pauses after each step for a few tens of microseconds. From the tip of the discharge a 'pilot streamer' having low luminosity and current of a few amperes propagates into the virgin air with a velocity of about 1×10^5 m/sec. The pilot streamer is followed by the stepped leader with an average velocity of about 5×10^5 m/sec and a current of some 100 A. For a stepped leader from a cloud some 3 km above ground shown in Fig. 8.1 it takes about 60 m/sec to reach the ground. As the leader approaches ground, the electric field between the leader and earth increases and causes point discharges from earth objects such as tall buildings, trees, etc. At some point the charge concentration at the earthed object is high enough to initiate an upwards positive streamer. At the instance when the two leaders meet, the 'main' or 'return' stroke starts from ground to cloud, travelling much faster ($\sim 50 \times 10^6$ m/sec) along the previously established ionized channel. The current in the return stroke is in the order of a few kA to 250 kA and the temperatures within the channel are 15 000°C to 20 000°C and are responsible for the destructive effects of lightning giving high luminosity and causing explosive air expansion. The return stroke causes the destructive effects generally associated with lightning.

The return stroke is followed by several strokes at 10- to 300-m/sec intervals. The leader of the second and subsequent strokes is known as the 'dart leader' because of its dart-like appearance. The dart leader follows the path of the first stepped leader with a velocity about 10 times faster than the stepped leader. The path is usually not branched and is brightly illuminated.

A diagrammatic representation of the various stages of the lightning stroke development from cloud to ground in Figs 8.2(a) to (f) gives a clearer appreciation of the process involved. In a cloud several charge centres of high concentration may exist. In the present case only two negative charge centres are shown. In (a) the stepped leader has been initiated and the pilot streamer and stepped leader propagate to ground, lowering the negative charges in the cloud. At this instance the striking point still has not been decided; in (b) the pilot streamer is about to make contact with the upwards positive streamer from earth; in (c) the stroke is completed, a heavy return stroke returns to cloud and the negative charge of cloud begins to discharge; in (d) the first centre is completely discharged and streamers begin developing in the second charge centre; in (e) the second charge centre is discharging to ground via the first charge centre and dart leader, distributing negative charge along the channel. Positive streamers are rising up from ground to meet the dart leader;

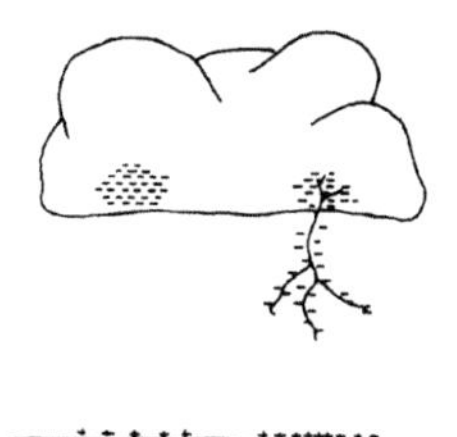

(a) Charge centres in cloud; pilot streamer and stepped leader propagate earthward; outward branching of streamers to earth. Lowering of charge into space beneath cloud.

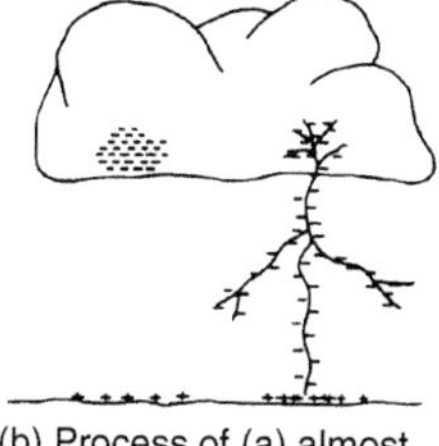

(b) Process of (a) almost completed; pilot streamer about to strike earth.

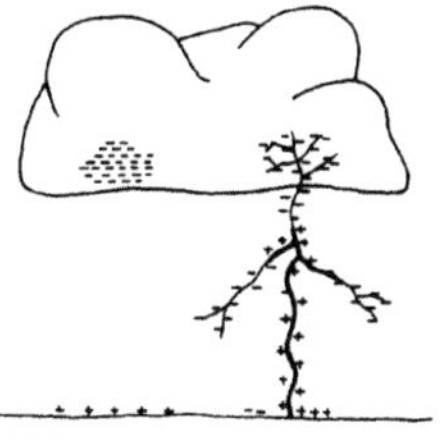

(c) Heavy return streamer; discharge to earth of negatively charged space beneath cloud.

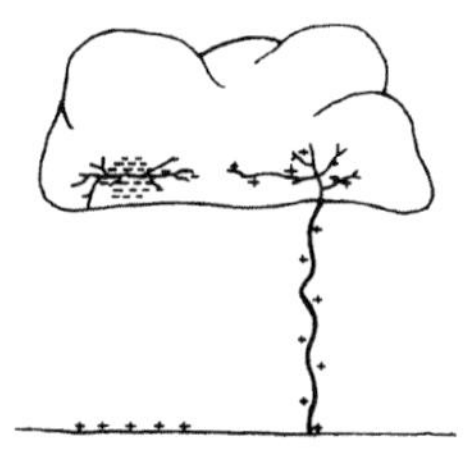

(d) First charge centre completely discharged; development of streamers between charge centres within cloud.

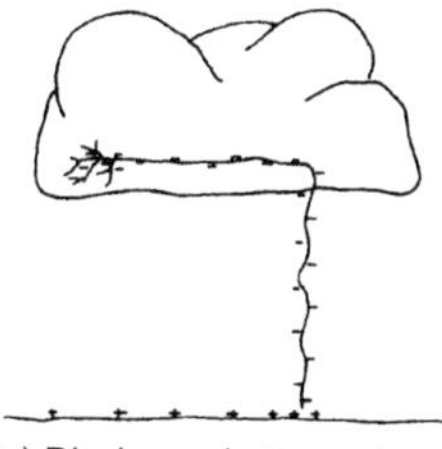

(e) Discharge between two charge centres; dart leader propagates to ground along original channel; dart leader about to strike earth; negative charge lowered and distributed along stroke channel.

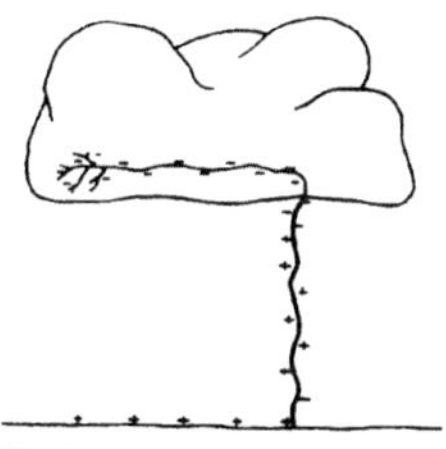

(f) Heavy return streamer discharge to earth of negatively charged space beneath cloud.

Figure 8.2 *Schematic representation of various stages of lightning stroke between cloud and ground*[6]

(f) contact is made with streamers from earth, heavy return stroke proceeds upwards and begins to discharge negatively charged space beneath the cloud and the second charge centre in the cloud.

Lightning strokes from cloud to ground account only for about 10 per cent of lightning discharges, the majority of discharges during thunderstorms take place between clouds. Discharges within clouds often provide general illumination known as 'sheath lightning'.

Measurements of stroke currents at ground have shown that the high current is characterized by a fast rise to crest (1 to 10 μsec) followed by a longer decay time of 50–1000 μsec to half-time. Figure 8.3 gives the probability distribution of times to crest for lightning strokes as prepared by Anderson.[7] There is evidence that very high stroke currents do not coincide with very short times to crest. Field data[3,20] indicate that 50 per cent of stroke currents including multiple strokes have a rate of rise exceeding 20 kA/μsec and 10 per cent exceed 50 kA/μsec. The mean duration of stroke currents above half value is 30 μsec and 18 per cent have longer half-times than 50 μsec. Thus for a typical maximum stroke current of 10 000 A a transmission line of surge impedance (say) $Z = 400\,\Omega$ and assuming the strike takes place in the middle of the line with half of the current flowing in each direction ($Z \cong 200\,\Omega$) the lightning overvoltage becomes $V = 5000 \times 400 = 2\,\text{MV}$. Based on many investigations the AIEE Committee[8] has produced the frequency distribution of current magnitudes, shown in Fig. 8.4, which is often used for performance calculations. Included in Fig. 8.4 is a curve proposed by Anderson.[7]

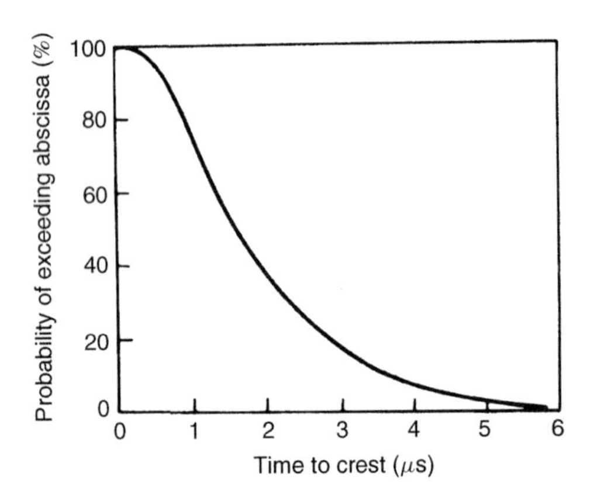

Figure 8.3 *Distribution of times to crest of lightning stroke currents (after Anderson[7])*

The data on lightning strokes and voltages has formed the basis for establishing the standard impulse or lightning surge for testing equipment in laboratories. The standard lightning impulse waveshape will be discussed later in this chapter.

8.1.1 Energy in lightning

To estimate the amount of energy in a typical lightning discharge let us assume a value of potential difference of 10^7 V for a breakdown between a cloud

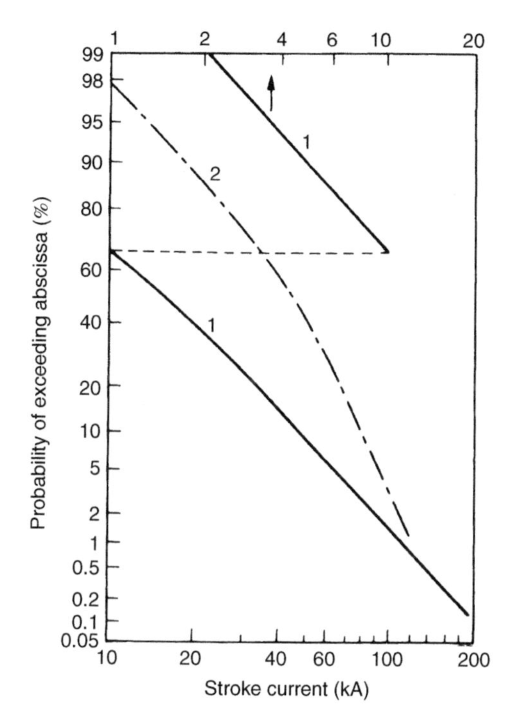

Figure 8.4 *Cumulative distributions of lightning stroke current magnitudes: 1. After AIEE Committee.*[8] *2. After Anderson*[6]

and ground and a total charge of 20 coulombs. Then the energy released is 20×10^7 Ws or about 55 kWh in one or more strokes that make the discharge. The energy of the discharge dissipated in the air channel is expended in several processes. Small amounts of this energy are used in ionization of molecules, excitations, radiation, etc. Most of the energy is consumed in the sudden expansion of the air channel. Some fraction of the total causes heating of the struck earthed objects. In general, lightning processes return to the global system the energy that was used originally to create the charged cloud.

8.1.2 Nature of danger

The degree of hazard depends on circumstances. To minimize the chances of being struck by lightning during thunderstorm, one should be sufficiently far away from tall objects likely to be struck, remain inside buildings or be well insulated.

A direct hit on a human or animal is rare; they are more at risk from indirect striking, usually: (a) when the subject is close to a parallel hit or other tall object, (b) due to an intense electric field from a stroke which can induce sufficient current to cause death, and (c) when lightning terminating on earth sets up high potential gradients over the ground surface in an outwards direction from the point or object struck. Figure 8.5 illustrates qualitatively the current distribution in the ground and the voltage distribution along the ground extending outwards from the edge of a building struck by lightning.[9] The potential difference between the person's feet will be largest if his feet are separated along a radial line from the source of voltage and will be negligible if he moves at a right angle to such a radial line. In the latter case the person would be safe due to element of chance.

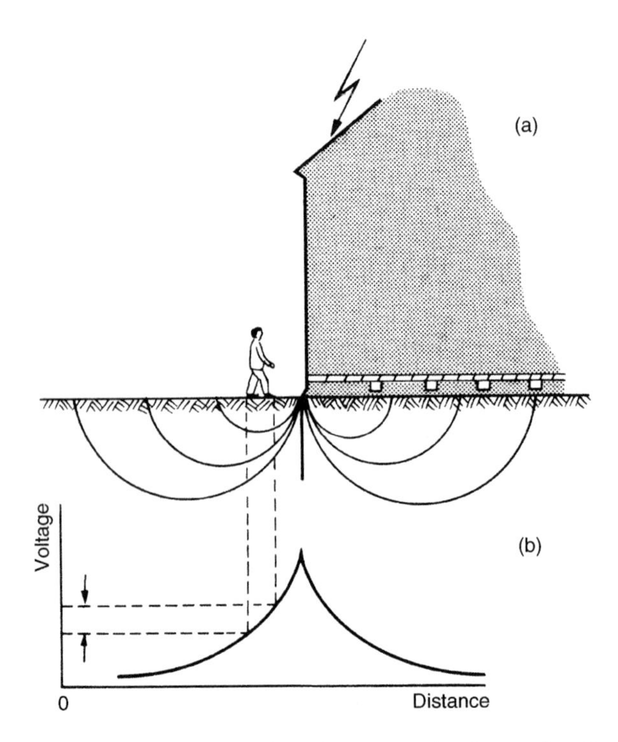

Figure 8.5 *Current distribution and voltage distribution in ground due to lightning stroke to a building (after Golde[9])*

8.2 Simulated lightning surges for testing

The danger to electric systems and apparatus comes from the potentials that lightning may produce across insulation. Insulation of power systems may be classified into two broad categories: external and internal insulation. External insulation is comprised of air and/or porcelain, etc., such as conductor-to-tower clearances of transmission lines or bus supports. If the potential caused

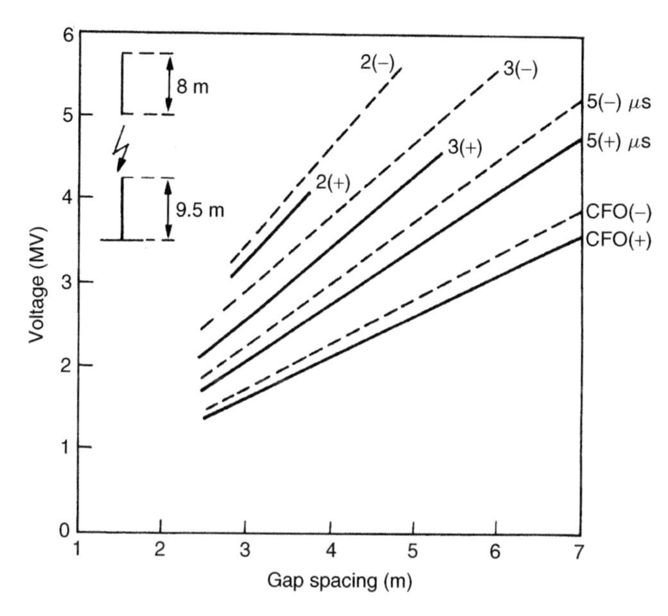

Figure 8.6 *Impulse (1.2/50 μsec) flashover characteristics of long rod gaps corrected to STP (after Udo[10])*

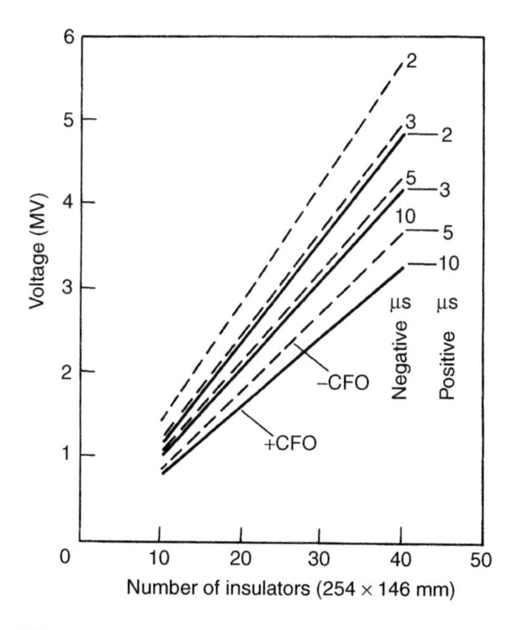

Figure 8.7 *Impulse (1.2/50 μsec) flashover characteristics for long insulator strings (after Udo[10])*

by lightning exceeds the strength of insulation, a flashover or puncture occurs. Flashover of external insulation generally does not cause damage to equipment. The insulation is 'self-restoring'. At the worst a relatively short outage follows to allow replacement of a cheap string of damaged insulation. Internal insulation most frequently consists of paper, oil or other synthetic insulation which insulates h.v. conductors from ground in expensive equipment such as transformers, generators, reactors, capacitors, circuit-breakers, etc. Failure of internal insulation causes much longer outages. If power arc follows damage to equipment it may be disastrous and lead to very costly replacements.

The system's insulation has to be designed to withstand lightning voltages and be tested in laboratories prior to commissioning.

Exhaustive measurements of lightning currents and voltages and long experience have formed the basis for establishing and accepting what is known as the standard surge or 'impulse' voltage to simulate external or lightning overvoltages. The international standard lightning impulse voltage waveshape is an aperiodic voltage impulse that does not cross the zero line which reaches its peak in 1.2 μsec and then decreases slowly (in 50 μsec) to half the peak value. The characteristics of a standard impulse are its polarity, its peak value, its front time and its half value time. These have been defined in Chapter 2, Fig. 2.23.

Extensive laboratory tests have shown that for external insulation the lightning surge flashover voltages are substantially proportional to gap length and that positive impulses give significantly lower flashover values than negative ones. In addition, for a particular test arrangement, as the applied impulse crest is increased the instant of flashover moves from the tail of the wave to the crest and ultimately to the front of the wave giving an impulse voltage–time ('$V-T$') characteristic as was discussed in Chapter 5, Fig. 5.45. Figures 8.6 and 8.7 show typical impulse sparkover characteristics for long rod gaps and suspension insulators obtained by Udo[(10)] at various times to flashover. These figures include the critical or long time flashover characteristics (CFO) occurring at about 10 μsec on the wave tail as well as the characteristics corresponding to shorter time lags near the wave crest. Data for both polarities are shown. The values plotted in Figs 8.6 and 8.7 have been corrected to standard atmospheric conditions.

8.3 Switching surge test voltage characteristics

In power transmission systems with systems voltages of 245 kV and above, the electrical strength of the insulation to switching overvoltages becomes important for the insulation design. A considerable amount of data on breakdown under switching surges is available. However, a variety of switching surge waveshapes and the correspondingly large range of flashover values make it difficult to choose a standard shape of switching impulses. Many tests have shown that the flashover voltage for various geometrical arrangements under unidirectional switching surge voltages decreases with increasing the front duration of the surge, reaching the lowest value somewhere in the range between 100 and 500 μsec. The time to half-value has less effect upon the breakdown strength because flashover almost always takes place before or at the crest of the wave. Figure 8.8 illustrates a typical relationship for a critical flashover voltage per metre as a function of time to flashover for a 3-m rod-rod gap and a conductor-plane gap respectively.(11) It is seen that the standard impulse voltages give the highest flashover values, with the switching surge values of crest between approx. 100 and 500 μsec falling well below the corresponding power frequency flashover values.

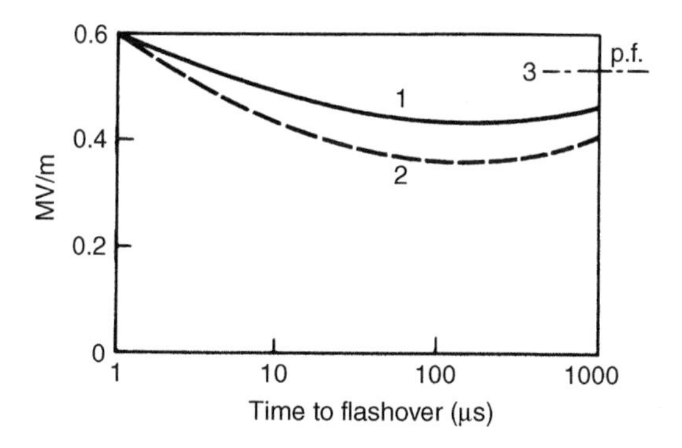

Figure 8.8 *Relationship between vertical flashover voltage per metre and time to flashover (3 m gap). 1. Rod-rod gap. 2. Conductor-plane gap. 3. Power frequency*

The relative effect of time to crest upon flashover value varies also with the gap spacing and humidity.(21) Figure 8.9 compares the positive flashover characteristics of standard impulses and 200/2000 μsec with power frequency voltages for a rod-rod gap plotted as flashover voltage per metre against gap spacing.(11) We observe a rapid fall in switching surge breakdown strength with increasing the gap length. This drastic fall in the average switching surge strength with increasing the insulation length leads to costly design clearances, especially in the ultra-h.v. regions. All investigations show that for nearly all gap configurations which are of practical interest, positive switching impulses result in lower flashover voltage than negative ones. The flashover behaviour of external insulations with different configurations under positive switching impulse stress is therefore most important. The switching surge voltage breakdown is also affected by the air humidity. Kuffel *et al.*(22) have reported that over the range from 3 to 16 g/m^3 absolute humidity, the breakdown voltage

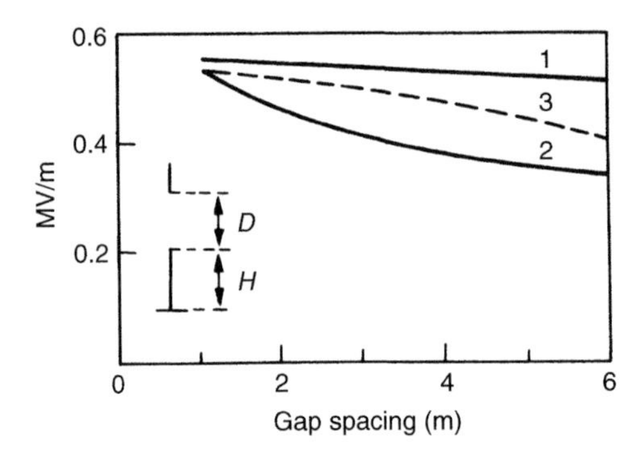

Figure 8.9 *Relationship between flashover voltage (MV/m) and gap length for 1: 1.2/50 μsec impulses, 2: 200/2000 μsec switching surges and 3: power frequency voltages*

of positive rod gaps increases approximately 1.7 per cent per 1 g/m^3 increase in absolute humidity.

For testing purposes the standard switching surge recommended by IEEE St-4-1995 Publication[12] and IEC Publication 60-1[13] 1998-11 has a front time $T_2 = 2500\,\mu\text{sec} \pm 20$ per cent and half-time value $T_2 = 2500\,\mu\text{sec} \pm 60$ per cent. The general designation for a standard switching impulse is given as 250/2500 μsec. The front is counted from the actual beginning of the impulse till the peak value is reached. Full characteristics of a standard switching test surge have been defined in Chapter 2, Fig. 2.24.

It was shown in Chapter 5, section 5.9 that in non-uniform field gaps the shape of both electrodes affects the formation and propagation of streamers and directly influences the flashover values. This explains the different flashover values observed for various insulating structures, especially under switching surges. Much of the laboratory flashover data for large gaps under switching surges have been obtained for rod-plane gaps. Subsequently, several attempts have been made to relate data for other structures to rod-plane gap data. Several investigators[14,15] have shown that the positive 50 per cent switching surge voltage of different structures in air in the range from 2 to 8 m follow the expression

$$V_{50} = k\,500\,d^{0.6}\ \text{kV} \tag{8.1}$$

where d is the gap length in metres and k is gap factor relating to the electrodes geometry. For rod-plane gaps the factor k is accepted as unity. Thus, the 'gap factor' k represents a proportionality factor of the 50 per cent flashover voltage of any gap geometry to that of a rod-plane gap for the same distance or

$$k = \frac{V_{50}}{V_{50}\ \text{rod-plane}} \tag{8.2}$$

Expression (8.1) applies to data obtained under the switching impulse of constant time to crest. A more general expression which gives minimum strength and applies to longer times to crest has been proposed by Gallet and Leroy[16] as follows:

$$V_{50} = \frac{k3450}{1 + \dfrac{8}{d}}\ \text{kV} \tag{8.3}$$

where k and d have the same meaning as in expression (8.1).

In expression (8.2) only the function V_{50} rod-plane is influenced by the switching impulse shape, while the gap factor k depends only on the gap geometry and hence upon the field distribution in the gap. The parameters influencing the gap factor (k) have been fully discussed by Schneider and Weck.[17] These authors have measured the gap factor (k) for different gap geometries and spacings using a large three-dimensional electrolytic tank and modelling scaled down gaps. Their data are included in Table 8.1. The corresponding geometric configurations are shown in Fig. 8.10(a) to (f).

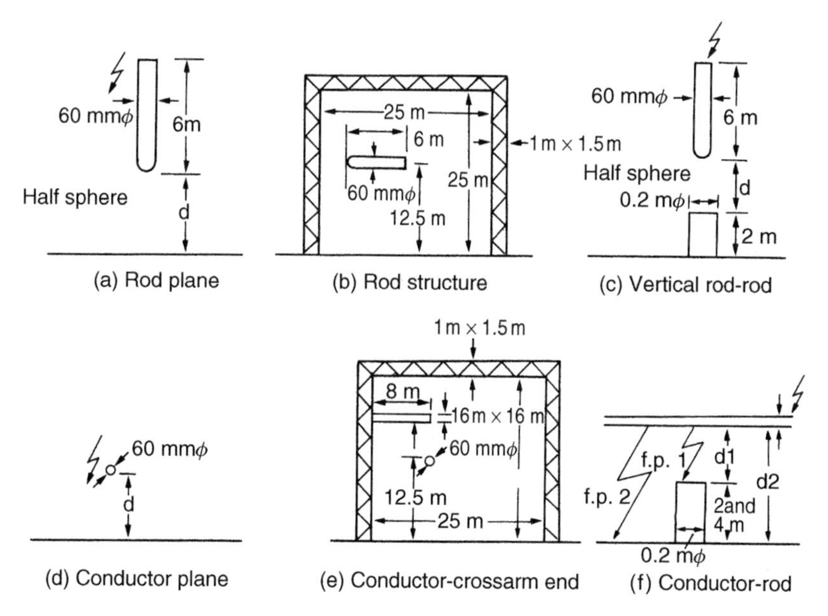

Figure 8.10 *Configuration (gap factor)*

Expressions (8.1) and (8.3) together with data presented in Table 8.1 can be used in estimating required clearances in designing e.h.v. and u.h.v. structures. Refinements to these expressions are being introduced as more data become available.

Table 8.1 *Geometric gap factor for various structures*

Configuration	*Figure*	$d = 2\,m$ k	$d = 3\,m$ k	$d = 4\,m$ k	$d = 6\,m$ k
Rod-plane	(a)	1	1	1	1
Rod-structure	(b)	1.08	–	1.07	1.06
Rod-rod vertical					
$H = 2\,\text{m}$	(c)	1.27	1.26	1.21	1.14
Conductor-plane	(d)	1.08	–	1.14	1.15
Conductor-cross					
arm end	(e)	1.57	1.68	1.65	1.54
Conductor-2m rod	(f)	1.47	–	1.40	1.25
Conductor-4m rod	(f)	1.55	–	1.54	1.40

8.4 Laboratory high-voltage testing procedures and statistical treatment of results

Practical high voltage insulation systems comprise various types of dielectrics, e.g. gases, liquids, solids or any combination of these. The result, following the application of a voltage stress to insulation, individually and also collectively, is a discharge or withstand, and has a random nature. Hence the parameters characterizing the behaviour of the insulation must be handled statistically.

Test methods and procedures adopted for the determination of the parameters characterizing insulation behaviour generally involve the repeated application of dielectric stress and the appropriate evaluation of the results. The aim of the statistical evaluation of the test methods is to establish procedures for relevant interpretation of the parameters characterizing the insulation behaviour and to determine confidence limits for the data obtained. Hence a brief treatment of the statistical methods generally used will be presented.

The documents addressing this issue are the IEEE Standard[12] and the IEC Publication 60-1 1989-11.[13]

8.4.1 Dielectric stress–voltage stress

A voltage stress when applied to a piece of insulation is completely defined when the applied voltage $V(t)$ is known during the time of stress $(t_O,\ t_M)$. Trying to correlate the behaviour of the insulation to even a slightly different value of $V(t)$ requires accurate knowledge of the physical processes occurring inside the insulation.

8.4.2 Insulation characteristics

The main characteristic of interest of an insulation is the disruptive discharge which may occur during the application of stress. However, because of the randomness of the physical processes which lead to disruptive discharge, the same stress applied several times in the same conditions may not always cause disruptive discharge. Also, the discharge when it occurs may occur at different times. In addition, the application of the stress, even if it does not cause discharge, may result in a change of the insulation characteristics.

8.4.3 Randomness of the appearance of discharge

Randomness of the appearance of discharge can be modelled by considering a large number of stress applications, a fraction p of which causes discharge, D, and the remaining fraction $q = (1 - p)$ being labelled as withstand, W. The value of p depends on applied stress, S, with $p = p(S)$ being the 'probability of discharge' and it represents one of the characteristics of the insulation. Recognizing that the time to discharge will also vary statistically, the probability of discharge will become a function of both the stress, S, and the time t.

$$p(V) = p(t, S) \tag{8.4}$$

8.4.4 Types of insulation

Insulations are grouped broadly into:

(i) Self-restoring (gases) – no change produced by the application of stress or by discharge, hence the same sample can be tested many times.
(ii) Non-self-restoring (liquids) – affected by discharge only, the same sample can be used until discharge occurs.
(iii) Affected by applied stress, insulation experiences ageing and in testing it becomes necessary to introduce a new parameter related to the sequential application of stress.

8.4.5 Types of stress used in high-voltage testing

For design purposes it is sufficient to limit the knowledge of the insulation characteristics to a few families of stresses which are a function of time $V(t)$ e.g. switching surge of double exponential with time to crest T_1 and to half value T_2 and the variable crest value V (see definitions in Chapter 2 for lightning and switching surges). For testing purposes, the family is further restricted by using fixed times T_1 and T_2, hence only one variable is left (V). The same applies to both types of surges. The behaviour of the insulation is then defined by the discharge probability as a function of crest voltage $p = p(V)$.

The most commonly used distribution function is the normal (Gaussian) distribution which has a particular shape (bell shape), plotted in Fig. 8.11. The equation for the normal distribution density function is

$$p(f) = \frac{1}{\sigma\sqrt{2\pi}} e^{-((f_k - f_{av})^2/2\sigma^2)} \tag{8.5}$$

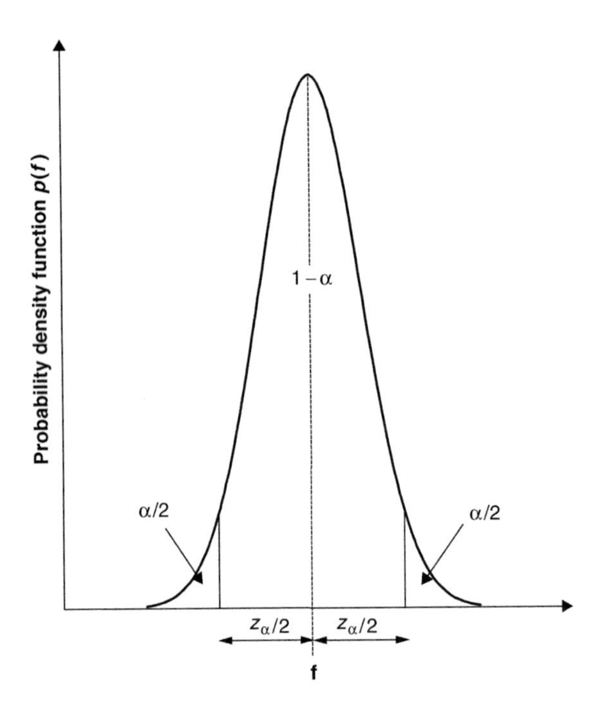

Figure 8.11 *Gaussian (normal) distribution curve with confidence limits*

where f_k is the k_{th} value of the variable, f_{av} is the average value and σ is the standard deviation. When the applied voltage, V, becomes the variable the Gaussian distribution function used takes the form

$$p(V) = \frac{1}{\sigma\sqrt{2\pi}} e^{-((V - V_{50})^2/2\sigma^2)} \tag{8.6}$$

where V_{50} is the voltage which leads to 50 per cent probability of discharge.

The knowledge of V_{50} and σ allows us to calculate the value of the probability $p(V)$ for any applied voltage.

Also shown In Fig. 8.11 are the confidence limits A and B. The confidence in our results when expressed in per cent is shown by the area $(1-\alpha)$ between the limits $-(\alpha/2)$ and $+(\alpha/2)$. A more convenient form of the normal distribution is the cumulative distribution function, the integral of eqn (8.12), which has the form

$$P(V) = \frac{1}{\sigma\sqrt{2\pi}} \int_{-\infty}^{\infty} e^{-(V-V_{50})^2/2\sigma^2}\,dx \tag{8.7}$$

A plot of this function is included in Fig. 8.12. When plotted on the probability scale a straight line results as shown in Fig. 8.13. In this figure are plotted the cumulative frequency P_{WS} of withstand voltage, the P_{FO} of flashover voltage and the parameter z, explained below, versus the breakdown voltage of a 1-m rod gap under positive switching impulse voltage in atmospheric air. We note

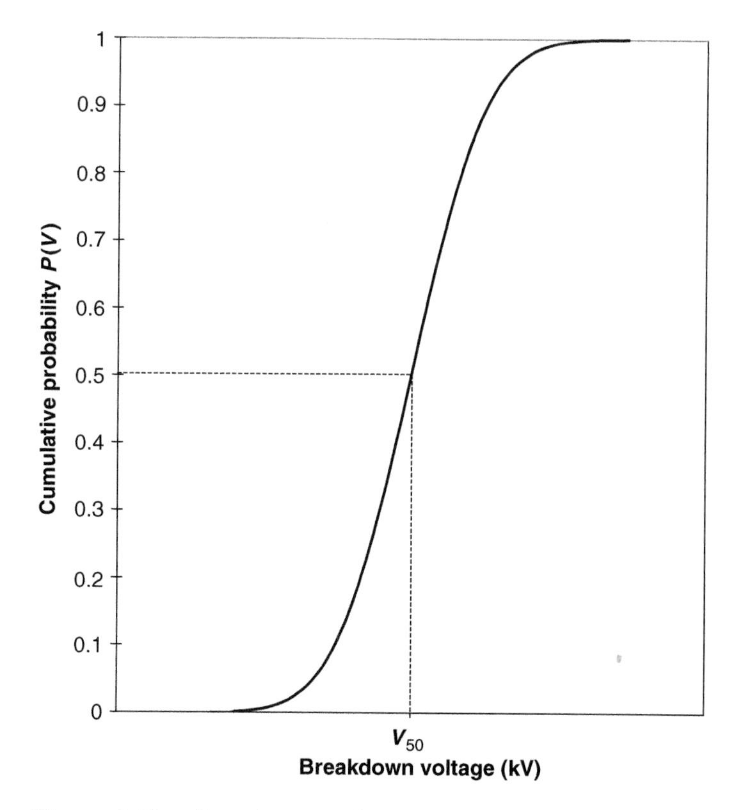

Figure 8.12 *Gaussian cumulative distribution function*

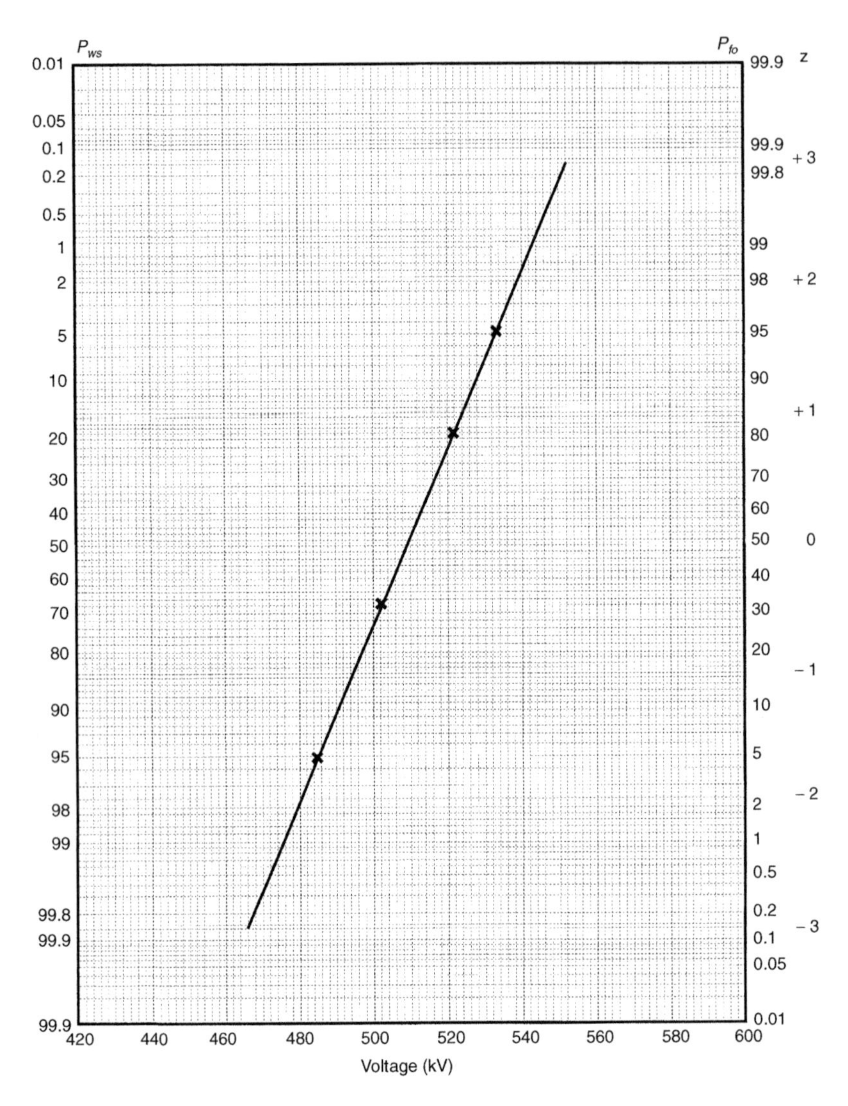

Figure 8.13 *Breakdown voltage distribution plotted on probability scale*

that there are three vertical scales, two non-linear giving directly the P_{WS} (l.h.s.), the P_{FO} (r.h.s.) and further to the right a linear scale given in units of dimensionless deviation z. The parameter z is convenient for analysis of normal distribution results. Equation (8.7) is rewritten in the form

$$P(\mathbf{z}) = \frac{1}{\sigma\sqrt{2\pi}} \int_{-\infty}^{z} e^{-(z^2/2)}\, dz \tag{8.8}$$

where

$$\mathbf{z} = \frac{V - V_{50}}{\sigma}$$

As noted earlier the distribution of flashover of the gap is characterized by two parameters:

(i) V_{50}, called the critical flashover (CFO),
(ii) σ, called the standard deviation.

Both can be read directly from the best fit line drawn through the experimentally determined points. Note, that CFO corresponds to $z = 0$ and σ is given by the difference between two consecutive integers of z. In practice the voltage range over which the probability of flashover is distributed is

$$\text{CFO} \pm 3\sigma \tag{8.9}$$

- $(\text{CFO} - 3\sigma)$ is known as the statistical withstand voltage (SWV) and represents the point with flashover probability 0.13 per cent;
- $(\text{CFO} + 3\sigma)$ is known as the statistical flashover voltage (SFOV) and represents the point with flashover probability 99.87 per cent.

The SWV and SFOV are used in insulation coordination and will be discussed later. For a complete description of insulation parameters, the time to breakdown must also be considered. The times to breakdown are represented by

$$P(t) = \frac{1}{\sigma\sqrt{2\pi}} \int_0^t e^{-(t-\bar{t})^2/2\sigma^2}\, dt \tag{8.10}$$

where
$\bar{t}$ = mean time to breakdown,
σ = standard deviation.

An example of the distribution of times to breakdown is included in Fig. 8.14.
In this example the range

$$\bar{t} \pm \sigma = \bar{t} \pm z$$

is shown by a straight line but not at the extremities. Nevertheless the method is often used to represent distribution of times to breakdown because of its simplicity.

Another frequently used distribution function for representing breakdown voltage probability is the Weibull function of the form:

$$P(V) = 1 - 0.5^{[1+((V-V_{50})^m/n\sigma)\ln 2]} \tag{8.11}$$

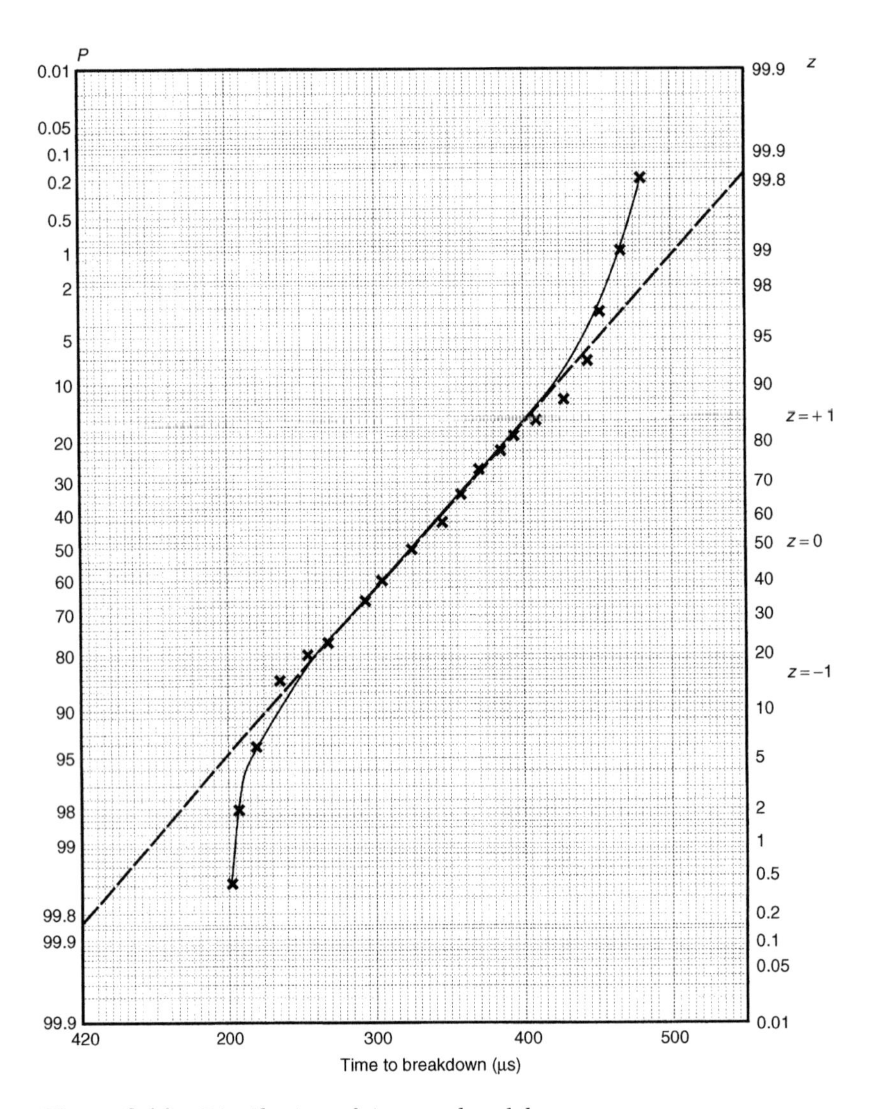

Figure 8.14 *Distribution of times to breakdown*

where

$P(V)$ = the probability of flashover,
V = the applied voltage,
V_{50} = the applied voltage which gives 50 per cent probability flashover,
σ = the standard deviation.

In the Weibull function n is not known but it determines the voltage $V_{50} - n\sigma$ below which no flashover occurs, or $P(V) = 0$ for $V \leq V_{50} - n\sigma$. For air n

lies in the range $3 \le n \le 4$. The value 3 is usually used resulting in

$$m = \frac{\ln \dfrac{\ln 0.84}{\ln 0.5}}{\ln \dfrac{n-1}{n}} = 3.4 \tag{8.12}$$

The adaptation of the Weibull function to normal distribution using the above values for n and m gives $P(V) = 0.5$ for $V = V_{50}$ and $P(V) = 0.16$ for $V = V_{50} - \sigma$. Both the Gaussian and the Weibull functions give the same results in the range $0.01 \le P(V) \le 0.99$.

8.4.6 Errors and confidence in results

In the determination of a parameter two types of error are present:

(i) error associated with the statistical nature of the phenomena and the limited number of tests (ε_S),
(ii) error in the measurement (ε_M).

The statistical error is expressed by means of two confidence limits C per cent. The total error is given by

$$\varepsilon_T = \sqrt{\varepsilon_M^2 + \varepsilon_S^2} \tag{8.13}$$

The various IEC recommendations specify the permissible measurement accuracy as 3 per cent. Hence, a statistical error of, say, 2 per cent will increase the total error by a factor of 1.2, while a statistical error of 1.5 per cent will increase the total error by 1.1.

The outcome of a test procedure and the analysis of the results is usually an average of a parameter z with C per cent confidence limits z_A and z_B (see Fig. 8.11). For a normal distribution the probability density of a function for a frequency of occurrence can be represented graphically in terms of area as shown in Fig. 8.11 $(1 - \alpha)$.

8.4.7 Laboratory test procedures

The test procedures applied to various types of insulation are described in national and international standards as already mentioned before.[12,13] Because the most frequently occurring overvoltages on electric systems and apparatus originate in lightning and switching overvoltages, most laboratory tests are conducted under standard lightning impulse voltages and switching surge voltages. Three general testing methods have been accepted:

1. Multi-level method.
2. Up and down method.
3. Extended up and down method.

1. Multi-level test method

In this method the procedure is:

- choose several test voltage levels,
- apply a pre-specified number of shots at each level (n),
- count the number (x) of breakdowns at each voltage level,
- plot $p(V)$ (x_j/n) against V (kV),
- draw a line of best fit on a probability scale,
- from the line determine V_{50} at $z = 0$ or $P(V) = 50$ per cent,
- and σ at $z = 1$ or $\sigma = V_{50\%} - V_{16\%}$

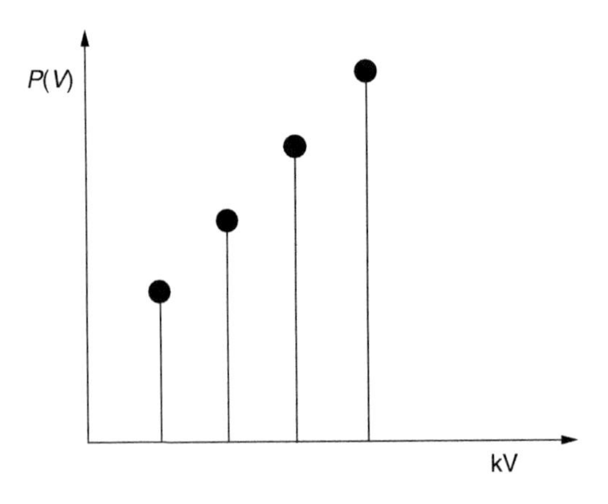

Figure 8.15 *Probability of breakdown distribution using the multi-level method*

The recorded probability of breakdown, x_j/n, is the number which resulted in breakdown from the application of n shots at voltage V_j. When x_j/n is plotted against V_j on a linear probability paper a straight line is obtained as shown in Fig. 8.15.

The advantage of this method is that it does not assume normality of distribution. The disadvantage is that it is time consuming, i.e. many shots are required.

This test method is generally preferred for research and live-line testing (typically 100 shots per level, with 6–10 levels).

2. Up and down method

In this method a starting voltage (V_j) close to the anticipated flashover value is selected. Then equally spaced voltage levels (ΔV) above and below the starting voltage are chosen. The first shot is applied at the voltage V_j. If breakdown occurs the next shot is applied at $V_j - \Delta V$. If the insulation withstands, the next voltage is applied at $V_j + \Delta V$. The sequential procedure of testing is illustrated in Fig. 8.16.

Figure 8.17 illustrates the sequence with nine shots applied to the insulation under test. The IEC Standard for establishing V_{50} (50 per cent) withstand voltage requires a minimum $n = 20$ voltage applications for self-restoring

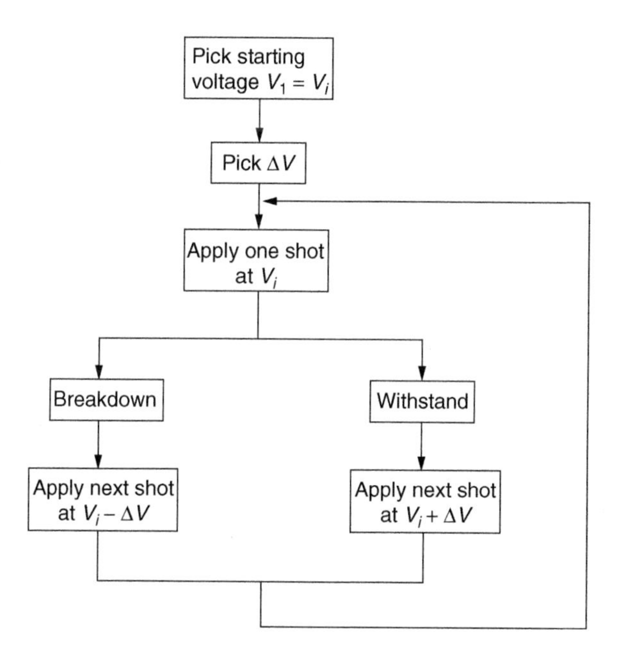

Figure 8.16 *Schematics of the sequential up and down procedure*

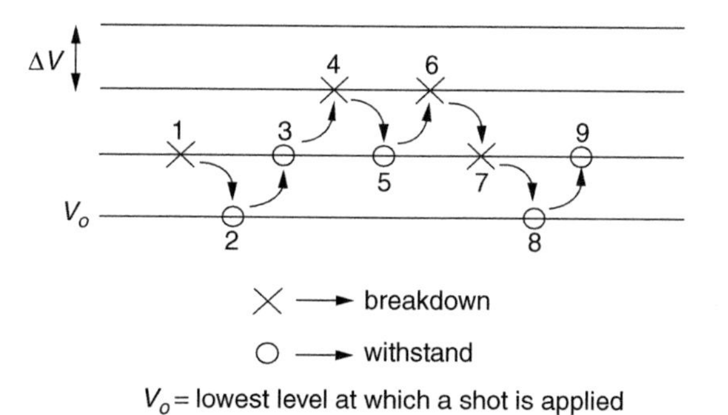

Figure 8.17 *Example illustrating the application of nine shots in the sequential up and down method. X = breakdown; O = withstand*

insulation. To evaluate the V_{10} (10 per cent) withstand voltage for self-restoring insulation with the up and down method with one impulse per group also requires a minimum of $n = 20$ applications.

In practice the points, expressing the probability of withstand, are plotted against the voltage V_j on a probability scale graph as was shown in Fig. 8.13. The best straight line is then plotted using curve fitting techniques. The 50 per cent and 10 per cent discharge voltages are obtained directly from the graph. This method has the advantage that it requires relatively few shots and therefore is most frequently used by industry. The disadvantage is that it assumes normality and is not very accurate in determining σ. Alternatively, the V_{10} can be obtained from the V_{50} using the formula

$$V_{10} = V_{50}(1 - 1.3z) = V_{50} \cdot 0.96 \tag{8.14}$$

From the sequentially obtained readings (Fig. 8.17), the values of V_{50} and σ can be also calculated analytically as follows.

In the example chosen (Fig. 8.17): total number of shots $n = 9$, total number of breakdowns $n_b = 4$, total number of withstands $n_w = 5$, and lowest level at which a shot is applied $= V_0$.

In calculating $V_{50\%}$ and σ,

if $n_b > n_w$ then n_i = number of withstands at level j

if $n_w > n_b$ then n_j = number of breakdowns at level j

(always use the smaller of the two). The expressions are:

$$V_{50} = V_0 + \Delta V\left[\frac{A}{N} \pm \frac{1}{2}\right] \Rightarrow \begin{cases} n_i = n_{bi} \text{ use negative sign} \\ n_i = n_{wi} \text{ use positive sign} \end{cases} \tag{8.15}$$

$$\sigma = 1.62AV\left(\frac{NB - A^2}{N^2} + 0.029\right) \tag{8.16}$$

where

$$N = \sum_{i=0}^{k} n_{iw} \quad \text{or} \quad \sum_{i=0}^{k} n_{ib}$$

$$A = \sum_{i=0}^{k} in_{iw} \quad \text{or} \quad \sum_{i=0}^{k} in_{ib}$$

$$B = \sum_{i=0}^{k} i^2 n_{iw} \quad \text{or} \quad \sum_{i=0}^{k} i^2 n_{ib}$$

with i referring to the voltage level, n_{iw} to the number of withstands and n_{ib} the number of breakdowns at that level.

3. *The extended up and down method*

This method is also used in testing self-restoring insulation. It can be used to determine discharge voltages corresponding to any probability p. A number of impulses are applied at a certain voltage level. If none causes discharge, the voltage is increased by a step ΔV and the impulses are applied until at least one causes breakdown, then the voltage is decreased. For an example of the extended up and down method procedure see Fig. 8.18.

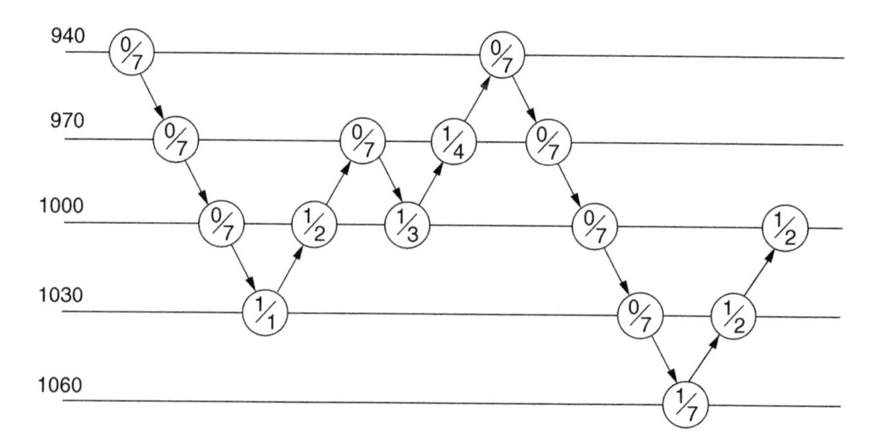

Figure 8.18 *Example of the extended up and down method*

The number n is determined such that a series of n shots would have 50 per cent probability of giving at least one flashover. The 50 per cent probability of discharge is given by

$$0.5 = 1 - (1 - p)^n$$

or

$$n = 0.5 = \ln(1 - p) \tag{8.17}$$

from which p becomes a discrete value. The value $n = 7$ impulses per voltage level is often used as it allows the determination of 10 per cent discharge voltage without the necessity to use σ. Substituting $n = 7$ into eqn (8.17) gives $p = 0.094$ or approximately 10 per cent.

The IEC switching withstand voltage is defined as 10 per cent withstand, hence the extended up and down method has an advantage. Other advantages include: discharge on test object is approximately 10 per cent the number of applied impulses rather than 50 per cent as applicable to the up and down method. Also the highest voltage applied is about V_{50} rather than $V_{50} + 2$. In the up and down method the $V_{10\%}$ may also be obtained from:

$$V_{10} = V_{50}(1 - 1.13z) = V_{50} \cdot 0.96 \quad (8.18)$$

In today's power systems for voltages up to 245 kV insulation tests are still limited to lightning impulses and the one-minute power frequency test. Above 300 kV, in addition to lightning impulses and the one-minute power frequency tests, tests include the use of switching impulse voltages.

8.4.8 Standard test procedures

1. Proof of lightning impulse withstand level

For self-restoring insulation the test procedures commonly used for withstand establishment are:

(i) 15 impulses of rated voltage and of each polarity are applied, up to two disruptive discharges are permitted,
(ii) in the second procedure the 50 per cent flashover procedure using either the up and down or extended up and down technique as described earlier.

From the up and down method the withstand voltage is obtained using eqn (8.18). In tests on non-self-restoring insulation, three impulses are applied at the rated withstand voltage level of a specified polarity. The insulation is deemed to have withstood if no failure is observed.

2. Testing with switching impulses

These tests apply for equipment at voltages above 300 kV. The testing procedure is similar to lightning impulses using 15 impulses. The tests are carried out in dry conditions while outdoor equipment is tested under positive switching impulses only. In some cases, when testing circuit isolators or circuit breakers which may experience combined voltage stress (power frequency and switching surge) biased tests using combined power frequency and surge voltages are used. The acceptable insulating capability requires 90 per cent withstand capability.

8.4.9 Testing with power frequency voltage

The standard practice requires the insulation to perform a one-minute test with power frequency at a voltage specified in the standards. For indoor equipment, the equipment is tested in dry conditions, while outdoor equipment is tested under prescribed rain conditions for which IEC prescribes a precipitation rate of 1–1.5 mm/min with resistivity of water.

8.4.10 Distribution of measured breakdown probabilities (confidence in measured P(V))

We apply at a level V_i, n shots and obtain x breakdowns. The outcome is breakdown or withstand, that is

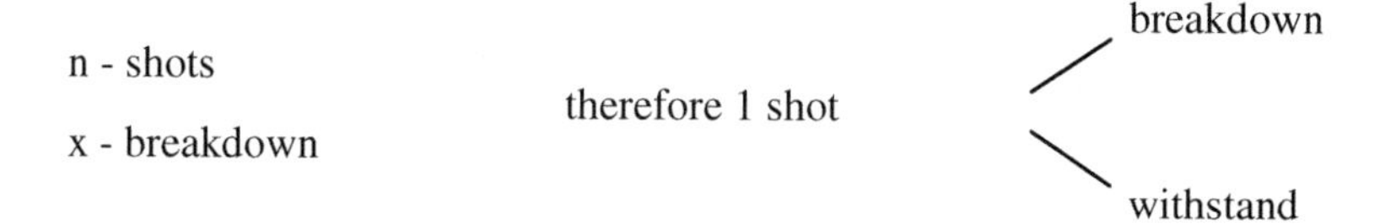

hence, the distribution of $P(V)$ is binomial around the expected value x/n. This distribution depends on x, n and q (the breakdown around which $P(x, n, \theta)$ is centre d) as shown in Fig. 8.19.

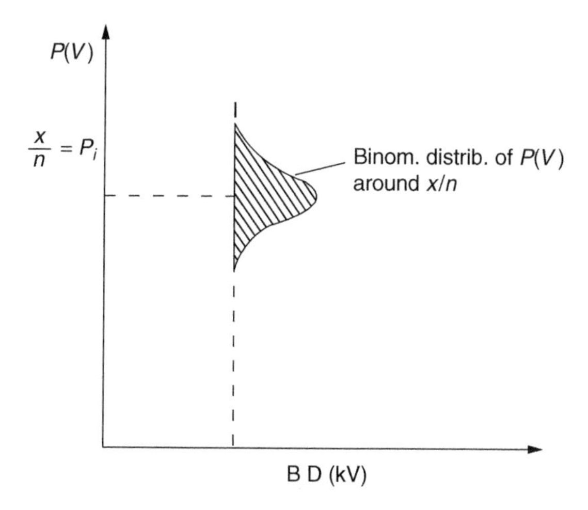

Figure 8.19 *Binomial distribution of $P(V)$ around the expected value x/n*

P_i is distributed around the value the point gives. For example, if we get: $V = 500\,\text{kV}$; P(FO) = 78 per cent, we do not really know that it is 78 per cent but we do know that it is distributed around 78 per cent.

The binomial distribution of P around x/n is given by

$$P(x, n, \theta) = \binom{n}{x} \theta^x (1 - \theta)^{n-x}$$

where

$$\binom{n}{x} = \frac{n!}{x!(n-x)!}$$

θ = true value of the most likely outcome (value around which the distribution is centred).

We do not know θ but we can replace it with the expected value x/n as was shown in Fig. 8.19.

Hence

$$P(x, n, \theta) = \binom{n}{x} \theta^x (1 - \theta) \tag{8.19}$$

with $x/n = 0.5$, $P(x)$ is symmetrical around x/n but at extremities ($x/n = 1$ per cent and 99 per cent). $P(x)$ is skewed as seen in Figs 8.20(a) and (b).

To obtain these distributions, leave x/n as the expected value and then vary x to obtain the corresponding $P(x)$.

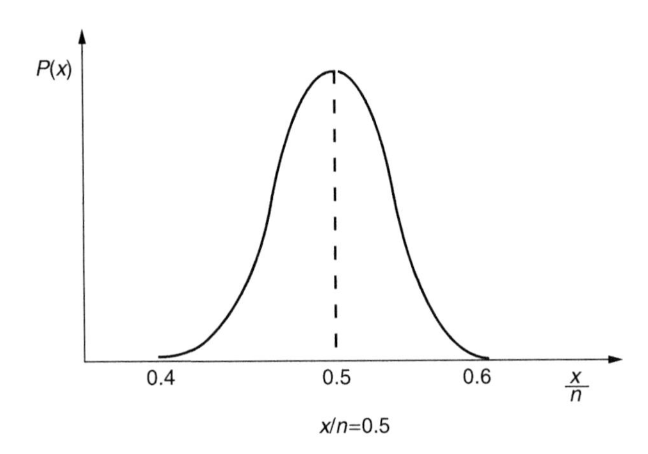

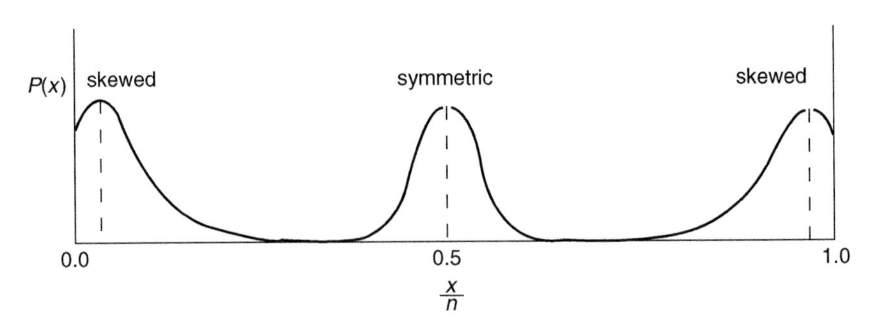

Figure 8.20 *Relation between P(x) and x/n: (a) Symmetric around x/n = 0.5. (b) Skewed*

Example

$n = 5$; $x = 2$, therefore $x/n = 0.4$: find $P(x = 4)$, using eqn (8.19)

$$P(x = 4) = \frac{5!}{4!(5-4)!}(0.4)^4(1-0.4)^{(5-4)} = 7.7\%$$

Using the eqn (8.19) we find that as n increases with x/n being constant we have greater confidence in $P(x)$ as seen in Figs 8.21(a) and (b), while for small values of n the results are spread.

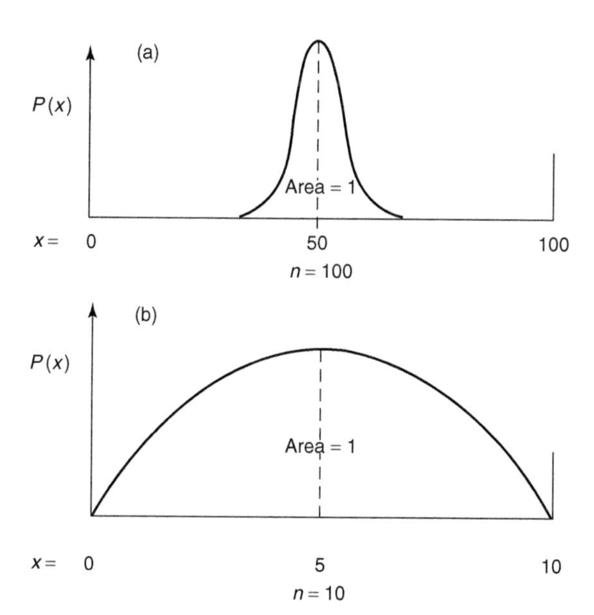

Figure 8.21 *Effect of number of shots on $P(x)$ distribution: (a) $n = 100$; (b) $n = 10$*

8.4.11 Confidence intervals in breakdown probability (in measured values)

The normalized value of the variable x in the binomial distribution is

$$\frac{X - n\theta}{\sqrt{n\theta(1-\theta)}} \tag{8.20}$$

For a given level of confidence $1 - \alpha$, where α is the level of significance as shown in Fig. 8.11, the confidence interval at a measured point is given by

$$-Z_{a/2} < \frac{x - n\theta}{\sqrt{n\theta(1-\theta)}} < +Z_{a/2} \tag{8.21}$$

The probability of breakdown with a confidence level $1 - \alpha$ is given by

$$P(V) = \frac{x}{n} \pm Z_{a/2}\sqrt{\frac{\frac{x}{n}\left(1 - \frac{x}{n}\right)}{n}} \tag{8.22}$$

Using this expression it can be shown on the linear probability scale that the confidence in the measured values of breakdown is at maximum at $x/n = 0.5$ and progressively decreases as the extreme values of breakdown probability are approached. $Z_{a/2}$ is obtained from tables of statistics or for convenience from the graph directly.

Example

$n = 10$; $x = 5$ for a confidence level of 90 per cent

$$\alpha = 1 - 0.9 = 0.1$$

using statistical tables,[25] we obtain for

$$\frac{a}{2} = 0.05, \quad Z_{a/2} = 1.64$$

hence

$$P(V) = \tfrac{1}{2} \pm (1.64)\sqrt{\frac{\frac{1}{2}\left(\frac{1}{2}\right)}{10}}$$

The confidence limit at near $P(V) = 50$ per cent is much smaller than the confidence limit for $P(V)$ approaching 1 per cent or 99 per cent. The solution is a non-linear distribution of n, that is we need to take many shots near the limits and a few in the middle. Confidence expressed in terms of kV is more convenient than confidence in probability as shown in Fig. 8.22.

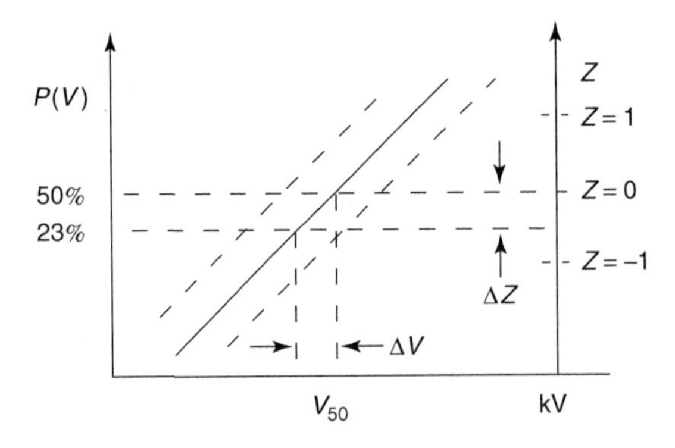

Figure 8.22 *Confidence expressed in kV*

Using the same example as before

$23\% \leq P(V_{50}) \leq 77\%$

to determine $\Delta_z = z_1 - z_2$

z_1 is determined from the value of $P(0.50)$

z_2 is determined from the value of $P(0.23)$

from Fig. 8.22

for $F(z) = 50\% = 0.5, \ z_1 = 0.0$

for $F(z) = 23\% = -0.77, \ z_2 = -0.74$

therefore

$$\begin{aligned} \Delta z &= z_1 - z_2 \\ &= 0 - (-0.74) \\ &= 0.74 \end{aligned}$$

the standard deviation σ is the run from $z = 0$ to $z = 1$ therefore

$$\begin{aligned} \text{rise} &= \text{slope} \\ &= \Delta z / \Delta V \\ &= 1/\sigma \end{aligned}$$

and therefore $\Delta V = \Delta z \sigma$

Thus the confidence in V is

$$\begin{aligned} V &= V \pm \Delta V \\ &= V \pm \Delta z \sigma \end{aligned} \tag{8.23}$$

confidence in

$$V_{50} = V_{50} \pm (0.74)\sigma \tag{8.24}$$

8.5 Weighting of the measured breakdown probabilities

Weights can be assigned to various data points to the measured breakdown probability and the number of impulses applied at each level.

8.5.1 Fitting of the best fit normal distribution

On probability paper the normal distribution best characterizing the data points will appear as the best fit straight line. An example of this is shown in Fig. 8.23.

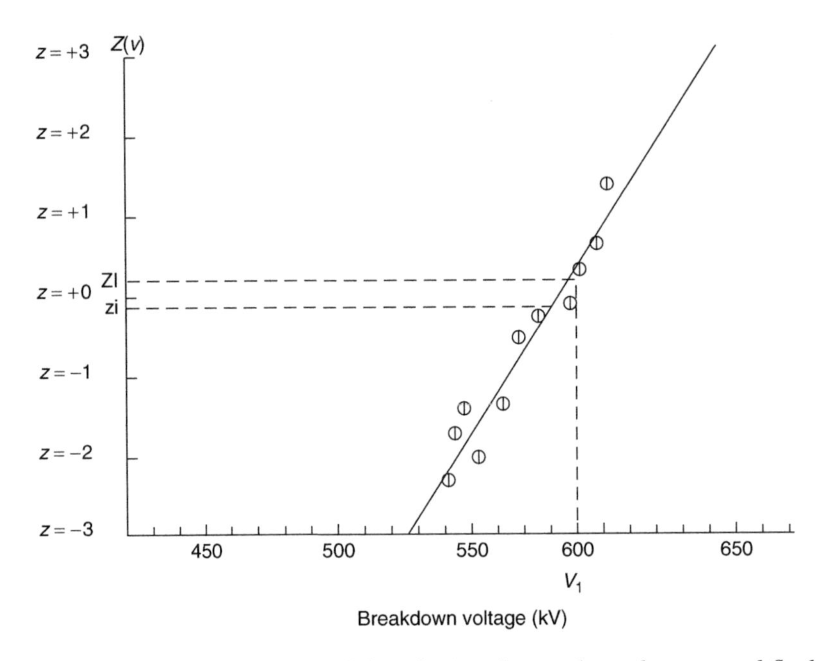

Figure 8.23 *Best fit normal distribution drawn through measured flashover probability points*

In order to obtain this best fit straight line, it is necessary to minimize the deviation of the data points around the line. The root mean square deviation for the case shown in Fig. 8.23 is given by

$$\sqrt{\frac{1}{m}\sum_{i=1} w_i\,(z_i - \xi_i)^2} \tag{8.25}$$

where x_i is the value of the measured breakdown probability on the probit scale at the voltage level V_i, x_i is the probit scale value of the breakdown probability as given by the best fit straight line for the same voltage level, and w_1 is the weighting coefficient assigned to the measurement, x_i. The expression given in eqn (8.25) is in terms of the dimensionless deviation z. This can be rewritten using

$$z_i = \frac{V_i - V_{50}}{\sigma} \tag{8.26}$$

to obtain

$$\sqrt{\frac{1}{m}\sum_{i=1} w_i\left(\frac{V_i - V_{50}}{\sigma} - \xi_i\right)^2} \tag{8.27}$$

Minimizing this expression is equivalent to minimizing

$$\sum_{i=1} w_i \left(\frac{V_i - V_{50}}{\sigma} - \xi_i \right)^2 \tag{8.28}$$

The minimum value of the above expression occurs when the quantity

$$\sum_{i=1} w_i \, (V_i - V_{50} - \sigma\xi_i)^2 \tag{8.29}$$

is at its minimum. The best fit straight line which is in fact the normal distribution best representing the breakdown probability can now be obtained by setting

$$\frac{\partial}{\partial V_{50}} \sum_{i=1} w_i \, (V_i - V_{50} - \sigma\xi_i)^2 = 0 \tag{8.30}$$

and

$$\frac{\partial}{\partial^{\sigma}} \sum_{i=1} w_i \, (V_i - V_{50} - \sigma\xi_i)^2 = 0 \tag{8.31}$$

and solving for V_{50} and σ. These values are found by carrying out the partial differentiation of eqns (8.30) and (8.31). This gives the following two simultaneous equations

$$\sum_{i=1} w_i \, v_i - \sum_{i=1} w_i V_{50} - \sigma \sum_{i=1} w_i \xi_i = 0 \tag{8.32}$$

and

$$\sum_{i=1} w_i V_i \xi_i - V_{50} \sum_{i=1} w_i \xi_i - \sigma \sum_{i=1} w_i \xi_i^2 = 0 \tag{8.33}$$

which can be solved to obtain

$$V_{50} = \frac{\sum_{i=1} w_i V_i - \sigma \sum_{i=1} w_i \xi_i}{\sum_{i=1} w_i} \tag{8.34}$$

and

$$\sigma = \frac{\sum_{i=1} w_i V_i \sum_{i=1} w_i \xi_i - \sum_{i=1} w_i \sum_{i=1} w_i V_i \xi_i}{\left(\sum_{i=1} w_i \xi_i \right)^2 - \sum_{i=1} w_i \sum_{i=1} w_i \xi_i^2} \tag{8.35}$$

Thus values for V_{50} and σ are obtained.

8.6 Insulation coordination

Insulation coordination is the correlation of insulation of electrical equipment with the characteristics of protective devices such that the insulation is protected from excessive overvoltages. In a substation, for example, the insulation of transformers, circuit breakers, bus supports, etc., should have insulation strength in excess of the voltage provided by protective devices.

Electric systems insulation designers have two options available to them: (i) choose insulation levels for components that would withstand all kinds of overvoltages, (ii) consider and devise protective devices that could be installed at the sensitive points in the system that would limit overvoltages there. The first alternative is unacceptable especially for e.h.v. and u.h.v. operating levels because of the excessive insulation required. Hence, there has been great incentive to develop and use protective devices. The actual relationship between the insulation levels and protective levels is a question of economics. Conventional methods of insulation coordination provide a margin of protection between electrical stress and electrical strength based on predicted maximum overvoltage and minimum strength, the maximum strength being allowed by the protective devices.

8.6.1 Insulation level

'Insulation level' is defined by the values of test voltages which the insulation of equipment under test must be able to withstand.

In the earlier days of electric power, insulation levels commonly used were established on the basis of experience gained by utilities. As laboratory techniques improved, so that different laboratories were in closer agreement on test results, an international joint committee, the Nema-Nela Committee on Insulation Coordination, was formed and was charged with the task of establishing insulation strength of all classes of equipment and to establish levels for various voltage classification. In 1941 a detailed document[(18)] was published giving basic insulation levels for all equipment in operation at that time. The presented tests included standard impulse voltages and one-minute power frequency tests.

In today's systems for voltages up to 245 kV the tests are still limited to lightning impulses and one-minute power frequency tests, see section 8.3. Above 300 kV, in addition to lightning impulse and the one-minute power frequency tests, tests include the use of switching impulse voltages. Tables 8.2 and 8.3 list the standardized test voltages for ≤245 kV and above ≥300 kV respectively, suggested by IEC for testing equipment. These tables are based on a 1992 draft of the IEC document on insulation coordination.

Table 8.2 *Standard insulation levels for Range I ($1\ kV < U_m \leq 245\ kV$) (From IEC document 28 CO 58, 1992, Insulation coordination Part 1: definitions, principles and rules)*

Highest voltage for equipment U_m *kV (r.m.s. value)*	*Standard power frequency short-duration withstand voltage kV (r.m.s. value)*	*Standard lightning impulse withstand voltage kV (peak value)*
3.6	10	20
		40
7.2	20	40
		60
12	28	60
		75
		95
17.5	38	75
		95
24	50	95
		125
		145
36	70	145
		170
52	95	250
72.5	140	325
123	(185)	450
	230	550
145	(185)	(450)
	230	550
	275	650
170	(230)	(550)
	257	650
	325	750
245	(275)	(650)
	(325)	(750)
	360	850
	395	950
	460	1050

Table 8.3 *Standard insulation levels for Range II (U_m > 245 kV) (From IEC document 28 CO 58, 1992, Insulation coordination Part 1: definitions, principles and rules)*

Highest voltage for equipment U_m kV (r.m.s. value)	Longitudinal insulation (+) kV (peak value)	Standard lightning impulse withstand voltage Phase-to-earth kV (peak value)	Phase-to-phase (ratio to the phase-to-earth peak value)	Standard lightning impulse withstand voltage kV (peak value)
300	750	750	1.50	850 950
	750	850	1.50	950 1050
362	850	850	1.50	950 1050
	850	950	1.50	1050 1175
420	850	850	1.60	1050 1175
	950	950	1.50	1175 1300
	950	1050	1.50	1300 1425
525	950	950	1.70	1175 1300
	950	1050	1.60	1300 1425
	950	1175	1.50	1425 1550
765	1175	1300	1.70	1675 1800
	1175	1425	1.70	1800 1950
	1175	1550	1.60	1950 2100

(+)Value of the Impulse component of the relevant combined test.
Note: The introduction of $U_m = 550$ kV (instead of 525 kV), 800 kV (instead of 765 kV), 1200 kV, of a value between 765 kV and the associated standard withstand voltages, are under consideration.

8.6.2 Statistical approach to insulation coordination

In the early days insulation levels for lightning surges were determined by evaluating the 50 per cent flashover values (BIL) for all insulations and providing a sufficiently high withstand level that all insulations would withstand. For those values a volt–time characteristic was constructed. Similarly the protection levels provided by protective devices were determined. The two volt–time characteristics are shown in Fig. 8.24. The upper curve represents the common BIL for all insulations present, while the lower represents the protective voltage level provided by the protective devices. The difference between the two curves provides the safety margin for the insulation system. Thus the

$$\text{Protection ratio} = \frac{\text{Max. voltage it permits}}{\text{Max. surge voltage equipment withstands}} \tag{8.36}$$

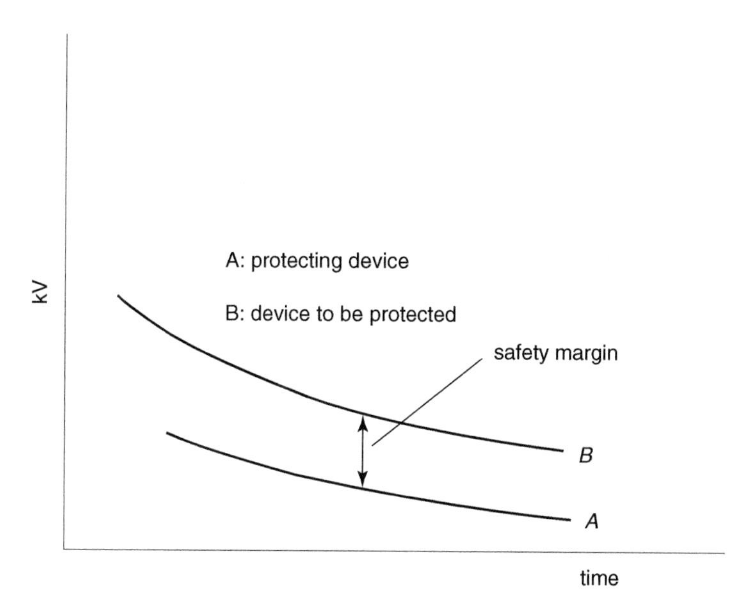

Figure 8.24 *Coordination of BILs and protection levels (classical approach)*

This approach is difficult to apply at e.h.v. and u.h.v. levels, particularly for external insulations.

Present-day practices of insulation coordination rely on a statistical approach which relates directly the electrical stress and the electrical strength.[11] This approach requires a knowledge of the distribution of both the anticipated stresses and the electrical strengths.

The statistical nature of overvoltages, in particular switching overvoltages, makes it necessary to compute a large number of overvoltages in order to determine with some degree of confidence the statistical overvoltages on a system. The e.h.v. and u.h.v. systems employ a number of non-linear elements, but with today's availability of digital computers the distribution of overvoltages can be calculated. A more practical approach to determine the required probability distributions of a system's overvoltages employs a comprehensive systems simulator, the older types using analogue units, while the newer employ real time digital simulators (RTDS).[24]

For the purpose of coordinating the electrical stresses with electrical strengths it is convenient to represent the overvoltage distribution in the form of probability density function (Gaussian distribution curve as shown in Fig. 8.11) and the insulation breakdown probability by the cumulative distribution function (shown in Fig. 8.12). The knowledge of these distributions enables us to determine the 'risk of failure'. As an example, let us consider a case of a spark gap for which the two characteristics in Figs 8.11 and 8.12 apply and plot these as shown in Fig. 8.25.

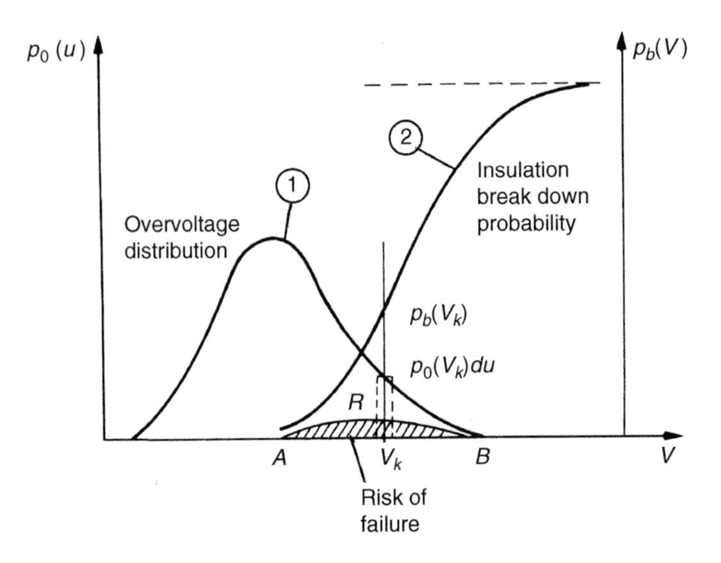

Figure 8.25 *Method of describing the risk of failure. 1. Overvoltage distribution–Gaussian function. 2. Insulation breakdown probability–cumulative distribution)*

If V_a is the average value of overvoltage, V_k is the kth value of overvoltage, the probability of occurrence of overvoltage is $p_0(V_k)\,du$, whereas the probability of breakdown is $P_b(V_k)$ or the probability that the gap will break down at an overvoltage V_k is $P_b(V_k)p_0(V_k)\,du$. For the total voltage range we obtain for the total probability of failure or 'risk of failure'

$$R = \int_0^\infty P_b(V_k)p_0(V_k)\,\mathrm{d}u. \tag{8.37}$$

The risk of failure will thus be given by the shaded area under the curve R.

In engineering practice it would become uneconomical to use the complete distribution functions for the occurrence of overvoltage and for the withstand of insulation and a compromise solution is accepted as shown in Figs 8.26(a) and (b) for guidance. Curve (a) represents probability of occurrence of overvoltages of such amplitude (V_s) that only 2 per cent (shaded area) has a chance to cause breakdown. V_S is known as the 'statistical overvoltage'. In Fig. 8.26(b) the voltage V_w is so low that in 90 per cent of applied impulses, breakdown does not occur and such voltage is known as the 'statistical withstand voltage' V_w.

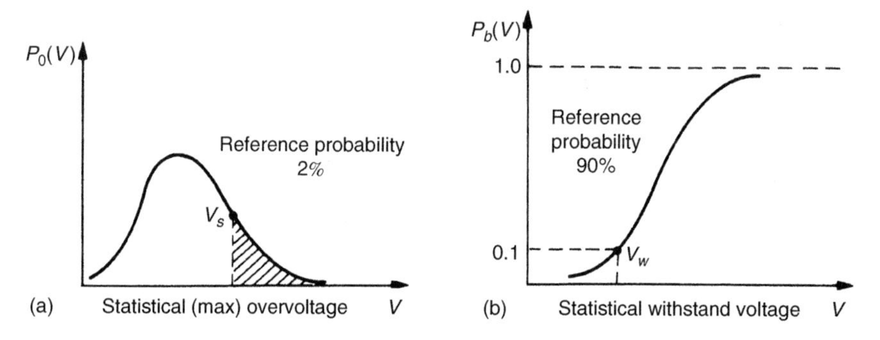

Figure 8.26 *Reference probabilities for overvoltage and for insulation withstand strength*

In addition to the parameters statistical overvoltage 'V_S' and the statistical withstand voltage 'V_W' we may introduce the concept of statistical safety factor γ. This parameter becomes readily understood by inspecting Figs 8.27(a) to (c) in which the functions $P_b(V)$ and $p_0(V_k)$ are plotted for three different cases of insulation strength but keeping the distribution of overvoltage occurrence the same. The density function $p_0(V_k)$ is the same in (a) to (c) and the cumulative function giving the yet undetermined withstand voltage is gradually shifted along the V-axis towards high values of V. This corresponds to increasing the insulation strength by either using thicker insulation or material of higher insulation strength. As a result of the relative shift of the two curves [$P_b(V)$ and $p_0(V_k)$] the ratio of the values V_w/V_s will vary. This ratio is known as the statistical safety factor or

$$\frac{V_w}{V_s} = \gamma \tag{8.37}$$

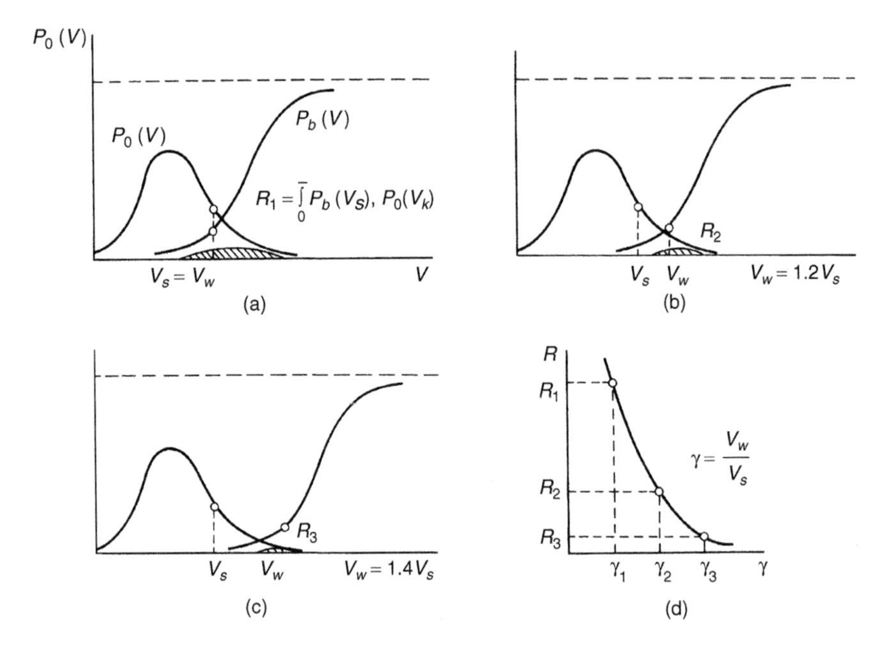

Figure 8.27 *The statistical safety factor and its relation to the risk of failure (R)*

In the same figure (d) is plotted the relation of this parameter to the 'risk of failure'. It is clear that increasing the statistical safety factor (γ) will reduce the risk of failure (R), but at the same time will cause an increase in insulation costs. The above treatment applies to self-restoring insulations. In the case of non-self-restoring insulations the electrical withstand is expressed in terms of actual breakdown values. The statistical approach to insulation, presented here, leads to withstand voltages (i.e. probability of breakdown is very small), thus giving us a method for establishing the 'insulation level'.

8.6.3 Correlation between insulation and protection levels

The 'protection level' provided by (say) arresters is established in a similar manner to the 'insulation level'; the basic difference is that the insulation of protective devices (arresters) must not withstand the applied voltage. The concept of correlation between insulation and protection levels can be readily understood by considering a simple example of an insulator string being protected by a spark gap, the spark gap (of lower breakdown strength) protecting the insulator string. Let us assume that both gaps are subjected to the same overvoltage represented by the probability density function $p_0(V)$, Fig. 8.28. The probability distribution curves for the spark gap and the insulator string are presented by $P_g(V)$ and $P_i(V)$ respectively in Fig. 8.28.

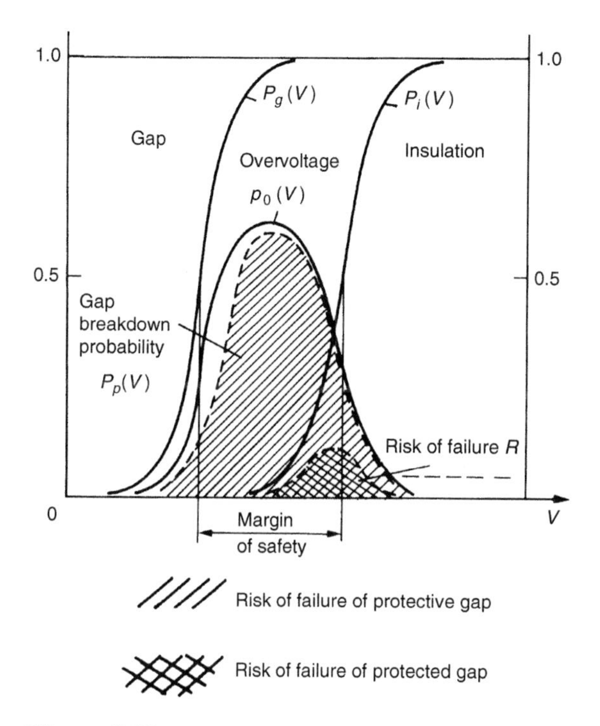

Figure 8.28 *Distribution functions of breakdown voltages for protective gap and protected insulation both subjected to an overvoltage $p_0(V)$*

The statistical electrical withstand strength of the insulator string is given by a curve identical to Fig. 8.26. The probability of breakdown of this insulation remains in the area R which gives 'risk of failure'. Since the string is protected by a spark gap of withstand probability $P_g(V)$, the probability that the gap will operate (its risk of failure) is obtained from integrating the product $P_g(V)p_0(V)\,dV$. In Fig. 8.28 this probability is denoted (qualitatively) by $P_P(V)$. As is seen the probability is much higher than the probability of insulation damage or failure R. In the same figure is shown the traditional margin of safety corresponding to the voltage difference between the 50 per cent flashover values of the protecting gap and the protected gap.

For overvoltages of the highest amplitude (extreme right of Fig. 8.28) the probability curves of insulation failure and that of protective spark gap breakdown overlap. In reality such cases will not arise. Figure 8.28 is simplified in that it contains information pertaining to the amplitude of the overvoltage, and ignores the effect of time of voltage application on the breakdown of both the protective gap and the insulation. In practice, the protective gap will in general break down before the insulation and will cause a reduction (to a safe limit) in overvoltage reaching the protected insulation.

8.7 Modern power systems protection devices

8.7.1 MOA – metal oxide arresters

The development of MOA (metal oxide arresters) represented a breakthrough in overvoltage protection devices. It became possible to design arresters without using gaps which were indispensable in the conventional lightning arresters, which utilized non-linear resistors made of silicon Carbide (SiC) and spark gaps. Figure 8.29 shows a block diagram of the valve arrangements in the two types of arrester.

In (a) the elements and the spark gaps are connected in series. In (b) the elements are stacked on top of each other without the need for spark gaps.

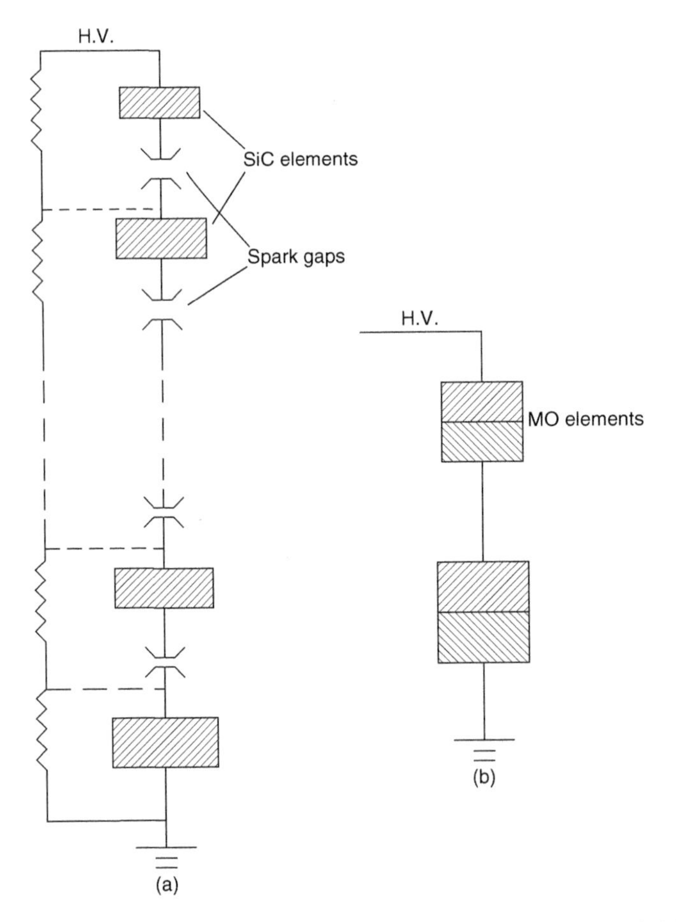

Figure 8.29 *Block diagram of valve arrangements in (a) SiC, (b) MOA*

An ideal lightning arrester should: (i) conduct electric current at a certain voltage above the rated voltage; (ii) hold the voltage with little change for the duration of overvoltage; and (iii) substantially cease conduction at very nearly the same voltage at which conduction started.[25] In Fig. 8.29(a) the three functions are performed by the combination of the series spark gaps and the SiC elements. In the (b) case the metal oxide valve elements perform all three functions because of their superior non-linear resistivity.

The volt–current characteristics for the two types of arresters can be represented by the following equations:

$$\text{For SiC valves: } I = kV^a \quad \text{where } a = 4\text{–}6 \tag{8.38}$$

$$\text{For ZnO valves: } I = kV^b \quad \text{where } k = \text{const},\ b = 25\text{–}30 \tag{8.39}$$

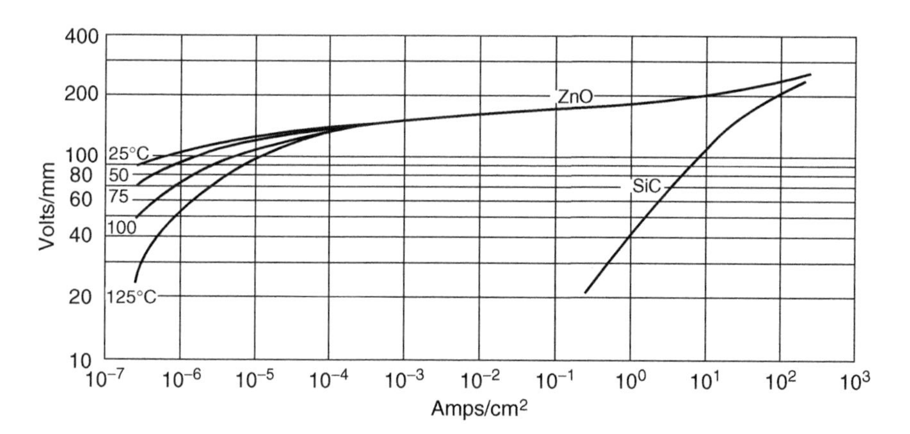

Figure 8.30 *Normalized volt–ampere characteristic of zinc oxide and silicon carbide valve elements*

Typical volt–current characteristics for the valve elements used in the two types of arresters are plotted in Fig. 8.30. The metal oxide varistors, which consist of compacted and sintered granules of zinc oxide with a small amount of other carefully selected metal oxide additives (Bi_2O_3, MnO, Cr_2O_3, Sb_2O_3) to improve the V–I non-linearity, were first introduced in the electronics industry in 1968 by Matsushita Electric Industrial Co. in Japan. The ZnO grains have a low resistivity, while the additives (oxides) which form the boundaries between the grains provide high resistance. The two are strongly bonded when sintered at high temperature. Figure 8.31 shows the microstructure of a metal oxide varistor.

Subsequently these were developed for use as a substitute for SiC valve blocks in surge arresters by General Electric Co.[26] From Fig. 8.30 it can be seen that for a change in current from 10^{-3} to 10^2 A/cm^2, the voltage increase

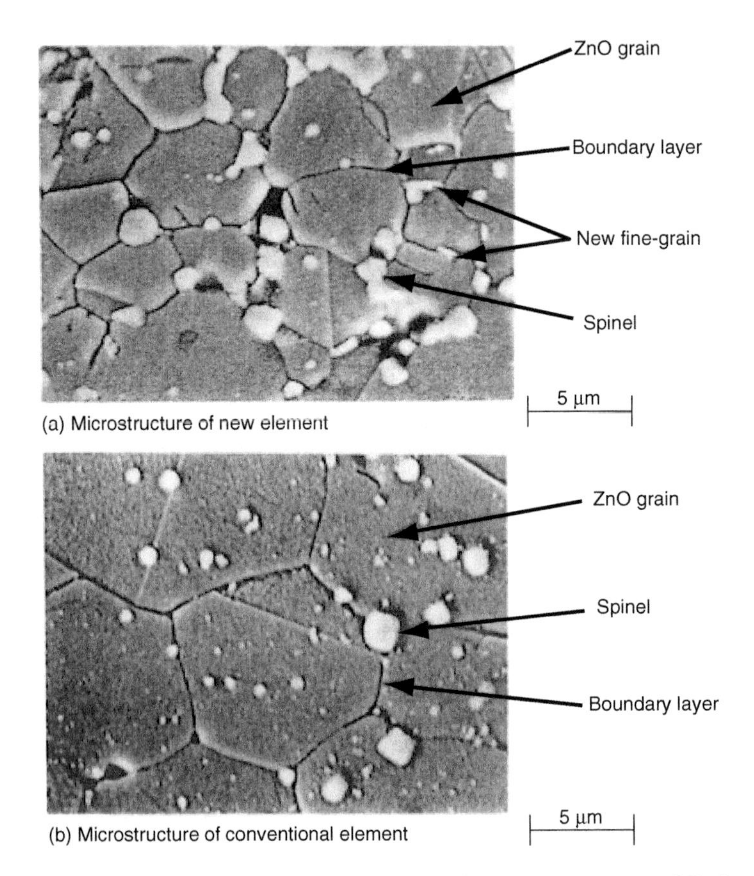

Figure 8.31 *Cross-section, showing the microstructure of ZnO elements. (a) Latest type (advanced). (b) Older conventional type (courtesy of Mitsubishi Elec. Co.)*

for ZnO is only 56 per cent.[25] With such a high degree of non-linearity it is entirely feasible to use these elements without series gaps in an arrester with a current of only tens of μA at operating voltage.

The elements are manufactured in the form of discs of several sizes. The disc voltage rating has been increasing with the improvement in the manufacturing technology and the microstructure composition, e.g. Fig. 8.32 compares the V–I characteristics of an older type ZnO element with that of a new type, both developed by Mitsubishi.[27]

It is noted that the voltage rating per unit valve has been approximately doubled. For higher voltage and current ratings the discs are arranged in series and in parallel. Figure 8.33 shows a schematic structure of a three-column arrangement of the arrester valves in an advanced MOA compact structure manufactured by Mitsubishi.[27]

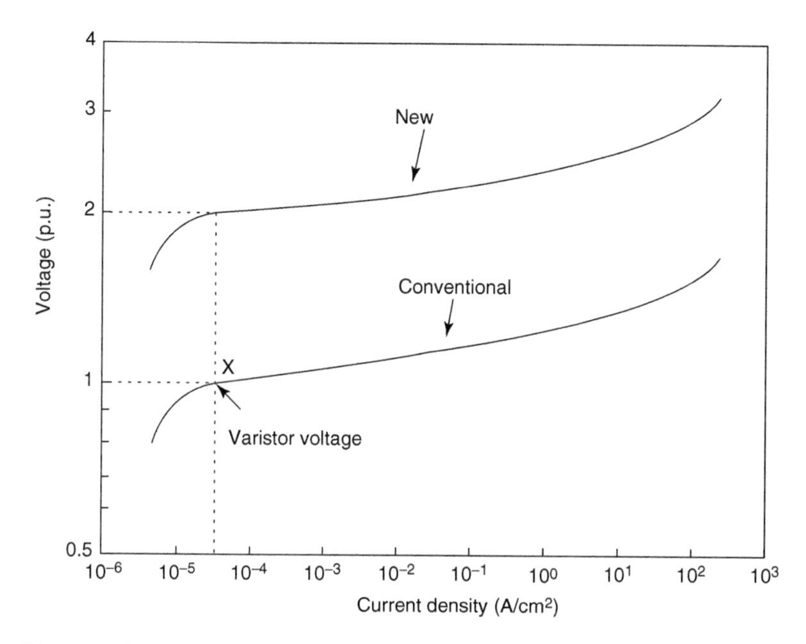

Figure 8.32 *Comparison of volt–current characteristics of (a) advanced MOA with (b) that of an older type MOA (courtesy of Mitsubishi Co.)*

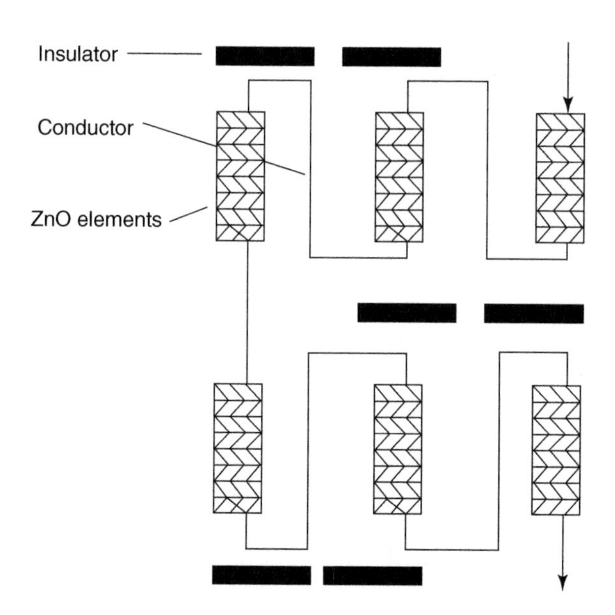

Figure 8.33 *Schematic structure of a three column series arrangement of elements in advanced MOAs*

In Fig. 8.34 is shown part of an assembled advanced 500 kV MOA. The percentages indicate the reduction in size by replacing the older type MOA with the advanced MOA elements whose V–I characteristics are shown in Fig. 8.32.

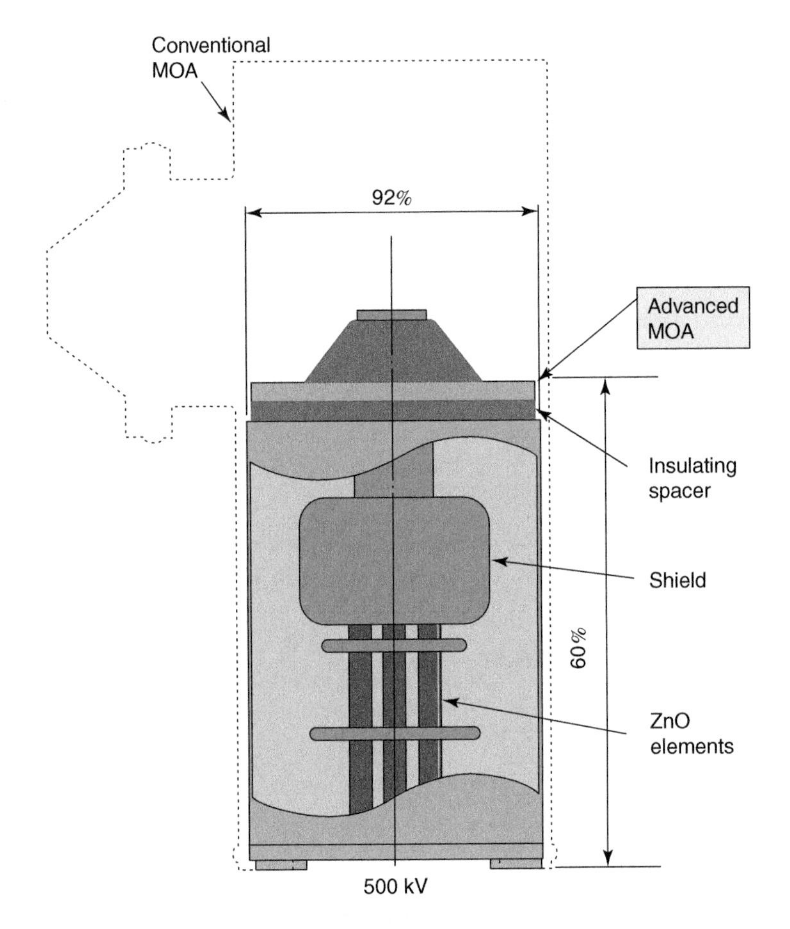

Figure 8.34 *Part of an assembled 500 kV MOA Arrester. (courtesy of Mitsubishi Co.)*

In this construction the individual surge arresters are interconnected by means of corona-free stress distributors. The modular design and the lightweight construction allow easy on-site erection and in the event of any units failing the individual unit may be readily replaced.

The advantages of the polymeric-housed arresters over their porcelain-housed equivalents are several and include:

- No risk to personnel or adjacent equipment during fault current operation.
- Simple light modular assembly – no need for lifting equipment.

- Simple installation.
- High-strength construction eliminates accidental damage during transport.
- The use of EPDM and/or silicon rubber reduces pollution flashover problems.

Thus the introduction of ZnO arresters and their general acceptance by utilities since late 1980s, and in 1990s in protecting high voltage substations, has greatly reduced power systems protection problems.

In the earlier construction the valve elements were mounted within a ceramic housing. The metal oxide elements were surrounded by a gaseous medium and the end fittings were generally sealed with rubber O-rings. With time in service, especially in hostile environments, the seals tended to deterio-

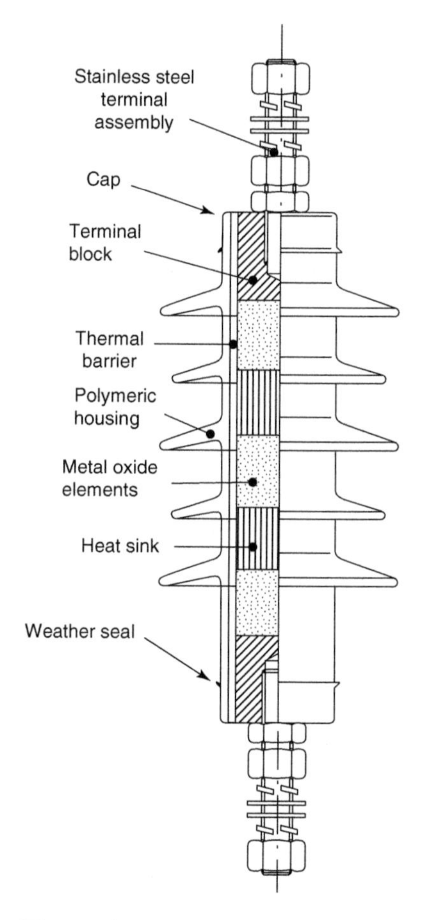

Figure 8.35 *Cross-section of a polymer-housed arrester (courtesy of Bowthorpe EMP)*

rate allowing the ingress of moisture. In the 1980s polymeric-housed surge arresters were developed. Bowthorpe EMP (UK)[28] manufactures a complete range of polymeric-housed arresters extending from distribution to heavy duty station arresters for voltages up to 400 kV. In their design the surface of the metal oxide elements column is bonded homogeneously with glass fibre reinforced resin. This construction is void free, gives the unit a high mechanical strength, and provides a uniform dielectric at the surface of the metal oxide column. The housing material is a polymer (EPDM)–Ethylene propylene diene monomer–which is a hydrocarbon rubber, resistant to tracking and is particularly suitable for application in regions where pollution causes a problem. A cross-section detailing the major features of a polymeric-housed arrester is given in Fig. 8.35.

The ZnO elements are separated by aluminium blocks which serve as heat sinks. To achieve higher voltages and higher current ratings a modular construction with the individual units mounted in series–parallel arrangement is shown in Fig. 8.36.

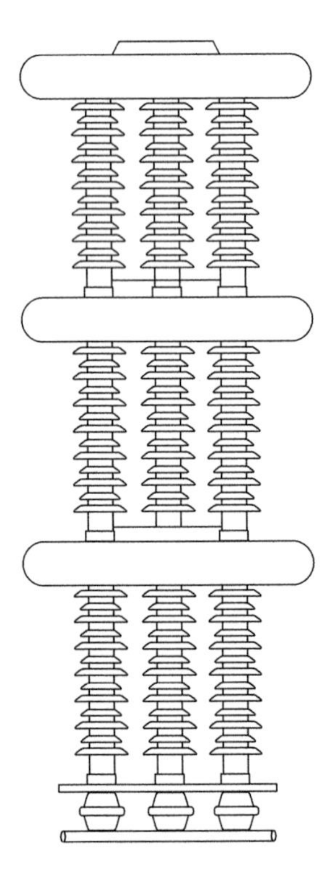

Figure 8.36 *Construction of a series–parallel polymeric-housed arrester. (courtesy of Bowthorpe EMP)*

References

1. Hydro-Quebec Symposium on Extra-High Voltage Alternating Current, Oct. 1973.
2. L.V. Bewley. *Travelling Waves on Transmission Systems*. Dover Publications, New York NY, 1963.
3. *Electrical Transmission and Distribution*. Westinghouse Electric Co., Pittsburgh, Penn., 1964.
4. W.W. Lewis. *The Protection of Transmission Systems against Lightning*. Dover Publications Inc., New York, 1965.
5. R.H. Golde (ed.). *Lightning*, Vol. I: *Physics of Lightning*; Vol. II: *Lightning Protection*. Academic Press, London/New York/San Francisco, 1977.
6. B.F.J. Schonland. Progressive lightning, IV. The discharge mechanism. *Proc. Roy. Soc.* Series A, **164** (1938), p. 132.
7. *E.H.V. Transmission Lines*. Reference Book General Electric Company, Edison Electrical Institute, New York, 1968, p. 288.
8. AIEE Committee Report. Method for Estimating Lightning Performance of Transmission Lines. *Trans. AIEE* Part III, **69** (1950), p. 1187.
9. R.H. Golde. A plain man's guide to lightning protection. *Electronics and Power*, March 1969.
10. T. Udo. Sparkover characteristics of long gaps and insulator strings. *Trans. IEEE* **PAS 83** (1964), p. 471.
11. W. Diesendorf. *Insulation Coordination in High Voltage Electric Power Systems*. Butterworths, 1974.
12. IEEE St-4-1995, Standard Techniques for High Voltage Testing. IEEE Inc. publication.
13. International Electrotechnical Commission, IEC Intern. Standard 61-1, 1989-11, High Voltage Test Techniques, Pt. 1: General Definitions and Test Requirements.
14. L. Paris. Influence of air gap characteristics on line to ground switching surge strength. *Trans. IEEE* **PAS 86** (1967), p. 936.
15. L. Paris and R. Cortina. Switching surge characteristics of large air gaps and long insulator strings. *Trans. IEEE* **PAS 87** (1968), p. 947.
16. G. Gallet and G. Leroy. Expression for switching impulse strength suggesting the highest permissible voltage for AC systems. IEEE-Power, Summer Power Meeting, 1973.
17. K.H. Schneider and K.H. Week. *Electra* No. 35 (1974), p. 25.
18. Standard Impulse, Basic Insulation Levels. A Report of the Joint Committee on Coordination of Insulation AIEE, EEI and NEMA. EEI Publication No. H-9, NEMA Publication #109, AIEE Transactions, 1941.
19. Dielectric Stresses and Coordination of Insulation. Brown Boveri Publication No. CH-A0500 20E No. 4, 1972.
20. K. Berger. Method und Resultate der Blitzforschung auf dem Monte San Salvatore bel Lugano in den Jahren 1963–1971.
21. W. Büsch. The effect of humidity on the dielectric strength of long air gaps of UHV-configurations subjected to positive impulses. Ph.D. thesis, ETH, Zurich, 1982 (see also: W. Büsch. *Trans. IEEE* **PAS 97** (1978), pp. 2086–2093).
22. J. Kuffel, R.G. van Heswijk and J. Reichman. Atmospheric influences on the switching impulse performance of 1-m gaps. *Trans. IEEE* **PAS 102**(7), July 1983.
23. International Electrotechnical Commission. IEC Intern. Standard IEC 71-2 1996 'Insulation Coordination Part 2. Application Guide'.
24. R. Kuffel, J. Giesbrecht, T. Maguire, R.P. Wierckx and P. McLaren. 'RTDS' A Fully Digital Power System Simulator Operating in Real Time, Proceedings of the First International Conference on Digital Power System Simulators, pp. 19–24, April 1995.
25. E.C. Sakshang *et al.* A new concept in Station Design. *IEEE Trans.* **PAS**, Vol. 96, No. 2, 1977, pp. 647–656.

26. A. Sweetana *et al.* Design, development and testing of 1,200 kV and 550 kV gapless surge arresters. *IEEE Trans.* **PAS** Vol. 101, No. 7, 1982, pp. 2319–2327.
27. Private communication.
28. Bowthorpe EMP Catalogue 030 1992.

Chapter 9

Design and testing of external insulation

Conventional air-insulated substations represent a large majority of installed high-voltage substations. They range in voltage from distribution levels to 765 kV systems. The external insulation generally utilized in these outdoor substations takes the form of insulators (posts, suspension and pin types) and housings. These types of apparatus are generally broken into classifications based on manufacturing and materials.

The two broad categories of insulators are ceramic and polymeric. Ceramic insulators include those constructed from porcelain and glass. Polymeric insulators, often referred to as non-ceramic insulators (NCIs), are made up of various designs, usually incorporating a fibre glass core encapsulated in rubber housings which afford protection to the fibre glass core from electrical stresses and moisture. Both categories are described in more detail in section 9.5.

In addition to the general review of the design and use of outdoor insulation, this chapter also presents a synopsis of the physical mechanism of insulator contamination flashover. Following this, methods used in the evaluation and testing of insulators operating in contaminated environments are discussed, from the laboratory and in-service perspectives. Finally, methods of mitigating contamination-related flashovers of in-service insulators are reviewed.

9.1 Operation in a contaminated environment

The environment in which an insulator is installed can have a significant impact on the unit's performance. When insulators are situated in areas where they are exposed to contamination, their performance can deteriorate significantly. This is likely the single greatest challenge encountered in the design and operation of substation insulation. In order to provide some insight into this topic, the processes of contamination accumulation and the flashover mechanism of polluted insulators will be presented, together with some remedies to control the problem.

Although the problem of contamination of insulators has been recognized for over 50 years most studies have been carried out within the last three decades, i.e. since the advent of compact transmission systems with reduced insulation. Contamination flashover has become the most important and often the limiting factor in the design of high-voltage outdoor insulation and hence became a subject of extensive studies. A number of empirical and theoretical models for the flashover mechanisms have been proposed.(1,2,3,4,5,6,7)

Contamination flashover (FO) requires both soluble salts and moisture. To a large extent differences in insulator behaviour arise due to the variety of environments and complex wetting mechanisms. Hence the performance of insulators in contaminated environments is best assessed by tests done under natural conditions.

However, for practical reasons artificial tests which can be performed in h.v. laboratories are required (see section 9.3). Artificial tests assess insulators on a relative basis, because they cannot account for the effect of the characteristics of shape on collection of contamination and self-cleaning properties.

The various sources of pollution that affect power system insulation include:

- Sea salt – salt from sea water is carried by winds up to 15–30 km inland or further.
- Industrial products which contain soluble salts.
- Road salts.
- Bird excrement.
- Desert sands.

9.2 Flashover mechanism of polluted insulators under a.c. and d.c.

Insulators in service become covered with a layer of pollution. When the surface is dry the contaminants are non-conducting; however, when the insulator surface is wetted by light rain, fog, or mist, the pollution layer becomes conducting with the following sequence of events:

- conducting layer build-up,
- dry band formation,
- partial arcing,
- arc elongation,
- eventual arc spanning the whole insulator followed by flashover.

The pollution layer in general is not uniform. When conduction starts, the currents are in the order of several milliamps, resulting in heating of the electrolyte solution on the insulator surface. The leakage current begins to dry the pollution layer and the resistivity of the layer rises in certain areas. This leads to dry band formation, usually in areas where the current density is highest. The dry band supports most of the applied voltage. The air gap flashes over, with the arc spanning the dry band gap which is in series with the wet portion of the insulator. The arc may extinguish at current zero and the insulator may return to working conditions. Dry band formation and rewetting may continue for many hours.

The current coinciding with the occurrence of dry band breakdown is in the order of 250 mA. The current at this stage is in surges, and the voltage is unaffected.

9.2.1 Model for flashover of polluted insulators[1,2,6]

Let us assume a uniform pollution layer with resistance r kΩ/mm as shown in Fig. 9.1. When the arc is burning in series with the pollution layer, the voltage across the insulator with an arc partially bridging the insulator will be given by:

$$V = V_{\mathrm{arc}}(I, x) + I(L - x)r \tag{9.1}$$

where the function $V_{\mathrm{arc}}(I, x)$ relates the arc voltage to the current I and the arc length x. In general, for a given resistance r the curve relating V to x/L has the form shown in Fig. 9.2.

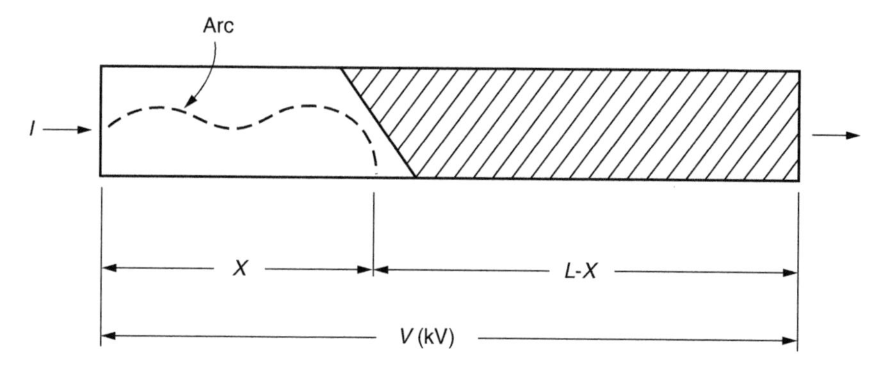

Figure 9.1 *Model of a single arc developing on a polluted surface (uniform pollution layer r, kΩ/mm)*

For an applied voltage V_a, x/L may have values no greater than x/L_a.The curve has a maximum critical voltage V_c, and for voltages equal to or greater than V_c, x/L may have values up to unity. When the applied voltage V_a is less than V_c, x/L cannot increase to unity and flashover cannot occur. Numerous empirical relations have been[1,4,5] proposed to solve eqn (9.1). For example, for vertical 33 kV and above

$$V_c = 0.067 r^{1/3} L_a^{2/3} L_s^{1/3} \ \mathrm{kV(r.m.s.)} \tag{9.2}$$

where L_a is the minimum arc length (mm) to the bridge insulator and L_s is the leakage path (mm) on the insulator surface.

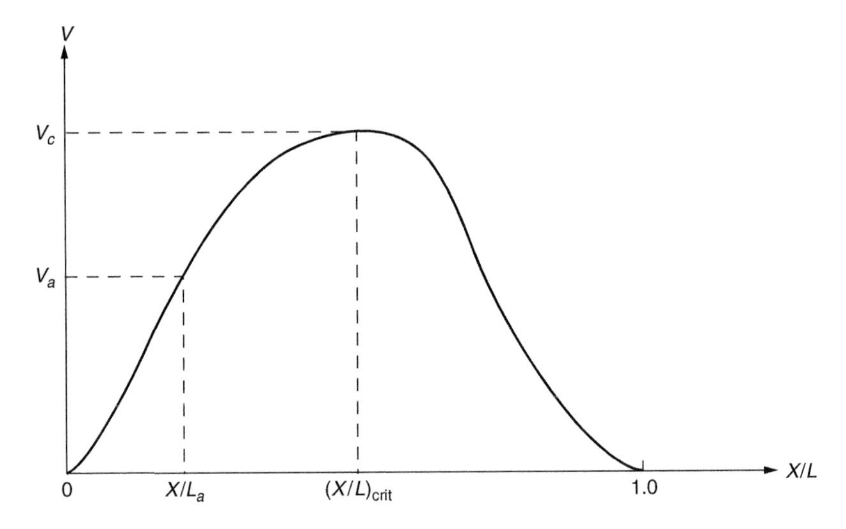

Figure 9.2 *Voltage versus x/L for an arc in series with a pollution layer of a fixed resistance per unit length*

For a cap and pin type insulator string,

$$V_c = 0.67^{1/3} N r^{1/3} L_S^{2/3} \lambda^{-1/3} \text{ kV(r.m.s)} \tag{9.3}$$

where λ is a constant and N is the number of insulators in the string.

For a given insulator type

$$L_s \propto L_a$$

or

$$L_a = kL_s \quad \text{with } k = \text{constant}$$

Therefore eqn (9.3) can be written as

$$V_C = 0.067^{2/3} k r^{-1/3} L_s \tag{9.4}$$

9.3 Measurements and tests

Assessments of the performance of insulators is based on laboratory and field tests which include:

(i) measurement of insulator dimensions;
(ii) measurement of pollution severity;
(iii) tests on polluted insulators.

9.3.1 Measurement of insulator dimensions

In order to effectively assess the degree of contamination present on an insulator surface the dimensions of the insulator must be taken into account. The relevant dimensions include the leakage path L_s, and the surface gradient expressed in kV/L_s (L_s in mm). For a definition of L_s see Fig. 9.3. The indentations X and Y are assumed filled with a conducting material.

The insulator surface area is required to determine the equivalent salt density deposit (ESDD) in mg/cm^2 (usually mean area based on maximum and minimum areas).

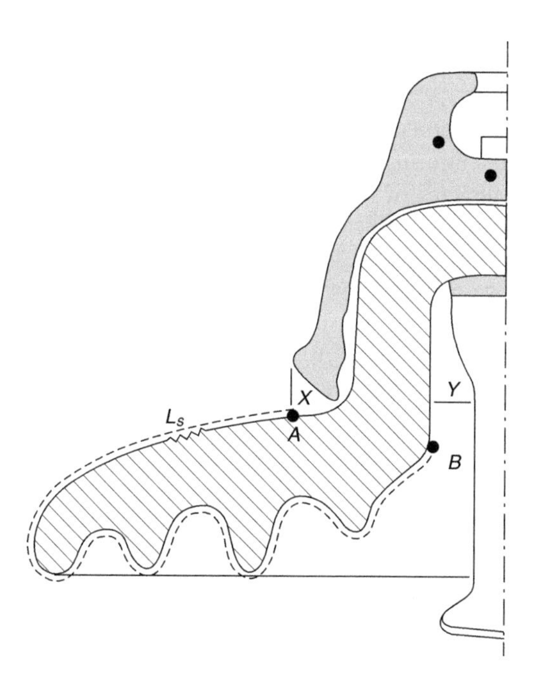

Figure 9.3 *Evaluation of insulator dimensions for a typical shape. L_s = leakage path length.*

The parameter relating resistance R of a polluted insulator in air to the surface resistivity σ is known as the form factor F, defined as:

$$F = \frac{R}{\sigma} \tag{9.5}$$

$$F = \int_0^{L_S} \frac{dL_S}{2\pi a} \tag{9.6}$$

where a is the radius corresponding to the path element dL_s.

In the laboratory, the resistance R can be measured with a low-voltage bridge and the average resistivity σ is determined from eqns (9.5) and (9.6). The average value of r is obtained from:

$$r = \frac{R}{L_S} \tag{9.7}$$

Therefore σ can be related to the minimum flashover voltage.

An example of the process of evaluating the profile for a simple insulator is included in Fig. 9.4.

9.3.2 Measurement of pollution severity

In general, the severity of the contamination present on the surface of insulators is classified according to ESDD. This information is used to designate various severity zones based on the characteristics of pollution present in the service environment where the insulators are used. The classifications are shown in Table 9.1. They are used as a guideline for choosing the leakage distance as a function of system voltage for a particular environment. It should be noted that the data in the table is meant for ceramic insulators, but in the absence of parallel information for composite insulators, it is also often used as a guideline for application of non-ceramic insulators.

Determination of equivalent salt (NaCl) deposit density

When an insulator is recovered from service, swabs are taken from the surface using a pre-cleaned cloth according to the following procedure:

- The conductivity of distilled water with the clean cloth submersed is measured.
- Swabs are taken from the top and bottom surfaces of the insulator independently.
- The cloth is rinsed and the conductivity is remeasured and the increase is noted.
- The solution is transferred into a standard volumetric flask.
- The solution is diluted with distilled water to a volume of 0.5 or 1 litre.
- The conductivity of the diluted solution is measured at two different temperatures and the conductivity corresponding to a temperature of 20°C is calculated through interpolation.
- The ESDD in mg/cm^3 is calculated using the following expression[15]

$$\text{ESDD} = \frac{0.42(\text{vol. in ml})}{\text{Area in cm}^2}(\sigma_{20^\circ\text{C}})^{1.039}$$

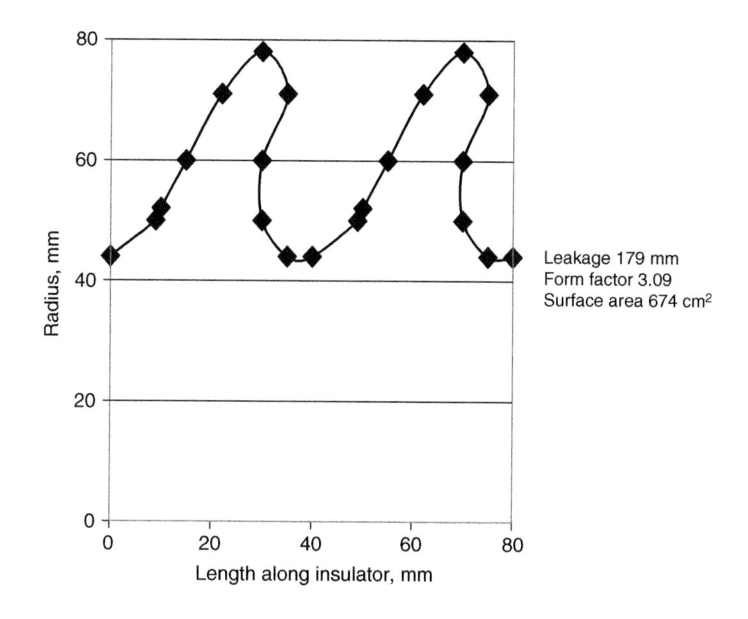

Dist (mm)	Rad (mm)	delta rad	Avg rad	dl	dl/avgr	dl*avgr
0	44	6	47	10.81665383	0.230141571	508.3827298
9	50	2	51	2.236067977	0.04384447	114.0394669
10	52	8	56	9.433981132	0.168463949	528.3029434
15	60	11	65.5	13.03840481	0.199059615	854.0155151
22	71	7	74.5	10.63014581	0.142686521	791.945863
30	78	−7	74.5	8.602325267	0.115467453	640.8732324
35	71	−11	65.5	12.08304597	0.184473984	791.4395113
30	60	−10	55	10	0.181818182	550
30	50	−6	47	7.810249676	0.166175525	367.0817348
35	44	0	44	5	0.113636364	220
40	44	6	47	10.81665383	0.230141571	508.3827298
49	50	2	51	2.236067977	0.04384447	114.0394669
50	52	8	56	9.433981132	0.168463949	528.3029434
55	60	11	65.5	13.03840481	0.199059615	854.0155151
62	71	7	74.5	10.63014581	0.142686521	791.945863
70	78	−7	74.5	8.602325267	0.115467453	640.8732324
75	71	−11	65.5	12.08304597	0.184473984	791.4395113
70	60	−10	55	10	0.181818182	550
70	50	−6	47	7.810249676	0.166175525	367.0817348
75	44	0	44	4.99	0.113409091	219.56
79.99	44	0	44	0.01	0.000227273	0.44
80	44					
			Sums	179.301749	3.091535268	10732.16199

Leakage (mm)	Form factor	Area (cm²)
179.301749	3.091535268	674.3210559

Figure 9.4 *Spreadsheet-based example for the evaluation of the relevant dimensions of a simple insulator*

Table 9.1 *IEC 815 contamination severity table*

Pollution level (max. ESDD)	*Examples of typical environments*	*Min. leakage distance*
I – Light (0.06 mg/cm^2)	Areas without industries and low density of houses equipped with heating plants. Areas with low density of industries or houses but subjected to frequent winds and/or rainfall. Agricultural areas (use of fertilizers can lead to a higher pollution level). Mountainous areas. *Note:* All these areas shall be situated at least 10 km to 20 km from the sea and shall not be exposed to winds directly from the sea.	16 mm/kV
II – Medium (0.20 mg/cm^2)	Areas with industries not producing particularly polluting smoke and/or with average density of houses equipped with heating plants. Areas with high density of houses and/or industries but subjected to frequent winds and/or rainfall. Areas exposed to wind from the sea but not too close to the coast (at least several km distant).	20 mm/kV
III – Heavy (0.60 mg/cm^2)	Areas with high density of industries and suburbs of large cities with high density of heating plants producing pollution.	

Table 9.1 *(continued)*

Pollution level (max. ESDD)	*Examples of typical environments*	*Min. leakage distance*
	Areas close to the sea or in any case exposed to relatively strong winds from the sea.	25 mm/kV
IV – Very heavy (>0.60 mg/cm^2)	Areas generally of moderate extent, subjected to conductive dusts and to industrial smoke producing particularly thick conductive deposits.	
	Areas generally of moderate extent, very close to the coast and exposed to sea-spray or to very strong and polluting wind from the sea.	31 mm/kV
	Desert areas, characterized by no rain for long periods, exposed to strong winds carrying sand and salt, and subjected to regular condensation.	

Note. The table is based on ceramic and glass insulators (ref. IEC 815). Its use for composite insulators is still to be verified.

9.3.3 Contamination testing

Contamination testing of insulators can be realized under field or laboratory conditions. Both of these are discussed below.

Field tests

Insulators are exposed to natural pollution at testing stations while subjected to operating voltage. During testing their performance is monitored throughout the measurement of parameters such as frequency of flashover and the levels, durations, and repetition rates of leakage current bursts. Subsequently they are removed and tested in the laboratory to establish flashover levels. Field test sites also provide an opportunity to characterize the pollution by removing it from the insulator and submitting it to chemical analysis. Typical pollutants which may be found on insulators include NaCl, $CaSO_4$, $MgCl_2$ and $CaCO_4$.

Laboratory tests

There are two laboratory testing procedures in common use. They are known as the salt fog and the clean fog tests.

(a) Salt fog test

The 'salt fog method' reflects the contamination mechanism prevalant along coastal areas; it is largely followed by European countries and Japan.

In this method a clean dry insulator is energized at its highest working voltage and is exposed to a salt fog produced by standardized nozzles. The highest fog salinity (kg of $NaCl/m^3$) which the insulator can withstand for three out of four one-hour tests is used to characterize the insulator's performance.

Alternatively the salinity of the fog is fixed at 80 kg/m^3 of solution and the voltage is raised from 90 per cent of flashover voltage in steps of 2 to 3 per cent each 5 minutes until flashover.

The standards in salinity vary in values from country to country. For example, Table 9.2 lists values from the Italian specification.

Table 9.2 *Example of salinity standards*

Severity of pollution at site	*Withstand salinity required* kg/m^3
Light	20–40
Moderate	40–80
Heavy	80–160
Very heavy	100–200

(b) Clean fog test

Clean fog tests reflect the contamination mechanism occurring in industrial areas. It is the most widely used method. Both methods are accepted as standards.[12,13,14] Studies show that when wetting conditions are properly controlled the most critical conditions for insulator flashover occur when the surface is totally contaminated and efficiently wetted by fog or dew with minimum washing.

9.3.4 Contamination procedure for clean fog testing

The insulator is dipped into a slurry consisting of kaolin, water and NaCl to give sufficient conductivity. The kaolin provides the mechanical matrix binding the conductive salt to the insulator. The insulator is allowed to dry, and is then tested. Alternatively, the contaminants may be deposited by spraying. This method gives poorer uniformity than dipping.

The deposit density should be recorded for various parts of the insulator. The clean fog method is applicable to suspension or post type ceramic, as well as polymer insulators. Figure 9.5 shows the relationship between ESDD and NaCl concentration of the slurry applied to the insulator under test.

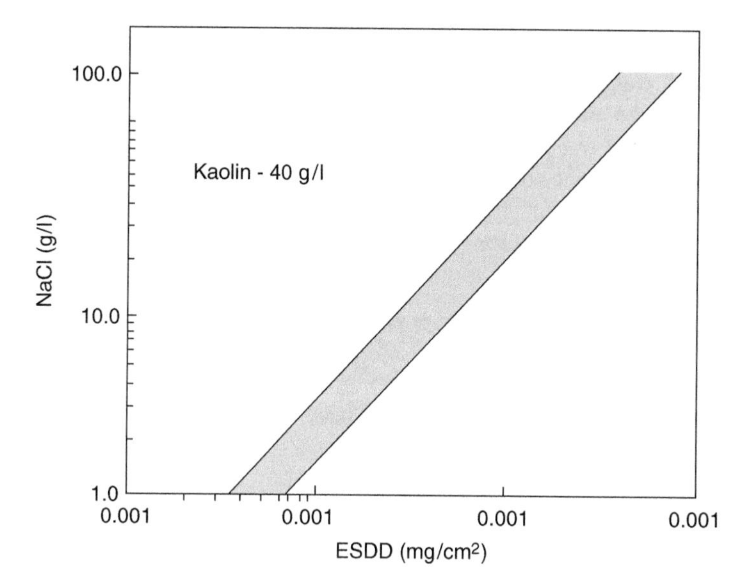

Figure 9.5 *Relationship between ESDD and contaminant NaCl concentration*

9.3.5 Clean fog test procedure

The dried precontaminated insulators are energized under constant voltage, and wetted by clean fog. The test voltage is maintained until FO or WS takes place. For WS the voltage is maintained for at least twice the time required for FO. Time to maximum wetting rate is determined by monitoring the leakage current. The maximum leakage current is measured after about 25–30 minutes of wetting. This is found to be the time required to reach maximum wetting. The critical flashover voltage is then determined using the up and down method.

Other important parameters which effect the recorded flashover voltage include the short-circuit current of the testing transformer[8] and the rate of fog temperature rise. The recommended minimum values are a short-circuit current of 5 A and a fog temperature rise of 0.8°C/min.[16]

9.3.6 Fog characteristics

The characteristics of the fog introduced to provide insulator wetting have an effect on the experimental results. The fog is characterized by:

1. Droplet size distribution.
2. Fractional liquid volume (1.8–6.2 g/m^3).
3. Fog temperature. (fog temperature rise).

1 and 2 are interrelated and contribute to wetting and washing of the exposed insulator surface by droplet impingement. For contamination testing it is essential that the natural washing conditions be simulated. Depending upon the source, both droplet size and the fractional liquid volume of natural fog vary over a wide range. At a fractional volume below 1.8 g/m^3 it is difficult to control temperature and density due to heat loss. The fog temperature must be significantly higher than ambient in order to ensure effective wetting through condensation on the insulator surface.

9.4 Mitigation of contamination flashover

There are a number of methods used to reduce or eliminate the possibility of contamination flashover of ceramic insulators. These include:

- use of insulators with optimized shapes,
- periodic cleaning,
- grease coating,
- RTV coating,
- resistive glaze,
- replacement of ceramic insulators with composite.

9.4.1 Use of insulators with optimized shapes

The shape and leakage distance of insulators can be varied to address environmental conditions. Generally the shapes are aerodynamically optimized to gather as little pollution as possible and to enhance self-cleaning through wind and rain. Standardized variation in shape parameters is available for service areas characterized by different environmental contamination processes. Special designs of varying shed profiles, diameters, spacings, leakage distance, etc. are available.

9.4.2 Periodic cleaning

In many installations high-pressure water systems, or corn and CO_2 pellet blasting, are utilized to periodically clean surface contaminants off insulators. Of these, high-pressure water cleaning is predominant and by far the cheapest.

Corn and CO_2 pellet blasting are far more effective for cleaning cement-like deposits that are difficult to remove. These procedures are generally applied on a repetitive basis linked to the pollution composition, severity, and deposition mechanism as well as local wetting conditions

9.4.3 Grease coating

Coating of insulator surfaces with petroleum gels or hydrocarbon greases is utilized in areas of heavy contamination. These coatings produce hydrophobic surfaces and the surface layer is able to encapsulate the contaminants into its bulk. The former characteristic was covered in the section on non-ceramic insulators, while the latter prevents the contaminants from going into solution upon initiation of the surface wetting mechanism. This approach has proven effective and has been in use for many years. As with washing, this is a maintenance-based solution, which must be periodically repeated. Usually the old grease is removed before new grease can be applied. In most instances, both the application of the new and removal of the old grease are manual operations. The process is slow and requires circuit outages.

9.4.4 RTV coating

Room temperature vulcanizing (RTV) silicone coatings are being applied with increasing frequency on both substation and line insulators. RTV coatings are applied over porcelain insulators and bushings to provide hydrophobic surfaces (described in detail in the section on non-ceramic insulators). Current information based on service experience and laboratory testing shows that these coatings perform well and will last for a number of years. The lifetime depends upon the coating composition, the application thickness and of course the pollution severity. RTV coatings are popular in that they represent a longer-term solution, which does not require replacement of the insulators.(19) They can be applied over existing insulators after adequate cleaning. A further advantage is that they can be applied to insulators on live circuits. Their mechanism of resistance to contamination flashover is based on surface hydrophobicity maintenance and contamination encapsulation. These processes are similar to those described in subsequent sections dealing with polymer insulators. As is the case with non-ceramic insulators, RTV coatings can rapidly deteriorate in the presence of electrical discharges, so care must be taken at higher voltage levels to ensure that the insulators are free of corona.

9.4.5 Resistive glaze insulators

In areas of heavy contamination resistive glaze insulators are often used to alleviate contamination flashover. Resistive glaze insulators utilize a specialized glaze, which is partially conductive. The glaze is formulated so as to provide steady state power frequency current flow along the insulator surface. Its use results in a uniform electric field distribution and surface heating. Both of these contribute to superior contamination performance. Surface heating inhibits wetting through condensation and aids in the drying process, whilst the more uniform electric field distribution acts to control dry band flashover. The conduction current of the glaze is generally designed to be approximately 1 mA and results in an insulator surface that is several degrees warmer than the ambient surroundings.

The improvement in contamination flashover performance through the use of resistive glaze was first demonstrated in the 1940s.(17) Since that time resistive glaze insulators have met with mixed success when applied in service. The technology has been commercially available since the 1950s. The 1970s saw production of both suspension and post type resistive glaze insulators. While they both provided excellent contamination flashover resistance, they suffered from glaze corrosion at the junction point where electrical contact was made between the metal portions of the insulator and the glaze. The problem was particularly severe in the case of suspension units where there is a high current density at the glaze/pin junction. The glaze corrosion resulted in a break in the conductive path between the insulator's line and ground end. This prevented the flow of resistive current and thereby eliminated the improvement in contamination performance. Manufacture of resistive glaze suspension insulators was halted after several years, but application of the technology to the production of post insulators and bushings continued. Over the years, the corrosion problem on post insulators has been studied and performance improved. Currently, resistive glaze post insulators and bushings are successfully utilized in many installations worldwide where the environmental conditions are severe. Recently, there has been renewed interest in resistive glaze suspension insulators. The process utilized in producing these units has undergone significant improvement and work performed by manufacturers suggests that the problems associated with severe glaze corrosion have been successfully addressed.(18)

9.4.6 Use of non-ceramic insulators

Application of non-ceramic insulators as a solution to contamination flashover problems has been growing since the early 1980s. Their application for this purpose and a number of others is given in the section on non-ceramic insulators.

9.5 Design of insulators

The two basic types of insulators in use are:

1. *Ceramic insulators.* The material includes porcelain or toughened glass, the connection is provided by a zinc-coated iron pin and cap-clevis-ball-socket; cement is used for mounting.
   Typical examples are shown in Figs 9.6(a) to (c); all are suspension type, but are of different shape. Of special interest is the insulator in Fig. 9.6(c) which has a significantly longer leakage distance of 17.3 inches compared with 11.5 in the previous two figures. This design is known as the antifog insulator. All are standard ANSI classes.
   The type shown in Fig 9.6(c) is also used in d.c. applications and in cases where the pollution is very severe.
2. *Non-ceramic insulators (NCI).* Composite, both suspension and post type, is displacing the earlier ceramic types, especially at the lower voltage levels. These were first introduced in the 1960s and 1970s with extensive applications only within the last two decades. Examples illustrating general design features are shown in Figs 9.7(a) and (b). The basic materials are polymers with metal end fittings. The inner part is a fibre glass reinforced resin rod which provides the mechanical strength. The outer sheds consist of polymer material, nowadays made of synthetic rubber EPDM or silicon rubber. Other materials such as epoxies have also been tried in the past, but the use is limited to indoor applications because of epoxy tracking and erosion under polluted-wet conditions.

The limiting factors as far as electrical withstand of insulators is concerned are the environmental conditions, particularly pollution. In dry and clean conditions there are seldom flashover problems experienced with outdoor insulators. Pollution sets the practical limits for insulator size and design.

The two basic categories of insulators will now be described in more detail.

9.5.1 Ceramic insulators

Insulators are made of ceramic materials which include porcelain and glass. Their initial use precedes the construction of power systems. They were first introduced as components in telegraph networks in the late 1800s.

There are a number of basic designs for ceramic insulators, examples were shown in Figs 9.6(a) to (c). Porcelain is used for the production of cap and pin suspension units, solid and hollow core posts, pin type, multi-cone and long rod insulators, and bushing housings. Glass, on the other hand, is used only for cap and pin suspension and multi-cone posts. Porcelain and glass insulators are well established, as might be expected based on their long history of use. Currently these types of insulators comprise by far the majority of in-service units. Continuous improvements in design and manufacturing processes have resulted in insulators, which are both reliable and long lasting. Porcelain units are coated with a glaze to impart strength to the surface. Today's glass insulators are predominantly manufactured from thermally toughened glass, which

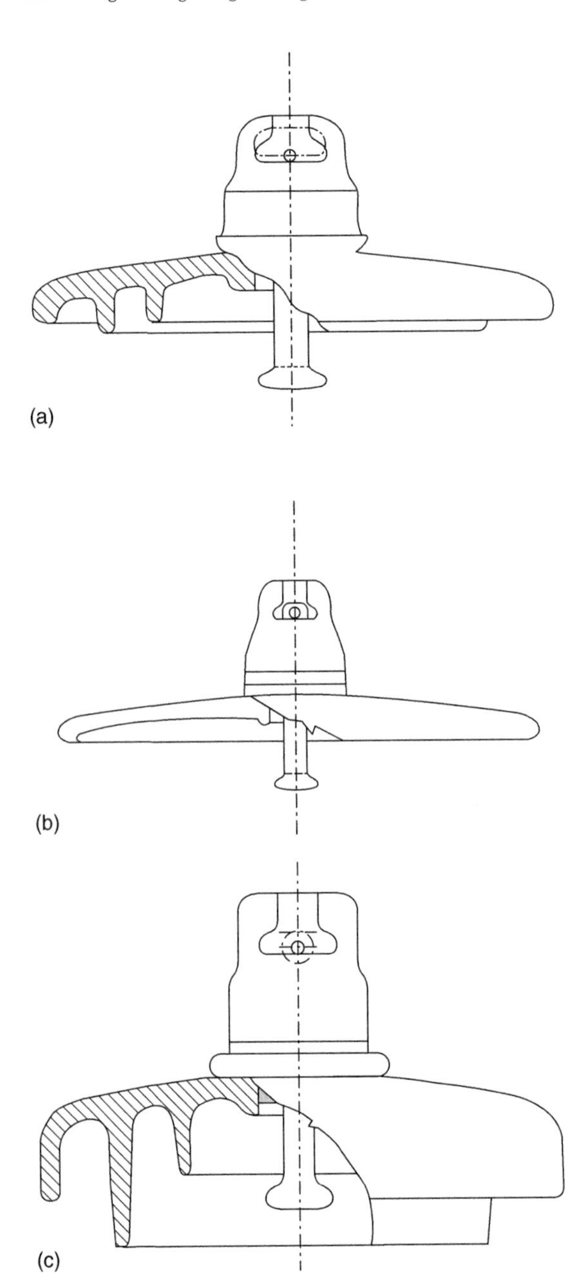

Figure 9.6 *Typical constructions of ceramic type suspension insulators. (a) Standard. (b) Open profile (self-cleaning). (c) Anti-fog and for d.c. applications*

prevents crack formation. Both of the materials have inert surfaces, which show very good resistance to surface arcing, and both are extremely strong in compression.

The manufacturing process for electrical porcelain is complex and involves numerous steps. With glass insulators, the manufacturing process is less complex, but still requires tight control. Failures of porcelain and glass insulators can usually be traced back to the manufacture, material or application of the units. If adequate caution and control in these areas is not maintained, the likelihood of an inferior product increases. However, as previously mentioned, when well made, both porcelain and glass insulators are highly reliable. The majority of bushings and lightning arresters installed in today's substations are contained within porcelain housings. Porcelain housings are, in essence, hollow core post insulators.

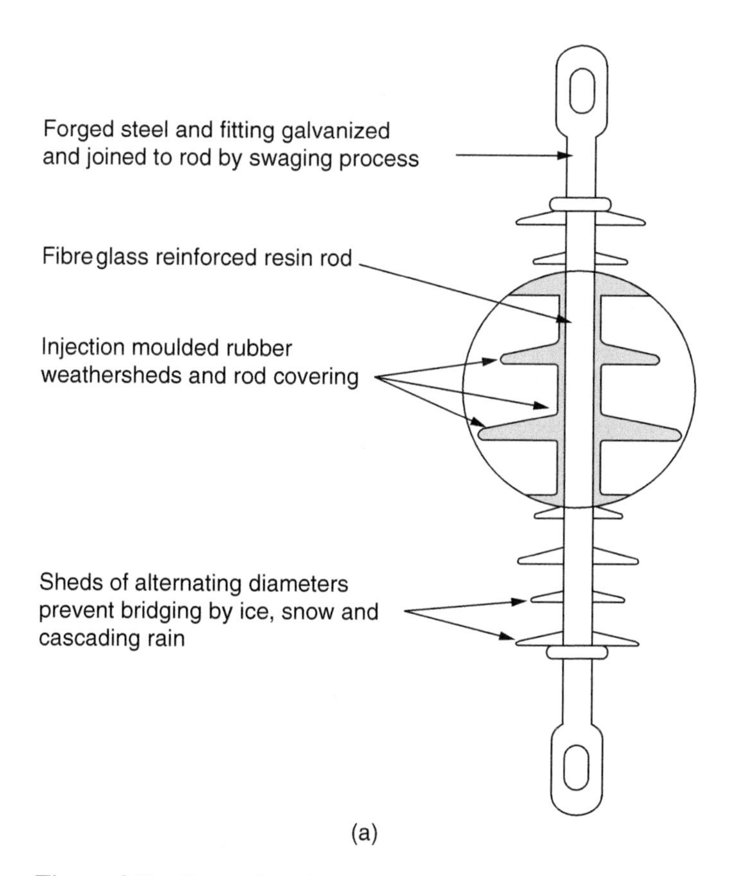

Figure 9.7 *Examples of non-ceramic insulators. (a) Suspension type. (b) Post type*

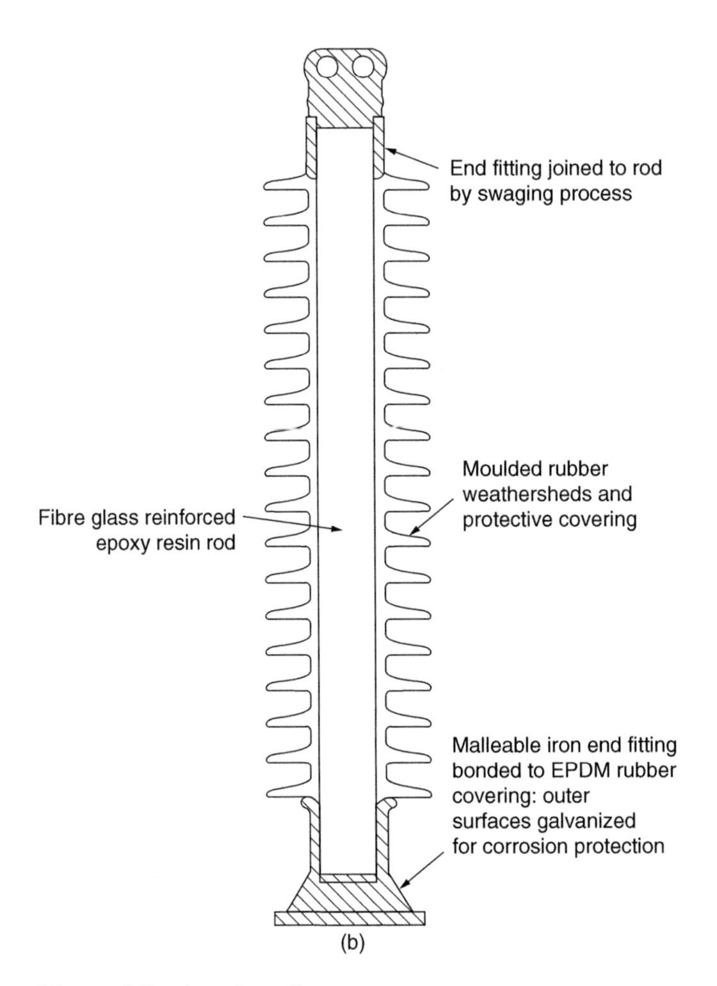

Figure 9.7 *(continued)*

9.5.2 Polymeric insulators (NCI)

Polymeric or non-ceramic insulators were first introduced in 1959. They were made from epoxy and when used outdoors or in contaminated environments, they were susceptible to problems associated with u.v. degradation, tracking, and erosion. NCIs were produced by various manufacturers through the 1960s and 1970s. Those early designs were primarily of the suspension/dead-end and post type. Certain fundamental aspects of the early designs formed the basis of today's production units. They utilized a pultruded fibre glass core as the strength member. The fibre glass core afforded protection against the environment through encapsulation in a rubber housing. The mechanical connections at the insulator ends were made using a variety of means. Some designs used glued fittings, others had a wedge type attachment, still other manufacturers utilized crimping. In all cases metal end fittings were attached to the fibre glass rod to give the insulator the mechanical strength the applications required.

Early advocates of NCIs claimed that they achieved up to 90 per cent weight reduction when compared to their ceramic equivalents. They also had superior resistance to shock loads due to conductor or hardware failure on adjacent spans. Another area in which they showed promise was their ability to withstand vandalism. Significant portions of ceramic insulator failures are due to vandalism involving shooting. When a bullet hits a ceramic unit, it breaks or shatters. NCIs do not fail immediately when shot because their components are not brittle. There are instances reported where NCIs have remained in service without problems for many years after being shot. A final advantage claimed by manufacturers and users of early NCIs was that they could be designed with extremely high leakage lengths which could be easily optimized to differing environmental conditions.

Early experience with NCIs was confined to short lines and trouble spots. The trouble spots were generally associated either with areas of environmental contamination or gunshot damage. The initial experience with these applications proved somewhat disappointing. A host of problems not previously experienced with ceramic units were encountered. Amongst these were tracking and erosion, u.v. damage, chalking and crazing, hardware separation, corona splitting, and water penetration. Many of these were associated with the use of inappropriate housing materials and manufacturing techniques, poor quality fibre glass rods, modular sheds and poor sealing between the rod, housing and end fittings. These operating problems resulted in a significant number of outages and line drops. Based on the initial field performance, NCIs saw limited use and therefore production.

By the 1980s the technology had evolved sufficiently to address the concerns generated through the early field experience. Understanding of the early failure mechanisms combined with improvements in materials and manufacturing technology resulted in the development of the NCIs available today. Generally, today's NCIs are characterized by a one-piece shed or housing structure. This one-piece external housing is obtained through single stage moulding or post assembly vulcanization. Tracking and erosion performance as well as resistance to u.v. has increased markedly. Most industry standards include tracking and erosion tests, and most of the insulators in production today utilize a track-free high-temperature vulcanized elastomer housing. The importance of sealing the exterior of the insulator against moisture has been well recognized, and is addressed in most current designs. Present experience with these insulators is beginning to indicate failure rates approaching that of ceramic units.

Today NCIs are utilized as standard products in many of the world's power delivery systems. Their main areas of application include distribution and transmission systems rated up to 345 kV. There is limited use above 345 kV all the way up to 765 kV.

Shed material

In today's manufacture of NCIs the most commonly used shed and housing materials are hydrocarbon and silicone elastomers. The hydrocarbon elastomers include ethylene-propylene rubbers such as ethylene-propylene monomer (EPM), ethylene-propylene diene monomer (EPDM), and a co-polymer of ethylene-propylene and silicone (ESP). The silicone elastomers include both high temperature and room temperature vulcanizing silicones. Both these families of materials utilize aluminatrihydrate (ATH) as a filler which enhances the materials' tracking performance. The silicone and hydrocarbon elastomer housing materials have been developed to the stage where the tracking and u.v. degradation encountered with older designs are no longer a concern. Both materials are utilized on distribution and transmission systems. The EPR materials have shown good performance in clean environments, whereas the silicone-based materials function well in both clean and contaminated applications.

One of the key characteristics affecting the contamination performance of NCIs is surface hydrophobicity. Hydrophobicity is a characteristic ability of a surface to 'bead' water which is deposited on it. As previously explained, contamination flashover of external insulation involves dry band arcing which develops due to heating and evaporation of electrically continuous liquid paths formed from the dissolution of surface contaminants in a layer of moisture present on the insulator surface. When a surface has a high degree of hydrophobicity, water deposited on it forms individual beads or droplets. This droplet formation inhibits the occurrence of leakage currents and the associated dry band arcing process. Simply put, an insulator with a highly hydrophobic surface will be characterized by significantly better contamination flashover performance than an identical one with a non-hydrophobic surface.

Most polymer insulator housings are hydrophobic when the insulators are first installed. Exposure to surface discharges, corona and certain chemicals (including water) reduces the hydrophobicity of polymer surfaces. With EPR-based housings, exposure to the operating environment results in the reduction and eventual permanent elimination of surface hydrophobicity. This is one of the more significant differences between the two housing materials. Unlike the EPR compounds, silicone housings have the ability to recover a highly hydrophobic surface state after it has been lost. In the silicone materials used, high and low weight molecular chains constantly break down and recombine. The material's initial hydrophobic state is due to the presence of the low molecular weight oils on the surface. The process of losing hydrophobicity involves the removal of these oils. In service this occurs primarily through exposure to surface arcing which can be present when the insulators are applied in areas of severe contamination. Typically even under extremely severe conditions, the duration of conditions that cause surface arcing is limited to tens of hours. When the arcing abates, the surface again becomes coated with the low molecular weight oil and the hydrophobicity is regained. This process of hydrophobicity regeneration takes somewhere between several hours and several days. The number of times that the process can repeat is not known, but given the thickness of the bulk material used, it is expected that the process can go on for the expected life of the insulators.

Fibre glass core

The mechanical strength of NCIs is provided through the use of a fibre glass core. For strain, dead-end and solid core post designs, the fibre glass rod is generally manufactured using a pultrusion process. These pultruded rods contain axially aligned electrical grade glass fibres in a resin matrix. Two types of resin are in common use. Epoxy resin is generally believed to give better performance, while polyester resin is a lower cost alternative. Potential problems associated with these types of pultruded rod include axial cracking due to poor handling or manufacturing procedures, and stress corrosion cracking otherwise termed brittle fracture. Brittle fracture is a process which culminates in the physical parting of the insulator under low mechanical loading and is therefore of significant concern. It is not fully understood, and is currently the focus of a significant amount of research.(11)

End fitting attachment and moisture ingress

Since their inception, several methods of attaching end fittings to solid core NCIs have been utilized. Some of the original designs had end fittings that were glued, while others used a wedge method of connection. The end fitting fulfils two very important requirements of an NCI. First, it has to be able to support mechanical loading of the insulator with no slippage. Second, it must be designed so as to ensure that moisture cannot reach the fibre glass core through the interface that exists where the end fitting is joined to the insulator. The importance of the first function is obvious; however, if long-term performance is to be achieved, the second requirement is more critical. Most of the end fittings used in today's designs are either swaged or crimped. This type of connection has proven to give the best performance from both the strength and the sealing aspects. Moisture sealing is achieved in three ways. RTV or some other sealant is applied over the end fitting/housing interface, the end fitting is installed using an interference or friction fit over the housing, or the housing material can be extruded over a portion of the end fitting during the moulding phase of the manufacturing process. The last of these appears to be the most effective, and the first has proven least reliable in preventing moisture ingress.

Hollow core NCIs

Hollow core NCIs (HCNCIs) are made of fibre glass filament tubes impregnated with glass epoxy resin. The housing is then generally extruded over the fibre tube, this extrusion process can result in the manufacture of the weathersheds, or alternatively, the weathersheds can be fitted over the housing and vulcanized. In substations they are used primarily as housings for lightning arresters, and for transformer, circuit breaker and wall bushings. There are also some applications where they are used as station post insulators for supporting buswork, switches, and other electrical equipment. When compared to conventional ceramic bushings and insulators they offer several advantages. Amongst these are light weight, superior contamination performance and increased reliability under earthquake conditions. For bushings, their use represents an important safety enhancement in that unlike porcelain bushings they do not fail explosively when internal power arcs occur.

9.6 Testing and specifications

All insulators are tested according to standard procedures outlined in various national and international publications. Ceramic and glass insulators are mechanically and electrically proof tested prior to shipment. In the case of NCIs, prior to leaving the factory each production piece is subject to mechanical but not to electrical proof testing. The primary reason for this difference is that ceramic and glass units are generally made of a number of smaller units in series. For example, a 230 kV station post would generally comprise two smaller posts bolted together to give the clearances required for 230 kV. This allows for piecewise testing of individual components. With NCIs a 500 kV insulator is manufactured as a single piece. Performing electrical tests on each unit would require significant time and investment in a sizeable h.v. test facility. In addition to mechanical and electrical proof tests, the raw materials used in the production of ceramic, glass, and polymer insulators are tested as a control on the production process.

With regards to qualification and application testing, the most widely used standards are those issued by IEC, ANSI, IEEE, CSA, and CEA. Porcelain, glass, and polymeric insulators are subjected to both electrical and mechanical tests. Depending upon the type of insulator, the electrical tests include wet and dry power frequency flashover, lightning impulse flashover, steep front impulse flashover, power arc, and RIV/corona tests. Mechanical tests include tension, thermal mechanical cycling, torsion, cantilever, and electrical–mechanical testing. Contamination performance tests are also performed on these insulators in accordance with the techniques discussed earlier.

Contamination flashover performance tests were described in earlier sections of this chapter. The two test methods mentioned earlier were both developed for ceramic and glass insulators. These methods and variations upon them are also being applied to NCIs, and at the same time other tests to characterize the operation of polymer insulators in contaminated environments are being developed.

9.6.1 In-service inspection and failure modes

Insulators are often periodically inspected to ensure their continued integrity. For regular porcelain, this includes visual inspection and in the case of suspension units, an in-service electrical test. For glass, monitoring usually comprises only a visual inspection. These simple actions are usually sufficient to detect any impending problems with ceramic or glass insulators. In-service monitoring of NCIs presents a greater challenge. Methods such as IR thermography, radio noise detection, corona observation, and electric field monitoring have proven somewhat effective as diagnostic tools for assessing the in-service condition of NCIs. The difference in approaches to NCI and ceramic/glass monitoring are due to the different failure mechanisms characterizing each type of insulator. Damage to ceramic insulators is generally noticeable due to surface cracks. In suspension units, there can be hidden electrical punctures through the insulator in the area between the cap and the pin. These are not visually detectable, but can be detected using a simple field instrument. With glass insulators, any significant physical damage usually results in destruction of the shed as the units are made of tempered glass. With NCIs the insulator can be seriously damaged inside with no indication on its exterior. The mode of failure predominant with NCIs involves mechanical or electrical failure due to rod breakage or surface or internal tracking. Because of this, monitoring of NCIs is more complex than monitoring ceramic or glass insulators. Up-to-date information on different approaches to the monitoring of NCIs in-service is summarized in reference 14.

References

1. L.L. Alston and S. Zolendziowski. Growth of Discharges on Polluted Insulation. *Proc. IEE* Vol. 110, No. 7, July 1963, pp. 1260–1266.
2. H. Boehme and F. Obenaus. Pollution Flashover Tests on Insulators in the Laboratory and Systems and the Model Concept of Creepage–path–flashover. CIGRE paper No. 7, June 1969, pp. 1–15.
3. B.F. Hampton. Flashover Mechanism of Polluted Insulation. *Proc. IEE* Vol. 111, No. 5, May 1964, pp. 985–990.

4. H.H. Woodson and A.J. Mcleroy. Insulators with Contaminated Surfaces, Part H: Modelling of Discharge Mechanisms. *IEEE Trans. on Power Apparatus and Systems* Nov./Dec. 1970, pp. 1858–1867.
5. F.A.M. Rizk. Mathematical Models for Pollution Flashover. *Electra* Vol. 78, 1981, pp. 71–103.
6. D.C. Jolly. Contamination Flashover Part II, Flat Plate Model Tests. *IEEE Trans.* Vol. PAS-90, No. 6, Nov. 1972, pp. 2443–2451.
7. R. Sudararajan and R.S. Gorur. Dynamic arc modeling of pollution flashover of insulators under d.c. voltage. *IEEE Trans. on Elec. Insul.* Vol. 28, No. 2, April 1993, pp. 209–219.
8. F.A. Chagas. Flashover Mechanism and Laboratory Evaluation of Polluted Insulators under d.c. Voltages. Ph.D. Thesis, Elec. Eng. Dpt., University of Manitoba, Canada, 1996.
9. ANSI/IEEE Std. 987-1985. IEEE Guide for Application of Composite Insulators.
10. IEEE Std. 1024-1988. IEEE Recommended Practice for Specifying Composite Insulators.
11. ANSI/IEEE Std. 987–Final draft balloted in 1999: IEEE Guide for Application of Composite Insulators, to be published in 2000.
12. IEEE Std. 4-1995. IEEE Standard Techniques for High-Voltage Testing.
13. IEC Publication 507 (1997). Artificial Pollution Tests on High Voltage Insulators (a.c.).
14. IEC Publication 1245 (1993). Artificial Pollution Tests on High Voltage Insulators (d.c.).
15. W.A. Chisholm, P.G. Buchan and T. Jarv. Accurate measurement of low insulator contamination levels *IEEE Trans. on PD* Vol. 9, July 1994, p. 1552.
16. J.N. Edgar, J. Kuffel and J.D. Mitz. Leakage Distance Requirements for Composite Insulators Designed for Transmission Lines. Canadian Electrical Association Report CEA No. 280 T 621, 1993.
17. S.T.J. Looms. *Insulators for High Voltages*. Peter Peregrinus Ltd, London, 1998.
18. R. Matsuoka, M. Akizuki, S. Matsui, N. Nakashima and O. Fuji. Study of Performance of Semi-conducting Glaze Insulators. *NGK Review Overseas Edition* No. 21, Dec. 1997.
19. Round Robin Testing of RTV SILIKON Rubber Coating for Outdoor Insulation. *IEEE Trans. on PD* Vol. 11, No. 4, Oct 1996, p. 1889. Paper prepared by IEEE Dielectric and Insulation Soc. Committee S-32-33.

Index

A.C. transmission systems 1
A.C. peak voltage measurement 111
A.C. to D.C. conversion 10
Active peak-reading circuits 117
A/D converter 179–80
A/D recorder 183
Ammeter, in series with
  high ohmic resisters 96
Amorphous dielectric 370
Anode coronas 349, 351
Anti-fog insulator 524
Aperture uncertainty 195
Apparent charge 433
Arithmetic mean value 9
Attachment coefficient (η) 306
  cross-section 306
  negative ion formation 304
Avogadro's number 282

Band pass filter 421
Basic insulation level (BIL) 492
Belt-driven generator 24
Bernoulli–l'Hopital's rule 142
Best fit normal distribution 490
Binomial distribution 485
Biphase, half-wave rectifier 13
Boltzmann's constant 282
Boltzmann–Maxwell distribution 285
Boundary element method (BEM) 270
Boyle and Mariotte Law 281
Breakdown and corona inception:
  in coaxial cylindrical systems 213
  voltage 204
Breakdown criteria 345
Breakdown field strength (E_b) 340
  potential relationships 342

Breakdown in solids 367
  in liquids 385
  in non-uniform fields 326
  strength of insulating materials 201
  tests on solid dielectric plate materials 233
  under impulse voltage 360
Breakdown probability 485, 487, 489
Breakdown voltage curves for N_2 358
Breakdown voltage (V_b), expression:
  for air 338, 342
  for SF_6 347–8
Breakdown voltage, of:
  rod gaps 94
  uniform field gaps 92
Bubble breakdown 391
Bushing, capacitor 235–6, 238, 241
  simple, arrangement 236

Capacitance of spark gaps 62–3
Capacitor, reservoir 11
Cathode corona 352
  processes 316
Capacitors:
  compressed gas 122–4
  high voltage 118
Carbon track 385
Cascade circuits 13–14
Cascaded transformers 21, 37, 38, 39
Cavity breakdown 383, 390
Ceramic insulator 523
Charge coupled device (CCD) 177
Charge density 25
Charge Q 11–12
Charge simulation method (CSM) 254
Charge transferred 12

Charging resistors 61
  voltage 52
Chubb–Fortescue methods:
  fundamental circuit 110–11
  voltage and current relation 112
CISPR Publication 16-1 438
Clean fog testing 518–9
Cloud chamber photographs 328
Coaxial cable:
  cylindrical and spherical fields 209–10
  with layers of different permittivities 231
  spherical terminations 212
Cockroft–Walton voltage doubler 14
Collision cross section 288
Collisions, elastic 283
Confidence interval 487
Contamination flashover 520
  test 517
Corona discharges 348
Crest voltmeter for a.c. measurement 113
  with discharge error compensation 117
Cross-section, effective 288
Current comparator bridge 417–9
Current density 312

Damped capacitor voltage divider 162–4
Damping resistors 32
D.C. cascade circuit with transformers 21
D.C. corona inception and breakdown in air 204
Deionization 302
Deltatron 22
Dielectric loss 411
Dielectric polarization 396
Dielectric refraction 232–3
Dielectric response analysis 410
Differential non-linearity 191
Differential PD bridge 449
Diffusion 313
  and mobility relationship 314
  coefficient 314, 331, 388
  equation 314
Digital transient recorder 175
Dimensioning of circuit elements 57–8
Diode 10, 13
Dirichlet boundaries 247, 251
Discharge in cavity 382
Discharge resistors 53, 64
Disruptive discharge 78
  of sphere gaps 83–7
Dissipation factor 406
Divider for D.C. 107
Drag force 388
Drift velocity 311, 391–2
Duhamel's integral 136
Dynamic error 183

EBS tube 178
Eddy motion 391
Edge breakdown 374
Effective ionisation, coefficient 300, 325
Efficiency of impulse generator 55–7
Electric power 1
Electric strength of highly purified liquids 387
Electric stress distribution:
  cylindrical conductors in parallel 221–2
  in sphere-to-sphere arrangement 216, 218
Electroconvection 391
Electrodynamic model of breakdown 391
Electromechanical breakdown 373
Electron affinity 304
Electron attachment 306, 345
  avalanche 297
  emission by photon impact 323
Electrostatic charging tendency (ECT) 394
Electrostatic fields 201
  generators 24
  voltmeters 94–6
Elliptical display 437
Energy balance equation 372
Energy functional 250–51

Energy level diagram 370
Energy in lightning 464
Energy transfer, collision 292–3, 301
Engetron 22
Environment, contaminated 509
Epoxy disc spacer 234
Equivalent salt deposit density (ESDD) 514
Erosion breakdown 381
Exciting winding 33

Fast digital transient recorder 175
Felici generator 28
Field computations by CSM with surface charges 268–9
Field distortions by:
  conducting particles 221
  space charge 326
Field distribution in non-uniform field gaps 343
Field efficiency factor η 202, 203, 214
Field emission 319
Field sensor 107
Field stress control 201
Fields in:
  homogeneous, isotropic materials 205
  multidielectric materials 225
Finite difference method (FDM) 242
Finite element method (FEM) 246
Flashover characteristics of:
  long rod gaps 467–8
  insulator strings 467–8
Force, electrostatic voltmeter 94–5
Form factor 513
Formative time lag in N_2 332
Four terminal network of measuring system 132–3
Fourier:
  series 133
  transform 134–5
Fowler–Nordheim equation 321
Free path λ 287
Free paths, distribution of 290
Frequency 9
Frequency domain 404
Frequency response of measuring system 133
Front-chopped impulses 50–51, 188
Front oscillation 189
Gap factors (k) 471–2
Gaussian probability distribution curve 474–5
Gay–Lussac's Law 281
Generation of high voltages 8
Generating voltmeters 107
  principle of 108
G.I.S. (gas insulated substations) 44, 66, 183, 212
Grading electrode 106
Grading rings 106
Grease coating 521

Half wave rectifier 11, 13
Harmonic currents 41
High energy breakdown criterion 371
High temperature breakdown 370
High voltage bridge with Faraday cage 420
High voltage capacitors 118
HVAC 9
HV output:
  loaded 16
  open circuit 14
HVDC technology 2, 9
  voltages and power transmitted 4

Ideal rectifiers 16
IEC Publication 60 9
IEEE Standard 4 9
Image intensifier photographs 329
Impedance of cable 131
  supply 41
Impedance, internal of diodes 11
Impulse generator:
  outdoor construction 67
  indoor construction 69
Impulse generators, design and construction of 66
Impulse thermal breakdown 378
  voltages 48

Impulse voltage generators 52
  wave components 55
  withstand level 493–4
Inductance of h.v. reactor 45
  nominal (L_n) 45–6
Insulation aging 409
Insulation coordination 492
  level 492
Integral non-linearity 194
Intrinsic breakdown 368
  electric strength 369
Ionisation constants 299–300
  cross-sections 295, 297
  processes 294

Kanal 326, 328
Kaolin 518–9

Laplace transform circuit, of impulse generator 53–4
Laplace's equation 245, 258
Laplacian field 246
Leakage path length 513
Lifetime stress relationship of polyethylene 384
Lightning current 461, 464–5
Lightning impulse voltage, definition of 50
Lightning mechanism 460
Lightning overvoltages 49, 460
Lightning stroke between cloud and ground, development 463
Lightning strokes 49, 461
Line charge, finite 258, 263
Loss measurement 411
Lossless transmission line 143, 163
Low voltage arm of the divider 171, 174

Main stroke 462
Malter effect 317
Marx generator 61
Matching impedance for signal cable 172–3
Mean free path 287
  of atom 289
  of electron 289
  of ions 309
Mean free time 311
Mean molecular velocities 286
Measurement of high voltages 77
Measurement system, computation of 139–40
Metastables 301
Minimum breakdown voltage 334, 336–7
Minimum sparking constants 337
Mixed resistor-capacitor dividers 156–7
MOA (metal oxide arrester) 500
Mobility 308, 391
  of electron 313
  of single charged gaseous ions 313
Multi-level test method 480
Multiplier circuit 13
Multiphase rectifiers 13
Multistage generator 60, 61, 64

Narrow band amplifiers 440–41
Narrow band PD detector circuits 437
Nearby earth objects, effect of 89
Non-ceramic insulator 522
Non-destructive insulation test 77, 395
Normalised amplitude frequency spectra of impulse voltage 135
Numerical methods 241

Operational amplifiers in crest voltmeters 117
Optimum, number of stages 19
Oscillating switching impulses 66
Output load 11, 16
Output voltage 15
Overvoltages 3, 460

Partial discharge currents 428–9
  equivalent circuit 423–4
  measurements 421
Paschen's curve for air 338
  Law 333, 339
Passive rectifier circuits for peak voltage measurement 113

Peak measurement 184
Peak reverse 11
Peak voltage, measurement of 78, 109
Peak voltmeter for impulse voltages 116
Peek's equation 344
Penning effect 339
Phase resolved PD measurement 454
Photoelectric emission 317
Photoexcitation 301
Photoionisation 301
Pilot streamer 462
Point-plane breakdown and corona inception in air 355
Poisson's equation 245, 258
Polarity effect of 354
Polluted insulator 510
Polution severity 514, 516
Polymer-housed arrestor 505
Polymeric insulator 526
Positive ion, emission by 317
Potential barrier 319
Potential coefficients 258, 263, 267
Potential distribution along gas capacitor 123
Potential related to point charge 260
Potentials 14
Power rating 31
Probability density function 496
Protection level 495, 498

Quality factor 42, 421, 436

Radio disturbance voltage (RDV) 438–9, 443
Radio interference voltage (RIV) 438, 443, 445–6
Rate of energy gain 371–2
Reactances of transformer windings 39–40
Reactor, prototype 47
Real time digital simulator (RTDS) 496
Recombination 302
Recombination, coefficient of 303
  in air 304
Recovery voltage 402, 410–11
Rectifier circuits 10
Rectifier half wave single phase 10
  regulation 19
Reference measuring system 91
Relative air density (RAD) 89–90, 339
Relaxation current 401–3
Resistance, time dependent 71
Resistive glaze insulator 521
Resistor, shielded 99–100
Resistor h.v.:
  equivalent network 101
  standard 99–100
  voltage dividers 149, 173
Resistors:
  wavefront 52
  wavetail 52
Resonance frequency 40
Response function of transfer network 133
Response time of impulse measuring system 137–8
Response time of resistor dividers 151–2, 166, 170
Return stroke 461
Return voltage 402
Ripple 12, 16, 17, 19
Risk or failure 496, 499
R.M.S. (voltage) 30
Rod gaps 93
Rod-to-plane electrode configuration 202
Rogowski's profile 207
Rotating barrel generators 28
RTV coating 521

Saha's equation 302
Sames generator 29
Sampling error 182, 185–6
Saturation current density 318
Schering bridge 412–13
Schottky's equation 321, 386
Secondary avalanches 330
Secondary electron emission 323
Self-cleaning insulator 524
Self restoring insulation 468

Series resonant circuits 40, 42–4
with variable test frequency 45–6
Simulation of dielectric boundary by discrete charges 266
Single stage generator circuits 52–3
Space charge 354
Space charge field
in negative point-plane gap 359
Sparking voltage-Paschen's Law 333–4
Sphere gap, clearances around 83
horizontal 81
peak voltage 84–7
vertical 80
Sphere gaps 79
Sphere shank 80
'Stacked' capacitor units 124
Standard capacitor for 1000 kV 124
Standard deviation 474
Standard insulation levels 493–4
Static electrification 393
Static error 179
Statistical overvoltage 496
safety margin 495
time lag 360
withstand voltage 484, 493–4
Straight PD detection circuits 431
Stray capacitances of standard capacitor 125
Streamer breakdown 373
Streamer spark criterion 329, 331
Streamer velocity 350–51
Streamer or "Kanal" mechanism of spark 326
Streamers under impulse voltage 350
Stress control by floating screens 235
Surge impedance 1, 163, 165, 172
Suspended solid particle mechanism 387
Switching impulse, standard 51–2
Switching impulse voltages, circuits for generating 64–5
Switching impulses, oscillating 66
Switching overvoltages 460
Switching surge voltage characteristics 468
Synchronisation and tripping 70
Tandem accelerator 27
Tank of transformer 34, 37
Taylor's series 243
Tertiary exciting windings 37
Testing transformers 32
single unit 33–4
with mid-point potential 34–5
Testing voltages 5, 479, 484
Testing with lightning impulses 5, 484
A.C. voltages 5, 484
D.C. voltage 6
switching impulses 6, 484
very low frequency 7
Thermal breakdown 369, 375
impulse 378
minimum 379
Thermal capacity 377
Thermal instability 376
Thermal ionisation 302
Thermionic emission 318
Thermal voltage, minimum 380–81
Time delay, built in 70
Time delay of lead τ_L 143–4
Time domain 398
Time lag 359
experimental studies 362–3
overvoltage relationship 364–5
Time to crest 65
Time to peak 51
Tolerances on sphere gaps 81
Townsend breakdown mechanism 324
Townsend criterion for spark 325
Townsend criterion for spark for non uniform field 342
Townsend first ionisation coefficient 295
Townsend second ionisation coefficient 321
Tracking 385
Transfer characteristics of measuring system 132
Transformer 11, 32, 35
Transformer current 12
Transformation of a square grid from W to Z plane 207–8
Transient digital recorder 176
Transmission voltage 2

Travel time τ_L 163
Treeing 374
Trichel pulses 353
Trigatron 72–3
Trigatron characteristics 73–4
Trigger electrode 72
Tripping 70, 74
Tunnel effect 319

Ultra-wide-band PD detection 447
Uniform field gaps 92, 206
  breakdown voltage 92
Up and down method 480, 483
USR for capacitor voltage divider 160–62
USR and definition of response time 137, 145–6
USR for low value resistor divider 168
USR for mixed dividers 157–8, 166
USR for resistor dividers 152, 169

Van de Graaf generator 24
Velocities, distribution of 284, 286
Virtual front time 50
Virtual time to half value 50
Voltage dividers 96, 130, 147, 149, 156, 159, 163, 171
Voltage dividing systems, impulse measurements 129
Voltage doubler 13
Voltage multiplier 13, 21
Voltage regulator transformer 42
Voltage stresses 3, 472
Volt–ampere characteristics 501, 503
Volt–time characteristics 361–2
$V_{r.m.s.}$ 30, 95

Wagner earth 415–16
Wave shaping network 52
Weibull function 477
Wide band PD detection circuits 434–5
Wimshurst machine 24
Work function for typical elements 317

ZnO element 501–2, 504

Printed and bound by CPI Group (UK) Ltd, Croydon, CR0 4YY
08/05/2025
01864805-0001